Engineering Design and Visualization Techniques

Engineering Design and Visualization Techniques

Edited by **Niceto Salazar**

Willford Press
New York

Published by Willford Press,
118-35 Queens Blvd., Suite 400,
Forest Hills, NY 11375, USA
www.willfordpress.com

Engineering Design and Visualization Techniques
Edited by Niceto Salazar

International Standard Book Number: 978-1-68285-292-7 (Hardback)

Printed in China.

Contents

Preface

Visualization paradigms consistently contribute to the progress of multiple engineering disciplines. This book aims to bring forth the scientific basis of visualization to facilitate the decision making process in engineering. It presents design rules, principles and guidelines. The text is focused on the evolving trends, state-of-the-art technologies and analysis techniques. The book outlines some innovative applications and future trends in scientific visualization. It covers theoretical as well as practical knowledge to help the readers understand the applications of this field. Students will find this text beneficiary.

The information contained in this book is the result of intensive hard work done by researchers in this field. All due efforts have been made to make this book serve as a complete guiding source for students and researchers. The topics in this book have been comprehensively explained to help readers understand the growing trends in the field.

I would like to thank the entire group of writers who made sincere efforts in this book and my family who supported me in my efforts of working on this book. I take this opportunity to thank all those who have been a guiding force throughout my life.

Editor

1

SMART: scalable and modular augmented reality template for rapid development of engineering visualization applications

Suyang Dong[1*] and Vineet R Kamat[2]

Abstract

Background: The visualization of civil infrastructure systems and processes is critical for the validation and communication of computer generated models to decision-makers. Augmented Reality (AR) visualization blends real-world information with graphical 3D models to create informative composite views that are difficult to create or replicate on the computer alone.

Methods: This paper presents a scalable and extensible mobile computing framework that allows users to readily create complex outdoor AR visual applications. The technical challenges of building this reusable framework from the software and hardware perspectives are described.

Results: SMART is a generic and loosely-coupled software computing framework for creating AR visual applications with accurate registration algorithms and visually credible occlusion effects. While SMART is independent of any specific hardware realization, ARMOR is built as a hardware implementation of SMART to test its algorithm correctness and its adaption to engineering applications. In particular, ARMOR is a modular mobile hardware platform designed for user position and orientation tracking, as well as augmented view display.

Conclusion: The framework has been validated in several case studies, including the visualization of underground utilities for excavator control and collision avoidance to demonstrate SMART's rapid creation of complex AR visual applications.

Introduction

Augmented Reality (AR) refers to a visualization technology that superimposes virtual objects on the real world so as to provide information beyond reality and enhance people's interaction with the environment. It possesses distinct advantages over other visualization methods in at least three aspects: 1) From the perspective of visualization, the integration of the real world can significantly mitigate the efforts to create and render contextual models for virtual scenes, and can provide a better perception of the surroundings than virtual reality alone (e.g., visualization of construction simulations) (Behzadan and Kamat 2007), and the visualization of architectural designs (Thomas et al. 1999); 2) From the perspective of information retrieval, AR supplements a user's normal vision with context-related or Georeferenced virtual objects (e.g., looking through walls to see columns (Webster et al. 1996), looking beneath the ground to inspect subsurface utilities (Roberts et al. 2002), or retrieving drawings and BIM models for on-site communication (Yeh et al. 2012); and 3) From the perspective of interaction, authentic virtual models can be deployed to evaluate the physical condition of real objects (e.g., evaluation of earthquake-induced building damage (Kamat and El-Tawil 2007), and automation of construction process monitoring (Golparvar-Fard et al. 2009)). There have been other proposed applications for AR in construction such as training platforms for heavy construction equipment operators (Wang and Dunston 2007). For a comprehensive and state-of-the-art review of applications of AR, especially in construction and built environments, the reader is referred to (Wang et al. 2013).

While the aforementioned AR engineering applications possess tangible economic and social values, some fundamental technical challenges have to be overcome before

* Correspondence: dsuyang@umich.edu
[1]Department of Civil and Environmental Engineering, University of Michigan, 2350 Hayward Street, Suite 1380, Ann Arbor, MI 48105, USA
Full list of author information is available at the end of the article

AR can be deployed in practical outdoor applications. The difficulties are associated with two requirements: 1) maintaining a constant spatial alignment between the virtual and real entities, which is also referred as registration; and 2) creating a sustained illusion that the virtual and real entities co-exist, which is also referred to as occlusion.

Since most engineering AR applications expect to encounter registration and occlusion problems, it is reasonable to solve these challenges and build the solutions into a scalable and extensible AR computing framework that is openly accessible to the community. Researchers who are interested in exploring AR for their specific application in construction or another domain can immediately have access to the core logic components without starting from scratch on developing solutions for the registration and occlusion issues. The existence of such a reusable framework can significantly shorten the lifecycle of developing AR engineering applications.

A real-time occlusion algorithm with robust depth sensing and frame buffer techniques for handling occlusion problems in ubiquitous AR applications has been presented in the paper (Dong and Kamat 2012b), and the designed software module has been built into the SMART framework. This paper primarily focuses on the fundamental challenge of achieving precise registration in AR visualizations from both the hardware and software perspectives.

Backgound: Registration challenges in unprepared environments

The fundamental challenge in Augmented Reality is the difficulty of aligning virtual objects with the real environment with the correct pose, a process which is called registration in the literature (Azuma 1997). It is difficult because the registration errors arise from both the spatial and temporal domains (Azuma 1997). In addition, different tracking technologies have their own registration error sources. This paper focuses on solving the registration challenge in an unprepared environment (i.e., outdoors, where sensor-based AR is by far the most reliable tracking method free from constraints put on the user).

Errors in the spatial domain are also referred to as static errors when neither the user nor virtual objects are in motion. The static errors of sensor-based AR include: 1) inaccuracy in the sensor measurement; 2) mechanical misalignments between sensors; and 3) an incorrect registration algorithm. The selection of high-accuracy sensors is crucial, because the errors contained in the measurement can rarely be eliminated. Rigid placement of sensors on the AR backpack and helmet is also essential; else it can further compromise the system accuracy. Some early, relatively fragile AR backpack design examples can be found in the touring machine (Feiner et al. 1997) and Tinmith-Endeavour (Piekarski et al. 2004). A more robust and ergonomic version is demonstrated by the Tinmith backpack 2006 version (Stafford et al. 2006), in which a GPS antenna and an InterSense orientation tracker are anchored on top of a helmet. However, the 50cm accuracy of its GPS receiver is not sufficient for centimeter-level-accuracy AR tasks.

Compared with static errors, dynamic errors—errors in temporal domain—are much more unpredictable. The differences in latency between data streams create pronounced "swimming" effect of dynamic misregistration, which is called relative latency by (Jacobs et al. 1997). Given modern machine computational power, relative latency mainly results from: 1) off-host delay: the duration between the occurrence of a physical event and its arrival on the host; 2) synchronization delay: the time in which data is waiting between stages without being processed. Two common mitigation methods for resolving relative latency are: 1) adopting multi-threading programming or scheduling system latency (Jacobs et al. 1997); and 2) predicting head motion using a Kalman filter (Liang et al. 1991; Azuma et al. 1999).

Contributions: Scalable and modular augmented reality template

The mobile computing framework presented in this paper provides a comprehensive hardware and software solution for centimeter-level-accuracy AR applications in the outdoor environment. The robustness of the framework has been presented with a case study in visualizing underground infrastructure as part of an excavation planning and safety project.

The Scalable and Modular Augmented Reality Template (SMART) builds on top of the ARVISCOPE software platform (Behzadan and Kamat 2009). Besides its engine that interprets visual simulation scripts, another contribution of ARVISCOPE is creating some basic modules communicating with peripheral hardware that can later be imported into other potential AR applications. SMART takes advantage of these modules, and constructs an AR application framework that separates the AR logic from the application-specific logic, and make the libraries hardware independent, thus providing a flexible and structured AR development environment for the rapid creation of complex AR visual applications.

The in-built registration algorithm of SMART affords high-accuracy static alignment between real and virtual objects. Efforts have also been made to reduce dynamical misregistration, including: 1) in order to reduce synchronization latency, multiple threads are dynamically generated for reading and processing sensor measurement immediately upon the data arrival in the host system; and 2) an adaptive lag compensation algorithm is designed to reduce the latency induced by the Finite Impulse Response (FIR) filter.

In order to test the robustness and effectiveness of SMART in specific engineering application, ARMOR is constructed as an hardware realization of SMART, The Augmented Reality Mobile OpeRation platform (ARMOR) evolves from the UM-AR-GPS-ROVER hardware platform (Behzadan et al. 2008). ARMOR improves the design of the UM-AR-GPS-ROVER in two distinct ways: rigidity and ergonomics. It introduces high-accuracy and lightweight devices, rigidly places all tracking instruments with full calibration, and upgrades the carrying harness to make it more wearable.

Methods: SMART software framework

SMART provides a default application framework for AR tasks, where most of its components are written as generic libraries and can be inherited in specific applications. The framework isolates the domain logic from AR logic, so that the domain developer only needs to focus on realizing application-specific functionalities and leaving the AR logic to the SMART framework. Furthermore, SMART is designed to be hardware independent, therefore developer can pick their own hardware profile, write modules that conforms to SMART interface, and make the hardware run on top of SMART.

The SMART framework follows the classical model-view-controller (MVC) pattern. Scene-Graph-Controller is the implementation of the MVC pattern in SMART: (1) the model counterpart in SMART is the CARScene that utilizes application-specific I/O engines to load virtual objects, and that maintains their spatial and attribute status. The update of a virtual object's status is reflected when it is time to refresh the associated graphs; (2) the CARGraph corresponds to the view and reflects the AR registration results for each frame update event. Given the fact that the user's head can be in constant motion, the graph always invokes callbacks to rebuild the transformation matrix based on the latest position and attitude measurement, and refreshes the background image; (3) the CARController—manages all of the UI elements, and responds to a user's commands by invoking delegates' member functions like a scene or a graph.

The framework of SMART that is based on a Scene-Graph-Controller set-up is constructed in the following way (Figure 1). The main entry of the program is CARApp, which is in charge of CARSensorForeman and CARSiteForeman. The former initializes and manages all of the tracking devices, like RTK rovers and electronic compasses. The latter one defines the relation among scene, graphs, and controller. After a CARSiteForeman object is initialized, it orchestrates the creation of CARController, CARScene, and CARGraph, and the connection of graphs to the appropriate scene. The controller keeps pointers to the scene and the graph(s).

Application for operations-level construction animation

As a study case of adapting SMART to a specific engineering application, ARVISCOPE animation functions have been re-implemented under the SMART framework as follows. In order to load the ARVISCOPE animation trace file (Behzadan and Kamat 2009), CARSiteForemanA contains CARSceneA, CARControllerA , and CARGraphA, all of which are subclasses inheriting from SMART's superclasses, and are adapted for animation function. For example, CARSceneA employs CARStatementProcessor and CARnimation classes as the I/O engine to interpret the trace file, and CARControllerA adds customized elements for controlling the animation such as play, pause, continue, and jump functions.

Static registration

Registration process

The registration process of Augmented Reality is very similar to the computer graphics transformation process: 1) positioning the viewing volume of a user's eyes in the world coordinate system; 2) positioning objects in the world coordinate system; 3) determining the shape of the viewing volume; and 4) converting objects from the world coordinate system to the eye coordinate system (Shreiner et al. 2006). However, unlike computer graphics where parameters needed for step 1 through 3 are coded or manipulated by the user, Augmented Reality rigidly fulfills these steps according to the 6 degrees of freedom measured by tracking devices and the lens parameter of the real camera. Figure 2 lists the registration process, the needed parameters, and their measuring devices.

Step 1 – viewing The viewing step computes the relative transformation from the world coordinate system to the eye coordinate system. The origin of the world coordinate system coincides with that of the eye coordinate system, which is the user's geographical location (Figure 3). The world coordinate system uses a right-handed system with the Y-axis pointing in the direction of the true north, the X-axis pointing to the east, and the Z-axis pointing upward. The eye coordinate system complies with the OpenSceneGraph (OSG) (Martz 2007) default coordinate system, using a right-handed system with the Z-axis as the up vector, and the Y-axis departing from the eye.

Yaw, pitch, and roll—all measured by the magnetic sensor—are used to describe the relative orientation between the world and eye coordinate systems (Figure 4). There are six possibilities of rotating sequences (i.e., xyz, xzy, zxy, zyx, yzx, and yxz), and zxy is picked to construct the transformation matrix between the two coordinate systems (Equation 1). Suppose the eye and world coordinate systems coincide at the beginning; the user's head rotates around the Z-axis by yaw angle $\Psi \in (-180, 180]$ to get the new axes, X' and Y'. Since the rotation is

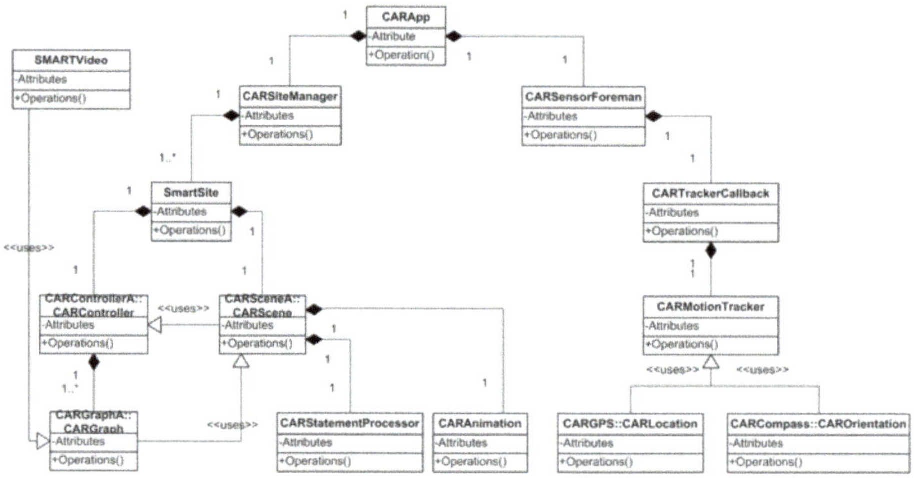

Figure 1 SMART framework architecture.

clockwise under the right-handed system, the rotation matrix is $R_z(-\Psi)$. Secondly, the head rotates around the X'-axis by pitch angle θ ϵ [−90, +90] to get the new axes, Y'' and Z', with counter-clockwise rotation of $R_{x'}(\theta)$. Finally, the head rotates around the Y''-axis by roll angle φ ϵ (−180, 180] with a counter-clockwise rotation of $R_{y''}(\varphi)$ to reach the final attitude.

Converting the virtual object from the world coordinate to the eye coordinate is an inverse process of rotating from the world coordinate system to the eye coordinate system, therefore the rotating matrix is written as: $R_z(\Psi)\ R_{x'}(-\theta)\ R_{''}(-\varphi)$ or $R_z(yaw)\ R_{x'}(-pitch)\ R_{y''}(-roll)$ (Equation 2). The rotation matrix can be further constructed as quaternion, a simple and robust way to express rotation, by specifying the rotation axis and angles. The procedure is explained as follows, and its associated equations are listed in sequence from Equation 3 to 5 (referenced in Figure 4): rotating around the Y''-axis by −φ degree, then rotating around the X'-axis by −θ degree, and finally rotating around the Z-axis by Ψ degree.

$$\begin{bmatrix} Xe \\ Ye \\ Ze \end{bmatrix} = \begin{bmatrix} \cos\Psi & \sin(-\Psi) & 0 \\ \sin\Psi & \cos\Psi & 0 \\ 0 & 0 & 1 \end{bmatrix} * \begin{bmatrix} 1 & 0 & 0 \\ 0 & \cos\theta & \sin\theta \\ 0 & \sin(-\theta) & \cos\theta \end{bmatrix} * \begin{bmatrix} \cos\varphi & 0 & \sin(-\varphi) \\ 0 & 1 & 0 \\ \sin\varphi & 0 & \cos\varphi \end{bmatrix} * \begin{bmatrix} Xw \\ Yw \\ Zw \end{bmatrix} \quad (1)$$

(X_w, Y_w, Z_w) is the point expressed in the world coordinate system; (X_e, Y_e, Z_e) is the same point but expressed in the eye coordinate system.

Step	Task	Illustration	Parameters and Device
Viewing	Position the viewing volume of a user's eyes in the world		Attitude of the camera (Electronic Compass)
Modeling	Position the objects in the world	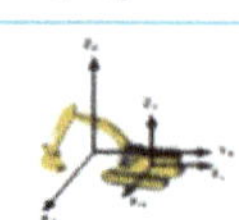	Location of the world origin (RTK GPS)
Creating Viewing Frustum	Decide the shape of the viewing volume	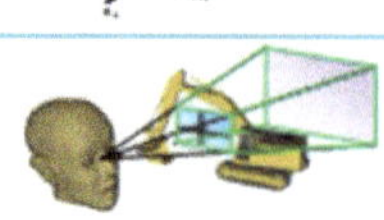	Lens and aspect ratio of camera (Camera)
Projection	Project the objects onto the image plane		Perspective Projection Matrix

Figure 2 The four steps of registration process.

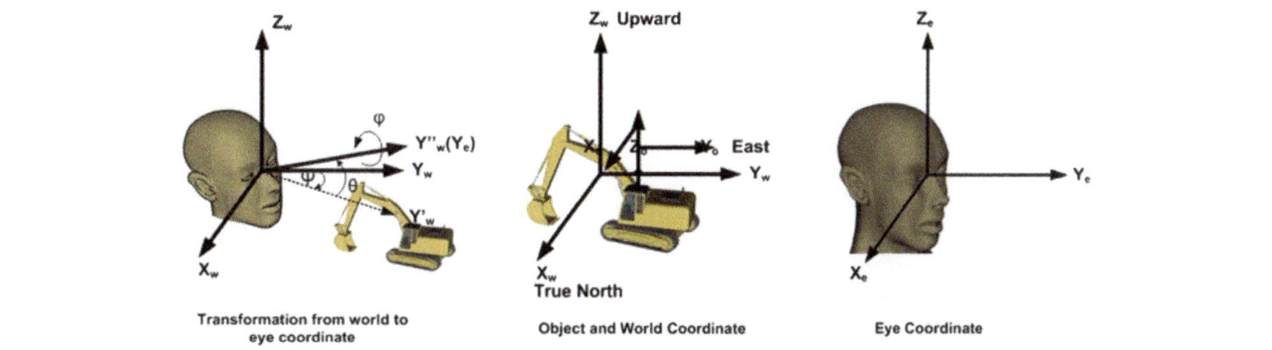

Figure 3 Definition of the object coordinate system, world coordinate system, and eye coordinate system.

$$Pe = Rz\Psi * Rx'\,(-\theta) * Ry''\,(-\varphi) * Pw \tag{2}$$

Equation 2 is a simplified version of Equation 1, while R_AB stands for rotation matrix around axis A by degree of B.

$$Z\text{-}axis = \begin{bmatrix} 0 \\ 0 \\ 1 \end{bmatrix} \tag{3}$$

$$X'\text{-}axis = \begin{bmatrix} \cos\Psi & \sin\Psi & 0 \\ \sin(-\Psi) & \cos\Psi & 0 \\ 0 & 0 & 1 \end{bmatrix} * \begin{bmatrix} 1 \\ 0 \\ 0 \end{bmatrix} = \begin{bmatrix} \cos\Psi \\ -\sin\Psi \\ 0 \end{bmatrix} \tag{4}$$

$$Y''\text{-}axis = \begin{bmatrix} 1 & 0 & 0 \\ 0 & \cos\theta & \sin(-\theta) \\ 0 & \sin\theta & \cos\theta \end{bmatrix} * \begin{bmatrix} \cos\Psi & \sin\Psi & 0 \\ \sin(-\Psi) & \cos\Psi & 0 \\ 0 & 0 & 1 \end{bmatrix} * \begin{bmatrix} 0 \\ 1 \\ 0 \end{bmatrix} = \begin{bmatrix} \sin\Psi \\ \cos\theta\cos\Psi \\ \sin\theta\cos\Psi \end{bmatrix} \tag{5}$$

Step 2 – modeling The modeling step positions the virtual objects in the world coordinate system. The definition of the object coordinate system is determined by the drawing software. The origin is fixed to a pivot point on the object with program-specified geographical location. The geographical location of the world coordinate origin is also given by the GPS measurement; the 3D vector between the object and world coordinate origins can thus be calculated. The methods to calculate the distance between geographical coordinates is originally introduced by (Vincenty 1975), and (Behzadan and Kamat 2007) proposed a reference point concept to calculate the 3D vector between two geographical locations. SMART adopts the same algorithm to place a virtual object in the world coordinate system using the calculated 3D vector. After that, any further translation, rotation, and scaling operations are applied on the object.

Steps 3 and 4 – viewing frustum and projection With the viewing frustum and projection, the virtual objects are ultimately projected on the image plane. The real world is perceived through the perspective projection by the human eye and the web camera. Four parameters are needed to construct a perspective projection matrix: horizontal angle of view, horizontal and vertical aspect ratio, and NEAR plane and FAR plane (any content is clipped if its distance falls outside the bound of NEAR and FAR place). All of them together form a viewing frustum and decide the amount of virtual content shown in

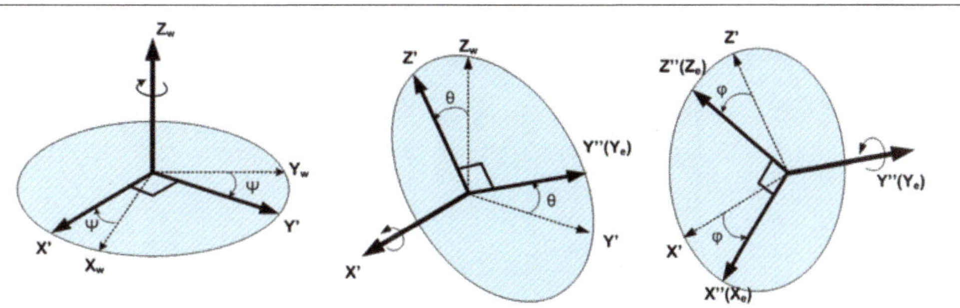

Figure 4 The relative orientation between the world and eye coordinate systems is described by yaw, pitch, and roll.

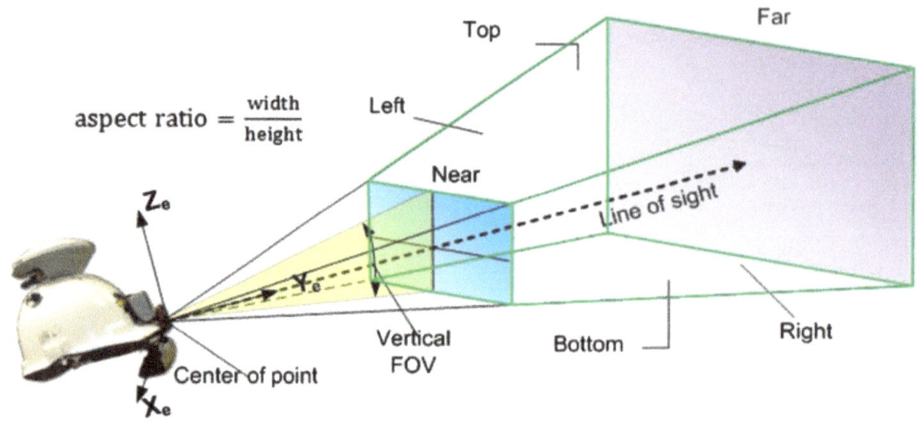

Figure 5 The viewing frustum defines the virtual content that can be seen.

the augmented space (Figure 5). Virtual objects outside of the viewing frustum are either cropped or clipped.

The NEAR and FAR planes do not affect how the virtual object appears on the projection plane. However, to maintain a high precision z-buffer, the principle is to keep the NEAR plane as far as possible, and the FAR plane as close as possible. The horizontal and vertical angle of view directly influence the magnification of the projected image and are affected by the focal length and aspect ratio of the camera. In order to ensure consistent perspective projection between the real and virtual camera, both of them need to share the same angle of view. This is achieved by calibrating the real camera and conforming the virtual camera's angle to view to that of the real camera.

Registration validation experiment

Calibration of the mechanical attitude discrepancy

The mechanical attitude discrepancy between the real camera and the electronic compass needs to be compensated by the following calibration procedure. The mechanical position discrepancy between the real camera and the GPS receiver can be calibrated in the similar way in the outdoor environment. A real box of size 12cm × 7 cm × 2 cm (length × width × height) is placed at a known pose. A semi-transparent 3D model

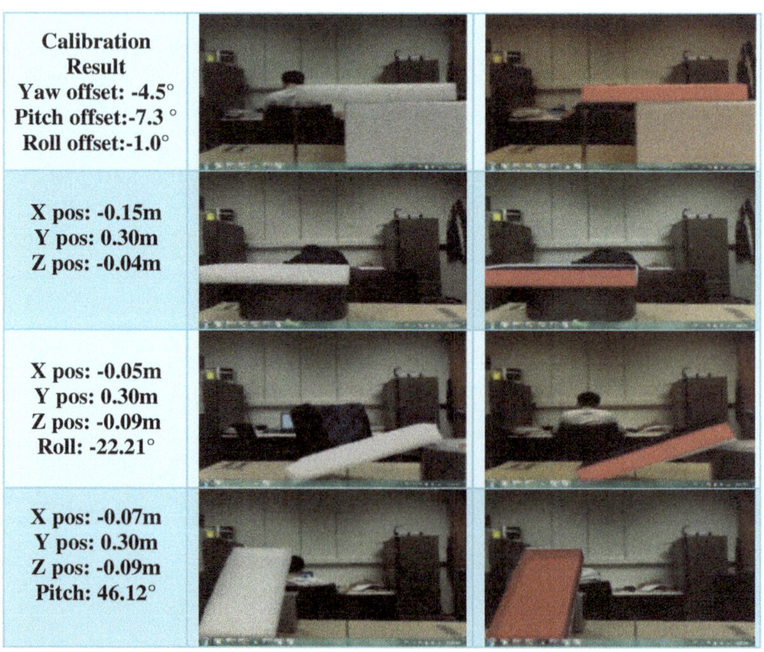

Figure 6 Mechanical attitude calibration result and validation experiment of registration algorithm.

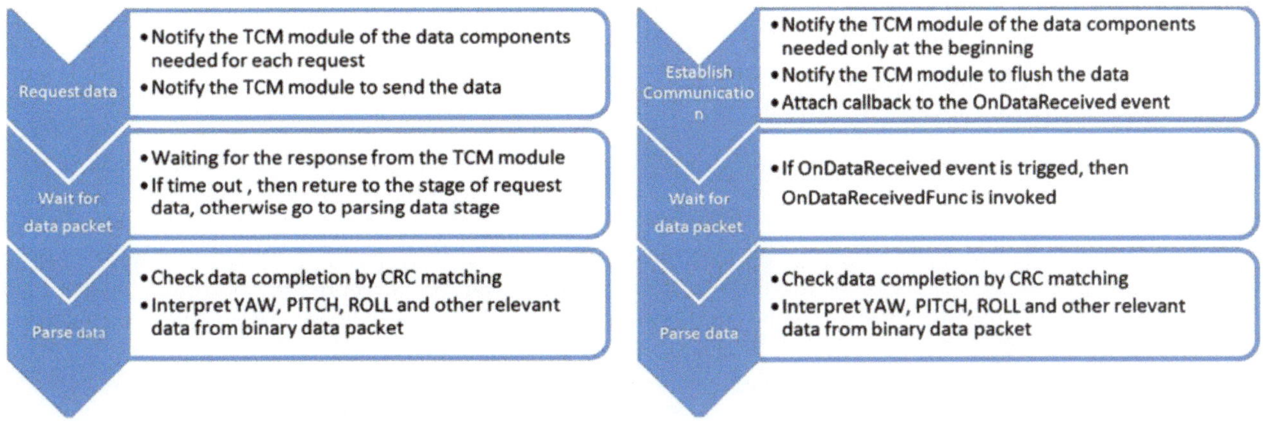

Figure 7 Communication stages in the PULL Mode (left) and the PUSH Mode (right).

of the same size is created and projected onto the real scene, so that the level of alignment can be judged. The virtual box is first projected without adjustments to the attitude measurement, and discrepancy is thus present. The virtual box is then shifted to align with the real one by adding a compensation value to the attitude measurement, as shown in Figure 6 Row 1.

Validation of the static registration algorithm A series of experiments are performed to validate the agreement between the real and virtual camera; if the static registration algorithm works correctly, the virtual box should coincide with the real box when moved together with 6 degrees of freedom. Overall, the virtual box accurately matches the real one in all tested cases, and a selected set of experiments are shown below in Figure 6 Rows 2~3.

Resolving the latency problem in the electronic compass
Due to the latency induced by the compass module itself, correct static registration does not guarantee that the user can see the same correct and stable augmented image when in motion. This section addresses the cause and solution for the dynamic misregistration problem.

Multi-threading to reduce synchronization latency
There are two options for communicating with the compass module: PULL and PUSH mode. PULL is a passive output mode for the compass module, and is used by

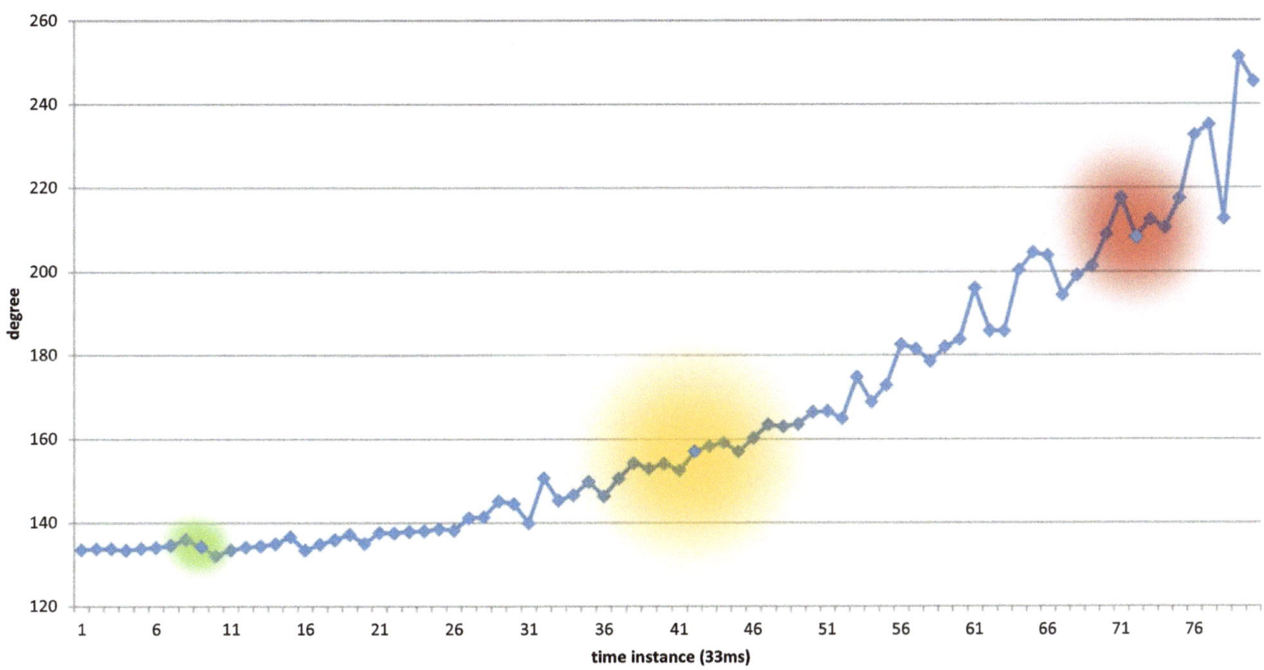

Figure 8 The data shows that the noise in the raw data increases as the motion accelerates.

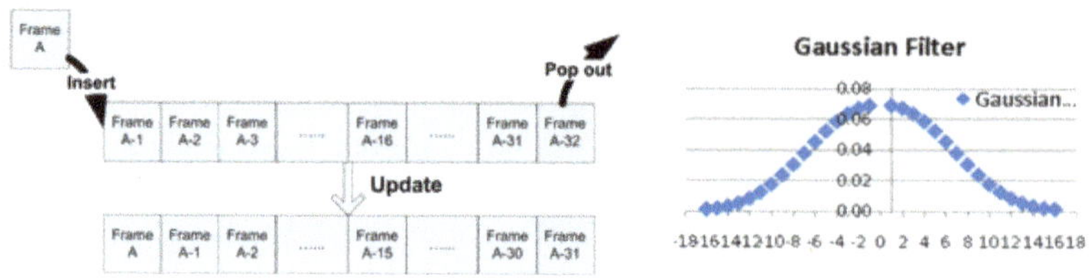

Figure 9 The Filter-induced latency when a 32 tap Gaussian filter is used.

the UM-AR-GPS-ROVER to pull data out of the module. Without separating I/O communication with the electronic compass as a background task, one loop of the pulling request costs 70ms on average, and significantly slows down program performance.

The PUSH mode is an active output mode for the compass module. SMART selects the PUSH mode as its data communication method to increase the program efficiency. If the PUSH mode is selected, the module outputs the data at a fixed rate set by the host system (Figure 7). If the fixed rate is set to 0, which is done by SMART, the module will flush the next data packet as soon as the previous one is sent out. The sampling and flushing happens at proximately 30 to 32 Hz. The biggest advantage of choosing the PUSH mode is that, once the initial communication is successfully established, and no FIR filtering is carried on the hardware board, the host system can acquire the observed orientation data in only 5 ms on average. Accordingly, SMART adopts an event-based asynchronous pattern to handle high frequency data packet arrival and process the data on a separate thread in the background without interrupting the main loop. To be specific, a dedicated listener is set up on a separate thread. It is active in listening to the data package posted from the module device. Once data package arrives, the listener triggers its attached delegate that computes the registration result. This multi-threaded processing accelerates the main function and also reduces the synchronization latency to the minimum possible value.

Filter-induced latency

Even though the PUSH mode is free of synchronization delay, there is still significant latency if the FIR filter is switched on inside of the compass module. This section explains the reason for this phenomenon. The calibration of the magnetometer can compensate for a local static magnetic source within the vicinity of the compass module. However, dynamic magnetic distortion still impacts the module in motion, and the noise magnification depends on the acceleration of the module.

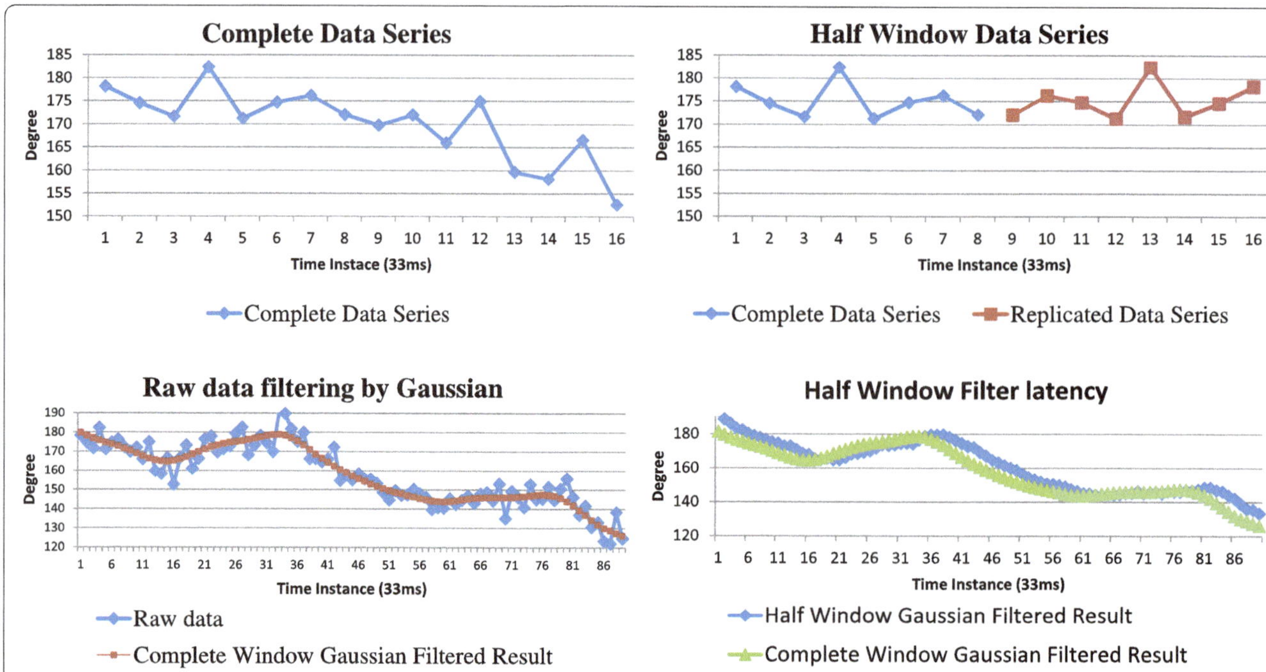

Figure 10 Half-Window Gaussian filter immediately generates estimation for time instance A, while the estimation still contains 4-5 frames of latency on average.

Usually the faster the acceleration is, the higher the noise is (Figure 8). Among the three degrees of freedom, heading is the most sensitive to the noise.

Except the high frequency noise that can be caused by abrupt vibration and can significantly distort FIR filtered results, other types of noise, especially jittering noise, can be removed/smoothed by a FIR Gaussian filter. The compass module comes with 5 options for filtering: 32, 16, 8, 4 and 0 tap filter. The higher the number is, the more stable the output, but the longer the expected latency. The reason can be illustrated by the case of selecting a 32 tap filter (Figure 9). When it is time to send out estimated data at time instant A, the module adds a new sample A to the end of the queue with the first one being dropped, and applies a Gaussian filter to the queue. However, the filtered result actually reflects

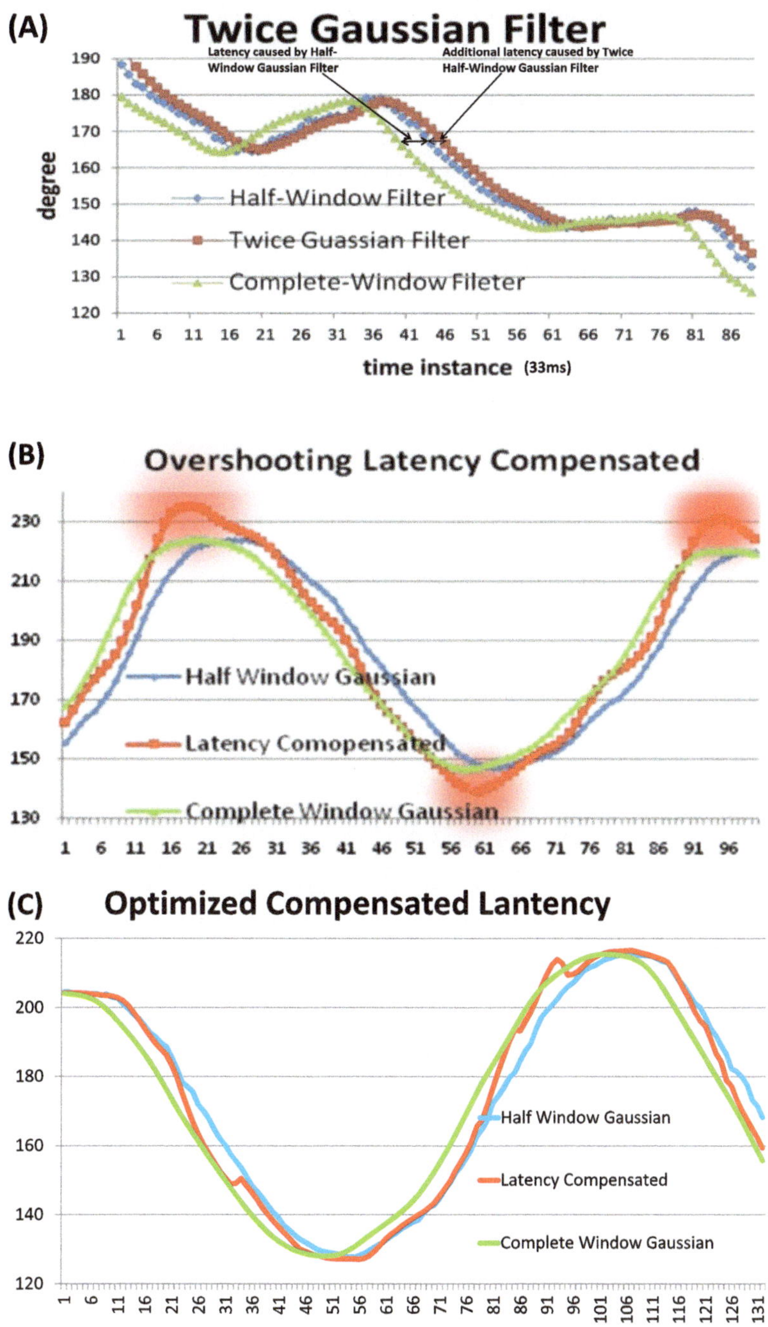

Figure 11 Twice Half-Window Gaussian Filter causes additional latency, however it is also a hint for making the Half-Window Gaussian filtered results more close to the 'true' value. (**A**) Shows the latency caused by Twice Gaussian Filter; (**B**) shows the compensation results with overshooting problem at the transitions; (**C**) shows the optimized compensation algorithm to alleviate the overshooting problem.

the estimated value at time instant (A–15). Since the module samples at approximately 30–32 Hz, it induces a 0.5 second delay for a 32 tap filter; a 0.25 second delay for 16 tap filter, and so on. This is called filter-induced latency, and it applies to both PULL and PUSH mode. A 0 tap filter implies no filtering, but significant jittering.

Half-window Gaussian filter

In order to avoid the filter-induced latency, the Gaussian FIR filter is moved from the hardware to the software, but with only half window size applied. For example, if a complete Gaussian window is used, it is not until time instant A+15 that the estimated value can be available for time instant A. However, Half-Window replicates the past data from time instant A-15 to time instant A as the future data from time instant A+1 to A+16, and immediately generates an estimated value for time instant A without any delay (Figure 10). Nevertheless, as is shown in the graph chart, half window still causes 4–5 frames of latency on average. Depending on the speed of module movement, the faster the speed, the longer latency it presents. We address this kind of latency as Half-Window induced latency.

Because the Half-Window Gaussian filter puts more emphasis on the current frame, it makes the estimated result more sensitive to noise contained in the current frame, and consequently the estimated result is more jittering than the one of the Complete-Window Gaussian filter. Therefore, a second Half-Window Gaussian is applied on the first Half-Window Gaussian filtered result for smoothing purposes, but this introduces 1–2 extra frames of latency. This extra latency can be found in Figure 9A, where the filtered result in red of Twice Half-Window Gaussian filter is pushed toward right for additional 2 time instances compared with the filtered result in blue of Half-Window filter (Figure 11A). However, unlike the Half-Window Gaussian latency which cannot be computed (since the 'true value' is unknown), the additional latency can be calculated as the subtraction between the Half-Window and Twice Half-Window Gaussian filter results. Furthermore, in observance that the additional latency can be is always smaller the original Half-Window latency. double the additional latency can thus be subtracted from the Twice Half-Window Gaussian filter result which makes the estimation closer to the 'true value'than the Half-Window Gaussian filter result. This approach works most of the time except at the transition state, and it leads to overshooting during change of direction, and during transitions from dynamic to static states as it is shown Figure 9B.

Adaptive latency compensation algorithm

In order to resolve the overshooting problem, the estimated result needs to be forced to the observed data when the module comes to a stop. This is possible because when the module is static, the observed data is very stable and close to the actual. Large collections of observed value show standard deviation is a good indicator of dynamic and static status of the sensor. Empirically, when the standard deviation is larger than 6, the module is in motion, otherwise it is in static or on the way to stopping (Figure 12). Therefore the adaptive algorithm computes the latency compensated value to the Twice Half-Window Gaussian filter in the following way: when the standard deviation is no smaller than 6, the compensated value is double of the subtraction of the Half-Window Gaussian filter by the Twice Half-Window Gaussian filter result; otherwise the compensated value is equal to the subtraction of the Twice Half-Window Gaussian Filter by the observed data (Figure 11C).if standard deviation >= 6

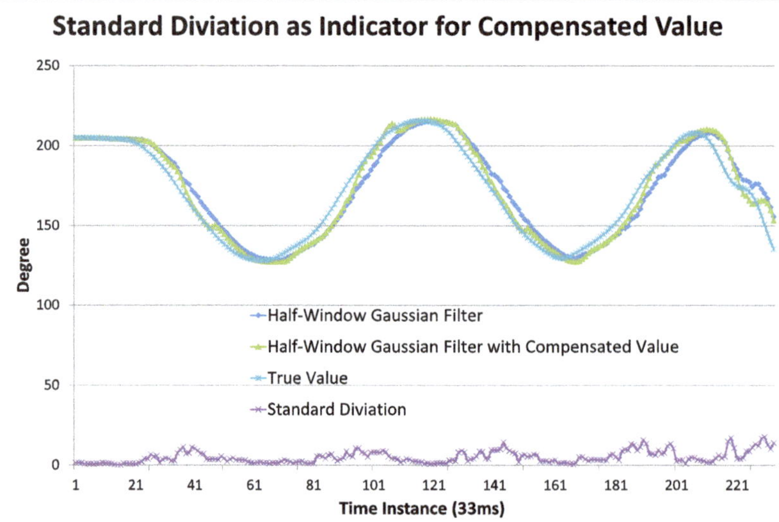

Figure 12 Standard deviation indicates the motion pattern.

Table 1 Comparison between the UM-AR-GPS-ROVER and ARMOR configuration

Device	UM-AR-GPS-ROVER	ARMOR	Comparison
Location Tracking	Trimble AgGPS 332 using OmniStar XP correction for Differential GPS method	Trimble AgGPS 332 using CMR correction broadcast by a Trimble AgGPS RTK Base 450/900	OmniStar XP provides 10~20 cm accuracy. RTK provides 2.5 cm horizontal accuracy, and 3.7 cm vertical accuracy
Orientation Tracking	PNI TCM 5	PNI TCM XB	Same accuracy, but ARMOR places TCM XB rigidly close to the camera
Video Camera	Fire-I Digital Firewire Camera	Microsoft LifeCam VX-5000	LifeCam VX-5000 is lightweight, small and uses less wire
Head-mounted Display	i-Glasses SVGA Pro video see-through HMD	eMagin Z800 3DVisor	Z800 3DVisor is lightweight with stereovision
Laptop	Dell Precision M60 Notebook	ASUS N10J Netbook	ASUS N10J is lightweight and smaller
User Command Input	WristPC wearable keyboard and Cirque Smart Cat touchpad	Nintendo Wii Remote	Wii Remote is lightweight and intuitive to use
Power Source	Fedco POWERBASE	Tekkeon myPower MP3750	MP3750 is lightweight and has multiple voltage outputs charging both GPS receiver and HMD
Backpack Apparatus	Kensington Contour Laptop Backpack	Load Bearing Vest	Extensible and easy to access equipment

Compensation Value = 2 * (Half-Window Gaussian Filter Result – Twice Half-Window Gaussian Filter Result)

else

Compensation Value = Twice Half-Window Gaussian Filter Result – Observed Data

ARMOR hardware architecture

To test SMART framework's correctness and robustness, ARMOR, an evolved product from UM-AR-GPS-ROVER (Behzadan and Kamat 2009), is designed as a hardware platform implementation of SMART. As a prototype design, the UM-AR-GPS-ROVER meets the need of a proof-of-concept platform for visual simulation. However, there are two primary design defects that are inadequately addressed: accuracy and ergonomics. First, the insecure placement of tracking devices disqualifies the UM-AR-GPS-ROVER from the centimeter-accuracy-level goal. Secondly, co-locating all devices, power panels, and wires into an ordinary backpack makes it impossible to accommodate more equipment like the Real Time Kinematic (RTK) rover radio. The weight of the backpack is also high for even distribution around the body.

Table 2 Power voltage demands of different devices

Component	External power supply	Voltage	Power
Head-mounted Display	MP3750	5 V	<1.25 W
Electronic Compass	AAA batteries	3.6 V~5 V	0.1 W
Web Camera	Through USB	Unknown	Unknown
RTK Rover Receiver	MP3750	10 V~32 V	4.2 W
RTK Rover Radio	Through RTK Rover Receiver	10.5 V~20 V	3 W
RTK Base Receiver	Integrated Internal battery	7.4 V	<8 W
User Command Input	AA batteries	3 V	Unknown
Laptop	Integrated Internal battery	11.1 V	17.76 W

ARMOR represents a significant upgrade from the UM-AR-GPS-ROVER. The improvements can be divided into three categories: 1) highly accurate tracking devices with rigid placement and full calibration; 2) lightweight selection of input/output devices and external power source, and 3) load-bearing vest to accommodate devices and distribute weight evenly around the user's body. An overview comparison between UM-AR-GPS-ROVER and ARMOR is listed in Table 1.

Tracking devices

Position tracking device — RTK-GPS

The UM-AR-GPS-ROVER uses the AgGPS 332 Receiver with OmniStar XP mode to provide a user's position (i.e., the position of camera center) with 10~20 cm accuracy. This level of accuracy is sufficient for creating animations of a construction process simulation, where slight positional displacement may not necessarily compromise the validation purpose. However, for precision critical applications, 10~20 cm accuracy is not sufficient for visual accuracy.

The AgGPS 332 Receiver used in UM-AR-GPS-ROVER is thus upgraded and three objectives are pursued: 1) The upgraded GPS must be able to support centimeter-level accuracy; 2) The hardware upgrade should have minimum impact on the software communication module; and 3) The existing device should be fully utilized given the cost of high-accuracy GPS equipment. Ultimately, the AgGPS RTK Base 450/900 GPS Receiver is chosen for implementing the upgrade for three reasons. First, it leverages RTK technology to provide 2.5 cm horizontal accuracy and 3.7 cm vertical

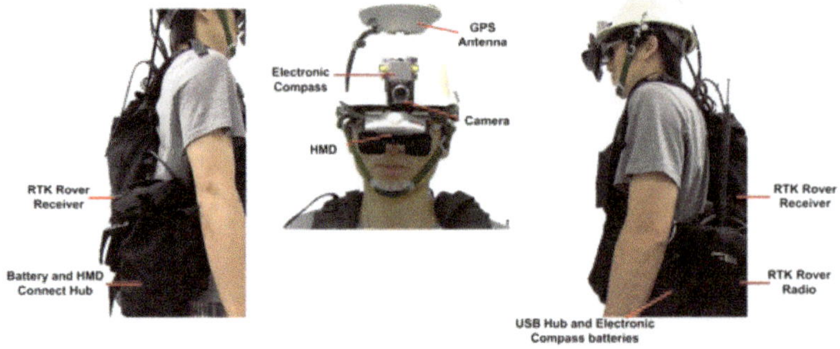

Figure 13 The profile of ARMOR from different perspectives.

accuracy on a continuous real-time basis. The RTK Base 450/900 Receiver is set up as a base station placed at a known point (i.e., control points set up by the government with 1st order accuracy), and tracks the same satellites as an RTK rover. The carrier phase measurement is used to calculate the real-time differential correction that is sent as a Compact Measurement Record (CMR) through a radio link to the RTK rover within 100 km (depending on the radio amplifier and terrain) (Trimble 2007). The RTK rover applies the correction to the position it receives and generates centimeter-level accuracy output. The second reason for choosing this receiver is that, despite the upgrade, the RTK rover outputs the position data in NMEA format (acronym for National Marine Engineers Association) (NEMA 2010), which is also used in OmniStar XP. No change therefore applies to the communication module. The third and final reason is that the original AgGPS 332 Receiver is retained as an RTK rover with its differential GPS mode being changed from OmniStar XP to RTK. A SiteNet 900 radio works with the AgGPS 332 Receiver to receive the CMR from the base station.

Improvement has also been made to the receiver antenna placement. The UM-AR-GPS-ROVER mounted to the receiver antenna on a segment of pipe that was tied to the interior of the backpack. This method proved to be inefficient in preventing lateral movement. Therefore ARMOR anchors the GPS receiver with a bolt on the summit of the helmet, so that the phase center of the receiver will never shift relative to the camera center. The relative distance between the receiver phase center and the camera center is calibrated beforehand and added to the RTK rover measurement.

ARMOR can work in either indoor or outdoor mode. Indoor mode does not necessarily imply that the GPS signal is totally lost, rather the qualified GPS signal is absent. The GPS signal quality can be extracted from the $GGA section of the NMEA string. The fix quality ranges from 0–8, for example 2 means DGPS fix, 4 means Real Time Kinematic, and 5 means float RTK. The user can define the standard (i.e., which fix quality is deemed as qualified) in the hardware configuration file. When a qualified GPS signal is available, the geographical location is extracted from the $GPGGA

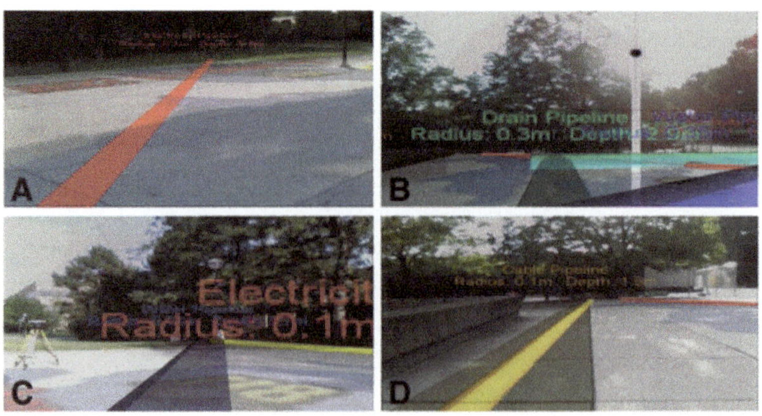

Figure 14 Labeling attribute information and color coding on the underground utilities. (**A**) Shows the proposed excavation area (the semi-transparent layer); (**B, C**) show the geometric and attribute information of the pipelines; (**D**) shows the buffering zone of the detected pipelines.

Figure 15 An x-ray view of the underground utilities.

section. Otherwise, a preset pseudo-location is used, and this pseudo-location can be controlled by user input.

Orientation tracking device—electronic compass

The TCM XB electronic compass is employed to measure the yaw, pitch, and roll that represent the relative attitude between the eye coordinate system and the world coordinate system. It measures the heading up to a full 360-degree range, and maintains the accuracy of 0.3°rms when the tilt (pitch and roll) is no larger than 65°, which is the common motion range of the human head.

Theoretically the electronic compass is applied to measure the orientation of the user's head. However, since a user's eyes are obstructed by a Video See-Through Head Mounted Display (HMD) and that the eyes' function is replaced by a video camera, the electronic compass is applied to measure the attitude of the camera instead of the eyes.

The UM-AR-GPS-ROVER places the electronic compass on the zenith of the helmet, which makes it very hard to align the camera and the electronic compass. ARMOR chooses to anchor the electronic compass rigidly close to the camera anchored on the visor of the helmet, and parallel to the line of sight, making the physical discrepancy calibration much easier. The calibration approach is described in Section 4.2.1. For safety reasons, the electronic

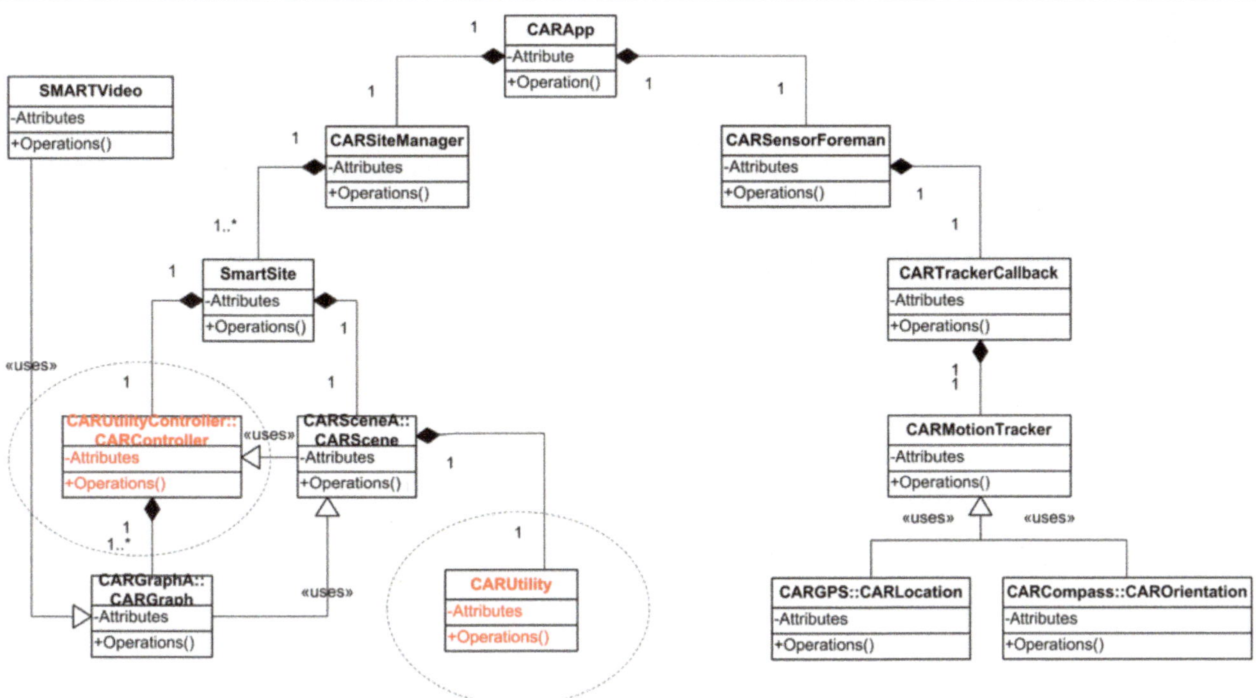

Figure 16 SMART adaption for visualization of underground utilities.

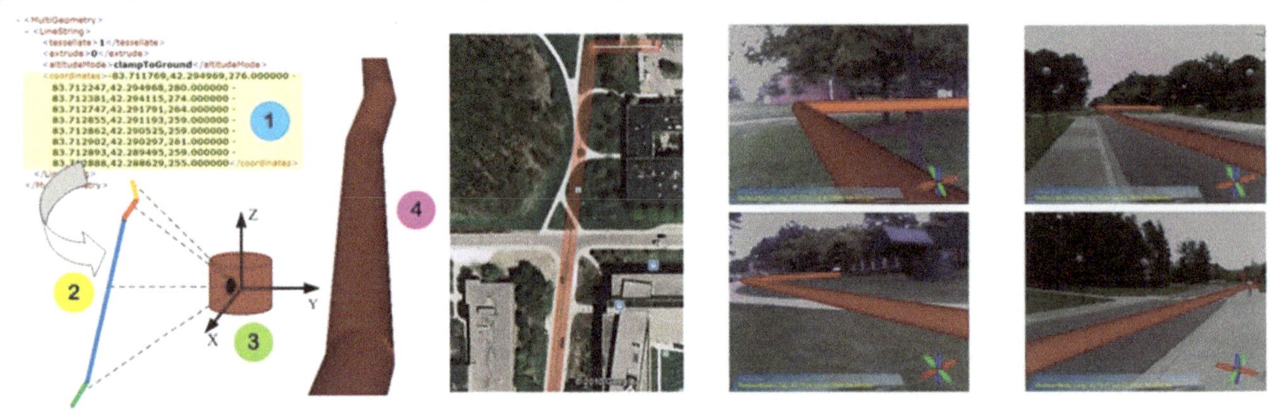

Figure 17 Conduit loading procedure, conduits overlaid on Google Earth and field experiment results.

compass is encapsulated in a custom-sized aluminum enclosure that is free of magnetic forces.

Magnetometer calibration also needs to be carried out for the purpose of compensating for distortions to the magnetic field caused by the host system and the local environment (PNI 2009). Given that ARMOR's entire periphery could have magnetic impact on the sensor—for example GPS receivers, HMDs, and web cameras—the TCM XB needs to be mounted within the host system that is moved as a single unit during the calibration.

Input/output devices and external power supply

Video sequence input: camera

The camera is responsible for capturing the continuous real-time background image. The ideal device should possess properties of high resolution, high-frequency sampling rate, and high-speed connection, with a small volume and lightweight. Microsoft LifeCam VX5000 is chosen for the following reasons—the size is only 45 mm × 45.6 mm and it does not compromise on resolution (640 × 480) or connection speed (480 Mbps). More importantly, it takes samples at 30 Hz, which is the same speed as the electronic compass.

Augmented view output: head-mounted display (HMD)

The augmented view generated by the video compositor is ultimately presented by the Video See-Through HMD. The eMagin Z800 3DVisor is chosen as the HMD component of ARMOR because it has remarkable performance in primary factors, including wide view angle, large number of colors, lightweight frame, and comfort. Furthermore, the stereovision capability is another important rendering effect that helps the user to better appreciate the 3D augmented space.

External power supply

External power supplies with variant voltage output are indispensible for powering devices without integrated internal batteries. ARMOR upgrades 'POWERBASE' of UM-AR-GPS-ROVER to 'Tekkeon myPower ALL MP3750' which shows improvements over 'POWERBASE' in four ways: 1) both the volume (17 cm x 8 cm x 2 cm) and weight (0.44 kg) of MP3750 are only 1/5 of POWERBASE's volume and weight; 2) the main output voltage varies from 10 V to 19 V for powering the AgGPS 332 Receiver (12 V), and an extra USB output port can charge the HMD (5 V) simultaneously (Table 2); 3) it features automatic voltage

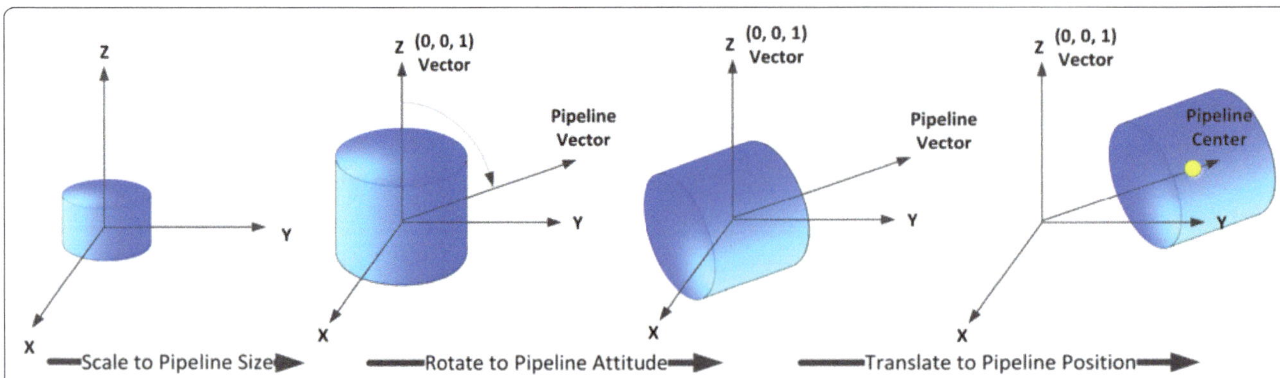

Figure 18 Process of creating pipeline segment from unit cylinder.

detection with an option for manual voltage selection; and 4) an extended battery pack can be added to double the battery capacity (Tekkeon 2009).

Load-bearing vest

The optimization of all devices in aspects of volume, weight, and rigidity allows the author to compact all components into one load-bearing vest. Figure 13 shows the configuration of the backpack and the allocation of hardware. There are three primary compartments: the back pouch accommodates the AgGPS 332 Receiver; the SiteNet 900 is stored in the right side pouch, and the left-side pouch holds the HMD connect interface box to a PC and the MP3750 battery; and the ASUS N10J netbook is securely tied to the inner part of the back. All of the other miscellaneous accessories—like USB to Serial Port hubs, or AAA batteries—are distributed in the auxiliary pouches.

The configuration of the vest has several overall advantages: 1) the design of the pouches allows for an even distribution of weight around the body; 2) the separation of devices allows the user to conveniently access and check the condition of certain hardware; and 3) different parts of the loading vest are loosely joined so that the vest can fit any body type, and be worn rapidly even when fully loaded. ARMOR has been tested by several users for outdoor operation that lasts continuously for over 30 minutes, without any interruption or significant discomfort.

Results and case study

The robustness of ARMOR and the SMART framework have been evaluated in an Augmented Reality application designed to visualize subsurface utilities during ongoing excavation operations for improved context awareness and accident avoidance.

Current practice of excavation damage prevention and its limitation

Every U.S. state's "One-Call Center" is a message handling system for underground utility owners. It collects excavation requests from contractors and distributes them to all of its members (utility owners). The standard procedure can be summarized as the following steps: 1) the contractor issues a text-based request to the "One-Call Center" describing planned excavation locations; 2) "One-Call Center" distributes the request to its members who may own buried assets in the proposed excavation area; 3) Each facility owner dispatches field locators to mark the approximate location of the underground facilities; 4) the excavator operators carry out the excavation activities based on the visual guidance marked on the ground.

Despite its effectiveness of reducing utility lines hit to a great extent, there are several weaknesses in the current practice of damage prevention to buried utilities that can be improved using Augmented Reality. First, the description about the proposed excavation is a text-based request ticket that can be vague and ambiguous, and it is up to the field locators to interpret the digging zone from the written request. In case of misinterpretation, the field locators either fail to cover the entire proposed excavation area, or take significant extra effort to mark the ground that is unnecessary. Second, even though the surface markings (paint, stakes, flags, etc.) serve as visual guidance for the excavation operator, they are vulnerable to heavy traffic, severe weather, and excavation activity which scrapes the surface and then removes the top soil, thus destroying the surface markings. This makes it challenging for an excavator operator to maintain spatial orientation, and must rely on memory and judgment to recollect marked utility locations as excavation proceeds (Talmaki and Kamat 2012).

Proposed practice for improvement of the "One-call" damage prevention practice with augmented reality

Augmented Reality can help to accurately visualize the proposed excavation area and digitize the located underground utilities, which bridges the communication among contractors, "One-Call Center", facility owners, field locators and excavation operators. First, the contractor can issue a "visual" ticket to the "One-Call Center" and facility owners by superimposing a semi-transparent layer above the precise proposed excavation range; second, dispatched locators come to the field, "see" the proposed digging area (Figure 14A) and can survey and mark precisely within that area. The markings are stored digitally and contain information about utility type and geometry attribute (Figure 14B, C) and buffer zone indicating the uncertainty of the survey result (Figure 14D); third, upon the completion of the digital marking, excavation operators can thus persistently "see" what utilities lie buried in a digging machine's vicinity, and consciously avoid accidental utility strikes (Figure 15).

Visualization of buried utilities based on SMART framework

The ultimate goal of SMART framework is to speed up the process of inventing engineering applications based on AR, so that domain experts can focus on building application logic and leave the AR logic to SMART framework. Figure 16 highlights the modifications compared with Figure 1 to adapt SMART for visualizing buried utilities. Essentially, two parts need to be overwritten: 1) The animation module (CARStatementProcessor and CARAnimation) is replaced by the utilities generation module (CARUtility), and the detail implementation of CARUtiliy is explained in the next section; 2) CARUtilityController replaces CARControllerA to respond to customized user inputs for creating, manipulating, visualizing, and archiving utilities.

Visualization logic of buried utilities

The workflow of CARUtility is briefly summarized here (Dong and Kamat 2012a). The digitized survey results are exported as KML (Keyhole Modeling Language) files. The following procedure interprets KML files and builds conduit (e.g. pipe) models in the augmented space (Figure 17):

(1) Extract the spatial and attribute information of pipelines from the KML file using libkml, a library for parsing, generating, and operating in KML (Google 2008). For example, the geographical location of pipelines is recorded under the Geometry element as "LineString" (Google 2012). A cursor is thus designed to iterate through the KML file, locate "LineString" elements, and extract the geographical locations.

(2) Convert consecutive vertices within one "LineString" from the geographical coordinate to the local coordinate in order to raise computational efficiency during the registration routine. The first vertex on the line string is chosen as the origin of the local coordinate system, and the local coordinates of the remaining vertices are determined by calculating the relative 3D vector between the rest of the vertices and the first one, using the Vincenty algorithm (Behzadan and Kamat 2007).

(3) In order to save storage memory, a unit cylinder is shared by all pipeline segments as primitive geometry upon which the transformation matrix is built.

(4) Scale, rotate, and translate the primitive cylinder to the correct size, attitude, and position (Figure 18). For simplicity, the normalized vector between two successive vertices is named as the pipeline vector. First, the primitive cylinder is scaled along the X- and Y-axis by the radius of the true pipeline, and then scaled along the Z-axis by the distance between two successive vertices. Secondly, the scaled cylinder is rotated along the axis—formed by the cross product between vector <0, 0, 1> and the pipeline vector—by the angle of the dot product between vector <0, 0, 1> and the pipeline vector. Finally, the center of the rotated pipeline is translated to the midpoint between two successive vertices. This step is applied to each pair of two successive vertices.

(5) Translucent rectangles (Figure 12D) are created and laid underneath the pipelines to represent the buffer zone for excavation.

(6) An inverted pyramid without bottom is created to represent the X-ray vision of the underground pipelines. Each cylinder segment and its associated rectangle (buffer zone) are tested against the surface of pyramid, and the parts falling outside are truncated (Figure 15).

Discussion and conclusion

This paper has presented the design and implementation of a robust mobile computing platform composed of the rigid hardware platform ARMOR and the application framework SMART with open access to the community. Researchers who are interested in improving the AR graphical algorithms or developing new AR engineering applications can take advantage of the existing AR framework implemented in this research, and can prototype their ideas in a much shorter lifecycle.

Targeting outdoor AR applications at centimeter-level accuracy, algorithms for both static and dynamic registration have been introduced. Several dynamic misregistration correction approaches are also described and evaluated. It is found that dynamic misregistration continues to be an open research problem and continues to be under investigation by the authors. Several efforts are being made in ongoing work, which include: 1) synchronizing the captured image and sensor measurements; and 2) optimizing the adaptive latency compensation algorithm with image processing techniques (e.g., optical flow can provide additional heuristics about the angular speed).

A video demo about SMART and ARMOR, and its application in excavation damage prevention can be found at <http://pathfinder.engin.umich.edu/videos.htm>. A video demo about its occlusion capability can also be found at the same site. SMART is an open source project that can be downloaded at <http://pathfinder.engin.umich.edu/software.htm>.

Competing interests

Both authors declare that they no competing interests.

Authors' contributions

SD coded the software package SMART, designed the hardware implementation ARMOR, and carried out the field experiment. VK co-authored the software package SMART, and co-designed the field experiment. All authors read and approved the final manuscript.

Author details

[1]Department of Civil and Environmental Engineering, University of Michigan, 2350 Hayward Street, Suite 1380, Ann Arbor, MI 48105, USA. [2]Department of Civil and Environmental Engineering, University of Michigan, 2350 Hayward Street, Suite 2340, Ann Arbor, MI 48105, USA.

References

Azuma, R. (1997). A survey of augmented reality. *Teleoperators and Virtual Environments, 6*(4), 355–385.

Azuma, R., Hoff, B., Neely, H., & Sarfaty, R. (1999). *A motion-stabilized outdoor augmented reality*. Houston, TX: Proceedings of the Virtual Reality. IEEE.

Behzadan, A. H., & Kamat, V. R. (2007). Georeferenced registration of construction graphics in mobile outdoor augmented reality. *Journal of Computing in Civil Engineering, 21*(4), 247–258.

Behzadan, A. H., & Kamat, V. R. (2009). Automated generation of operations level construction animations in outdoor augmented reality. *Journal of Computing in Civil Engineering, 23*(6), 405–417.

Behzadan, A. H., Timm, B. W., & Kamat, V. R. (2008). General-purpose modular hardware and software framework for mobile outdoor augmented reality applications in engineering. *Advances Engineering Informatics, 22*, 90–105.

Dong, S., & Kamat, R. V. (2012a). *Scalable and Extensible Augmented Reality with Applications in Civil Infrastructure Systems*. Ann Arbor: University of Michigan.

Dong, S., & Kamat, R. V. (2012b). Real-Time Occlusion Handling for Dynamic Augmented Reality Using Geometric Sensing and Graphical Shading. *Journal of Computing in Civil Engineering*. in press.

Feiner, S., Macintyre, B., & Hollerer, TA. (1997). A Touring Machine: Prototyping 3D Mobile Augmented Reality Systems for Exploring the Urban Environment. In *Proceedings of 1997 International Symposium on Wearable Computers* (pp. 74–81). Piscataway, NJ: IEEE.

Golparvar-Fard, M., Pena-Mora, F., Arboleda, C. A., & Lee, S. (2009). Visualization of construction progress monitoring with 4D simulation model overlaid on time-lapsed photographs. *Journal of Computing in Civil Engineering, 23*(6), 391–404.

Google. (2008). *Introducing libkml: a library for reading, writing, and manipulating KML*. http://google-opensource.blogspot.com/2008/03/introducing-libkml-library-for-reading.html.

Google. (2012). *KML Documentation Introduction*. https://developers.google.com/kml/documentation/.

Jacobs, M. C., Livingston, M. A., & State, A. (1997). Managing Latency in Complex Augmented Reality Systems. In *Proceedings of the 1997 Symposium on Interactives 3D Graphics* (pp. 49–54). New York, NY: ACM.

Kamat, V. R., & El-Tawil, S. (2007). Evaluation of augmented reality for rapid assessment of earthquake-induced building damage. *Journal of Computing in Civil Engineering, 21*(5), 303–310.

Liang, O., Shaw, C., & Green, M. (1991). On temporal-spatial realism in the virtual reality environment. In *Proceeding of 1991 Symposium on User Interface Software and Technology*. New York, NY: ACM.

Martz, P. (2007). *OpenSceneGraph Quick Start Guide*.

NEMA. (2010). *NEMA data*. http://www.gpsinformation.org/dale/nmea.htm.

Piekarski, W., Smith, R., & Thomas, B. H. (2004). Designing Backpacks for High Fidelity Mobile Outdoor Augmented Reality. In *Proceedings of the 2004 IEEE and ACM International Symposium on Mixed and Augmented Reality* (pp. pp. 280–281). Piscataway, NJ: IEEE.

PNI. (2009). *User Manual Field Force TCM XB*.

Roberts, G., Evans, A., Dodson, A. H., Denby, B., Cooper, S., & Hollands, R. (2002). The Use of Augmented Reality, GPS and INS for Subsurface Data Visualization. In *Proceedings of the 2002 FIG XIII International Congress* (pp. 1–12). Copenhagon, Denmark: International Federation of Surveyors, FIG.

Shreiner, D., Woo, M., Neider, J., & Davis, T. (2006). *OpenGL Programming Guide* (6th ed.). Boston, MA: Addison-Wesley.

Stafford, A., Piekarski, W., & Thomas, H. B. (2006). Implementation of God-like Interaction Techniques for Supporting Collaboration Between Outdoor AR and Indoor Tabletop Users. In *Proceedings of the 2006 IEEE and ACM International Symposium on Mixed and Augmented Reality* (pp. 165–172). Piscataway, NJ: IEEE.

Talmaki, S., & Kamat, V. R. (2012). Real-Time Hybrid Virtuality for Prevention of Excavation Related Utility Strikes. *Journal of Computing in Civil Engineering*.

Tekkeon. (2009). *MP3450i/MP3450/MP3750 datasheets*.

Thomas, B., Piekarski, W., & Gunther, B. (1999). Using Augmented Reality to Visualise Architecture Designs in an Outdoor Environment. In *Proceedings of the Design Computing on the Net*.

Trimble. (2007). *AgGPS RTK Base 900 and 450 receivers.: Trimble*.

Vincenty, T. (1975). Direct and inverse solutions of geodesics on the ellipsoid with application o fnested equations. *Survey Reviews, XXIII*, 88–93.

Wang, X., & Dunston, P. (2007). Design, strategies, and issues towards an augmented reality-based construction training platform. *Journal of Information Technology in Construction (ITcon), 12*, 16.

Wang, X., Kim, M. J., Love, P. E. D., & Kang, S.-C. (2013). Augmented Reality in built environment: Classification and implications for future research. *Automation in Construction, 32*, 13.

Webster, A., Feiner, S., Macintyre, B., & Massie, W. (1996). Augmented Reality in Architectural Construction, Inspection, and Renovation. In *Proceedings of 1996 ASCE Congress on Computing in Civil Engineering*. New York, NY: ASCE.

Yeh, K.-C., Tsai, M.-H., & Kang, S.-C. (2012). On-site building information retrieval by using projection-based augmented reality. *Journal of Computing in Civil Engineering, 26*(3), 14.

2

Combining photogrammetry and robotic total stations to obtain dimensional measurements of temporary facilities in construction field

Ming-Fung Francis Siu, Ming Lu* and Simaan AbouRizk

Abstract

Background: Three-dimensional (3D) modeling and visualization of temporary site facilities is instrumental in revealing potential space conflicts and refining time and cost estimates. This research focuses on implementation of photo-based 3D modeling in a time-dependent, dynamically-changing context.

Methods: We propose a cost-effective modeling technique to obtain dynamic dimension measurements of a moving object. The methodology resulting from integrating photo-based 3D modeling and robotic total station tracking technologies better caters to the application needs of visualization and measurement in construction that are critical to operational safety and structural accuracy. The computational foundation of photogrammetry is first addressed then the modeling procedure and the system design described.

Results: In a module assembly yard, a rigging system being lifted by a mobile crane was identified as the moving object. The length and the length changes of twelve slings on a newly-engineered rigging system at different stateswere measured in order to ensure quality and safety.

Conclusion: The proposed technique relies on utilizing two robotic total stations and three cameras and provides a simple, safe and effective solution to monitor the dimensional changes of a temporary facility in the construction field.

Keywords: Time-dependent modeling, Photogrammetry, Robotic total station, Synchronization, Integration, Rigging system, Lifting frame, Augmented reality, Visualization

Background

Photogrammetric surveying has been widely applied in medical surgery, 3D modeling, engineering, manufacturing and map production. Virtual reality (VR) and augmented reality (AR) researches have resorted to photos as the most straightforward and cost-effective means for field data collection in the construction management domain (Kamat et al. 2011). Due to the computational complexity of photogrammetric surveying, construction engineering and management researchers attempted to reduce the number of photos used by imposing geometric constraints and automating the modeling process based on pattern recognition and feature detection (Golparvar-Fard et al. 2009). (El-Omari and Moselhi 2008) integrated both photogrammetry and laser scanning to reduce time required for collecting data and modeling. (Golparvar-Fard et al. 2011) used site photos to generate the point cloud, then match and pair the images to generate the as-planned and as-built models enabled by the technique of structure from motion (SfM). (Golparvar-Fard et al. 2011) further reduced the modeling time and cost by generating the point cloud from both photos and SfM. Successful point cloud applications were demonstrated in (Golparvar-Fard et al. 2011) and (Bhatla et al. 2012). The point cloud based application is aimed to reduce the effort in 3D as-built modeling. However, the model object must be stationary during the laser scanning process while removing the noise data (redundant or irrelevant information) requires considerable time and expertise. Thus, "point-cloud" based techniques are not suitable for modeling a particular moving object on a near real-time basis in the field.

Research has extended photogrammetry into videogrammetry; for instance, (Fathi and Brilakis 2012) measured

* Correspondence: mlu6@ualberta.ca
Department of Civil & Environmental Engineering, The University of Alberta Markin/CNRL Natural Resources Engineering Facility, 9105 116th St, Edmonton, Alberta T6G 2W2, Canada

dimensions of a roof using two video cameras based on SfM. However, extensive video post-processing effort is necessary to match the time stamp of each video frame recorded by each camera. More than three stationary site control points are required in each photo frame for 3D modeling. To reduce the minimal quantity of the required control points from three to two, a simplified photogrammetry-enabled AR approach (Photo-AR) was applied to assist in crane allocation and bored pile construction sequencing on a congested building site (Siu and Lu 2009, 2010, 2011). In short, considering a life-size moving object found in the construction field (like the rigging system used in the case study of the present research), the inclusion of multiple fixed control points in each photo frame is infeasible. The above limitations identified in current 3D as-built methods have inspired us to develop an alternative solution which directly tracks and surveys target points by use of two synchronized RTS units, thus automatically providing the geo-referenced inputs as needed for photo-based 3D modeling.

At present, using total station is the common practice to determine positions in the site space, instead of applying traverse and leveling methods in traditional surveying (Kavanagh 2009). Mainstream surveying research focuses on improving the accuracy of collected data by computing, for instance, through the least square adjustment algorithm (King 1997). On the other hand, the state-of-the-art robotic total station (RTS) adds tracking and automation capabilities to enhance positioning and surveying applications in the field, including building settlement monitoring, bridge deflection and tunnel boring machine guidance (Shen et al. 2011). Currently, the drawbacks of robotic total stations include the high investment and application costs and its limited capability to track only one point at a particular time.

Physical dimensions of static building products that are not safely assessable can be surveyed through use of photogrammetry for quality control purposes (Dai and Lu 2010). It takes significant effort to process images to build 3D models of a static object, while dynamic changes of the object's geometric configurations over time are ignored. A Photo-AR scene, consisting of existing plant facilities plus virtual models of temporary facilities, can be linked with a scheduled event on an industrial construction project. The AR scene is instrumental in revealing and analyzing potential workface space conflicts and refining the estimate of productivity for construction scheduling. In previous research, the Photo-AR requires at least two control points with known coordinates fixed on the site explicitly visible in each photo. The scale of the AR scene can be fixed by using the two stationary control points (Siu and Lu 2009).

This paper reports our development and application of a modeling methodology for measuring physical dimensions of a moving object and checking any changes in those dimensions during a dynamic process in the construction field. On site, the stationary control points are usually located on the ground. It is difficult to include at least two ground-fixed points in each photo frame, especially when the large object being tracked is being lifted and moved from a source location to a destination location in the field. Therefore, this research is intended to model a moving object based on tracking a minimum quantity of dynamic control points on the object. As such, reliable dimensional measurements at one particular moment can be obtained with the least cost of equipment purchase and use. Photo-AR has been further enhanced through synchronizing cameras and RTS to track dynamic points on a moving object. The proposed time-dependent 3D modeling methodology is cost effective specifically for dynamic applications: physical dimensions of the moving object being modeled can be determined simply by processing photos from multiple cameras supplemented by point tracking results by two robotic total station units at a particular time. This would better cater to the application needs in industrial construction in terms of modeling dynamic temporary facilities which are critical to construction safety and productivity performances. The application background for our field experiments is given as follows.

As Canada's leading producer of oil, gas, and petrochemicals, Alberta is home to four ethane-cracking plants, including two of the world's largest, with combined annual capacity to produce 8.6 billion pounds of ethylene. In the foreseeable future, new refining capacity will be added to produce ethane from bitumen upgrading, which will directly source feedstock from downstream oil sands mining (Government of Alberta 2010). New construction and turnaround activities at industrial process plants consume substantial resources and involve diverse stakeholders who work closely towards delivering a project under a finite time window and a tight cost budget. In general, work items such as a pipe spool, a valve or a storage tank undergo a sequence of tasks which take place in a fabrication shop, at a module yard and on an industrial site. Each task is conducted by a specialist crew in a confined work space with the assistance of temporary facilities and equipment such as scaffolding, rigging system and cranes. A newly engineered rigging system was designed by a major industrial contractor to handle super modules with a maximum 160-ton lift capacity. The rigging frame system is made of steel and subjected to bending under loadings (Westover et al. 2012). Length-adjustable slings connect the rigging frame and an overhead plate to form a rigging system. The sling length measurements are critical to balance the frame. This ensures the load is evenly spread and carried by each sling. However, direct measurement of sling lengths such as using measurement tape is not feasible due to safety hazards and the dynamic movement of the rigging system.

In the remainder of this paper, the computing foundation of photogrammetry is briefly addressed. Then we describe the modeling procedure, the system design integrating the use of multiple cameras and two robotic total stations, and the field implementation to check sling lengths in modeling a rigging system in a module assembly yard. Field testing results are deliberated and analyzed. Conclusions are drawn based on discussions of the experimental findings and future research.

Computing foundation of photogrammetry

Three-point absolute modeling approach

This present research focuses on the implementation of close range photogrammetry in a time-dependent, dynamically-changing context. The collinearity equations, given in Eqs. (1) and (2), constitute the mathematical foundations of photogrammetry to determine (1) internal parameters of camera, (2) image and object coordinates, and (3) error adjustment. The direct linear transform algorithm was formalized by (Abdel-Aziz and Karara 1971) based on the collinearity equations, and has been further developed to simplify the transformation between the image pixel frame and the object space coordinates in digital photogrammetry (Bhatla et al. 2012; Mikhail et al. 2001; McGlone et al. (2004)). Basically, the camera's position and orientation parameters with respect to an object space coordinate system is determined by solving six unknowns, namely: three perspective center coordinates and three orientation parameters: $(X_C, Y_C, Z_C, \omega, \varphi, \kappa)$. As two collinearity equations can be written for one particular point, relating its imaging point (x, y) in the photo frame to its three coordinates (X_P, Y_P, Z_P) in the space frame, three different points with known (x, y) and (X_P, Y_P, Z_P) can define six equations, thus resulting in unique solutions of six unknowns $(X_C, Y_C, Z_C, \omega, \varphi, \kappa)$ (Figure 1). After the camera's position and orientation parameters are determined, any new point whose (x, y) in the photo frame are known while (X_P, Y_P, Z_P) in the object space are unknown,

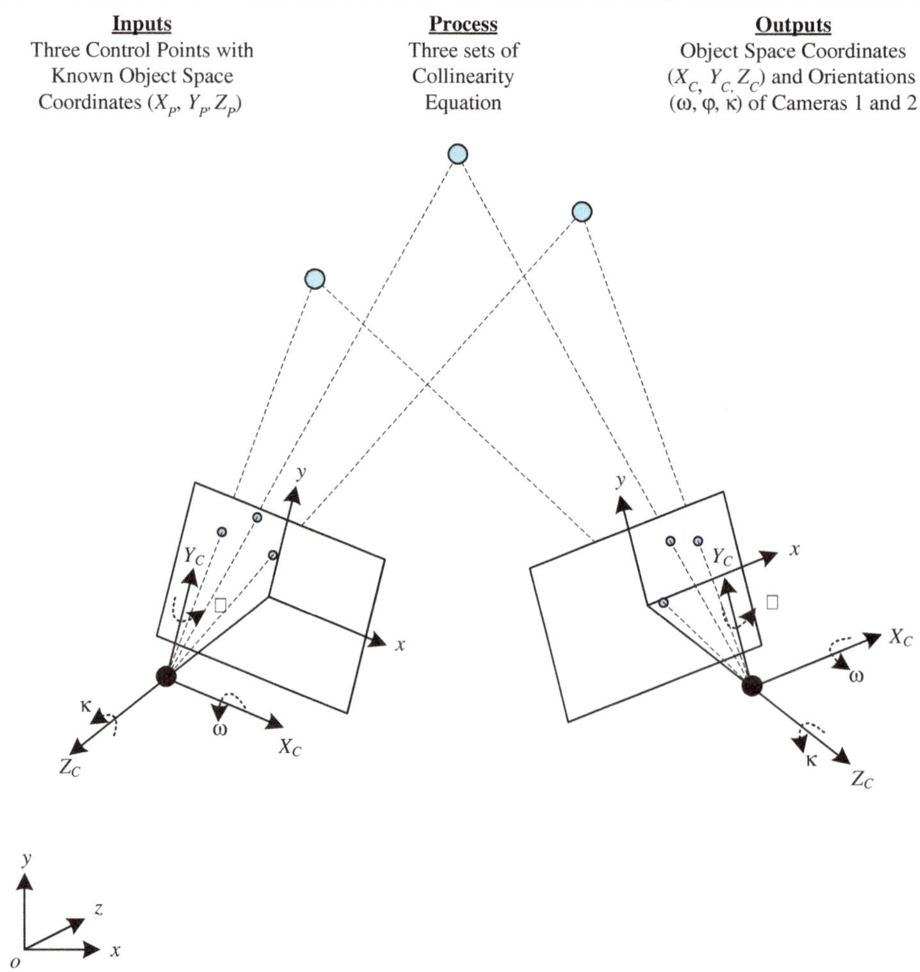

Figure 1 Determine camera object coordinate and orientation (absolute modeling approach).

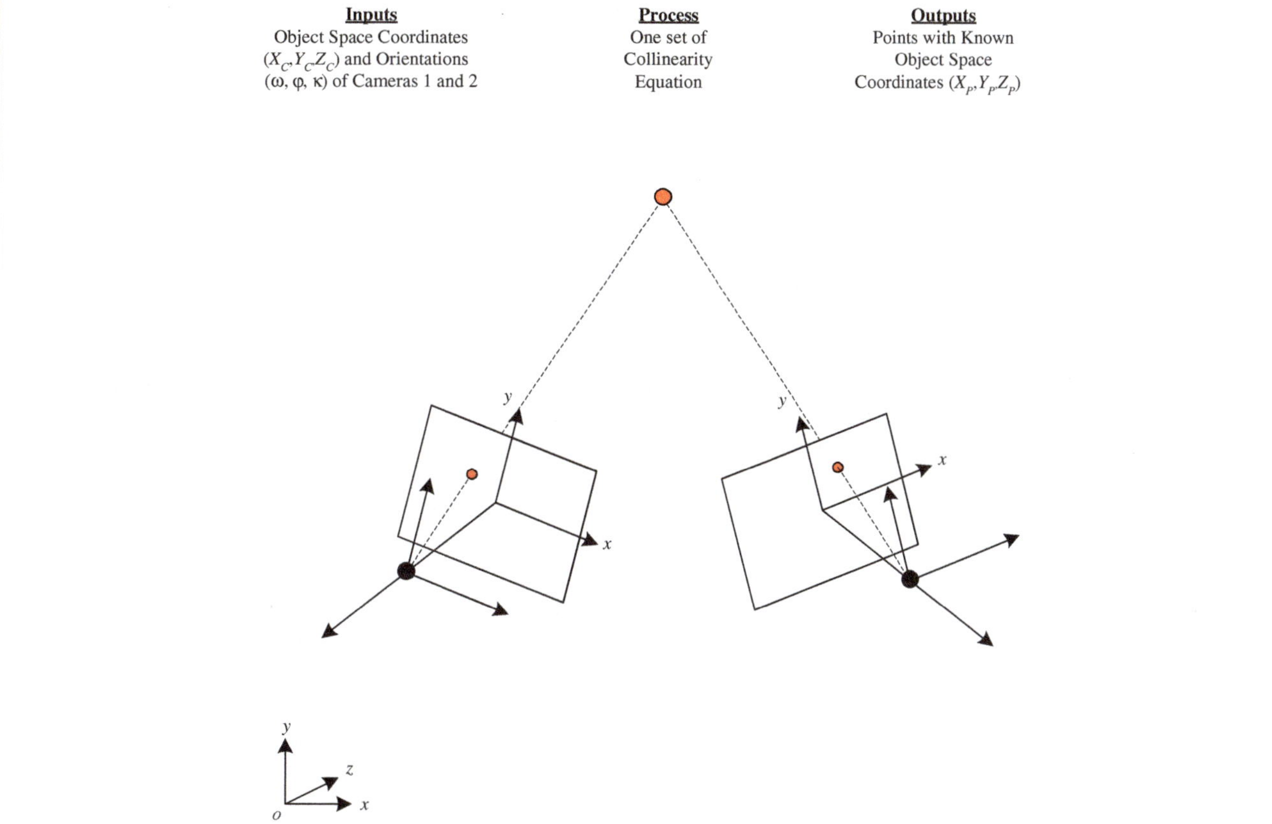

Figure 2 **Determine unknown object point coordinate (absolute modeling approach).**

(X_P, Y_P, Z_P) can be expressed as functions of (x, y) by transforming Eqs. (1) and (2); if two photos taken from different perspectives both capture the same point, the position and orientation of the two camera stations are all determined, then four equations having three unknowns can be solved by least square adjustment techniques in order to derive the most likely values of the object space coordinates for the new point (Figure 2).

$$x + \delta x - x_0 = -c_x \frac{m_{11}(X_P - X_C) + m_{12}(Y_P - Y_C) + m_{13}(Z_P - Z_C)}{m_{31}(X_P - X_C) + m_{32}(Y_P - Y_C) + m_{33}(Z_P - Z_C)} \quad (1)$$

$$y + \delta y - y_0 = -c_x \frac{m_{11}(X_P - X_C) + m_{12}(Y_P - Y_C) + m_{13}(Z_P - Z_C)}{m_{31}(X_P - X_C) + m_{32}(Y_P - Y_C) + m_{33}(Z_P - Z_C)} \quad (2)$$

Where:

$$\begin{bmatrix} m_{11} & m_{12} & m_{13} \\ m_{21} & m_{22} & m_{23} \\ m_{31} & m_{32} & m_{33} \end{bmatrix} = \begin{bmatrix} \cos\phi\cos\kappa & \cos\omega\sin\kappa + \sin\kappa\sin\phi\cos\kappa & \sin\omega\sin\kappa - \cos\omega\sin\phi\cos\kappa \\ -\cos\phi\sin\kappa & \cos\omega\cos\kappa - \sin\omega\sin\phi\sin\kappa & \sin\omega\cos\kappa + \cos\omega\sin\phi\sin\kappa \\ \sin\phi & -\sin\omega\cos\phi & \cos\kappa\cos\phi \end{bmatrix}$$

x, y, is image coordinate

x_0, y_0, is principal point coordinate

c_x, c_y, is principal distance scaled by λ in x and y directions ($c_x = c\lambda_x$ and $c_y = c\lambda_y$)

c, is principal distance

λ, is scale factor

X_P, Y_P, Z_P, is object space coordinate of the point

X_C, Y_C, Z_C, is object space coordinate of the perspective center

ω, ϕ, κ, is rotated angles with respect to x, y and z axis

δx, δy, is total lens distortion in x and y directions determined from camera calibration.

Five-point relative modeling approach

When the only objective of a particular field application is to take measurements and build a three-dimensional (3D) model of an object, instead of fixing the exact position of the object in space, then the three points with known coordinates in the field space are not needed. Instead, the relative orientation parameters can be determined from the (x, y) coordinates of a minimum of five points on the object in a minimum of two photos based on Eqs. (1) and (2), without requiring the coordinates of any point in the field. The five control points give five sets of collinearity equation with five degrees of freedom. The coordinates and orientations of two cameras $(X_C, Y_C, Z_C, \omega, \phi, \kappa)$ are determined with respect to the model coordinate system (Figure 3), whose origin o aligns with the principal point of one of the cameras. As such, the coordinates (X_P, Y_P, Z_P) of the object in the model frame can be determined by pairing imaging points (x, y) on two photos. The 3D model in relative measurements can be built (Figure 4). A scale bar, which can be the absolute measurement of a line section on the object, is used to scale the relative measurement of a dimension of the object in an absolute unit of measurement. Elaboration of the complete mathematical algorithms for the above five-point-two-photo approach can be found in (Dai and Lu 2013).

The proposed dynamic modeling approach essentially follows the "five-point relative modeling approach" with the assistance of two synchronized RTS units, at a particular moment, a scale bar is automatically fixed on the object by tracking two points. In other words, the length of the scale bar is subject to change over time. In the long run, the proposed methodology and system design can be readily extended to implement a time-dependent "three point absolute modeling" approach by synchronizing three RTS units to track the absolute coordinates of three points on the moving object.

Methods

Modeling procedure

The time-dependent dynamic modeling procedure is given in Figure 5. A minimum of two cameras plus two robotic total stations are required. Note the real-time kinematic based global positioning system (RTK-GPS) enabled by satellite navigation offers an alternative technology to RTS, but its reliability and accuracy in positioning a specific moving object is questionable. In contrast, the robotic total station (RTS) is the most accurate on-site survey instrument at present and is capable of providing millimeter-level accuracy for coordinate measurement of target points (Leica Geosystems 2012). It is possible to track multiple points on a moving object with multiple RTS units. As each total station can only track one target at a time, two RTS units are synchronized to simultaneously track and position two points through wireless automation command control.

The price of a robotic total station is around 80–100 times the price of a camera. More than two cameras can be utilized to enhance the quality of photogrammetry modeling. Two cameras and two RTS units are initialized by synchronizing their internal clocks. The images are taken once the object falls entirely into the fields of vision of the cameras. Each RTS unit is programmed to automatically track and survey one control point on the moving object being modeled. Via wireless communication networks, time stamped survey data and digital photos are collected in the field for immediate processing and modeling on a laptop. At time point T_0, the photos and the coordinates of two points on a moving object are captured by the two cameras and two RTS units, respectively. Time stamp checking is then performed to assure synchronization quality, which guarantees the data are captured at approximately the same time. In case the time differences between the photos and the control point survey data exceed the preset allowance (e.g. 1 second in the current case), then the photos and the survey data are discarded for ensuing 3D modeling. Two images taken from different angles are used to build the 3D model, by pairing and matching five common points in the two photos. The 3D model coordinate system is set up by selecting an arbitrary origin. The photo-based 3D model is set on an arbitrary scale. The image and object coordinates are then evaluated based on the collinearity equations. The scale of the model is determined based on the coordinates of the two points fixed by the two robotic total station units. Dimensions of the object can then be obtained at T_0. Next, the above tracking process is repeated on the next event until the tracking process is terminated. The details of the available software for building the photo-based models are discussed in the following section.

The time-dependent dynamic tracking method is proposed to monitor a moving object in the field space. Changes in measurements between the two time events (such as length, area and volume) can be evaluated from the resulting models for construction application purposes. As illustrated in Figure 6, a photo-based 3D model is generated with an arbitrary scale at time T_0. The origin o is specified at the corner of the box. The coordinates of the points a (0, 0, 10) and b (10, 0, 10) are fixed by employing robotic total station units. The length between a and b (10 cm) can be calculated and then used to fix the model scale. The geometric properties of the object, such as the area (A_{01}: 100 cm^2, A_{02}: 200 cm^2) and volume (V_0: 2000 cm^3), can be determined. At T_1, the object is remodeled following location and dimension changes as shown in Figure 7. Similarly, point coordinates of a (10, 10, 10) and b (30, 10, 10) are used to scale the photo-based 3D modal at T_1. The geometric properties (A_{11}: 200 cm^2; A_{12}: 200 cm^2; V_1: 4000 cm^3), along with any changes ($\Delta A_1 = 100\ cm^2, \Delta A_2 = 0\ cm^2, \Delta V = 2000\ cm^3$), from T_0 and T_1 can be evaluated.

The advantages in applying photogrammetric techniques compared to traditional surveying include (1) non-contact measurements, (2) relatively short turnaround cycle of data processing and modeling and (3) measuring multiple

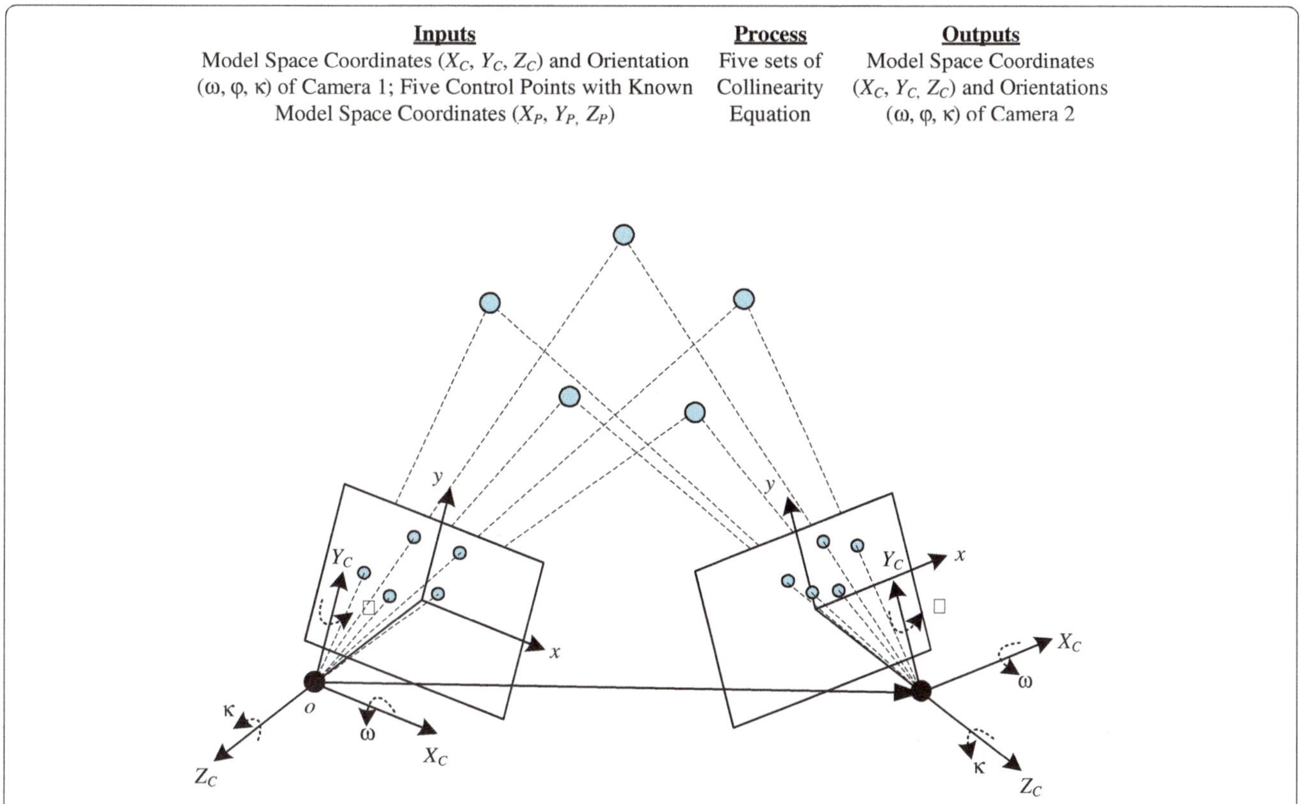

Figure 3 Determine camera model coordinate and orientation (relative modeling approach).

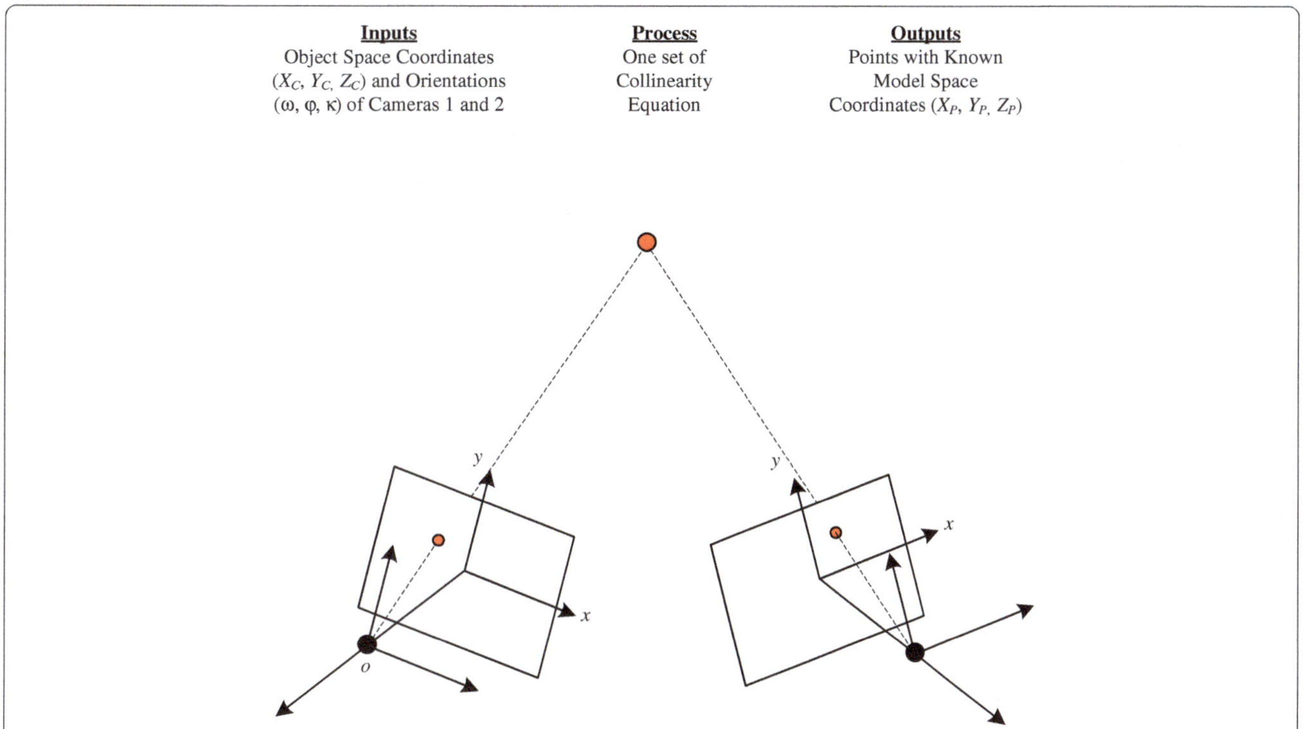

Figure 4 Determine unknown model point coordinate (relative modeling approach).

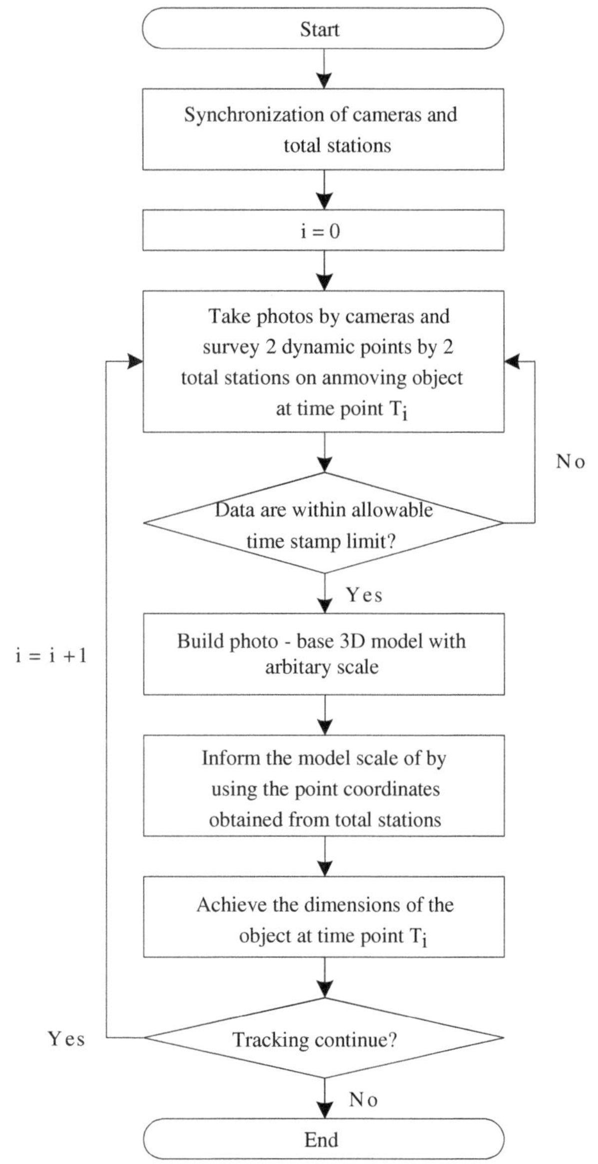

Figure 5 Time-dependent modeling flowchart.

points at one time. This application is particularly useful when direct-contact-based measurement is not viable due to the dynamic nature of a process and potential safety hazards. The following section presents the field application of the proposed time-dependent dynamic 3D modeling method on a rigging system engineered for lifting heavy industrial modules.

Cameras and photomodeler software

The system solution requires the integration of hardware and software. The software, PhotoModeler 2012 (EOS Systems 2011), was used to produce the photo-based 3D model with an arbitrary scale by importing three photos from different angles by three cameras and the coordinates of two control points on the rigging system, all taken simultaneously.

To achieve high quality photogrammetry modeling, three Canon T3i cameras were purchased and calibrated (CAD $699.98 each in May 2012). The internal clocks of the three cameras were synchronized such that the time stamps of the photos taken at the same time fell within a small margin of difference. In order to shorten the turnaround time during actual site testing, Eye-Fi cards were used to facilitate the wireless photo transfer from the cameras to a laptop. The card models are Connect×2 4GB (CAD $39.99) and Pro×2 8GB (CAD $99.99). The Eye-Fi card functions as a normal SD card for camera image storage with Wi-Fi signal receiver and transmitter embedded (Eye-Fi 2012). In the field testing, the time of

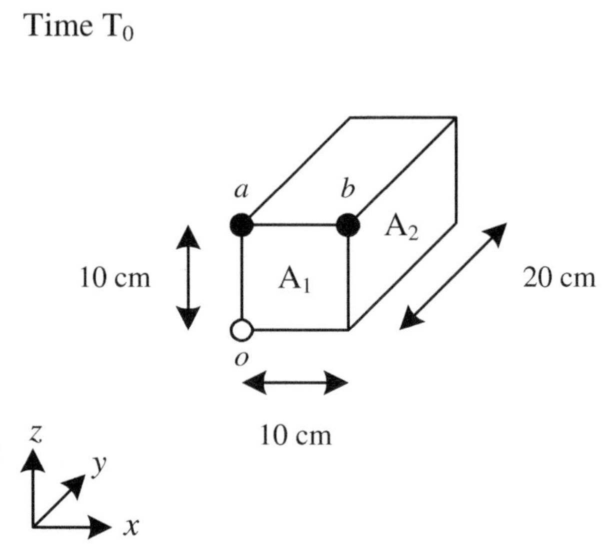

Figure 6 Tracking a moving object at T_0.

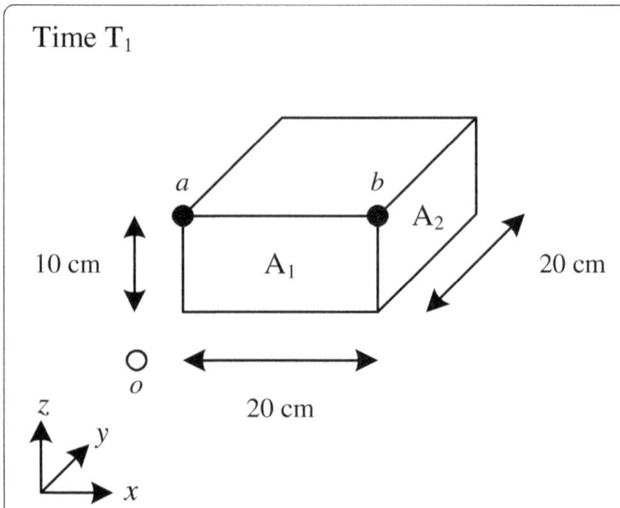

Figure 7 Tracking a moving object at T_1.

transferring three high resolution photos (5184×3456; 4.58 MB each) was under 30 seconds with the Internet or under 60 seconds with an "ad hoc" wireless network; the communication range of the Eye-Fi card was 27 m outdoors and 13 m indoors, respectively.

Camera calibration is essential to determine lens distortions (radial and decentering distortion), the principle distance and the location of principle point in the photo frame, referring to (δx, δy), c and (x_O, y_O) in Eqs. (1) and (2), respectively. System errors due to lens distortions can be regulated and corrected through camera calibration. High quality multiple-sheet calibration methods provided by PhotoModeler were used, as shown in Figure 8. There were 125 coded targets (5 coded targets on each calibration sheet, and 25 sheets were printed) and coded targets were automatically detected by PhotoModeler. Three cameras were successfully calibrated and the parameters were stored in PhotoModeler for system error adjustment.

Once object images taken from different perspectives are imported to PhotoModeler, building the model involves human interaction, to pick and match feature points in different photos. Human error in the modeling process is accessed in terms of residual error. The computational algorithm used in PhotoModeler, called bundle adjustment, essentially applies the colliearity equations (Eqs. 1 and 2) to simultaneously fix (1) camera orientations, (2) object and image point coordinates and (3) the residual error, with an objective to minimize the residual error.

Robotic total stations

The conventional approach to fix the scale of the photo-based 3D model is to use a scale bar mounted or marked on the object being modeled. However, the scale bar may not be visible in each photo if the object is moving. The reliability and accuracy of using real-time kinematics global positioning system (RTK-GPS) to position a specific moving object are questionable. In contrast, the robotic total station (RTS) is the most accurate survey instrument and is capable of providing point survey data of millimeter-level accuracy (Leica Geosystems 2012). As one total station can only track and measure one target at a time, two RTS units need to be synchronized in order to simultaneously survey two points on a moving object through wireless automation command control.

The models of the two RTS units used in field testing were Leica TS15I and Leica TCRP1203+. A tablet program was developed to synchronize the operations of the two RTS units during site experiments. The two RTS units were synchronized by the built-in protocols (COM_4 and COM_6) through an application programming interface (API) provided by Leica. In addition to synchronization, the two RTS units must be initialized in the object coordinate system. To set up the coordinate system, RTS_1 is assigned as the origin of the coordinate system. The direction pointing from RTS_1 to a fixed point in the ground is defined as north. The east direction can be established as a cross product of the north and the zenith (perpendicular to the ground). Each RTS unit automatically locks on a reflective glass prism and track its location. The coordinates of the two control points are simultaneously taken, which is controlled by the tablet program. The coordinates of the two control points at T_i are prerequisite inputs to calculate the distance between the two points by Eq. (3), which is used to take a distance measurement on the photo-based model in an absolute unit of measure.

$$L = \sqrt{(X_1 - X_2)^2 + (Y_1 - Y_2)^2 + (Z_1 - Z_2)^2} \quad (3)$$

The three cameras were operated manually by three persons who tracked the moving object and

Figure 8 Camera calibration.

continuously took photos in our field testing. In the near future, automation control of multiple cameras can be implemented in a way similar to the method used for the robotic total stations.

The focal lengths of the cameras were set to the allowable minimum value (18 mm) in order to capture the whole structure in the yard. The cameras were placed as far away from one another as possible in order to enhance the quality of photogrammetry modeling and reduce the residual error due to human factors. In general, more reliable results can be obtained if the angle between the perspectives of two cameras is close to 90°. Figures 9 and 10 show two different states of the rigging at T_0 and T_1, respectively. The photo-based 3D models can be built by using three photos simultaneously taken from different perspectives. The two RTS units track the two end points of a particular sling on the rigging system, as denoted by red arrows. In the end, the lengths of all the slings can then be fixed from the resultant models at T_0 and T_1, respectively.

Field testing

In a module assembly yard, a rigging system being lifted by a mobile crane was identified as the testbed object of practical size and complexity (Figure 11). The dimensions of the entire structure measured approximately 20 m wide, 20 m tall and 5 m deep. On each side of the rigging, six slings connected a horizontal beam and an overhead connection plate. The objectives of the testing include (1) measuring sling lengths of the moving rigging frame to ensure the main beam is leveled geometrically at a particular time (T_i), and (2) checking changes in sling lengths to examine if the rigging frame is stable during lifting. In addition, the quality of sling fabrication can be checked by comparing field measures against design specifications. In the field, sling lengths can also be adjusted by a limited magnitude in order to match the design. This will ensure proper tensions in all the slings when the rigging is used to handle the heavy load of a full module.

Surveying markers and glass prisms were placed on the rigging system to facilitate photogrammetric modeling and automatic RTS tracking. The prism placement on the rigging system and identifications of each sling are shown in Figure 12. In the field experiments, we placed two mini prisms at the two end points of sling B(L.) on the "Y-plate" and the beam, respectively.

Site constraints

The weather during field experiments was cloudy and windy. On site, temporary facilities such as a dumpster and the mobile crane were present likely to block the view

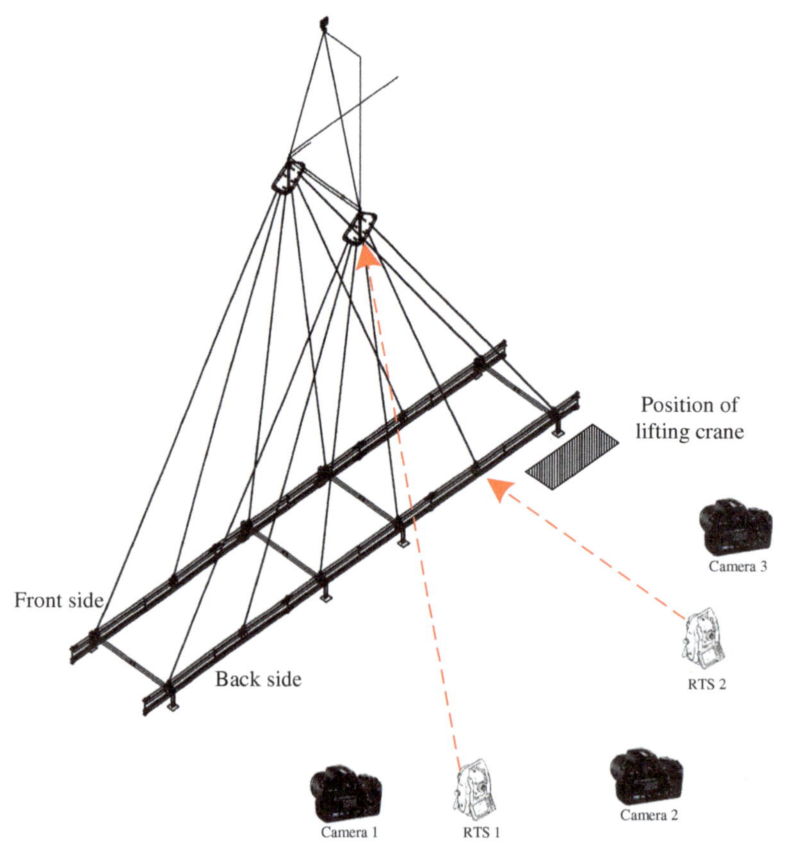

Figure 9 Moving object at T_0.

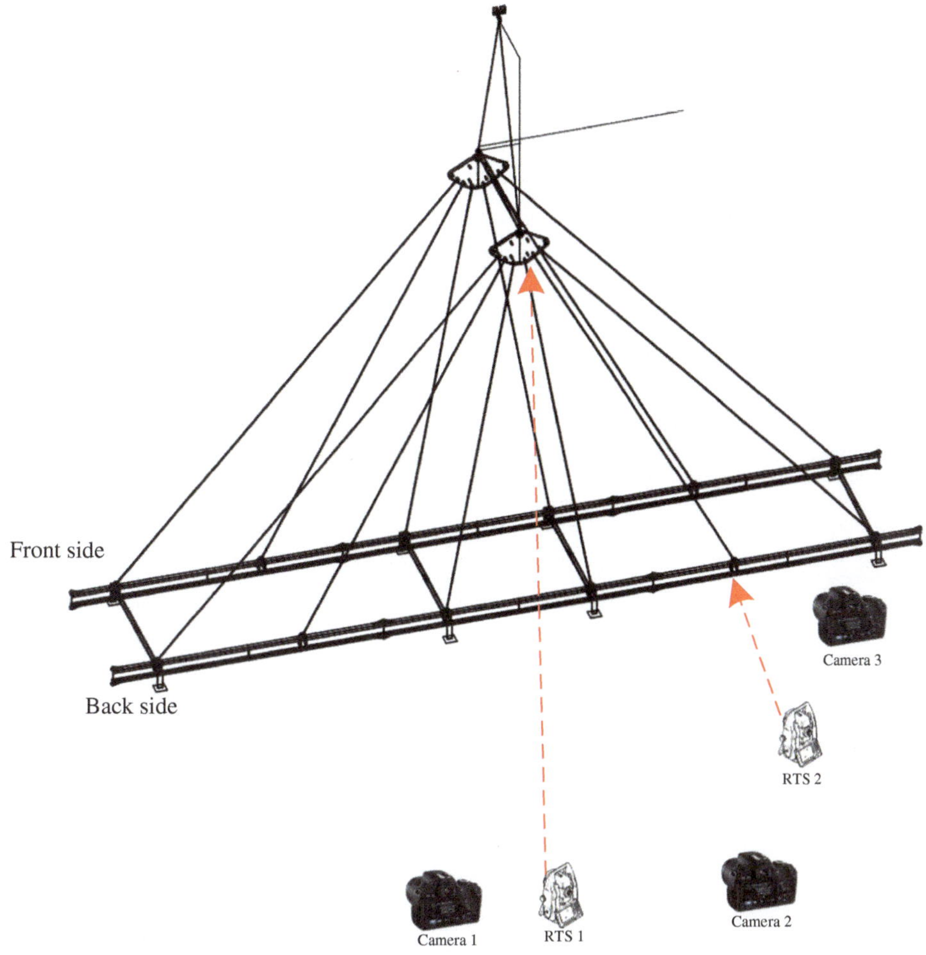

Figure 10 Moving object at T_1.

Figure 11 Rigging system.

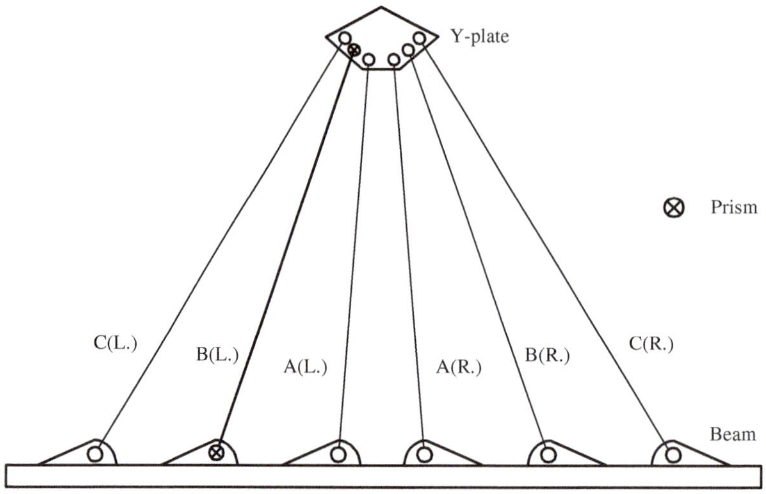

Figure 12 **Prism positions.**

of the rigging frame. Therefore, to allow full coverage of the whole frame structure in all the photos, 115 feet (35 meters) clearance between the cameras and the rigging frame was required. Additionally, to satisfy perspective-angle requirements between two photos in photogrammetry survey, a certain distance was maintained between the three cameras. Photos were collected using wireless network transfer and were selected for ensuing modeling if time stamp data recorded in the photos were checked to fall within one second of the RTS survey records. Then, on a laptop in the field, the photo-based models were built by manually matching the conjugate center points of markers on the rigging frame in the PhotoModeler software platform.

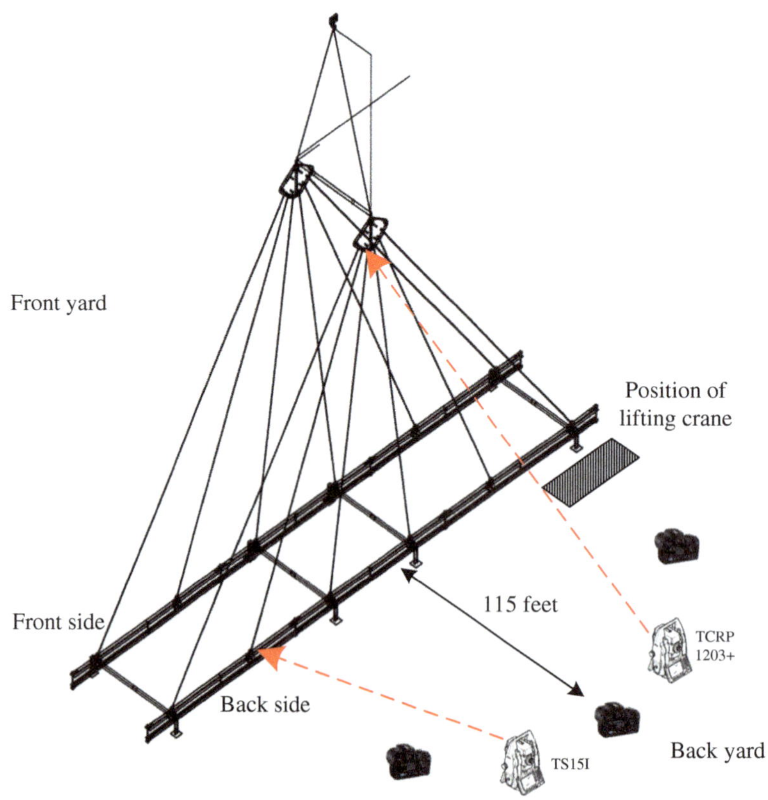

Figure 13 **Clearance without obstacles between the cameras and the rigging frame.**

Figure 14 **Photo-based modeling for site testing before removing the loading from the frame.**

The rigging frame swung under the wind load, making it difficult for total station units to track the position of the markers. By synchronizing robotic total station (RTS) units, two positions on a moving object were automatically tracked and accurately fixed at one particular time point: one RTS unit (Model: TS15I) was responsible for surveying the prism on the Y-plate, and the other (Model: TCRP1203+) locked on the prism on the beam (Figure 13). The surveying results at one particular time point were converted into the sling length by Eq. (3), which was required for model scaling. Note the sling length is assumed to be the straight-line distance between the two end points.

Results

Field testing results

Six models were built in the field for six particular moments in time; three moments (T_0, T_1, T_2) occurred when unbalanced weights were loaded near the left end of the frame, while the other three (T_3, T_4, T_5) represented the state of the rigging system after the unbalanced weights were removed. The models at T_0 and T_3 are shown in Figures 14 and 15, respectively. Each was built in PhotoModeler based on three photos taken from three perspectives, with an arbitrary scale. The two point coordinates obtained from the two synchronized RTS units were used to fix the length of sling B(L.), given in Figures 14 and 15.

Figure 15 **Photo based modeling for site testing after removing the loading from the frame.**

Sling length measurement data for T_0, T_1, T_2, and for T_3, T_4, T_5 are given in Table 1 and Table 2, respectively. The sling length changes $\Delta L(T_{i,\ i+1})$ between two time events, i and i+1, are summarized in Table 3. The measurement units are all in millimeters. The relatively large values of length change on A(L.) and A(R.) between any two time moments (Table 3) can be attributed to the longer distance between the sling and the center point of the rigging beam; while the significant changes on all the slings between T_2 and T_3 are largely caused by removing the "unbalanced" load from the rigging frame. It should be noted that a positive value in Table 3 denotes the increase of the distance between the two end points of a sling.

In the current site testing, we had only two RTS available; the problem definition is to measure the dimensional changes of the slings for a rigging system to ensure quality and safety. Hence implementing the "five point relative modeling approach" by use of two RTS is sufficient and cost effective in comparison with the "three point absolute modeling" using three RTS units. The objective of implementing the proposed technique to analyze the dimensional changes on the moving rigging system has been successfully fulfilled. The small magnitude of the sling length changes given in Table 3 indicates the rigging frame is relatively stable during lifting. In addition, the lengths of the sling assembled on site were adjusted to match the design based on Eq. (4). The proposed technique provides a simple, safe and effective solution to monitor the dimensional changes of a moving object, such as the sling length in this case study, and hence provides timely decision support making.

As the sagging effect of a loose sling is ignored in the current study, the length change measurement approximately indicates the sling length extension. In fact, the photo-based approach allows the modeling of the sagging effect on the sling. The determination of the exact sling length based on the end-to-end distance and the sagging effect will be investigated in the follow-up research.

$$\Delta L = L_{\text{As-Design}} - L_{\text{Surveyed}} \tag{4}$$

Table 1 Sling lengths before removing the loading from the frame during lifting

Slings	[08.22.06 am] T_0	[08.22.12 am] T_1	[08.22.19 am] T_2
A(L.)	21472	21469	21480
B(L.)	18193	18193	18193
C(L.)	16257	16251	16250
C(R.)	16270	16258	16264
B(R.)	18217	18217	18213
A(R.)	21475	21551	21469

Table 2 Sling lengths after removing the loading from the frame during lifting

Slings	[10.18.37 am] T_3	[10.18.45 am] T_4	[10.19.42 am] T_5
A(L.)	21462	21455	21451
B(L.)	18197	18189	18178
C(L.)	16273	16258	16259
C(R.)	16265	16260	16259
B(R.)	18197	18184	18223
A(R.)	21508	21435	21560

In general, the value of residual error is determined as an indicator of the quality of photogrammetry modeling which entails manually matching feature points on the object. In the field experiments, one object point P was identified and marked in the x-y frame of each photo taken by three cameras. Its coordinates in an arbitrary modeling space frame were then calculated based on the least-squares adjustment technique by applying the "five-point relative modeling approach." After the point coordinates were determined, the corresponding image point (x, y) were re-calculated by using collinearity Equations (1) and (2) in the corresponding image frame. The pixel differences between the manually marked (x, y) coordinate and recalculated (x, y) coordinate are defined as the residual error (EOS Systems 2011). As a rule of thumb, if the largest residual is less than 10 pixels, the quality of a photogrammetry modeling project is high; otherwise, re-modeling is needed. Note the residual errors on the models resulting from field experiments as described in this paper were controlled within 3 pixels.

To quantitatively assess the modeling accuracy of photogrammetry in construction field applications, Dai and Lu (2008) conducted case studies, in which the object dimensions ranged from 35mm to 5720mm, while the distance from the camera to the object ranged from 1m to 6m. Their experiment results showed that there was a 95% likelihood the photogrammetric modeling measurement and tape measurement would differ from −15.30 mm to +11.39 mm. A separate laboratory experiment was conducted to compare the results between the proposed technique and tape measurement. Note a particular sling's length can differ from its original design due to errors resulting from fabrication and assembly

Table 3 Changes of end-to-end lengths ΔL ($T_{i,\ i+1}$)

Slings	$\Delta L(T_{0,\ 1})$	$\Delta L(T_{1,\ 2})$	$\Delta L(T_{2,\ 3})$	$\Delta L(T_{3,\ 4})$	$\Delta L(T_{4,\ 5})$
A(L.)	−3	+11	−18	−7	−4
B(L.)	0	0	+4	−8	−11
C(L.)	−6	−1	+23	−15	+1
C(R.)	−12	+6	+1	−5	−1
B(R.)	0	−4	−16	−13	+39
A(R.)	+76	−82	+39	−73	+125

processes. And the true measurements of the rigging system used at a particular loading state in the field testing are unknown. It was impractical to manually gauge those slings' lengths due to large sizes of the rigging system, accessibility constraints and safety concerns. Therefore, a laboratory experiment was designed to examine the measurement error of the proposed technique against direct manual measurements based on a simplified model of the rigging system in reduced scale. In the laboratory, two cameras were set up to take photos of the mock rigging object (60 cm wide, 50 cm tall and 20 cm deep) at one particular instant. The distance between the cameras and the system was approximately 2m. It was found that the differences between photogrammetric and manual measurements were within 0.4 mm on average. As of the rigging system in the real world, the measurement error would be expected to be much larger as the size of the object and the camera-object distance were increased by roughly tenfold. The absolute measurement error is estimated to be in the order of 5 cm. Considering the sling length is about 20 m, the relative measurement error would be approximately 0.25% on the current field testing. Nonetheless, formal systematic analysis of accuracy of the proposed methodology and the further enhancement of the accuracy through real time computing will be left for future research.

Discussion

Main contributions of this research include (1) proposing cost-effective time-dependent dynamic modeling methodology to measure the dimensions and check dimensional changes on a moving object of practical size, by integrating photogrammetry and robotic total station based survey techniques. Note a minimum of two moving control points on an object being modeled are required to be shown in each photo frame; (2) Elucidating on the fundamentals of photogrammetric surveying techniques instead of using commercial available software as "black box". The algorithmic differences between the "three-point absolute" and "five-point relative" modeling approaches are also discussed; (3) Providing guidelines as regarded the selection of available software and hardware, the equipment set up and the expected site constraints and the modeling errors for implementing the technique on a construction site; (4) Providing a real-world construction case study to show that valuable measurement data can be obtained and used in support of decision making in the field.

Conclusions

This research study proves the potential and the feasibility of the proposed time-dependent dynamic modeling methodology based on synchronization of multiple cameras and two robotic total station units for checking the physical dimensions and changes in dimension of a moving object in the construction field. By building the photo-based 3D model at one particular time point, multiple dimensions of a dynamic object can be measured in absolute units of measure in a safe, cost-effective way. The methodology was successfully implemented to monitor the sling lengths of a rigging system engineered for lifting heavy modules in industrial construction. The data quality was further analyzed in terms of residual error. The measurement error is discussed based on a rigging system prototype of a reduced scale. The time-dependent dynamic modeling methodology provides a cost-effective solution to keep track of the physical dimensions and changes of a moving object over time. The proposed methodology would better cater to the application needs of modeling, visualization and measurement of dynamic objects critical to industrial construction.

Further research will be conducted in regard to (1) fully integrating and automating the modeling methodology using robust, reliable image processing methods such as pattern recognition and feature detection, so the manual operations of picking and matching points in photogrammetry modeling can be eliminated while the quality of modeling, which is quantified and evaluated by the residual error, is guaranteed; (2) extending synchronization of RTS units from two units to three units to track three control points on a dynamic object. The proposed dynamic modeling approach essentially follows the "five-point relative modeling approach" with the assistance of two synchronized RTS units at a particular moment, a scale bar is automatically fixed on the object by tracking two points. In other words, the length of the scale bar is subject to change over time. In the long run, the proposed methodology and system design can be readily extended to implement a time-dependent "three point absolute modeling" approach by synchronizing three RTS units to track the absolute coordinates of the three points on the moving object. As such, the coordinates of any point of the object in the field coordinate system can be fixed in real-time. This would be valuable to materialize AR on site level and lend timely, relevant decision support to construction engineers to improve quality, safety and productivity as field operations unfold.

The major barrier of implementing three RTS units lies in the cost of RTS. The price of a robotic total station is around 80–100 times the price of a camera. Therefore, this research is intended to model a moving object based on tracking a minimum quantity of dynamic control points on the object. As such, the reliable dimensional measurements at one particular moment can be obtained with the lowest equipment cost. In the current site testing, we had only two RTS units available; the problem definition is to measure sling lengths and their changes for a rigging system in order to ensure quality and safety. Hence implementing the "five point relative modeling" by use of two RTS was considered sufficient and cost effective in comparison with the "three point absolute modeling" using three RTS units; (3) analyzing dynamic structural loading effects based on real-time measurements of sling sagging magnitude, which can be made possible through photogrammetry modeling. As the sagging effect of a loose sling is ignored in the current study, the end-to-end length measurement gives only an approximation of the sling length. In fact, the photo-based approach allows direct modeling of the sagging effect on the sling. The determination of the exact sling length based on the end-to-end distance and the sagging effect will be investigated in the follow-up research.

Abbreviations

3D: Three-dimensional; VR: Virtual reality; AR: Augmented reality; RTS: Robotic total station; RTK-GPS: Real-time kinematic based global positioning system; CAD: Canadian dollar; API: Application programming interface.

Competing interests

The authors declare that they have no competing interests.

Authors' contributions

Siu conducted literature view, designed lab testing, participated in field testing, processed field collected data, and performed data analysis. Lu proposed the research, supervised the first author in doing research, and led a research team to design and perform field experiments. AbouRizk co-supervised the first author, advised on the design of field experiments and facilitated the implementation of field testing. All authors read and approved the final manuscript.

Acknowledgements

The presented research is substantially funded by a National Science and Engineering Council of Canada (NSERC) Collaborative Research and Development Grant (CRDPJ 414616-11). The authors are grateful to Ulrich Hermann (Manager of Construction Engineering PCL Industrial Management Inc.) for providing insight and advice on design and implementation of the field experiments. We also acknowledge Xuesong Shen and Sheng Mao for operating robotic total stations during field experiments.

References

Abdel-Aziz, YI, & Karara, HM. (1971). Direct linear transformation from comparator coordinates into object space coordinates in close-range photogrammetry. *Proceedings of the symposium on close-range photogrammetry*, 1–18.

Bhatla, A, Choe, S, Fierro, O, & Leite, F. (2012). Evaluation of accuracy of as-built 3D modeling from photos taken by handheld digital cameras. *Automation in construction, 28*, 116–127.

Dai, F, & Lu, M. (2010). Assessing the accuracy of applying photogrammetry to take geometric measurements on building products. *Journal of construction engineering and management, 136*(2), 242–250.

Dai, F, & Lu, M. (2013). Three-dimensional modeling of site elements by analytically processing image data contained in site photos. *Journal of construction engineering and management*. in press.

El-Omari, S, & Moselhi, O. (2008). Integrating 3D laser scanning and photogrammetry for progress measurement of construction work. *Automation in construction, 18*(1), 1–9.

EOS Systems. (2011). *PhotoModeler – Quick start guide*. Eos Systems Inc.

Fathi, H, & Brilakis, I. (2012). A videogrammetric as-built data collection framework for digital fabrication of sheet metal roof panels. In *Proceedings of the 19th EG-ICE Workshop on Intelligent Computing in Engineering, 4–6 July 2012*. Herrsching, Germany.

Golparvar-Fard, M, Peña-Mora, F, & Savarese, S. (2009). D^{4}AR- A 4-dimensional augmented reality model for automating construction progress data collection, processing and communication. *Journal of information technology in construction, 14*, 129–153.

Goodrum, P, Haas, C, Caldas, C, Zhai, D, Yeiser, J, & Homm, D. (2011). Model to predict the impact of a technology on construction productivity. *Journal of construction engineering and management, 137*(9), 678–688.

Government of Alberta. (2010). *Alberta's energy industry: an overview 2009*. Government of Alberta.

Kamat, VR, Martinez, JC, Fischer, M, Golparvar-Fard, M, Peña-Mora, F, & Savarese, S. (2011). Research in visualization techniques for field construction. *Journal of construction engineering and management, 137*(10), 853–862.

Kavanagh, BF. (2009). *Surveying with construction applications* (7th ed.). Prentice Hall.

King, BA. (1997). Some considerations for the statistical testing of least squares adjustments of photogrammetric bundles. *The photogrammetric record, 15*(90), 929–935.

Leica Geosystems. (2012). *Product overview – total stations*. Leica Geosystems AG.

McGlone, JC, Mikhail, EM, Bethel, JS, & Mullen, R. (2004). *Manual of photogrammetry* (5th ed.). American society for photogrammetry and remote sensing.

Mikhail, EM, Bethel, JS, & McGlone, JC. (2001). *Introduction to modern photogrammetry*. Wiley.

Shen, XS, Lu, M, & Chen, W. (2011). Tunnel-boring machine positioning during microtunneling operations through integrating automated data collection with real time computing. *Journal of construction engineering and management, 137*(1), 72–85.

Siu, MF, & Lu, M. (2009). Augmenting site photos with 3D as-built tunnel models for construction progress visualization. In *Proceedings of 9th international conference on construction applications of virtual reality (CONVR 2009), 187–196, Nov. 5–6, 2009*. Sydney, Australia.

Siu, MF, & Lu, M. (2010). Bored pile construction visualization by enhanced production-line chart and augmented-reality photos. In *Proceedings of 10th international conference on construction applications of virtual reality (CONVR 2010), 165–174, Nov. 4–5, 2010*. Sendai, Miyagi, Japan.

Siu, MF, & Lu, M. (2011). *Augmented-reality visualizations of bored pile construction*. Ottawa, Canada: Proceedings of the 2011 CSCE annual general meeting and conference (CSCE 2011), 10 pages, Jun. 14–17, 2011.

Westover, L, Olearczyk, J, Hermann, U, Adeeb, S, & Mohamed, Y. (2012). *Module rigging assembly dynamic finite element analysis, Proceedings of the 2011 CSCE annual general meeting and conference (CSCE 2011), Jun 6–9, 2012*. Alberta, Canada: Edmonton.

3

A visual energy performance assessment and decision support tool for dwellings

Amit Mhalas*, Mohamad Kassem, Tracey Crosbie and Nashwan Dawood

Abstract

Background: The target for carbon dioxide (CO_2) emissions reduction in the UK is set at 20% by 2020 and 80% by 2050. The UK housing stock is one of the least energy efficient in Europe. The energy used in homes accounts for more than a quarter of energy use and carbon dioxide emissions in Great Britain. Therefore, it is imperative to improve the energy performance of the existing housing stock and fully exploit energy efficiency and renewable energy interventions. The UK has developed several policies and initiatives to improve the energy performance of the housing stock and there are a number of databases that hold information about the condition of the housing stock. However, existing approaches and tools do not allow decision makers to assess the environmental and economic effectiveness of CO_2 reduction strategies at the neighbourhood level.

Methods: This research presents a methodology that integrates these energy databases with visualisation systems and multi-criteria decision analyses to enable the evaluation of the environmental and financial implications of various energy efficiency and renewable energy interventions at both building and neighbourhood levels. The methodology is prototyped in a proof-of-concept tool which is validated and tested in an empirical case study with local authorities and social housing providers.

Results: The validation study compared the energy performance of the dwellings estimated by the proposed methodology with the energy performance calculated from actual survey and confirmed that the results are consistent. The case study demonstrated that the methodology and the prototype can be reliably utilised to evaluate the environmental and financial implications of various energy efficiency and renewable energy interventions.

Conclusion: The findings illustrate that the tool is particularly useful for town planners, local authorities and social housing providers. They can make informed decisions about the implementation of energy policies and initiatives along with energy suppliers, building engineers and architects. The tool developed in the research and presented in this paper can contribute to meeting CO_2 emission reduction targets.

Keywords: Geographic Information System (GIS); Domestic energy assessment; Standard Assessment Procedure (SAP)

Introduction

There is a rising interest in tackling climate change. Subsequent to the 1992 Kyoto Protocol there is a growing incentive to reduce CO_2 emissions through increased use of renewable energy sources and reducing energy demand. The UK's commitment under the protocol is for a reduction in greenhouse gas emissions of 12.5% from 1990 levels by 2012. The UK government in its Climate Change Act is committed to reduce its CO_2 emissions by 80% by 2050 over its 1990 baseline (H.M. Government 2008). The UK government is also committed to meet the EU target to reduce its CO_2 emissions by 20% and obtain 15% of energy from renewable sources by 2020 (House of Lords 2008).

Buildings contribute almost a half of all CO_2 emissions in the UK. Of those emissions 17% come from approximately 26 million residential dwellings and 18% come from 2 million non-domestic buildings (All Party Urban Development Group 2008). It is expected that about

* Correspondence: a.mhalas@tees.ac.uk
Technology Futures Institute, Teesside University, Middlesbrough TS1 3BA, UK

75% of the existing domestic stock will be still present in 2050 (Wright 2008). The UK housing stock is one of the oldest and the least efficient in Europe. This poor quality housing stock means space heating consumed about 66% of the total delivered energy in 2006 (Palmer and Cooper 2011). Over 30% of the dwellings in England are thought to be 'non-decent' i.e. they are unhealthy, in disrepair, in need of modernisation or providing insufficient thermal comfort, with 80% of these failing to meet the criteria for comfort (Communities and Local Government 2012). The reduction of CO_2 emissions from the existing built environment is likely to be a key component of meeting the overall 80% CO_2 emissions reduction target (Jones et al. 2007).

A range of improvements through energy efficiency and renewable energy measures is promoted through Government policies and initiatives including Carbon Emissions Reduction Targets (CERT), Community Energy Savings Programme (CESP), Energy Company Obligation (ECO) and the Green Deal (DECC 2009). These initiatives include grants and advice programmes to achieve short and long term emission goals. These initiatives aim to reduce energy consumption, improve living standards and eliminate fuel poverty (DECC 2011a, b). The local development framework requires local governments to involve local community, utility providers, environmental groups and housing corporations amongst others in their appraisal and management process of the framework (Office of the Deputy Prime Minister 2010). Therefore, energy and carbon models which can undertake predictions and evaluate the potential of different energy efficiency and renewable energy interventions for the housing stock are essential for implementation of these policies and initiatives (Cheng and Steemers 2011).

This paper presents a methodology and a proof-of-concept tool that together integrates energy databases with visualisation systems and multi-criteria decision analyses to enable the evaluation of the environmental and financial implications of various energy efficiency and renewable energy interventions at both building and neighbourhood levels. The proof-of-concept tool is based on a GIS platform and makes use of aerial and terrestrial imagery, digital maps and information from various national statistics and databases. First, the paper presents the gaps identified through literature review of the existing dwelling models. Second, the paper illustrates the methodology and tool developed and their validation in an empirical case study with the involvement of a local authority and a social housing provider. Finally, the discussion of the case study results is presented conducted by comparing the tool outputs with the actual energy performance data from the housing provider and estimating the potential of energy saving and CO_2 emission reduction.

Background

The techniques to model energy consumption in the residential sector can be broadly classified into 'top-down' and 'bottom-up' approaches (Tuladhar et al. 2009). The approaches have a vast diversity in terms of their level of detail, their complexity, the data input required by the user, the time periods covered and their geographical coverage (Hourcade et al. 2006).

Top down approaches

The top-down approaches work on a macro-economic scale to model energy supply and energy demand. The development and use of these approaches grew significantly during the energy crisis in the late 1970s. The models require few details of the consumption process and treat dwellings as an energy sink and regress or apply factors that affect consumption to determine the trends (Swan and Ugursal 2009). This approach aims at fitting historical time series of national energy consumption or CO_2 emissions data on an aggregated level. Top-down models investigate the inter-relationship between the energy sector and the economy at large (Kavgic et al. 2010).

The strength of the top-down approach is that only aggregated data is required, which is widely available. The weakness of this approach is that it cannot model energy consumption at the building level and has no inherent capability to model discontinuous advances in technology (Swan and Ugursal 2009). Model developed by (Hirst et al. 1977) is sensitive to major demographic and economic factors, however these factors need to be continually updated to improve quality of the results. Haas and Schipper (1998) in their study identified *"non-elastic response due to irreversible improvements in technical efficiency."* It has been pointed out that a reliance on the past energy energy-economy interactions to predict future scenarios may not be appropriate while dealing with issues such as climate change; as the environmental, social and economic conditions may be significantly different to those experienced in the past (Kavgic et al. 2010). Several economists using top-down modelling approaches rely overly on the Autonomous Energy Efficiency Index leading to implementation costs for measures to mitigate CO_2 emissions being over estimated (Jaccard et al. 1996).

Due to these limitations, the top-down approach is clearly not suitable for identifying key areas for improvements in the demand side energy consumption at the building level (Swan and Ugursal 2009).

Bottom-up approaches

Bottom-up approaches consist of models that use input data from highly disaggregate components. Energy consumption from units such as individual houses, or groups of houses are considered and then extrapolated to regional or national levels. Based on the data input, these models can be further categorised into statistical models and building physics based models.

Statistical modelling relies on the availability of large quantities of energy supplier billing data. The data is however private information and may not be widely available (Swan and Ugursal 2009). Regression analysis is typically used within statistical modelling to determine the coefficients of the model corresponding to the input parameters to estimate the energy consumption of dwellings (Fung 2003). Though these models assess the energy consumption of residential sector, they have limited abilities to assess the impact of scenarios of reduction in energy consumption (Fung 2003).

The Building physics based models calculate the energy consumption based on physical characteristics of the buildings or its components. The energy calculation requires quantitative data on physically measurable variables. These include the efficiency of space heating systems and their characteristics, information on the areas of the different dwelling elements (walls, roof, floor, windows, doors) along with their thermal characteristics (U-values), internal temperatures and heating patterns, ventilation rates, energy consumption of appliances, number of occupants, external temperatures, etc. (Johnston 2003). The combination of building physics and empirical data from housing surveys, national data sets and assumptions about buildings operation, give modellers the means to estimate energy consumption in dwellings for the past, present, and future (Wilson and Swisher 1993).

The building physics based models consider detailed information about the building and hence estimate energy consumption with most clarity (Larsen and Nesbakken 2004). Furthermore, they do not depend upon historical values; however, the historical data can be used to calibrate the models. The major advantage of building physics based models are the modular structure of their algorithms. This means the users of this approach can easily modify the algorithms to suit particular needs (Kavgic et al. 2010). Building physics based models are the only methods that can fully estimate energy consumption of a sector without historical energy consumption information and evaluate the impact of new technologies (Swan and Ugursal 2009). The policies and initiatives such as CERT, CESP, ECO and Green Deal require practical decisions and are directed towards the level of the physical factors which influence energy use. Bottom-up approaches and in particular the building physics based models help in addressing these needs and hence is the preferred approach in this study.

Several building physics based models have been developed in the UK over a number of years to estimate the current and future residential demand. Some of the notable models include:

- Building Research Establishment's Housing Model for Energy Studies (BREHOMES) (Shorrock and Dunster 1997)
- The Johnston Energy and CO_2 Emission Model (Johnston 2003)
- The UK Domestic Carbon Model (UKDCM) (Boardman et al. 2005)
- The DECarb Model (Natarajan and Levermore 2007)
- The Energy and Environmental Prediction (EEP) Tool (Jones et al. 2007)
- The Community Domestic Energy Model (CDEM) (Firth et al. 2010)
- The Domestic Energy Carbon Counting and Carbon Reduction Model (DECoRuM) (Gupta 2009)

All these models have the same energy calculation engine which is the BREDEM (Building Research Establishment Domestic Energy Model) modified to varying degrees. The disaggregation levels vary significantly amongst the seven models. The transparency of models in terms of data sources and model structures is recognised by most authors as a crucial issue for the future deployment of the models. Furthermore, no access is available to the raw input data and core calculation algorithms of almost all the models, including the modified BREDEM-type modules (Kavgic et al. 2010; Natarajan and Levermore 2007). Except for EEP all other models, these tools lack the ability to be used by stakeholders for implementation of policy or initiatives.

Discussions were undertaken with stakeholders such as social housing providers, local councils, town planners and energy companies to understand how these models are currently used. Those questioned indicated that none of the models described above are used by them as the models cannot simulate scenarios for performance improvement for specific geographic areas. Further, the models rely on standard archetypes for simulation or use drive-by surveys to determine archetypes. In the case of the EEP tool it was necessary to collect data for 55,000 dwellings via a drive-by survey which took 18 months (Jones et al. 2007). Drive-by surveys are time consuming and costly. Furthermore, none of the models take into consideration the engagement of local community, utility providers, environmental groups and housing corporations as part of their decision making process (Kassem et al. 2012).

This research addresses these gaps through developing a methodology and proof-of-concept tool that integrates energy databases with visualisation systems and multi-criteria analyses. For this purpose, the Standard Assessment Procedure (SAP) was selected as a main element of the proof-of-concept tool. SAP is underpinned by BREDEM and is the recommended tool by the Department of Energy and Climate Change (DECC) for assessing and comparing the energy and environmental performance of dwellings. SAP is now the UK's National Calculation Methodology, meeting one of the requirements of the EU Energy Performance of Buildings Directive (DECC/BRE 2010). SAP provides indicators of energy performance through energy consumption per unit floor area, an energy cost rating (the SAP rating), an Environmental Impact rating based on CO_2 emissions (the EI rating) and Dwelling CO_2 Emission Rate (DECC/BRE 2010).

The methodology and the proof-of-concept tool enable stakeholders assess the baseline energy performance of dwellings on a neighbourhood level through use of visualisation techniques and databases. This eliminates the need for drive-by surveys and users can create customised archetypes representing the actual characteristics of dwellings rather than relying on standard archetypes. The tool permits stakeholders to develop tailor-made scenarios of energy efficiency and renewable energy interventions for individual or multiple dwellings thus assisting in the policy implementation stages. The tool informs the stakeholders of the environmental benefits in terms of increased SAP rating of dwellings, energy saved in kWh and amount of CO_2 reduced in kg for tailor-made energy performance improvement scenarios. Most importantly, the tool incorporates the use of multi-criteria decision analysis technique which assists stakeholders in meeting the requirements of the local development framework (Office of the Deputy Prime Minister 2010) through incorporation of a trade-off mechanism based on environmental, technical, economic and social criteria.

Methods

Within the SAP model the calculation is based on the energy balance of dwellings and considers the following factors:

- Materials used for construction of the main aspects of the dwelling such as walls, windows and roofs;
- Thermal insulation characteristics of the building fabric;
- Ventilation characteristics of the dwelling and ventilation equipment;
- Efficiency and control of the heating system(s)
- Gains into the dwellings from solar radiation, metabolism, cooking and lights and appliances;
- Fuel used to provide space and water heating, ventilation and lighting;
- Energy for space cooling, if applicable;
- Renewable energy technologies.

SAP requires an input of over 80 different items of data for each dwelling. These data are easily available for new developments; however, for existing dwellings, most of these data has to be gathered through site surveys. A detailed property survey by a trained assessor can last for at least 30 minutes (Rylatt et al. 2003). Thus collecting this data for each dwelling and then aggregating for locality, town, city, region, etc. can be hugely time consuming and expensive. To overcome this challenge, the research makes use of information from aerial and terrestrial imagery, published databases such as Homes Energy Efficiency Database (HEED), household surveys such as English House Condition Survey (EHCS), census and the Office of National Statistics (ONS). Data from these sources form input for the core SAP calculation engine. The framework used for developing the proof-of-concept tool is depicted in Figure 1.

Models of the dwellings in a neighbourhood are developed as a first step in this process. Next, energy performance calculations are undertaken for each of these dwelling models based on SAP algorithms. Based on the existing characteristics of the dwelling, the potential for carbon reduction is quantified. Finally, the energy improvement measures are ranked using a decision support system.

The prototype tool is built on a GIS platform as it helps in integrating and managing vast and various formats of data and can connect various data sets together by common locational data e.g. address (Goodchild 2009). The database created can be shared for various purposes including modelling and simulating scenarios. For the present research, ArcGIS 10 was used. It is one of the most commonly used GIS platforms and supports the data from the identified sources. ArcGIS 10 provides a geoprocessing functionality which allows personalizing tasks through inherent programming capabilities (Environmental Systems Research Institute 2010).

Create domestic dwelling models

The principal requirement of the prototype is to develop domestic dwelling models which act as source for the energy performance calculations. An investigation of the SAP algorithms reveals that the energy calculations and the data input work on three sub models. These sub-models are described below and the working mechanism is depicted in Figure 2.

- Geometric sub-model consisting of details on foot-print, floor height, exposed perimeter and wall area and roof-area;
- Physics sub-model consisting of details on ventilation and U-values of walls, windows, roof and floors;

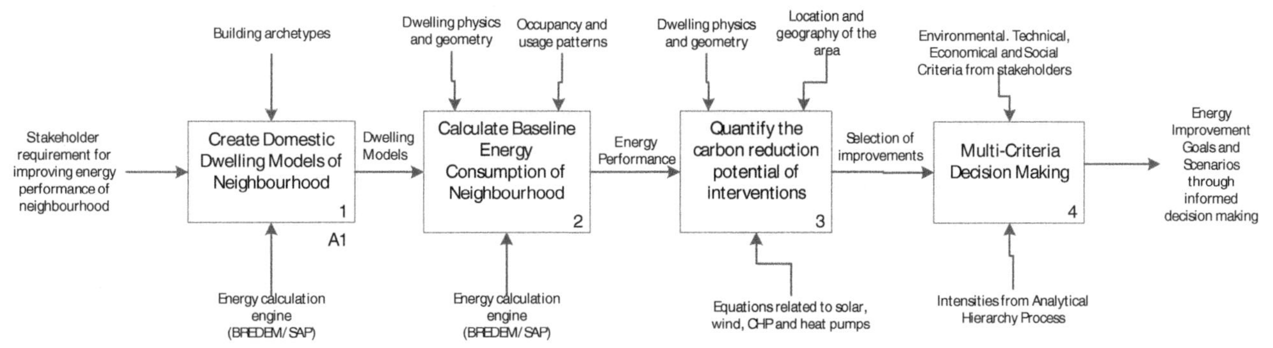

Figure 1 Framework of prototype development.

- Usage sub-model consisting of details on the type and use of heating system, heating controls and electrical appliances.

The detailed list of parameters required for the development of these models and the source where they can be obtained from is presented in Table 1.

The dwelling vector maps from Ordinance Survey are polygon features with address, area and footprint information included as attributes which builds the geometry model of dwellings (OS MasterMap Address Layer 2010; OS MasterMapTopography Layer 2010). Using the developer capabilities of ArcGIS, a user form is developed using Visual Basic for Applications (VBA) (Figure 3) which acts as an interface to include the information related to building physics and usage.

One of the significant advantages of using GIS based software is that the data entry process can be replicated for several dwellings that have similar characteristics. This is particularly useful in undertaking the energy assessment for terraced, semi-detached and detached houses built during similar time periods. The dwellings built in particular time period typically present similar physical characteristics as they were built to meet the needs of the relevant building specifications of that period.

Calculate baseline energy consumption

Once the dwelling models for a neighbourhood are developed with the attributes related to geometry, physics and usage of the dwellings stored, the next stage is to use these attributes for the calculation of baseline energy consumption. The ArcGIS developer capabilities allow for empirical equations such as those in SAP algorithms to be formulated into calculation tools that can source data from the recently stored attributes. The resultant data are values such as energy demand for space and water heating and electricity demand for lights and appliances. Based on the total energy used

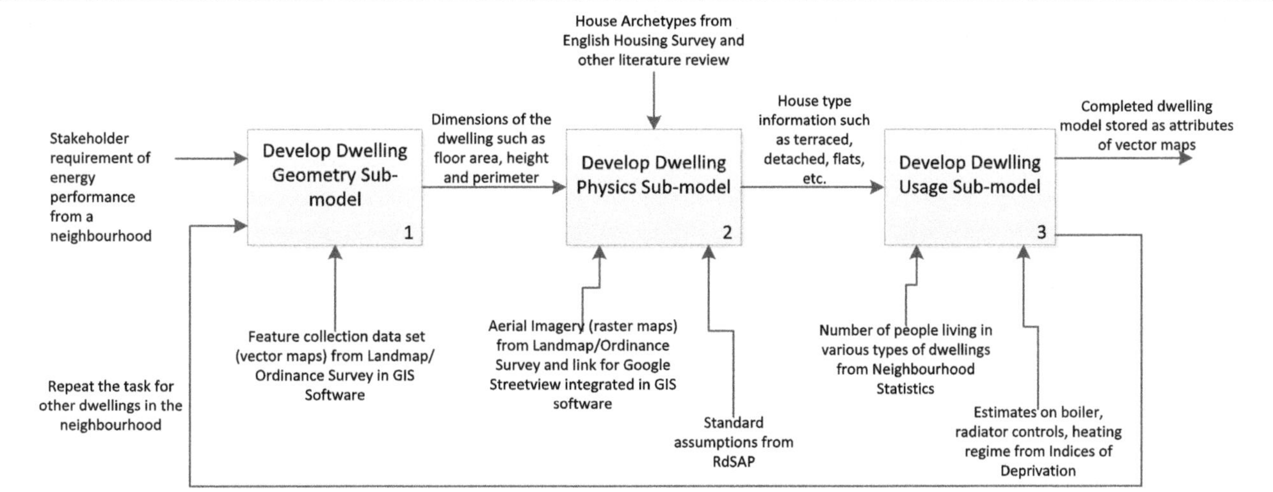

Figure 2 Creating domestic dwelling model.

Table 1 Data and their sources for model development

Information required	Data source
Dwelling geometry sub-model	
Number of storeys	Terrestrial imagery from Google Maps
Floor area and perimeter	Vector map of the area from Ordinance Survey
Height of each storey	Vector map from Landmap
Area of the roof	Vector map of the area from Ordinance Survey
Area of the exposed walls and windows	Vector map of the area from Ordinance Survey
Dwelling physics sub-model	
Degree day region	Vector map of the area from Ordinance Survey
Height above sea level (m)	Vector map of the area from Ordinance Survey
Mean wind speed	Data tables provided in SAP
Level of over-shading	Raster imagery from Ordinance Survey and Google Maps
Dwelling detachment (mid or end terraced, semidetached, detached, flat, etc.)	Vector map of the area from Ordinance Survey
Dwelling Age (Before 1900, 1900–1929, 1930–1949 1950–1966, 1967–1975, 1976–1982, 1983–1990, 1991–1995, 1996–2002, 2003–2006, 2007 – Onwards)	Vector map of the area from Landmap, raster imagery from Ordinance Survey and Google Maps
U Value for walls and windows	Inferred from age of the building and raster imagery
U Value for floor and roof	Inferred from age of the dwelling
Draught proofing	Inferred from economic deprivation data from ONS
Type of window and door frame	Terrestrial imagery from Google Maps
Orientation of windows	Raster imagery from Ordinance Survey and Google Maps
Number of flues, chimneys	Raster imagery from Ordinance Survey and Google Maps
Number of fans and vents	Inferred from age of the dwelling
Presence of mechanical ventilation	Assumed no mechanical ventilation
Type of water heater (gas, oil or solid fuel boiler, electric immersion.)	HEED
If hot water tank present its volume, thickness of insulation, thermostat and insulation of primary pipework	Default SAP values depending on age of the dwelling
Dwelling usage sub-model	
Heating system type e.g. open, sealed, etc. and fuel e.g. gas, oil, solid fuel, electric, etc.	HEED and EHCS
Efficiency of the heating system	Inferred based on HEED and EHCS data
Type of heating controls (programmers, thermostats)	Inferred based on HEED and EHCS data
Type of cooking system (electric cooker, cooking range, etc.)	Inferred based on HEED and EHCS data
Total number of lighting outlets and proportion of low energy lights	Inferred based on economic deprivation data from ONS, HEED and EHCS data
Number of occupants	Census data from ONS
Heating periods	Inferred based on economic deprivation data from ONS
Demand temperatures	Inferred based on economic deprivation data from ONS
Level of use of hot water, lights and cooking (average, below average and above average)	Inferred based on economic deprivation data from ONS

and the fuel supplied for space and domestic water heating, the total amount of CO_2 is computed. Based on the level of energy consumption, ArcGIS allows developing thematic maps where areas with higher energy consumption can be identified as hotspots. An example of such thematic energy consumption is presented in Figure 4.

The lowest geographical level on which the data from the Office of National Statistics is aggregated is the Lower Level Super Output Area (LLSOA). LLSOAs have a minimum population of 1,000 equating to around 400 households. An energy calculation output for one such LLSOA based on the framework described is depicted in Figure 5.

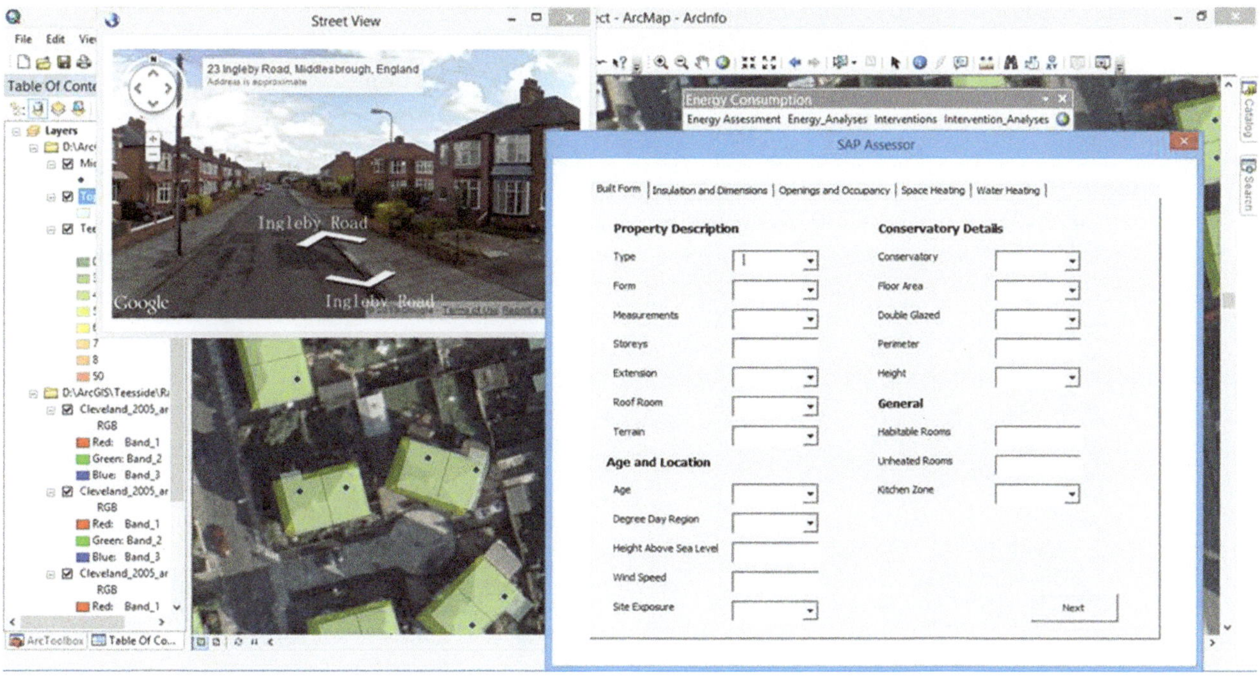

Figure 3 Userform to input data and create dwelling models.

Quantify carbon reduction potential

Subsequent to identifying the baseline energy consumption and associated CO_2 levels, the next task is to identify which energy efficiency and renewable energy interventions are applicable to the area and what are their energy saving / energy generation potentials. The most common energy efficiency and renewable energy interventions within UK are floor, roof and wall insulation, solar photo voltaic and solar thermal panels, condensing combination boilers, wind turbines, micro-combined heat and power systems and air and ground source heat pumps (Boardman 2007).

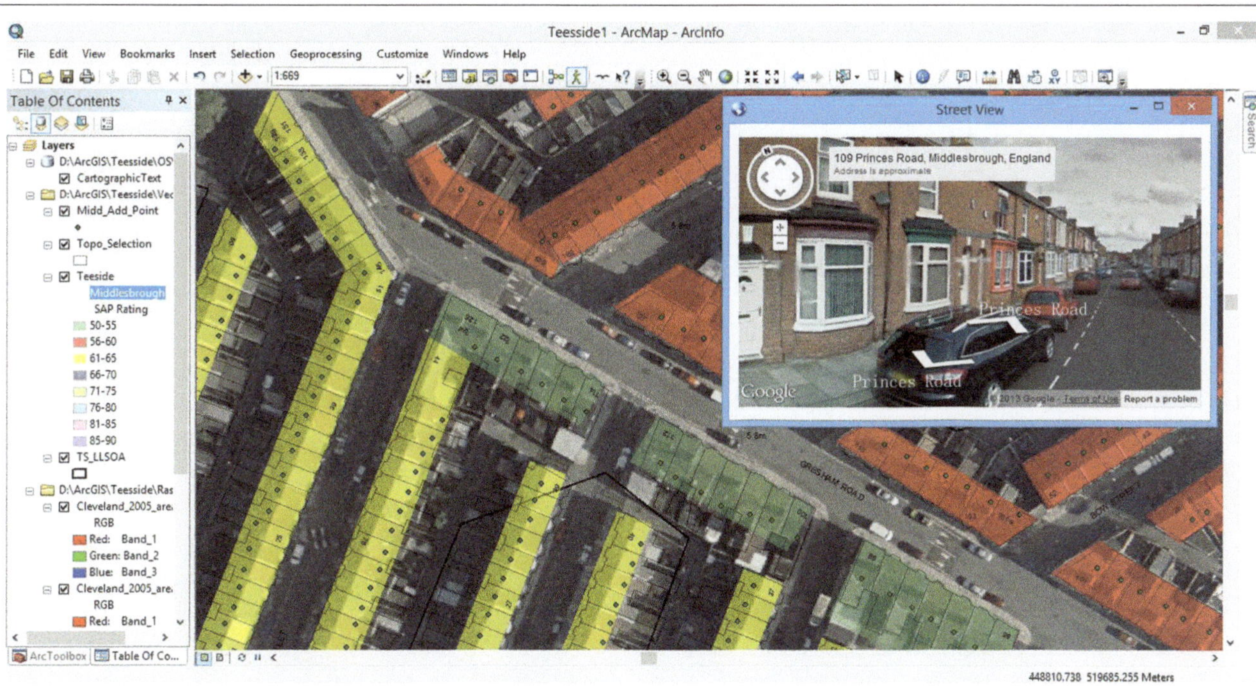

Figure 4 Thematic display of dwelling energy consumption levels.

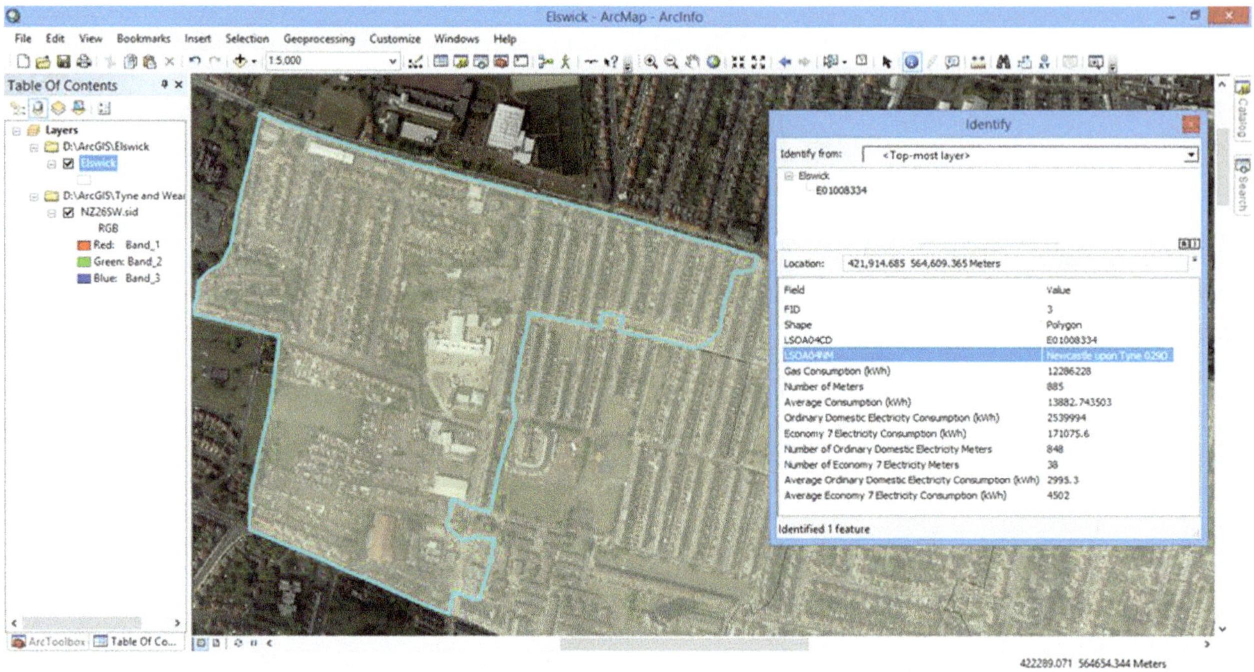

Figure 5 Baseline energy consumption calculation on LLSOA level.

Using the UK wind speed database available from the DECC (2013) as shown in Figure 6; decisions can be made if the areas meet the space and wind speed requirements. Based on the wind speed, and the size of the wind turbine that could be installed, the wind potential of the area is identified. The dwelling models contain information on the orientation and area of roof. Using ArcGIS querying capabilities, dwellings having south facing roof are identified (Figure 7). SAP algorithms provide equations for calculation of solar potential based on roof orientation and area.

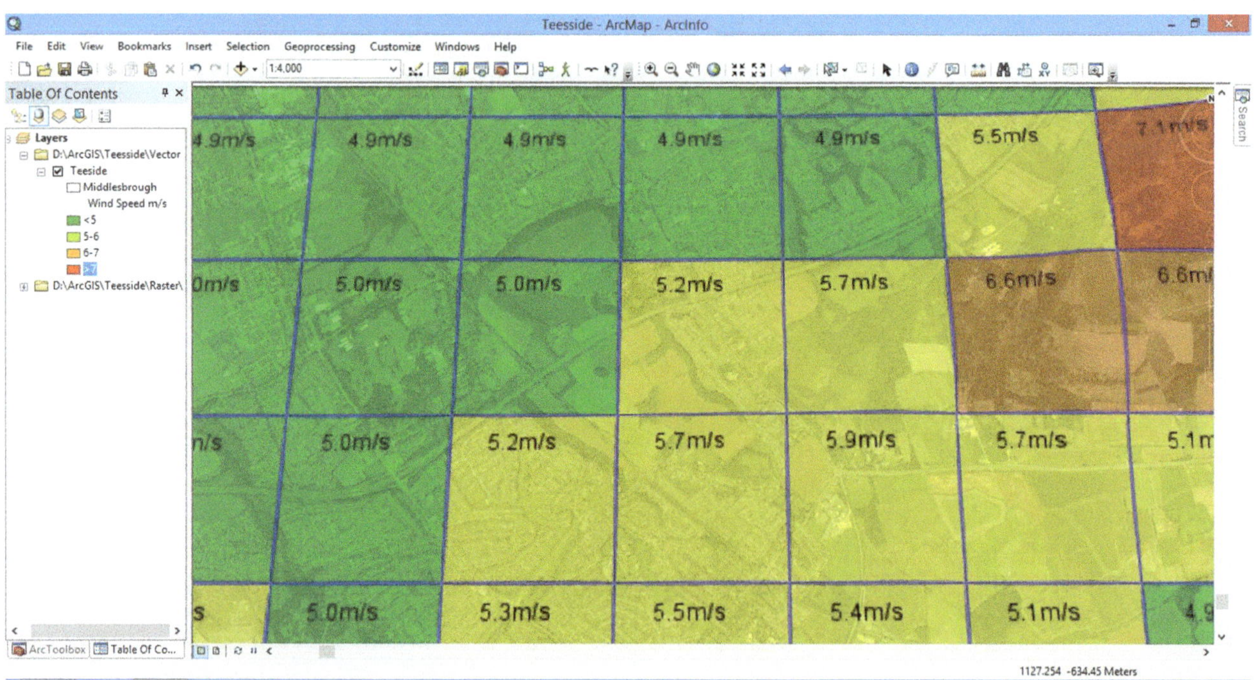

Figure 6 UK wind speed database.

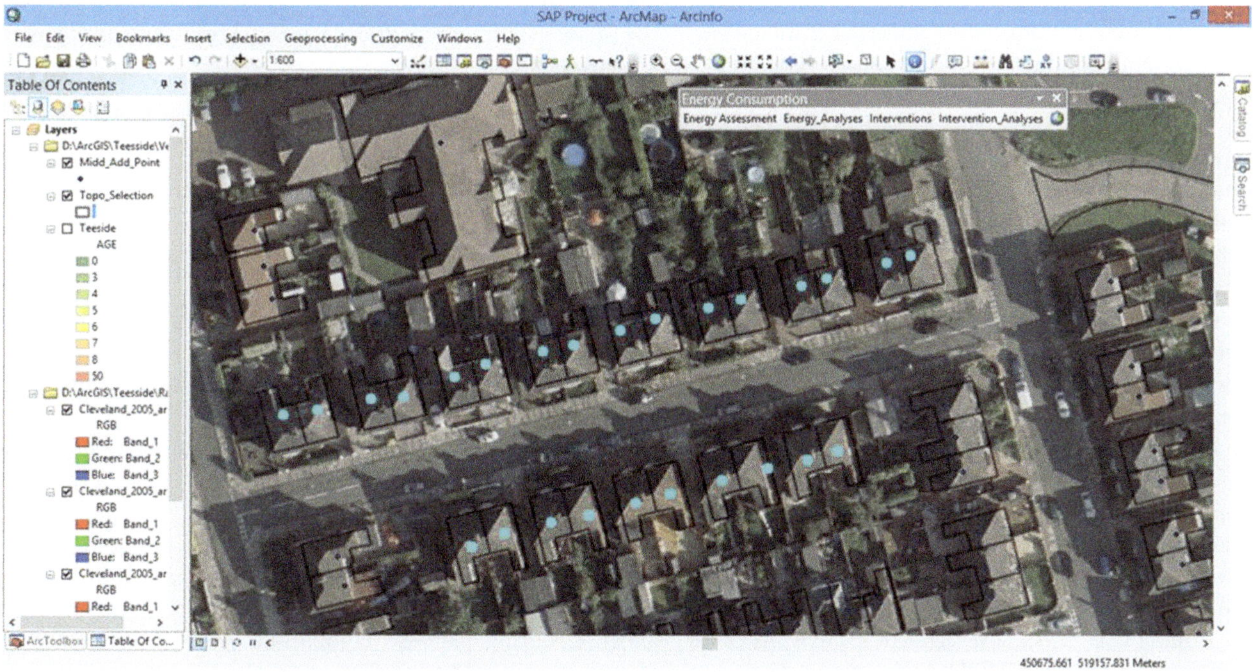

Figure 7 Houses with south facing roofs for solar panels.

The Ordinance Survey maps along with the actual building foot-print also show the measurements of area around the property. This information is used to select dwellings having large and accessible backyards as they have a potential for installation of ground source heat pumps (Figure 8). As the heat demand of the dwellings is now known, the potential for a heat pump can be calculated using equations from SAP algorithms.

Currently equations are developed to estimate the energy savings (kWh/annum), reduction in CO_2 (kg/annum) and

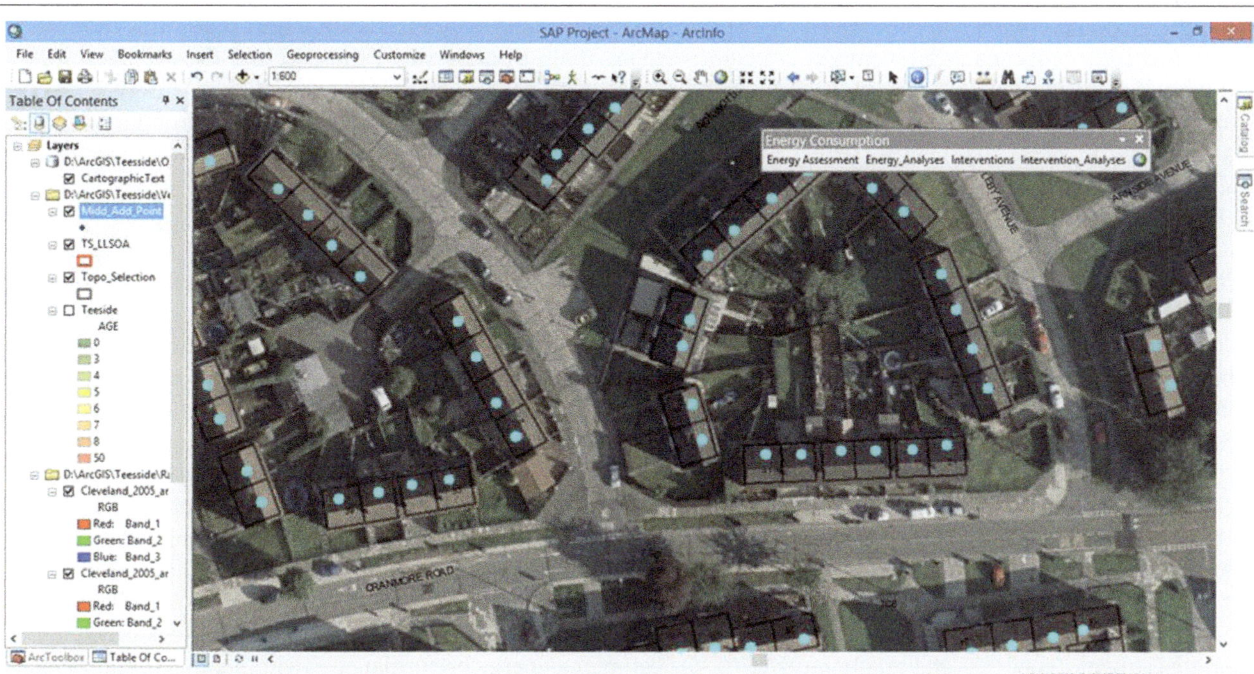

Figure 8 Dwellings with potential for heat pumps.

savings (£/annum) for standalone interventions. Energy savings from some interventions will be different if they were installed as combination, for e.g. an air source heat pump will provide more savings for a house without eternal wall insulation or filled cavity wall. Where the house is well insulated, the heat pumps will not require much energy to produce and hence the savings will be less and payback time will be higher. Work is currently being undertaken to address this issue.

Decision support system

Energy planning decisions are complex as several parameters are involved in the process thus necessitating a decision support system. Furthermore, as the case with most energy-related decisions, various groups of decision makers are involved. Multi-criteria decision making (MCDM) deals with making decisions in presence of multiple stakeholders' criteria and alternatives (Wang et al. 2009). MCDM can be further divided into Multi-Objective Decision Making (MODM) and Multi-Attribute Decision Making (MADM). Several methods exist in each of the above categories. In MODM, the alternatives are not predetermined, but instead a set of objective functions is optimised to a set of constraints. In MADM a small number of alternatives are evaluated against a defined set of attributes. In either case, a criterion is the basis for a decision which can be measured and evaluated quantitatively or qualitatively (Pohekar and Ramachandran 2004). Analytical Hierarchy Process (AHP) is one such attribute-based process developed by Saaty (1980). AHP decomposes a complex problem into a hierarchy with a goal (objective) at the top and criteria and sub-criteria at subsequent lower levels. Elements at a given hierarchy are compared in pairs to assess their relative preference with respect to each of the elements at the next higher level. A scale of 1–9 is used to assess the intensity with 1 indicating equal importance and 9 extremely high importance (Saaty, 1980). More importantly, the AHP supports trade-offs with intangible and tangible values. As this approach is considering the social perspective along with environmental, technical and economic factors, the AHP is selected as the trade-off method of the present research.

AHP has been used for energy source assessment in various studies undertaken to select between interventions such as wind farm, solar energy, geothermal, hydroelectric, etc. based on multiple criteria (Akash et al. 1999; Gamboa and Munda 2007; Georgopoulu et al. 2003; Erol and Kilkis 2012). In this research, AHP is used to rank following alternatives:

- Annual Reduction in CO_2 levels
- Initial Investment (fixed cost and grants received)
- Return on investment (annual running cost to user and savings made through feed-in-tariff)
- Social acceptability
- Ease of implementation (access to resources and timeline)

A pairwise comparison is then undertaken for the following improvement measures for each of the above criteria (second hierarchy).

- Solar photovoltaic (PV)
- Wind turbines
- Micro-combined heat and power pump (μ-CHP)
- Condensing boiler
- Air source heat pumps (ASHP) and ground source heat pumps (GSHP)
- Fabric change (roof, floor and wall insulation and double glazed windows

A user-form similar to the one presented in Figure 3 is prepared for decision support. It allows users to input intensities during pairwise comparison. The intensities assigned during pairwise comparison are converted into square matrices. The matrices are evaluated to identify the eigenvectors, which represent the weighting. The weighting for each alternative for each criterion (second hierarchy) is then multiplied by the weighting of the criteria (first hierarchy). A sum of these products across all criteria provides the final ranking of improvement measures (Saaty, 1980).

Thus, the tool enables stakeholders to assess the interventions that they are considering. So for example if a social housing provider is trying to decide if it would be more effective in terms of reducing CO_2 emissions to install solar PV or ground source heat pump this tool will allow them to understand the cost and CO_2 implications along with other criteria for each of these interventions. The stakeholders can then engage with appropriate building engineering companies and architects in installing these improvement measures.

Results and discussion

To validate the framework, baseline energy calculations were undertaken on a set of dwellings owned by a social housing provider in Middlesbrough, UK. The social housing providers have a regulatory requirement to maintain an energy performance certificate (EPC) for their properties. Therefore, the energy performance of the dwellings estimated by the approach of the present research can be compared with the energy performance calculated from actual survey of the property. Properties of various age and detachment were selected to ensure adequate representation of different archetypes. The results from the validation are presented in Table 2 and indicate that the estimated energy performance is within a maximum range of ±8% of the actual energy performance provided by the social housing provider. The average error over the 34

Table 2 Data and their sources for model development

No.	Actual Energy (kWh/m^2)	Estimated Energy (kWh/m^2)	Error	No.	Actual Energy (kWh/m^2)	Estimated Energy (kWh/m^2)	Error
1	298.7837	303.00	1.41%	18	166.1845	166.07	−0.07%
2	191.6192	188.00	−1.89%	19	147.8607	147.35	−0.35%
3	187.522	189.00	0.79%	20	165.0032	166.53	0.93%
4	242.8596	241.00	−0.77%	21	215.8848	205.09	−5.00%
5	174.995	182.88	4.51%	22	189.8272	196.54	3.54%
6	176.6468	177.18	0.30%	23	285.2347	274.00	−3.94%
7	190.0774	194.00	2.06%	24	162.7286	174.91	7.49%
8	163.5331	171.82	5.07%	25	265.662	272.02	2.39%
9	283.2615	282.75	−0.18%	26	217.2638	228.00	4.94%
10	249.1428	249.04	−0.04%	27	243.0846	254.39	4.65%
11	286.2131	272.42	−4.82%	28	258.9484	261.97	1.17%
12	221.8962	233.04	5.02%	29	260.5743	261.97	0.54%
13	361.7491	359.91	−0.51%	30	304.2866	313.82	3.13%
14	272.4004	278.90	2.39%	31	212.298	213.46	0.55%
15	267.3812	275.58	3.07%	32	362.5949	357.44	−1.42%
16	265.7164	279.75	5.28%	33	264.5559	261.41	−1.19%
17	167.6343	166.07	−0.93%	34	183.3866	185.82	1.33%

properties is just about 1% and the standard deviation of the error is 0.03. The results are clearly within a close proximity of the actual energy performance and hence the approach is reliable.

Subsequent to the validation, a case study is undertaken for an LLSOA in Middlesbrough. This LLSOA consists of 756 dwellings with a majority of terraced houses followed by a small number of semi-detached and detached houses. The annual energy consumption details for these house types estimated using the approach developed in this research is presented in Table 3. The results show an annual energy consumption of just over 16.5 GWh for the 756 dwellings in the LLSOA. This computes to an average energy consumption of 21.85 MWh per dwelling per year within the LLSOA. The national average energy consumption estimated for domestic dwellings is 19.8 MWh (OFGEM 2011). The estimated energy performance is approximately 10% higher than the national average. The results can however be considered consistent as most of the dwellings in this LLSOA are built prior to world war two and have low insulation standards (HEED 2012).

The prototype was further used to estimate the potential energy saving from the dwellings in the same LLSOA. The average wind speed in this area of Middlesbrough was less than the required 5 m/s and was thus not considered suitable (DECC 2013). Since most dwellings in this area are terraced houses, space is a constraint for GSHP, hence only ASHP are considered. Table 4 shows the results from the analysis of various interventions including changes to building fabric and installation of Solar PV, μ-CHP, condensing boiler and ASHP.

Table 3 Energy consumption in an LLSOA

Dwelling type	Number	Energy consumption (kWh/Annum)
Terraced	719	15,948,608
Semi-detached	23	357,700
Detached	14	265,387
Total	756	16,571,695

Fixed cost is based on the costs for individual dwellings provided by the Energy Savings Trust. Grants are typically available to energy suppliers, councils and social housing providers under the CESP and CERT that can significantly contribute towards the fixed costs and can help in bring down the costs indicated in Table 4 (DECC 2011a). Furthermore, under the UK Government's feed-in-tariff, payments are made to dwellings for each kWh of energy generated using micro-generation technologies which includes Solar PV and μ-CHP. This payment is made for 20 years from date of installation. Contribution from this feed-in-tariff has been included in the annual savings presented in Table 4. The lifetime of Solar PV is considered as 25 years and μ-CHP is considered as 20 years although these technologies can last typically longer. The tariffs for ASHP were expected to be available in September 2012; however, at the time of writing this paper, the tariffs were not available and are hence not included in the annual savings. The lifetime for all technologies other than Solar PV and μ-CHP is considered as 30 years.

The results indicate that for this LLSOA, fabric insulation and use of μ-CHP offer the most CO_2

Table 4 Analyses of interventions for an LLSOA in Middlesbrough

Intervention	Fixed cost	Energy saved (MWh/Year)	CO_2 saved (Tons/Year)	Annual savings	Lifetime savings
Fabric Change	£5,973,156	9,084	1,795	£281,090	£8,432,708
Solar PV	£4,309,200	631	127	£217,123	£5,428,080
μ-CHP	£1,814,400	1,311	678	£259,308	£5,186,160
Condensing Boiler	£1,209,600	585	115	£24,158	£483,179
ASHP (Under-floor)	£6,804,000	4,005	793	£123,984	£3,719,520
ASHP (Radiator)	£4,536,000	−1,037	−205	-£32,130	−£963,900

reduction potential for investment made. The use of ASHP using radiators as heat emitting source currently offer no savings in energy used due to their low efficiencies. The ASHP using under-floor heating as heat emitting source appear to have good impact on CO_2 reduction however is currently cost intensive. Further analyses of the results show that, just by improving the fabric of the building through insulation of solid walls and roof and installation of low emissivity double glazed windows can reduce the energy demand by 9,084 MWh or just over 41.5%. Installation of condensing boilers can contribute towards reducing energy demand and installation of μ-CHP and solar PV contribute towards electricity generation, thus reducing the demand from the national grid.

To test the decision making process, the information presented in Table 4 was presented to a planner representing one of the stakeholders. The planner was asked to undertake pairwise comparison for criteria and improvement measures. The intensities assigned were developed as matrices to arrive at final weighting. The prototype has interfaces to assign intensities and then visualise the final rankings.

Table 5 presents the summary of the pairwise comparison of the criteria and the alternatives weighted against each criterion. The results indicate that fabric insulation is ranked as the highest followed by μ-CHP, solar PV, condensing boiler and ASHP. Based on the rankings achieved, changes to the building fabric can be undertaken as a priority followed by installation of μ-CHP. In this way the tool enables informed decisions to be made regarding implementation of energy policies. The choice of the energy planner is consistent with the findings in Table 4 that fabric insulation provides the largest opportunity for reducing energy demand. This scenario may however be different for another LLSOA where housing fabric may be of higher standard and other means of energy improvement have to be identified.

Conclusion

A methodology that integrates visual systems, energy related databases and multi-criteria decision analysis to enable energy assessment and evaluation of various energy efficiency and renewable energy interventions for the housing stock was presented. The methodology was prototyped in a proof-of-concept tool, validated and tested with the involvement of a local authority on a housing stock composed of 756 dwellings. The prototype made use of a GIS platform, aerial and terrestrial imagery, digital maps and information from various national statistics and databases. The validation has showed that the developed methodology and prototype provide reliable estimates of energy consumption and enables a systematic analysis of various energy efficiency and renewable energy interventions. The developed methodology and prototype fill an important gap in the literature as there are no tools that enable both the evaluation of energy consumption and the assessment of energy efficiency and renewable energy interventions. In addition, the use of the tool showed the potential of reducing energy consumption and CO_2 emission while considering the intangible criteria such as social acceptance and ease of implementation associated with each improvement strategy. Therefore, the tool presents stakeholders with opportunity not only to identify the baseline energy performance of their housing stock at the neighbourhood level, but also to assess the impact of different scenarios for CO_2 reduction.

The findings presented in this paper clearly illustrate that the methodology and proof-of-concept tool can be used by local authorities, town planners and social housing providers

Table 5 AHP used in decision making process

Alternative	CO_2 reduced	Initial invest.	Return on invest.	Social accept.	Ease of implemen.	Goal
Fabric Change	0.1403	0.0550	0.0674	0.0359	0.0219	32.05%
Micro CHP	0.1103	0.0801	0.0517	0.0282	0.0125	28.29%
Solar PV	0.0663	0.0582	0.0349	0.0170	0.0086	18.50%
Boiler	0.0359	0.0561	0.0136	0.0092	0.0048	11.95%
ASHP	0.0218	0.0451	0.0093	0.0056	0.0104	9.21%

to make informed decisions with regard to the implementation of energy policies and initiatives. Building engineering companies and architects can use the methodology and prototype to undertake dwelling specific analysis during installation of interventions. These can altogether significantly contribute to meeting CO_2 emission reduction targets. Added to this it greatly reduces costs and simplifies the process by eliminating the need for drive-by surveys, saving a large amount of time and resources.

Future development of the tool will involve identification of the impact of interventions and the development of a more accurate calculation of the rate of return of investment based on inflation indices.

Competing interests

The authors declare that they have no competing interests.

Authors' contribution

AM is the main contributor to this article. The article is part of an outcome of the research undertaken by AM as a part of his PhD. MK was involved in compilation and proof-reading of the article. TC and ND supervised this research undertaken by AM. All authors read and approved the final manuscript.

References

Akash, B. A., Mamlook, R., & Mohsen, M. S. (1999). Multicriteria selection of electric power plants using analytical hierarchy process. *Electric Power Systems Research, 52*(1), 29–35.

All Party Urban Development Group. (2008). *Greening UK cities' buildings*. London, UK: All Party Urban Development Group.

Boardman, B. (2007). *Home Truths: A low carbon strategy to reduce UK housing emissions by 80% by 2050*. Oxford, UK: University of Oxford's Environmental Change Institute.

Boardman, B., Darby, S., Killip, G., Hinnels, M., Jardine, C. N., & Palmer, J. (2005). *40% house*. Oxford, UK: University of Oxford's Environmental Change Institute.

Cheng, V., & Steemers, K. (2011). Modelling domestic energy consumption at district scale: A tool to support national and local energy policies. *Environmental Modelling & Software, 26*, 1186–1198.

Communities and Local Government. (2012). *English Housing Survey 2010–11*. London, UK: Department of Communities and Local Government.

DECC. (2009). *Community Energy Saving Programme*. Department of Environment and Climate Change Consultation.

DECC. (2011a). *Evaluation synthesis of energy supplier obligation policies*. London, UK: Department of Energy & Climate Change.

DECC. (2011b). *The Green Deal and Energy Company Obligation*. London, UK: Department of Energy & Climate Change.

DECC. (2013). Wind Speed Database. http://tools.decc.gov.uk/en/windspeed/default.aspx Accessed Jan 25, 2013.

DECC/BRE. (2010). *The Government's Standard Assessment Procedure for Energy Rating of Dwellings*. Watford: Building Research Establishment.

Environmental Systems Research Institute. (2010). A quick tour of what's new in ArcGIS 10. http://help.arcgis.com/EN/ArcGISDesktop/10.0/help/index.html#//00qp0000000p000000.htm. Accessed December 19, 2011.

Erol, O., & Kilkis, B. (2012). An energy source policy assessment using analytical hierarchy process. *Energy Conversion and Management, 63*, 245–252.

Firth, S. K., Lomas, K. J., & Wright, A. J. (2010). Targeting household energy-efficiency measures using sensitivity analysis. *Building Research and Information, 38*(1), 24–41.

Fung, A. S. (2003). *Modeling of National and regional energy consumption and associated greenhouse gas emissions. PhD Thesis*. Dalhousie University.

Gamboa, G., & Munda, G. (2007). The problem of windfarm location: a social multi-criteria evaluation framework. *Energy Policy, 35*(3), 1564–1583.

Georgopoulu, E., Sarafidis, Y., Mirasgedis, S., Zaimi, S., & Lalas, D. P. (2003). A multiple-criteria decision-aid approach in defining national priorities for greenhouse gases emissions reduction in the energy sector. *Europen Journal of Operational Research, 146*, 199–215.

Goodchild, M. F. (2009). *Geographic information systems and science: Today and tomorrow* (pp. 1037–1043). Xuzhou: Procedia Earth and Planetary Science.

Gupta, R. (2009). Moving towards low-carbon buildings and cities: experiences from Oxford, UK. *International Journal of Low-Carbon Technologies, 4*, 159–168.

H.M. Government. (2008). *Climate Change Act*. London: The Stationery Office Limited.

Haas, R., & Schipper, L. (1998). Residential energy demand in OECD-countries and the role of irreversible efficiency improvements. *Energy Economics, 20*(4), 421–442.

HEED. (2012). *Area Summary Report (Middlesbrough, Middlesbrough South and East Cleveland)*. Energy Savings Trust.

Hirst, E., Lin, W., & Cope, J. (1977). A residential energy use model sensitive to demographic, economic, and technological factors. *The Quarterly Review of Economics and Finance, 17*(2), 7–22.

Hourcade, J. C., Jaccard, M., Bataille, C., & Gershi, F. (2006). Hybrid modeling: new answers to old challenges. *Energy Journal*(Special Issue), 1–12.

House of Lords. (2008). *The EU's Target for Renewable Energy: 20% by 2020*. London: The Stationery Office Limited.

Jaccard, M., Bailie, A., & Nyboer, J. (1996). CO_2 emission reduction costs in the residential sector: Behavioral parameters in a bottom-up simulation model. *Energy Journal, 17*(4), 107–134.

Johnston, D. (2003). *A physically based energy and carbon dioxide emission model of the UK housing stock*. Leeds, UK: Leeds Metropolitan University.

Jones, P., Patterson, J., & Lannon, S. (2007). Modelling the built environment at an urban scale - energy and health impacts in relation to housing. *Landscape and Urban Planning, 83*, 39–49.

Kassem, M., Dawood, N., & Mitchell, N. (2012). A decision support system for the selection of curtain wall systems at the design development stage. *Construction Management and Economics, 30*, 1039–1053.

Kavgic, M., Mavrogianni, A., Mumovic, D., Summerfield, A., Stevanovic, Z., & Djurovic-Petrovic, M. (2010). A review of bottom-up building stock models for energy consumption in the residential sector. *Building and Environment, 45*, 1683–1697.

Larsen, B. M., & Nesbakken, R. (2004). Household electricity end-use consumption: results from econometric and engineering models. *Energy Economics, 26*(2), 179–200.

Natarajan, S., & Levermore, G. J. (2007). Domestic futures-Which way to a low-carbon housing stock? *Energy Policy, 35*(11), 5728–5736.

Office of the Deputy Prime Minister. (2010). *Planning Policy Statement 12: Local Development Frameworks*. London: Stationary Office.

OFGEM. (2011). *Typical domestic energy consumption figures*. London, UK: OFGEM.

OS MasterMap Address Layer. (2010). *Ordinance Survey, GB*. Middlesbrough: Coverage.

OS MasterMapTopography Layer. (2010). *Ordinance Survey, GB*. Middlesbrough: Coverage.

Palmer, J., & Cooper, I. (2011). *Great Britain's housing energy fact file*. London, UK: Department of Energy and Climate Change.

Pohekar, S. D., & Ramachandran, M. (2004). Application of multi-criteria decision making to sustainable energy planning - A review. *Renewable and Sustainable Energy Reviews, 8*, 365–381.

Rylatt, R. M., Gadsden, S. J., & Lomas, K. J. (2003). Methods of predicting urban domestic energy demand with reduced datasets : a review and a new GIS based approach. *Building Services Engineering Research and Technology, 24*(2), 93–102.

Saaty, T. L. (1980). *The analytic hierarchy process*. New York: Macgraw-Hill.

Shorrock, L. D., & Dunster, J. E. (1997). The physically-based model BREHOMES and its use in deriving scenarios for the energy use and carbon dioxide emissions of the UK housing stock. *Energy Policy, 25*(12), 1027–1037.

Swan, L. G., & Ugursal, V. I. (2009). Modeling of end-use energy consumption in the residential sector: A review of modeling techniques. *Renewable and Sustainable Energy Reviews, 13*, 1819–1835.

Tuladhar, S. D., Yuan, M., Bernstein, P., Montgomery, W. D., & Smith, A. (2009). A top-down bottom up modelling approach to climate change policy analysis. *Energy Economics, 31*(Supp. 2), S223–S234.

Wang, J. J., Jing, Y. Y., Zhang, C. F., & Zhao, J. H. (2009). Review on multi-criteria decision analysis aid in sustainable energy decision-making. *Renewable and Sustainable Energy Reviews, 13*, 2263–2278.

Wilson, D., & Swisher, J. (1993). Exploring the gap. Top-down versus bottom-up analyses of the cost of mitigating global warming. *Energy Policy, 21*(3), 249–263.

Wright, A. (2008). What is the relationship between built form and energy use in dwellings? *Energy Policy, 36*, 4544–4547.

4

High-precision vision-based mobile augmented reality system for context-aware architectural, engineering, construction and facility management (AEC/FM) applications

Hyojoon Bae[1]*, Mani Golparvar-Fard[2] and Jules White[3]

Abstract

Background: Many context-aware techniques have been proposed to deliver cyber-information, such as project specifications or drawings, to on-site users by intelligently interpreting their environment. However, these techniques primarily rely on RF-based location tracking technologies (e.g., GPS or WLAN), which typically do not provide sufficient precision in congested construction sites or require additional hardware and custom mobile devices.

Method: This paper presents a new vision-based mobile augmented reality system that allows field personnel to query and access 3D cyber-information on-site by using photographs taken from standard mobile devices. The system does not require any location tracking modules, external hardware attachments, and/or optical fiducial markers for localizing a user's position. Rather, the user's location and orientation are purely derived by comparing images from the user's mobile device to a 3D point cloud model generated from a set of pre-collected site photographs.

Results: The experimental results show that 1) the underlying 3D reconstruction module of the system generates complete 3D point cloud models of target scene, and is up to 35 times faster than other state-of-the-art Structure-from-Motion (SfM) algorithms, 2) the localization time takes at most few seconds in actual construction site.

Conclusion: The localization speed and empirical accuracy of the system provides the ability to use the system on real-world construction sites. Using an actual construction case study, the perceived benefits and limitations of the proposed method for on-site context-aware applications are discussed in detail.

Background

Automated, on-demand, and inexpensive access to project information on-site has significant potential to improve decision-making during construction or facility management activities. This information, which is usually in the form of specifications, drawings, or schedule information, enables prompt identification, processing, and communication of discrepancies between actual and expected performance. Fast access to this information also helps project managers to proactively decide on corrective actions and minimize the cost and delays due to performance discrepancies (Golparvar-Fard et al. 2012). Despite the importance of on-site information access, most of current approaches to jobsite progress monitoring include manual and time consuming data collection, non-systematic analysis and visually/spatially complex reporting (Golparvar-Fard et al. 2012; Navon and Sacks 2007). As a part of data collection and analysis, field personnel have to carry large stacks of specifications and drawings on jobsites and spend significant amount of time to write down an actual progress on paper and compare it to relevant cyber-information (Khoury and Kamat 2009). Such inefficiencies in site analysis and information gathering can cause downtime or rework and ultimately lead to schedule delays or cost overruns. In addition, the quality and timing of information access and exchange can either delay or facilitate successful execution of on-site activities (Chen and Kamara 2011).

*Correspondence: hjbae@vt.edu
[1] Bradley Department of Electrical and Computer Engineering, Virginia Polytechnic Institute and State University, Blacksburg, VA, USA
Full list of author information is available at the end of the article

To minimize these inefficiencies, we have proposed a new context-aware vision-based mobile augmented reality system, Hybrid 4-Dimensional Augmented Reality (HD^{4}AR), which identifies location and orientation of field personnel solely based on a site photograph (Bae et al. 2012). As described by Bae et al. (2012), HD^{4}AR allows field personnel to query and access semantically-rich 3D cyber-information and see it precisely overlaid on top of real-world imagery. HD^{4}AR does not need RF-based tracking technologies or inertial measurements to find a user's position. Rather, the system takes a photograph from jobsite as input and computes the location and orientation of the user's camera using a set of computer vision algorithms. As a result, HD^{4}AR can be used with any camera-equipped mobile device, such as a smartphone or tablet, to provide accurate on-site localization of a field engineer and thus is practical and inexpensive to use on a construction site. As shown in Figure 1, the proposed vision-based system can successfully localize a photograph even with large changes in the viewpoint of a user. Moreover, as shown in the bottom-right screenshot, the system successfully recognizes the target building from the photograph, which includes the cellular phone displaying the building, and overlays the cyber-information precisely. This photograph has different color histogram and pixel values from the photograph taken at the actual site, which can simulate the different illumination conditions of the target scene.

This paper extends our prior work on HD^{4}AR (Bae et al. 2012) in the following ways: 1) the localization speed is further increased using direct 2D-to-3D matching, 2) different image feature description methods are implemented and tested to investigate the impact of those descriptors on performance of 3D reconstruction and localization, 3) a new homography-based 3D content-creation (annotation) method, which allows a field engineer to draw and mark any building elements within the photograph, is described. The enhanced localization speed and impact of feature descriptors will be further discussed in Section 'Results and discussion', while 3D annotation functionality will be examined in Section 'High-precision augmentation with HD^{4}AR'.

The remainder of this paper is organized as follows: After brief summary of related works in Section 'Related work', Section 'Method: Hybrid 4-dimensional augmented reality' presents an overview and technical approach of the HD4AR system. The details of the 3D reconstruction process that generates a 3D point cloud model from a set of unordered photographs are discussed in Section '3D reconstruction with HD^{4}AR'. Section 'High-precision augmentation with HD^{4}AR' presents the localization and augmentation process using a generated 3D point cloud. The new feature of the HD4AR, e.g., 3D annotation, is also discussed in this section. Section 'Results and discussion' presents empirical results from experiments with HD4AR and also compares the performance to other state-of-the-art Structure-from-Motion based (SfM-based) localization solutions. Finally the perceived benefits and limitations are described in Section 'Conclusion'. Video demos and detailed performance data of HD^{4}AR can be found at http://www.magnum.ece.vt.edu/index.php/research-projects/100-hd4ar and http://raamac.cee.illinois.edu/hd4ar.

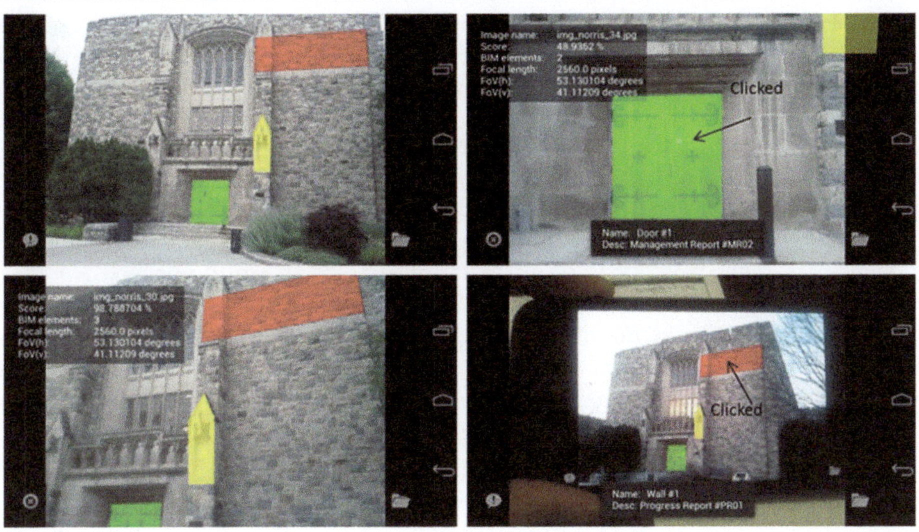

Figure 1 Screenshots of the Android HD^{4}AR client. 3D BIM information is precisely overlaid on photos From different viewpoints. (adopted from Bae et al. 2012).

Related work

The advantages of using augmented reality system on Architectural, Engineering, Construction (AEC) application has been extensively studied by many researchers (*e.g.*, Behzadan and Kamat 2007; Behzadan et al. 2008; Chen et al. 2011; Chi et al. 2013; Dunston et al. 2003; Golparvar-Fard 2009a, 2009b; Hammad et al. 2009; Hou and Wang 2010; Kuula et al. 2012; Schall et al. 2009; Wang 2008; Wang and Dunston 2006; Woodward et al. 2010; Yeh et al. 2012). They have shown that augmented reality indeed improve physical task performance and can reduce mental workload of engineers for AEC tasks (Wang and Dunston 2006). They have also indicated that augmented reality improves design activities as well as design visualization by providing better spatial cognition (Dunston et al. 2003). On-site building information retrieval using a wearable device, proposed by Yeh et al. (2012), also validates that the proper displaying of user-required information on-site leads to shorter task completion time and higher correctness than traditional approach.

To exploit these benefits, many research projects have focused on providing cyber-information to field personnel through mobile devices and/or augmented reality systems (*e.g.*, Akula et al. 2011; Anumba and Aziz 2006; Behzadan et al. 2008; Shin and Dunston 2008; Hakkarainen et al. 2008; Khoury and Kamat 2009; Irizarry et al. 2012; Pasman and Woodward 2003). These works have primarily focused on using Global Positioning Systems (GPS), Wireless Local Area Networks (WLAN), or Indoor GPS for accurately positioning the user within congested construction environments. The main drawback of these Radio Frequency based (RF-based) location tracking technologies is their high degree of dependency on pre-installed infrastructure, which makes their application either difficult or impractical for construction sites. Application of fiducial markers is also suggested by several researchers (*e.g.*, Feng and Kamat 2012; Hakkarainen et al. 2008; Lee and Akin 2011; Yakubi et al. 2011). The systems are also infrastructure-dependent and require the markers to be attached to various surfaces on construction sites, which challenges their applications for large-scale implementations.

On the other hand, some researches have focused on developing infrastructure-independent location tracking systems (*e.g.*, Akula et al. 2011; Ojeda and Borenstein 2007). These systems are typically based on inertial measurements and make use of highly accurate accelerometers and gyroscopes. Given their independence from an existing infrastructure, however, their application may result in accumulated drift error which grows with the distance traveled by the users.

Recent advances in image processing and computer vision have led to new research on the application of image-based reasoning for various construction management tasks and techniques that can manually, semi-automatically, and automatically interpret them (Cheng and Chen 2002; Carozza et al. 2012; Golparvar-Fard et al. 2010,2010; Kiziltas et al. 2008). These researches have shown that a set of overlapping images can be used to extract accurate 3D geometry of stationary objects such as buildings under construction. After the physical models (*e.g.*, the generated 3D point cloud models) and the cyber models (*e.g.*, Building Information Modeling (BIM)) are aligned, they can be compared to determine the actual state of the physical elements on construction site versus the expected state. Some researchers have proved that the fused (aligned) cyber-physical model is accurate to within millimeters (Golparvar-Fard et al. 2011) and can be used to predict the actual construction progress versus the planned cyber model with high accuracy, even when visual obstructions are present (2010). Recent works such as Carozza et al. (2012) extend marker-less augmented reality systems for urban planning purposes. Despite the fact that these systems rely on tracking the camera position and orientation, and does not require additional infrastructure, such systems still require a large amount of matching to be conducted at each step. As reported by Carozza et al. (2012), the overall tracking is likely not as efficient as tracking image features.

Although this body of computer vision research has shown the potential and high-accuracy of image-based reasoning, the speed of 3D reconstruction/localization and the lack of on-site localization methods make these systems difficult to use on worksites. Generating a 3D point cloud model from a set of construction photographs requires non-linear multi-dimensional optimization as well as exhaustive matching of the photographs in the data set and can take hours or days. A specific aim of HD^{4}AR was to overcome these challenges, speeding up overall time of 3D reconstruction and localization by optimizing and enhancing each process.

Method: Hybrid 4-dimensional augmented reality

Overview

HD^{4}AR combines user localization and AR visualization to target on-site query and view of project information on top of real-world imagery. For user localization, HD^{4}AR uses a computer vision-based and model-based method, which obtains detailed information from pre-reconstructed 3D point cloud models built from daily construction photos, and estimates the location of a field engineer's camera using these models. Using 3D point cloud models additionally permits the system to estimate the complete pose of the camera and therefore can support high-accuracy applications such as construction progress monitoring where millimeter-level precision is needed. Because HD^{4}AR relies on 3D point cloud models, it requires that users first take overlapping photos of the target scene to produce the initial 3D point cloud used for localization. This initial 3D reconstruction is based on a Structure-from-Motion (SfM) algorithm that triangulates the 3D position of image features in photographs through feature extraction, matching, and an optimization process called Bundle Adjustment.

Once the 3D reconstruction is done, a field engineer can take a new photo at a random location and his/her location and orientation are determined by comparing the new image to the generated 3D point cloud. Specifically, the system attempts to estimate extrinsic camera parameters, *e.g.* rotation matrix and translation vector of the camera, to find the relative position of the camera. After recovering the complete pose of the user's camera, HD^4AR decides what cyber-information, such as elements of the BIM, should appear in the field engineer's photograph.

Finally, HD^4AR allows a field engineer to select physical objects in the photograph by touching on them in order to retrieve more information associated with each object. Moreover, a field engineer can create new BIM elements by simply drawing a polygon on the photograph. The user-created 2D BIM elements are then automatically back-projected to cyber 3D space and attached to the existing cyber-physical model. Once user-created elements are successfully back-projected, they can be accurately overlaid on other photographs, which are taken from significantly different viewpoints. This simple 3D annotation functionality is one of the distinct features of HD^4AR. Figure 2 summarizes the overall procedures of HD^4AR, from initial 3D reconstruction process to localization/augmentation process.

Technical approach

As aforementioned, HD^4AR is based on a set of computer vision algorithms. However, due to exhaustive computations including non-linear multi-dimensional optimization processes in the SfM algorithm, model-based approaches are often considered impractical solutions for user localization. For example, the Bundler package (Snavely et al. 2007), a widely-used software package that implements SfM for 3D reconstruction, takes from hours to days to generate a 3D point cloud even for small number of base images. In addition, it uses the SIFT (Scale Invariant Feature Transformation) descriptor (Lowe 2004) for feature extraction, which have good invariance properties but require multiple layers of computation for each spatial scale, and thus is time consuming. Therefore, we designed and implemented a new parallelized 3D reconstruction module that operates across cores in a multi-core CPU and GPU. HD^4AR uses a client-server architecture with the mobile phone as the client that uploads photos to the server for localization and the major image processing load located on the server-side. The entire system consists of following components:

- A 3D reconstruction component runs on the server on a multi-core CPU and GPU. This component generates a 3D point cloud from the initial base images through feature extraction, matching, and the SfM procedure.
- A user localization component runs on the server. This component takes a single photograph taken from a mobile device as input and derives the 3D position and orientation of the mobile device with respect to the 3D point cloud by solving a Direct Linear Transform equation followed by a Levenberg-Marquardt optimization against the underlying point cloud model.
- A client component, which is a small program that runs on Android and iOS smartphones, sends user-captured images to the server. It also has the

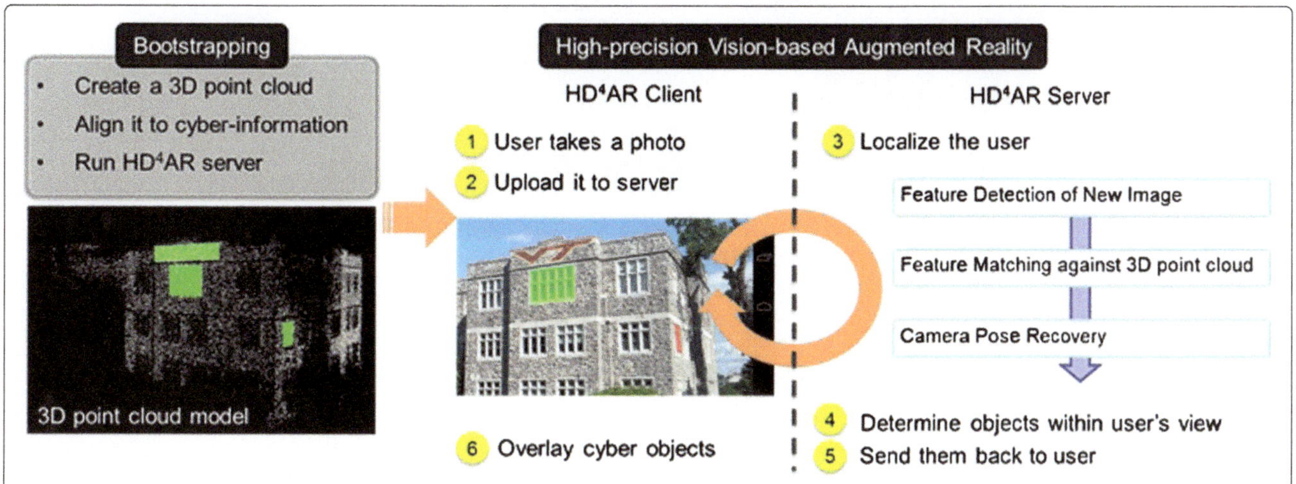

Figure 2 The overall procedures of HD^4AR system.

capability of drawing cyber objects on top of the photograph once it gets localization results from the server.

3D reconstruction with HD^{4}AR

Most engineering workstations today have a multi-core CPU with 2–16 cores and a GPU with anywhere from 4 to 128 cores. Exploiting this hardware parallelism is key to the performance and scalability of HD^{4}AR. We parallelize all the steps for 3D point cloud generation to obtain performance gains and implement GPU-based SURF (Speeded-Up Robust Features) descriptors (Bay et al. 2008) and CPU-based FREAK (Fast REtinA Keypoint) descriptors (Alahi et al. 2012) for fast feature extraction. The 3D reconstruction procedure in HD^{4}AR mostly follows the original steps in the SfM algorithm of the Bundler package except that it 1) uses different feature detectors and descriptors, 2) introduces new optimization parameters for reducing noise in the 3D point cloud to improve localization accuracy, and 3) exploits multi-core CPU and GPU hardware for faster processing speeds. Figures 2 and 3 show the steps of image-based 3D reconstruction from a high-level perspective. Each step can be summarized as follows:

Feature Detection and Extraction

To find a set of image feature points, a feature detection and extraction algorithm is executed on each base image. Two different state-of-the-art feature descriptors, *e.g.*, SURF and FREAK, are implemented and tested in the HD^{4}AR 3D reconstruction pipeline. In contrast to SIFT, SURF creates a stack of integral images without downsampling for higher levels in the pyramid and it filters the stack using a box filter approximation of second-order Gaussian partial derivatives to speed up the processing (Bay et al. 2012). On the other hand, FREAK uses retinal sampling patterns to compare image intensities and produces a cascade of binary strings (Alahi et al. 2012). Both SURF and FREAK are invariant to image scale and rotation, but provide faster feature extraction than SIFT. Therefore, the HD^{4}AR pipeline now supports SURF and FREAK, in addition to SIFT, to speed up feature extraction.

Feature Matching

The next step is finding correspondences between each image pair (*e.g.*, pair-wise matching). For each image pair, HD^{4}AR creates a *kd*-tree of the descriptors and runs the Approximate Nearest Neighbors (ANN) algorithm (Arya et al. 1998) to find the two nearest neighbors of each descriptor. Then the HD^{4}AR performs a distance ratio-test (Lowe 2004) to remove erroneous matches. In addition, if more than one feature descriptor matches the same feature in the opposite image, the HD^{4}AR removes all of those matches. Finally, the HD^{4}AR robustly estimates a Fundamental matrix with the eight-point algorithm (Hartley and Zisserman 2004) loop and removes matching outliers for every image pair. This filtering process removes false matches using an epipolar geometry constraint given by the estimated Fundamental matrix. To shorten the overall matching time, each image pair is processed on different CPUs with parallelized I/O tasks.

Structure-from-Motion (SfM)

1) *Camera Registration and Point Triangulation*: The SfM algorithm estimates a set of camera parameters, such as the focal length, rotation matrix, and translation vector, for each image and triangulates 3D positions of feature points observed in each image. Similar to the Bundler package, the HD^{4}AR

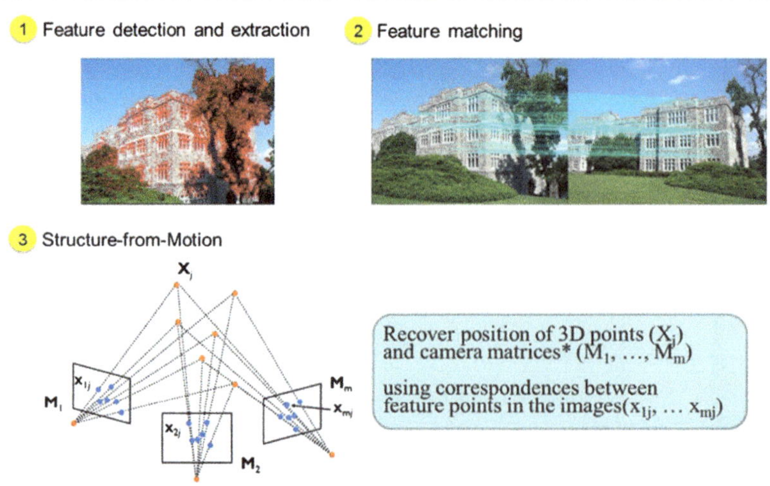

Figure 3 The sequence of HD^{4}AR 3D reconstruction.

uses an incremental approach, *e.g.*, recovering a few cameras at a time. The HD^{4}AR starts with initial image pair to recover camera parameters using Nistér's five-point algorithm (Nistér 2004), and triangulates their feature points. As discussed by Snavely et al. (2007), the initial pair should have a large number of matched feature points, but also have a long separation distance between the cameras to avoid converging in a local minimum during the optimization process. After estimating the camera parameters of the initial image pair, the HD^{4}AR attempts to calibrate the camera parameters of each additional base image using the already triangulated 3D points and matching information between the images. If the system successfully recovers camera parameters of an additional base image, it registers the new camera and triangulates the points seen by the newly registered camera. This registration fails in the event that an additional base image does not have any matched feature points against the previously registered images. In the HD^{4}AR, these camera registration and point triangulation steps are well-parallelized to exploit multi-core CPUs.

2) *Incremental Bundle Adjustment*: While the base images are being added (registered), the 3D reconstruction pipeline is run through a GPU-based sparse Bundle Adjustment module to minimize the overall re-projection error, *e.g.*, the difference between predicted 2D positions of the feature points in the photographs given their triangulated 3D positions and the locations of where the feature points are actually extracted in the images. The HD^{4}AR adopts Parallel Bundle Adjustment (Wu et al. 2011) to significantly enhance the speed of this optimization.
3) *Noise Reduction*: Bundle Adjustment is an optimization process that tries to minimize the overall re-projection error of all 3D points at the same time. It is possible that some 3D points have high re-projection error while other 3D points have a very small re-projection error, resulting in an overall small Minimum Mean-Square Error (MMSE). Since the ultimate purpose of the 3D point cloud generation is user localization, not the visual representation of target scene in 3D, it is very important to reduce the noise in the 3D point cloud by removing 3D points with high re-projection errors. To achieve this, HD^{4}AR uses a double-threshold scheme. The first threshold is for controlling the target MMSE of Bundle Adjustment. We set this threshold as 1.0 pixel^2 so that the average re-projection error of entire 3D point cloud is not greater than 1.0 pixel. Another threshold, which we call an absolute re-projection threshold, is for removing individual 3D points from point cloud. This threshold is set to be 4.0 pixels so that no 3D points in final point cloud have a re-projection error greater than 4.0 pixels.

Due to our algorithmic enhancements and parallelization, 3D reconstruction with HD^{4}AR is up to 35 times faster than the Bundler package. In Section 'Results and discussion', the experimental results of 3D reconstruction are discussed in detail. Figure 4 shows some examples of 3D point clouds generated by HD^{4}AR using real-world construction site photos and existing building photos in Blacksburg, VA.

High-precision augmentation with HD^{4}AR

Localization and augmentation

Once the HD^{4}AR has the 3D point cloud of the target construction site or building, the system can accurately localize and augment new photographs captured on a mobile device. Figures 2 and 5 summarize this process from a high-level perspective. In this use case, a field engineer first takes a picture of the building elements, which he/she wishes to query the information about, and uploads the photograph to the HD^{4}AR server. Upon receiving the photo from user's device, the server starts to run feature detection on the received image, finding correspondences between the image and 3D point cloud, and camera calibration to identify the relative pose of the camera. If the server successfully estimates the camera pose information, it determines what cyber-information is within the camera's field of view and where the information should appear. This decision is done by first projecting each vertex of cyber-information:

$$s\begin{bmatrix} x \\ y \\ 1 \end{bmatrix} = \begin{bmatrix} f_x & 0 & c_x \\ 0 & f_y & c_y \\ 0 & 0 & 1 \end{bmatrix} \begin{bmatrix} R \mid T \end{bmatrix} \begin{bmatrix} X \\ Y \\ Z \\ 1 \end{bmatrix} \quad (1)$$

where $[X, Y, Z]$ is a 3D vertex point of cyber-information, $[R|T]$ is the estimated rotation matrix and translation vector, $[f_x, f_y]$ is the focal lengths expressed in pixel units, $[c_x, c_y]$ is a principal point, and $[x, y]$ is the resulting projected point in pixels. Then the simple visibility test is performed to determine whether the cyber-information appears in current image or not:

$$V(x,y) = \begin{cases} 1, & 0 \le x \le W, 0 \le y \le H \\ 0, & otherwise \end{cases} \quad (2)$$

where W is image width and H is image height. The visible cyber-information is then sent back to user's device with

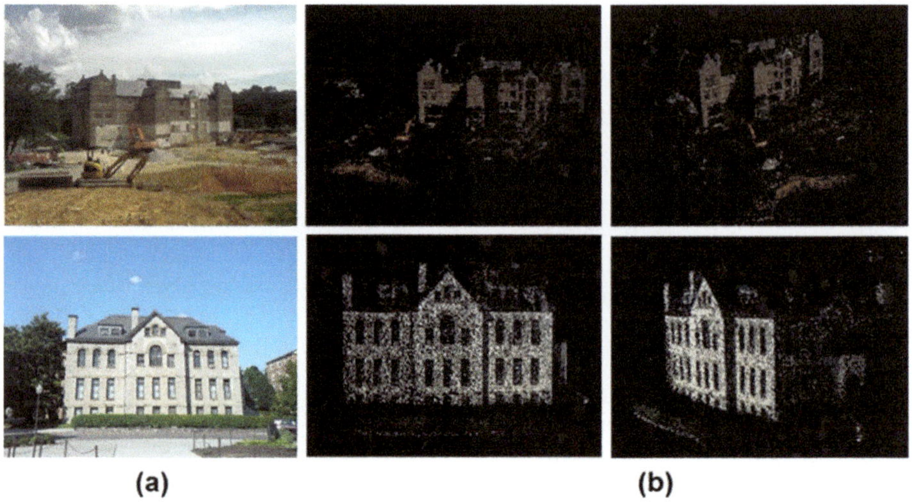

Figure 4 Example of HD^{4}AR 3D reconstruction. (**a**) Initial base images. (**b**) 3D point-cloud from HD^{4}AR. Resulting 3D point clouds well-represent the target construction site and building.

positional information and semantics. Finally, the user's device renders the returned visible cyber-information on the top of captured-image. As shown in Figure 6, HD^{4}AR can precisely localize and augment photographs with various test cases and it implies that HD^{4}AR remains stable under different illumination conditions and totally different viewpoint of the user's device.

One of the improvements over our previous work is that finding correspondences between image and the 3D point cloud is further accelerated using direct 2D-to-3D matching. HD^{4}AR only compares feature descriptors of the image to that of each 3D point in the point cloud to find 2D-to-3D correspondences. With our previous approach, however, HD^{4}AR matched feature descriptors of the image to an entire set of feature descriptors from all base images, which incurs unnecessary descriptor comparisons. Consequently, the localization time did depend on the number of base images as we discussed in our previous work (Bae et al. 2012). As we will see in Section 'Results and discussion', the new direct 2D-to-3D matching approach further speeds up the localization by an average factor of 2.79.

3D annotation

Upon successful localization of a new photograph, a field engineer can easily create and add a new 3D BIM element with HD^{4}AR by drawing a polygon on the localized photograph. Since HD^{4}AR keeps all the base images that are registered during the 3D reconstruction, it can estimate homography matrices between the localized photograph and each base image using a RANSAC algorithm. HD^{4}AR then utilizes these estimated homographies to find correspondences of a user-created element for each base image. As shown in Figure 7a and 7b, a window drawn by the user is correctly found in base images by the system. Finally, HD^{4}AR triangulates each vertex of the user-created polygon using camera parameters of localized photograph and registered base images. After running Bundle Adjustment to minimize the reprojection error of a triangulated polygon, the resulting 3D element is well-aligned with the existing 3D point cloud as shown in Figure 7c. Once the user-created element has 3D positional information, it can be precisely overlaid on other photographs from different viewpoints as shown in Figure 7d. This simple and robust 3D annotation/tagging functionality makes

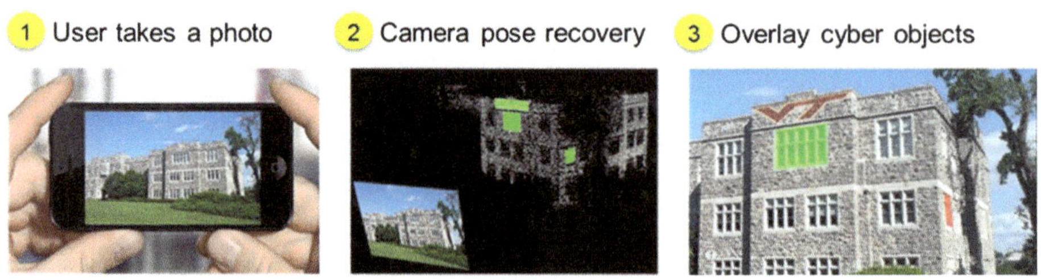

Figure 5 The sequence of HD^{4}AR localization and augmentation.

Figure 6 Example of HD^{4}AR localization and augmentation. Cyber-information is precisely overlaid on user's photograph despite the significant change of viewpoints.

it easier to create 3D content associated with building elements on-site, and is one of distinct features of HD^{4}AR.

Results and discussion

This section presents experimental results of 3D reconstruction and user localization with HD^{4}AR. To assess the ability of HD^{4}AR to produce the initial 3D point clouds, 3D reconstruction is performed on several data sets, which were randomly collected from actual construction sites and existing buildings. For user localization, test images were taken at random locations and localized on-site for validating the correctness. The details of data set specification and experimental results will be discussed in following subsections.

Platform specification and data sets

The server side of HD^{4}AR was running on a desktop computer with 8 gigabytes of 667 MHz DDR3 RAM, and a 4-core Intel i7 CPU 870 (@2.93 GHz) processor running Ubuntu version 12.04. The NVIDIA GeForce GTX 560 Ti graphic card was used for GPU computations. The image data sets used to create 3D point clouds came from the actual construction sites and existing buildings on Virginia Tech campus. Table 1 shows the summary of data sets that we used for 3D reconstruction.

Several Android smartphones were used to run the HD^{4}AR client for localization tests. For fast data transfer, the client-side communication was based on Wi-Fi 802.11g connection rather than using the cellular network.

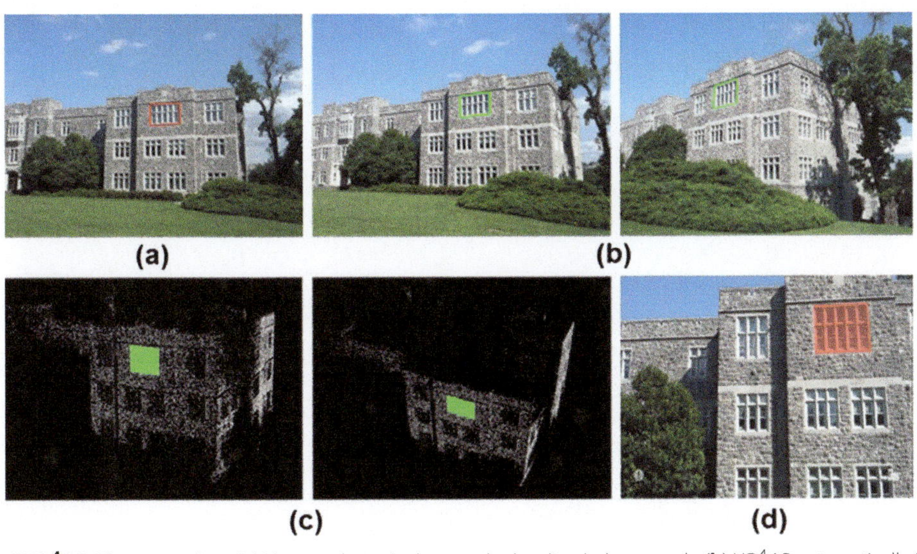

Figure 7 Example of HD^{4}AR 3D annotation. (**a**) User marks a window on the localized photograph, (**b**) HD^{4}AR automatically finds correspondences of window for each base image, (**c**) The system triangulates the window using camera information of base images and the localized photograph, (**d**) A user-created window element is precisely overlaid on other photographs from different viewpoints.

Table 1 Data sets for 3D reconstruction

Data set	Camera	Number of base images	Image size
Parking garage	Google Nexus S	143	2560 × 1920
Center for the arts	Nikon D300S	125	2144 × 1424
Norris hall	Google Nexus S	50	2560 × 1920
Patton hall	Samsung Galaxy Nexus	44	2592 × 1944

Performance of 3D reconstruction

An entire 3D reconstruction procedure with HD^4AR was run on each data set to produce the initial 3D point clouds. The performance of the Bundler package was also measured and compared to that of HD^4AR to demonstrate the performance gains of HD^4AR's optimizations. In addition, we tested two different descriptors, i.e. GPU-based SURF and CPU-based FREAK, to investigate the impact of feature descriptors on the performance of 3D reconstruction. Table 2 compares the overall elapsed time and number of recovered cameras for 3D reconstruction on each data set. The results show that HD^4AR with FREAK descriptor obtains the maximum performance gain of 3,471%. However, this only illustrates the tendency of time cost since there are many factors that influenced on the performance, such as number of base images, image sizes, and the texture of target scenes. Nevertheless, HD^4AR with FREAK outperformed HD^4AR with SURF and the Bundler package in all tested data sets. This result is due to the fact that the FREAK descriptor is a binary descriptor which uses simple Hamming distance calculations for descriptor matching, while SURF and SIFT descriptors are a vector of real numbers and must be compared using Euclidean norms. Compared to HD^4AR with SURF, however, HD^4AR with FREAK has fewer registered images in the Parking Garage and Patton Hall data sets. This outcome implies that the FREAK descriptor may not be as robust as SURF or SIFT descriptors for 3D reconstruction. Having a smaller number of registered images means that there is less 3D camera information in a point cloud and therefore it may affect the localization success-ratio. The success-ratio will be discussed further in the next subsection. Figure 8 shows results of HD^4AR 3D point cloud reconstruction for all test cases.

Performance of localization

In order to measure the reliability of the reconstructed 3D point clouds for localization of new photographs, localization tests were performed on each 3D point cloud. In this paper, the success in localization means that HD^4AR was able to solve the camera calibration equation, *e.g.* Direct Linear Transform equation followed by a Levenberg-Marquardt optimization, using a given image and 3D point cloud. More extensive quantification and measurement of the accuracy of localization with different feature descriptors will be investigated in future work to determine the feasibility of precise measurement using HD^4AR. As observed in Figure 6 and 9, however, the augmented photographs show that recovered camera parameters were accurate enough to precisely

Table 2 Performance of 3D point cloud reconstruction

Data set	System	Number of registered images/ Number of base images	Elapsed time
Parking garage	HD^4AR with FREAK	125 / 143	36.73 mins (× 17.25)
	HD^4AR with SURF	138 / 143	42.01 mins (× 15.08)
	Bundler package*	143 / 143	633.63 mins (× 1)
Center for the arts	HD^4AR with FREAK	125 / 125	9.25 mins (× 19.67)
	HD^4AR with SURF	125 / 125	12.49 mins (× 14.57)
	Bundler package*	125 / 125	181.95 mins (× 1)
Norris hall	HD^4AR with FREAK	50 / 50	2.62 mins (×27.53)
	HD^4AR with SURF	50 / 50	3.27 mins (×22.06)
	Bundler package*	50 / 50	72.13 mins (× 1)
Patton hall	HD^4AR with FREAK	37 / 44	3.78 mins (×34.71)
	HD^4AR with SURF	43 / 44	4.80 mins (×27.34)
	Bundler package*	44 / 44	131.22 mins (× 1)

*A widely-used software package for SfM (Snavely et al. 2007).

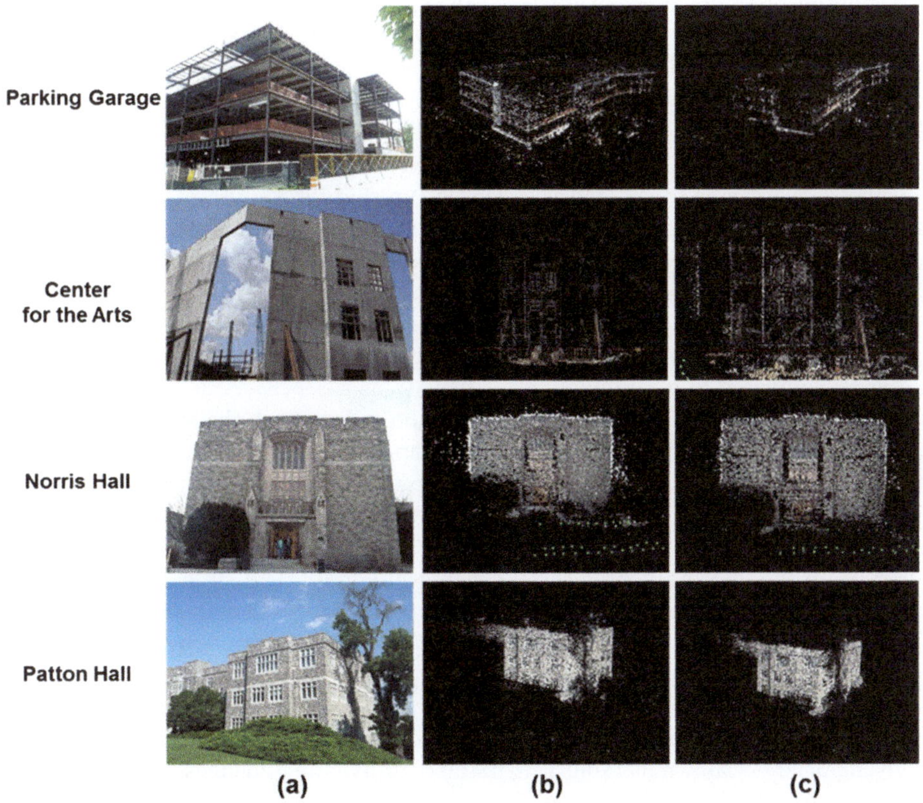

Figure 8 Results of HD^{4}AR 3D reconstruction. (**a**) Initial base images. (**b**) HD^{4}AR with SURF. (**c**) HD^{4}AR with FREAK.

Figure 9 Results of HD^{4}AR localization. From construction sites to existing building, HD^{4}AR provides high-precision of cyber-information visualization.

overlay the cyber-information on photographs from different viewpoints.

Table 3 shows the localization success-ratio and average localization time for each data set. As we expected, HD^4AR with FREAK has the lowest localization success-ratio, but the fastest localization speed. The localization success-ratio of 95.38% in worst case can be considered as a high-level of success for a construction site and therefore HD^4AR with FREAK is a good candidate for fast 3D reconstruction and localization that provides a reasonable level of localization success.

Another interesting result is that we achieved an additional speedup in localization, compared to our previous work. In our previous work, HD^4AR matched a given image to an entire set of base images for finding correspondences between image feature points and 3D points. As a result, the localization time depends on number of base images and increases as the number of base images increases (Bae et al. 2012). With the improvements presented in this paper, HD^4AR now directly compares feature descriptors of the image to that of 3D points in a point cloud, which reduces matching time significantly. The localization time does not depend on the number of base images anymore, but depends on the number of 3D points or texture of the given image. As shown in Table 3, our new approach further increases the speed of localization by an average factor of 2.79. Compared to the Bundler package, HD^4AR with FREAK is now up to 30 times faster in user localization.

All experimental results as observed in Table 3 and Figure 9 prove that HD^4AR can successfully localize a user solely based on an image and within few seconds. With HD^4AR using FREAK descriptor, a field engineer has to wait only 3–6 seconds after he or she takes a photo of target objects to retrieve related cyber-information. This is very promising for the use of HD^4AR in practice. Information retrieval using HD^4AR now can be done in much less time than our previous work or the traditional means (traveling back to trailer to lookup cyber-information or carrying large stacks of drawings on site and looking up for information on demand).

Discussion and research challenges

This paper presented a high-precision vision-based mobile augmented reality system for context-aware applications. The experimental results demonstrate the applicability of the proposed system to construction sites and existing buildings. The system can successfully localize the user solely based on an image, without using any external location tracking modules. Once the user's camera is accurately localized, the overlays or cyber-information can be overlaid on top of real-world imagery. The results, shown in Figures 6 and 9, indicate the robustness of the method to dynamic changes of illumination, viewpoint, camera resolution, and scale in the image, which are typical for unordered construction photo collections. While this paper presented the initial works toward vision-based localization and AR visualization for the purpose

Table 3 Performance of localization

Data set	System	Localization success-ratio	Average localization time
Parking garage	HD^4AR with FREAK	100% (50 / 50)	5.42 sec (×19.94)
	HD^4AR with SURF	100% (50 / 50)	6.42 sec (×16.84)
			18.45 sec (×5.86)**
	Bundler package*	100% (50 / 50)	108.10 sec (×1)
Center for the arts	HD^4AR with FREAK	95.38% (62 / 65)	3.17 sec (×29.40)
	HD^4AR with SURF	98.46% (64 / 65)	3.47 sec (×26.86)
			16.02 sec (×5.82)**
	Bundler package*	100% (65 / 65)	93.20 sec (×1)
Norris hall	HD^4AR with FREAK	100% (25 / 25)	4.98 sec (×7.51)
	HD^4AR with SURF	100% (25 / 25)	10.74 sec (×3.48)
			13.31 sec (×2.81)**
	Bundler package*	100% (25 / 25)	37.38 sec (×1)
Patton hall	HD^4AR with FREAK	100% (25 / 25)	6.33 sec (×5.10)
	HD^4AR with SURF	100% (25 / 25)	10.07 sec (×3.20)
			24.56 sec (×1.31)**
	Bundler package*	100% (25 / 25)	32.26 sec (×1)

*A widely-used software package for SfM (Snavely et al. 2007).

**The result using our previous approach (Bae et al. 2012).

of context-aware applications, several challenges remain. Some of the open research problems include:

- Quantifying the accuracy of image-based localization in terms of re-projection error to validate how cyber objects are precisely overlaid on top of real-world photograph.
- Quantifying the quality of 3D point cloud, which will guide users to take a minimal number of images from various sites for initial bootstrapping, *e.g.* 3D reconstruction.
- Further increasing the speed of localization by using supplemental information such as mobile GPS available in mobile devices to reduce data set to be matched. Minimizing the image resolution to reduce matching time is also in our focus.

Conclusion

The current practice of construction progress monitoring still has significant opportunities for improvement through the integration of cyber-information into regular site operations. The HD^4AR system was designed to provide such cyber-information on worksites using existing and already available camera-equipped mobile devices. HD^4AR takes vital project information, such as the expected quality of building elements, or location of elements, project schedule, and cost information, which traditionally has been difficult to access on a jobsite, and makes it mobile, accessible to on-site users. In addition, it provides an easy and intuitive method to create 3D information using 2D jobsite photographs. This content authoring capability may further facilitate the accurate exchange of project information among field personnel.

Using a set of computer vision algorithms, HD^4AR allows users to leverage any camera-equipped mobile device to take pictures for accurate on-site localization. This vision-based and location tracking-free system can support a range of promising context-aware AEC/FM applications since it does not require the installation of new technological components on the jobsite. HD^4AR uses image feature points as the basis for user localization and a SfM algorithm to build and match a 3D geometric model from regular smartphone camera images. Users can use a smartphone outfitted with a camera, screen, and wireless communication to upload a captured image, localize it, and then overlay the returned cyber-information on the physical objects in the photograph to which it pertains. The performance of HD^4AR, with a localization success-ratio of 95.38% (in worst case), implies that it is possible to develop a near real-time augmented reality systems using site photographs. It takes 3–6 seconds for localization and less than an hour for point cloud generation. With everyday data collection and application of HD^4AR, 3D point clouds can be produced very quickly, allowing AEC/FM practitioners to easily monitor construction progress by quickly and accurately accessing relevant information. In future work, we plan to use the full IFC-based (Industry Foundation Class) BIM rather than using manually created elements to completely test HD^4AR in an actual construction site. Enhancing localization speed to real-time with the aid of GPS information available in smartphone is also one of our focuses.

Competing interests

The authors declare that they have no competing interests.

Authors' contributions

All authors contributed extensively to the work presented in this paper. H.B. designed and implemented infrastructure-independent mobile augmented reality system (MARS), gathered and analyzed experimental data, and prepared the manuscript. M.G and J.W. designed the concept and study of infrastructure-independent MARS, supervised the overall project, and edited the manuscript. All authors read and approved the final manuscript.

Acknowledgment

This material is based upon work supported by the National Science Foundation under NSF CMMI-1200374 award. Any opinions, findings, and conclusions or recommendations expressed in this publication are those of the authors and do not necessarily reflect the views of the National Science Foundation.

Author details

[1]Bradley Department of Electrical and Computer Engineering, Virginia Polytechnic Institute and State University, Blacksburg, VA, USA. [2]Department of Civil and Environmental Engineering, Department of Computer Science, University of Illinois at Urbana-Champaign, Urbana, IL, USA. [3]Department of Electrical Engineering and Computer Science, Vanderbilt University, Nashville, TN, USA.

References

Akula, M, Dong, S, Kamat, VR, Ojeda, L, Borrell, A, Borenstein, J. (2011). Integration of infrastructure based positioning systems and inertial navigation for ubiquitous context-aware engineering applications. *Automation in Construction, 25*(4), 640–655.

Alahi, A, Ortiz, R, Vandergheynst, P (2012). FREAK: Fast Retina Keypoint, *Proceedings of 2012 IEEE Conference on Computer Vision and Pattern Recognition (CVPR 2012)* (pp. 510–517).

Anumba, C, Aziz, Z. (2006). Case studies of intelligent context-aware services delivery in AEC/FM. *Lecture Notes in Computer Science, 4200*, 23–31.

Arya, S, Mount, DM, Netanyahu, NS, Silverman, R, Wu, AY. (1998). An optimal algorithm for approximate nearest neighbor searching in fixed dimensions. *Journal of the ACM, 45*(6), 891–923.

Bae, H, Golparvar-Fard, M, White, J (2012). Enhanced HD^4AR (Hybrid 4-Dimensional Augmented Reality) for Ubiquitous Context-aware AEC/FM Applications, *Proceedings of 12th international conference on construction applications of virtual reality (CONVR 2012)* (pp. 253–262).

Bay, H, Ess, A, Tuytelaars, T, Gool, LV. (2008). Speeded-up Robust Features (SURF). *Computer Visioin and Image Understanding, 110*(3), 346–359.

Behzadan, A H, Kamat, VR. (2007). Georeferenced registration of construction graphics in mobile outdoor augmented reality. *Journal of Computing in Civil Engineering, 21*(4), 247–258.

Behzadan, A H, Aziz, Z, Anumba, CJ, Kamat, VR. (2008). Ubiquitous location tracking for context-specific information delivery on construction sites. *Automation in Construction, 17*(6), 737–748.

Carozza, L, Tingdahl, D, Bosché, F, Van Gool, L. (2012). Markerless Vision-Based Augmented Reality for Urban Planning. *Journal of Computer-Aided Civil and Infrastructure Engineering, 2012.* doi:10.1111/j.1467-8667.2012.00798.x.

Chen, Y, Kamara, J. (2011). A framework for using mobile computing for infomation management on construction sites. *Automation in Construction, 15*, 605–613.

Chen, Y C, Chi, HL, Hung, WH, Kang, SC. (2011). Use of tangible and augmented reality models in engineering graphics courses. *Journal of Professional Issues in Engineering Education & Practice*, *137*(4), 267–276.

Cheng, M Y, Chen, JC. (2002). Integrating barcode and GIS for monitoring construction progress. *Automation in Construction*, *11*, 23–33.

Chi, H L, Kang, SC, Wang, X. (2013). Research trends and opportunities of augmented reality applications in architecture, engineering, and construction. *Automation in Construction*. doi:10.1016/j.bbr.2011.03.031.

Dunston, P, Wang, X, Billinghurst, M, Hampson, B, *et al.* (2003). Mixed reality benefits for design perception. *National Institue of Standards (NIST) Special Publication*, 191–196.

Feng, C, Kamat, VR (2012). Augmented reality markers as spatial indices for indoor mobile AECFM applications, *Proceedings of 12th international conference on construction applications of virtual reality (CONVR 2012)* (pp. 235–242).

Golparvar-Fard, M, Peña Mora, F, Savarese, S. (2012). Automated model-based progress monitoring using unordered daily construction photographs and IFC as-planned models. *ASCE Journal of Computing in Civil Engineering*, 2012. in press.

Golparvar-Fard, M, Peña Mora, F, Savarese, S. (2011). Integrated sequential as-built and as-planned representation with D^4AR tools in support of decision-making tasks in the AEC/FM industry. *Journal of Construction Engineering and Management*, *137*(12), 1099–1116.

Golparvar-Fard, M, Savarese, S, Peña Mora, F (2010). Automated model-based recognition of progress using daily construction photographs and IFC-based 4D Models, *Proceedings of 2010 construction research congress* (pp. 51–60).

Golparvar-Fard, M, Peña Mora, F, Arboleda, CA, Lee, S. (2009a). Visualization of construction progress monitoring with 4D simulation model overlaid on time-lapsed photographs. *Journal of Computing in Civil Engineering*, *23*(6), 391–404.

Golparvar-Fard, M, Peña Mora, F, Savarese, S. (2009b). Application of D4AR–A 4-Dimensional augmented reality model for automating construction progress monitoring data collection, processing and communication. *ITcon, Special Issue on Next Generation Construction IT:TechnologyForesight, Future Studies, Roadmapping, and Scenario Planning*, *14*, 129–153.

Hakkarainen, M, Woodward, C, Billinghurst, M (2008). Augmented assembly using a mobile phone, *Proceedings of 7th IEEE/ACM international symposium on mixed and augmented reality (ISMAR 2008)* (pp. 167–168).

Hammad, A, Wang, H, Mudur, SP. (2009). Distributed augmented reality for visualizing collaborative construction tasks. *Journal of Computing in Civil Engineering*, *23*(6), 418–427.

Hartley, R, Zisserman, A (2004). *Multiple view geometry in computer vision*, Cambridge, UK: Cambridge University Press .

Hou, L, Wang, X (2010). Application of augmented reality technology in improving assembly task proficiency, *CD proceedings of the 10th international conference on construction applications of virtual reality* (pp. 4–5).

Irizarry, J, Gheisari, M, Williams, G, Walker, BN. (2012). InfoSPOT: A mobile Augmented Reality method for accessing building information through a situation awareness approach. *Automation in Construction*, 2012. in press.

Khoury, H, Kamat, VR. (2009). High-precision identification of contextual information in location-aware engineering applications. *Advanced Engineering Informatics*, *23*(4), 483–496.

Kiziltas, S, Akinci, B, Ergen, E, Tang, P, Gordon, C. (2008). Technological assessment and process implications of field data capture technologies for construction and facility/infrastructure management. *Journal of Informan Technology in Construction (ITcon), Special Issue on Sensors in Construction and Infrastructure Management*, *13*, 134–154.

Kuula, T, Piira, K, Seisto, A, Hakkarainen, M, Woodward, C (2012). User requirements for mobile AR and BIM utilization in building life cycle management, *Proceedings of 12th international conference on construction applications of virtual reality (CONVR 2012)* (pp. 203–211).

Lee, S, Akin, Ö. (2011). Augmented reality-based computational fieldwork support for equipment operations and maintenance. *Automation in Construction*, *20*(4), 338–352.

Lowe, D G. (2004). Distinctive image features from scale-invariant keypoints. *International Journal of Computer Vision*, *60*(2), 91–110.

Navon, R, Sacks, R. (2007). Assessing research issues in Automated Project Performance Control (APPC). *Automation in Construction*, *16*(4), 474–484.

Nistér, D. (2004). An efficient solution to the five-point relative pose problem. *IEEE Transactions on Pattern Analysis and Machine Intelligence*, *26*(6), 756–777.

Ojeda, L, Borenstein, J (2007). Personal dead-reckoning system for GPS-denied Environments, *Proceedings of the 2007 IEEE international workshop on safety, security and rescue robotics (SSRR): 27–29 September; Rome, Italy* (pp. 1–6).

Pasman, W, Woodward, C (2003). Implementation of an Augmented Reality System on a PDA, *Proceedings of the second IEEE and ACM international symposium on mixed and augmented reality (ISMAR 2003)* (pp. 276–277).

Schall, G, Mendez, E, Kruijff, E, Veas, E, Junghanns, S, Reitinger, B, Schmalstieg, D. (2009). Handheld augmented reality for underground infrastructure visualization. *Personal and Ubiquitous Computing*, *13*(4), 281–291.

Shin, D H, Dunston, PS. (2008). Identification of application areas for augmented reality in industrial construction based on technology suitability. *Automation in Construction*, *17*(7), 882–894.

Snavely, N, Seitz, SM, Szeliski, R. (2007). Modeling the world from internet photo collections. *International Journal of Computer Vision*, *80*(2), 189–210.

Wang, X (2008). Improving human-machine interfaces for construction equipment operations with mixed and augmented reality. In Balaguer, C, Abderrahim, M (Eds.), *Robotics and Automation in Construction* (pp. 349–362). Vienna, Austria: InTech.

Wang, X, Dunston, PS. (2006). Compatibility issues in Augmented Reality systems for AEC: An experimental prototype study. *Automation in Construction*, *15*(3), 314–326.

Woodward, C, Hakkarainen, M, Korkalo, O, Kantonen, T, Aittala, M, Rainio, K, Kähkönen, K (2010). Mixed reality for mobile construction site visualization and communication, *Proceedings of 10th international conference on construction applications of virtual reality (CONVR 2010)* (pp. 35–44).

Wu, C, Agarwal, S, Curless, B, Seitz, SM (2011). Multicore bundle adjustment, *Proceedings of 2011 IEEE conference on computer vision and pattern recognition (CVPR 2011)* (pp. 3057–3064).

Yabuki, N, Miyashita, K, Fukuda, T. (2011). An invisible height evaluation system for building height regulation to preserve good landscapes using augmented reality. *Automation in Construction*, *20*(3), 228–235.

Yeh, K C, Tsai, MH, Kang, SC. (2012). On-site building information retrieval by using projection-based Augmented Reality. *Journal of Computing in Civil Engineering*, *26*(3), 342–355.

5

The "BIM-sustain" experiment – simulation of BIM-supported multi-disciplinary design

Iva Kovacic*, Lars Oberwinter†, Christoph Müller† and Christoph Achammer

Abstract

Background: The AEC practice using BIM technology in Central European (CE) context is still very young; the previous experiences demonstrated a number of upcoming problems with BIM implementation on technical- (heterogonous data, interfaces, large data volumes) but even more so on process-level (question of responsibilities and work-load distribution, lacking standards or conventions on building-representation and in general lack of experience and knowledge on integrated practice).
The optimal data management, transfer and synchronization within inhomogeneous software context, as is often the case within inter-firm construction projects, require enormous organization, coordination and communication effort in the earliest design-phases. The BIM implementation implies therefore necessity of fundamental rethinking of the conventional design process, which in CE context is still predominantly based on sequential, segmented practice.

Methods: At the Vienna University of Technology a BIM-supported multi-disciplinary planning process with students of architecture, structural engineering and building physics, using several BIM-software tools was simulated. From the qualitative and quantitative evaluation of this BIM-supported multi-disciplinary collaboration will enable the compilation of guidelines for efficient use of BIM in design and planning process for the planners and standardization bodies.

Results: First insights on process-quality, such as team-, process- and technology satisfaction, as well as conflict- and stress levels will be presented in this paper. We were able to identify numerous technical problems related to the data transfer and inconsistencies in translation, which resulted in participant dissatisfaction and significant increasing of work-loads.

Conclusion: The first results imply at the importance of process-organization techniques such as face-to face communication, coordination and work-load allocation between the team-members in order to conduct the efficient BIM-supported process; as well as at urgent need for advancement of the tools in terms of data transfer and exchange. In the next step, using mandatory protocols and timesheets, a detailed statistical analysis of the people-process-technology issues will be conducted, as well as comparison of „Open-Platform-BIM" to „One-Platform-BIM" model.

Keywords: BIM; Collaboration; Integrated Planning; Exploratory Research; Experiment

* Correspondence: iva.kovacic@tuwien.ac.at
†Equal contributors
Department for Industrial Building and Interdisciplinary Planning, Vienna University of Technology, Karlsplatz 13, 1040 Vienna, Austria

Background

The consolidations for realization of sustainable built environment result in increasing complexity of planning and construction process. The increasing number of partaking disciplines uses wide spectrum of specialised visualisation, simulation or calculation tools. The practice calls for more integrated design and planning process, which would enhance simultaneous collaboration of various disciplines for sharing and creation of new common knowledge. Building Information Modelling (BIM) is believed to bear large potential for inducing a shift from the conventional, fragmented practice, which still largely dominates the AEC (architecture, engineering and construction) industry (Fellows and Liu 2012), towards more integrated design practice (Prins and Owen 2010). BIM as method is supportive of collaborative planning, facilitating communication and information exchange among various planning process participants (Rizal and van Berlo 2010). However, Rekkola et al. (2010) argue that "integrated design" is still handled rather loosely in the practice – often is the creation of BIM model sufficient for the project to be referred to as "integrated project", regardless of actual interdisciplinary data sharing and model use. BIM, in our understanding is much more about how (design of design process), than about what (building model and its properties).

This paper will focus on introduction of BIM-supported planning in the CE and particularly Austrian AEC (architecture, engineering and construction) market, where the application of this technology is still novel, as well as the integrated planning approach. Austrian market is characterised by a large number of very small planning offices (average size of architectural office of 2,7 employees (Forlati et al. 2006)) as well as construction companies largely coming from the small- or medium size sector. The traditional design and planning process is carried out by small scaled, highly segmented large number of experts working in sequential manner, using various kinds of tools and software. Therefore the standards, but also the knowledge for BIM-supported planning process is largely lacking.

The further problem that most of the Austrian offices are facing is the high fluctuation of the employees and of the related know-how loss; which is a common characteristic of the most of the small project-oriented firms. Owen et al. (2010) point out the need for enhancement of skills of project members, which are often highly specialised in their own fields of expertise, but seldom trained to work in integrated project environment. The organizations also seldom support this kind of professional development. The introduction of the new BIM-tools therefore mostly means more than simple CAD-tools shift, since the adoption is mostly related to the reorganization of the processes and management strategy of the project-based organization.

Seen in this light, in the practical BIM operation and use a number of problems on different levels can be met. On the technological level, the questions of the interfaces in the data transfer of the multi-disciplinary models arises, as well as of the heterogeneous data-structure of the different software, and of management of ever larger data-volumes. On the semantical level, it can be noticed that each discipline needs individual information; the professional languages differ strongly as well as the means and methods to represent a building (Bazjanac and Kiviniemi 2007). The spectrum reaches from diverse lists for project management and quantity surveys, over reduced slab models for structural engineering for earthquake simulation, to complete spatial representation of architectural models in the full geometric complexity.

The optimal management, filtering and reliable synchronisation of such highly differentiated information in the context of in the AEC industry, still dominated by the heterogeneous software-structure, requires high effort in organisation, administration interdisciplinary communication and know-how; especially in a market that lacks a tradition and knowledge of integrated planning practice. A standard solution offering the complete software package for this large spectrum does not yet exist, and it is questionable if such solution is viable, due to the prototypic nature of construction projects.

The BIM-based software-packages that would fully support and enhance the integrated, interdisciplinary planning practice and life-cycle data integration are still rather seldom. The one-stop-shop tools for modelling of architecture, structural and MEP (mechanical and electrical) engineering, energy simulation, life cycle costing and assessment are still not available according to the requirements of planning practice and building policy. The intra-firm project-constellation and mostly changing project-stakeholders with each new project, represent challenges for interoperability of new software-tools combinations with each new project. A pre-requisite for a successful implementation of life-cycle oriented planning and management is therefore a smooth and efficient data exchange without information losses.

If the early BIM research was mainly focused on the problem-solving of the software-interoperability and efficient data exchange, the current research efforts are focusing on the change of the planning practice towards integrated design and delivery, which is not only related to the handling of technical, but even more over so to the process-related issues (Succar 2009; Penttilä and Elger 2008; Gu and London 2010). The process-reorganization addresses both the intra- and inter-firm project organization and standardization of the workflows, role descriptions and related responsibilities of the stakeholders, as well as the general commitment towards collaborative planning attitude. Rekkola et al. (2010) argue, that the actual BIM-benefit lies in the domain of workflow and business practices therefore process-knowledge, beyond the technological issues. Within a case study of university building, they identified problems and benefits of BIM-supported integrated process by creating categories: people (competence or knowledge problem), process (work-flows, timing, contracts, roles) and technology (software). They argue that a) for enhanced integrative practice a participative process is necessary and b) that the slow BIM-adoption in the practice is caused by the difficulty of interrelation (triangulation) of the people-process-technology problems.

Therefore, the greatest challenge especially within markets still dominated by sequential design and planning method, either for holistic concepts such as Building Life-cycle Management (BLCM) (von Both 2011) or Integrated Design and Delivery Solutions (IDDS) Modell (Prins and Owen 2010), remain with the people (planning process stakeholder) and process – the process of model building of an integrated, interdisciplinary building model requires close cooperation and coordination of the planners, contractors, industry and facility managers, a highly skilled project team as well as detailed conventions on an inter-organizational level (et al. 2010, Plume and Mitchell 2007; Arayici et al. 2011).

Methods

Although a large bulk of literature is implying on the benefits of BIM (Azhar 2011; McGraw-Hill Construction 2010; Becerik-Gerber and Rice 2010; Gilligan and Kunz 2007), the over-all measurement of BIM-related benefits for planning networks and practice is still difficult to justify, due to the high level of complexity of employed tools and of the process, but also due to the lack of a standardized measuring methodology (Jung and Joo 2011; Barlish and Sullivan 2012). The issue of how to measure BIM benefits is especially important in the emerging markets, such as Austrian is, in order to enhance the adoption of the technology and more over the process in the industry. For the adoption in the AEC market the closer research of interrelations within the triangle: technology (operability) – people - process is necessary, in order to create a guideline for BIM adoption, assessment, usability, risks, and evaluation (Gu and Kerry London 2010).

In order to evaluate BIM-performance within an integrated planning process in relation to the technology-people-process triangle, we conducted an experiment with students, simulating a multi-disciplinary planning process for sustainable building within a design-studio class in the winter semester of 2012/13. The experiment is a part of an on-going research project "BIM-Sustain: Process Optimisation for BIM-supported Sustainable Design" involving co-operation of university research and BIM-software vendors and developers. This interdisciplinary collaboration of academy and industry enables development of customised strategic concepts for the individual BIM-settings within multi-disciplinary planning context. The final aim of the project is compilation of guidelines for BIM-supported design and planning. The guidelines will include the conventions for efficient data-exchange and a road-map for the standardization process at Austrian Standardisation Institute (standardization body), recommendations for the planners for the inter- and intra-firm organization of BIM-supported design process, and finally proposals for improvement of interoperability for the software-vendors; based on experiment-findings. Similar guidelines were

	Architect	Structural Engineer		Building Science (Simulation in TAS)	
	CAD	CAD	FEM	CAD	Calculation
1	Allplan	Allplan	Scia Engineer	Allplan	Allplan
2	Autodesk Revit Architecture	Autodesk Revit Structure	SOFiSTiK	Autodesk Revit MEP	PLANCAL
3	GRAPHISOFT ARCHICAD 16	TEKLA	Dlubal	PLANCAL	PLANCAL
4	GRAPHISOFT ARCHICAD 16	Allplan	Dlubal	PLANCAL	PLANCAL
5	Autodesk Revit Architecture	Allplan	Scia Engineer	PLANCAL	PLANCAL
6	GRAPHISOFT ARCHICAD 16	Allplan	Dlubal	Autodesk Revit MEP	PLANCAL
7	Allplan	TEKLA	SOFiSTiK	Autodesk Revit MEP	PLANCAL
8	Autodesk Revit Architecture	TEKLA	Scia Engineer	Allplan	Allplan
9	GRAPHISOFT ARCHICAD 16	Autodesk Revit Structure	Dlubal	PLANCAL	PLANCAL
12	GRAPHISOFT ARCHICAD 16	Allplan, TEKLA	Dlubal	Autodesk Revit MEP	PLANCAL
13	GRAPHISOFT ARCHICAD 16	TEKLA	SOFiSTiK	Autodesk Revit MEP	PLANCAL

Figure 1 Software-combinations used by the teams.

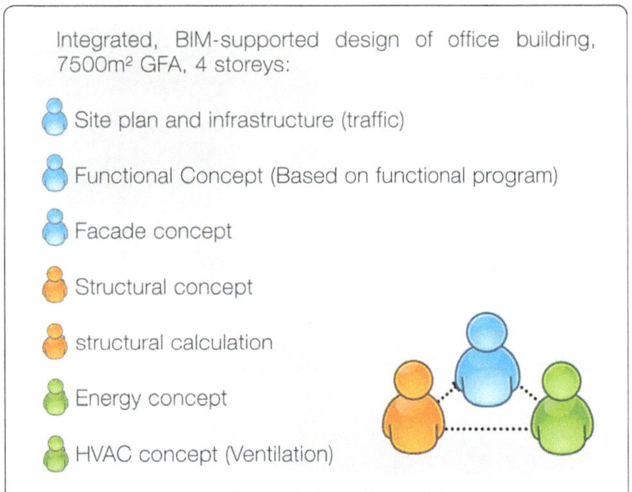

Figure 2 Assignments for each discipline of the team.

compiled by the Penn State within the Computer Integrated Construction Research Program (2012); or the Integrated Project Delivery For Public and Private Owners (2010).

The empirical research by experiment has often been employed to test of BIM-performance and capabilities. Plume and Mitchell (2007) conducted in 2004 an experiment with 23 students in a design studio setting, testing the IFC-model performance in multi-disciplinary collaboration (architecture, landscape architecture, MEP, statutory planning, sustainability and construction management.) They focused primarily on operational issues, such as building model (representation of a building model in different tools) and IFC –server data sharing issues. They conclude that the original architectural model needs significant adaptation for the use of other disciplines or their tools. Further issue needing closer attention is model management – tracing of the changes and updates carried out on the common model. Sacks et al. (2010) carried out the "Rosewood experiment", comparing the BIM-supported versus the traditional 2D CAD the planning and fabrication process of the pre-cast façade. BIM proved to be more efficient by 57%, however IFC proved not mature enough causing data inconsistency in transfer between architectural and engineering system. Losses in translation can be assigned to object-semantic, a similar problem addressed by the Plume and Mitchell (2007).

Sturts Dossick and Neff (2011) observed the collaboration of several teams on three real projects using a BIM-technology supported design process, focusing on people and process issues. They concluded that technology can even hinder the innovation of the design process through a too rigid corset of work-flow and knowledge exchange, hindering the exchange of tacit, informal knowledge. Their concept of "messy talk" – the informal, unstructured information exchange as often practiced in architecture and construction engineering is tested within student experiment, where geographically distributed teams work using BIM on a project in a virtual environment. They conclude that "...messy talk requires both the flexible, active, and informal setting described in the 2011 study as well as mutual discovery, critical engagement, knowledge exchange, and synthesis." (Dossick et al. 2012).

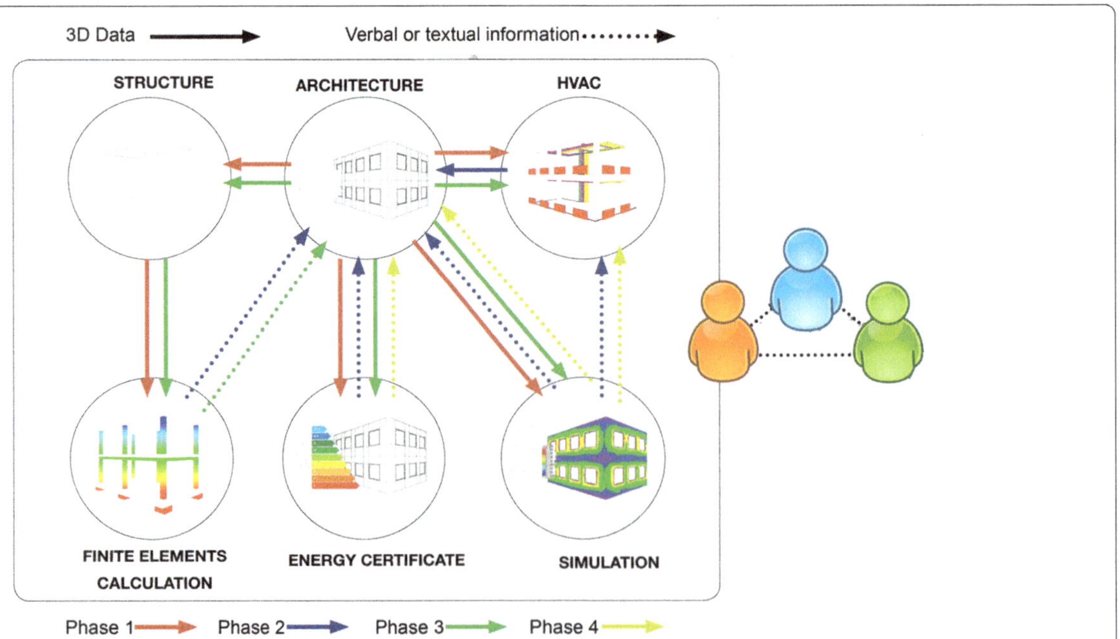

Figure 3 Modelling and data-transfer in multi-disciplinary team.

Peterson et al. (2011) conducted a simulation of integrated project management within two classes at Stanford and TU Twente, using pre-made BIM models in Revit, Tekla and AutoCAD 2006 and 2007 (which in our understanding cannot be considered as BIM models due to the lacking of parametric characteristics), the models were imported in the various cost and scheduling software such as Primavera or Vico.

The formerly mentioned experiments and research of BIM-supported planning practice focus on evaluation of singular issues - some primarily focus on the technology performance (interoperability, building model semantics), such as Plume and Mitchell (2007) experiment and Sacks et al. (2010). Sturts Dossick and Neff (2011) on the other hand focus mainly on the process issues. The student classes carried out at Stanford and TU Twente apply the holistic evaluation, however examine the BIM-supported project management, which in terms of data exchange displays lower complexity than multi-disciplinary design, involving structural and thermal simulation, which both are based on exact transfer of geometry.

In our research, we have addressed the triangulation of the technology, people and process parameters, in order to identify how they are correlated. Therefore, through the experiment the data on a) BIM-performance in terms of data-transferability in different software-constellations will be collected through protocols and revision of delivered models and b) the team performance using different BIM-tools will be assessed through protocols and recorded feedback workshop. The executed experiment is the first one to have a holistic approach on the evaluation of people-process-technology triangle, testing a large number of software tools (all together thirteen) and software combinations on the transfer of complex geometrical data, but also on usability.

Through exploratory research – an experiment within an interdisciplinary design class involving 40 students, the collaborative, multi-disciplinary BIM-supported planning for an energy-efficient office building is simulated. The multi-disciplinary teams consisting of: architect, structural engineer, building physicist (BS) were formed by the means of a pre-questionnaire, which questioned skill-level, experience and preference of the software. Upon the results of the questionnaire a matrix of software-combinations used by each team was compiled (Figure 1).

In the course of the experiment (design class) basically two work-flow models can be identified: One-Platform BIM (proprietary) and Open-Platform BIM (using IFC exchange format). The experiment began in September 2012, the latest available software versions were used. The Open-Platform BIM teams (Figure 1, Teams 3–13) use different, for each discipline typical (custom)

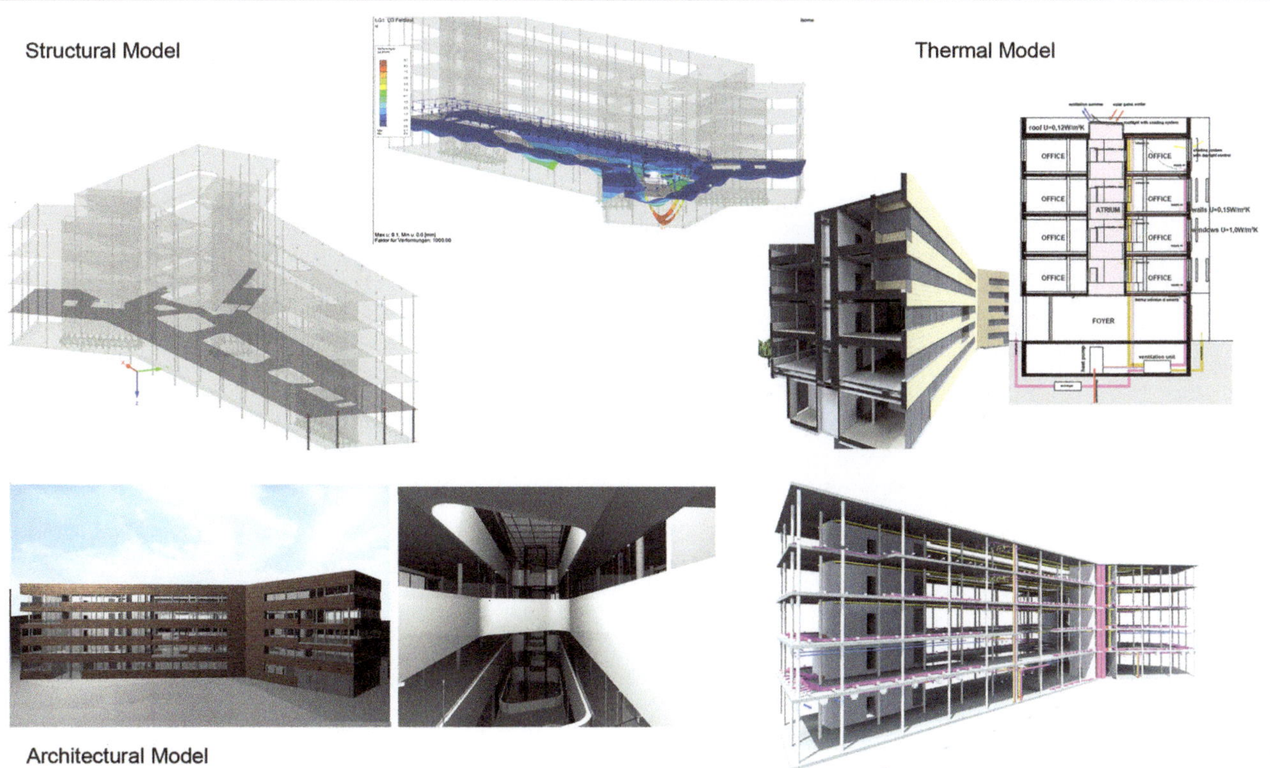

Figure 4 **Project example including architectural, structural and ventilation model, Team 3.**

software, and work with central architectural model, exchanging the data using the IFC. By central architectural model we mean the physical, architectural model as the point of origin for the further transfer into a) structural, analytical software or b) into the thermal analysis software. The One-Platform BIM teams (Figure 1, Teams 1 und 2) work with one software family Nemetschek Allplan (2012) or Autodesk Revit (2012) using proprietary standards, again starting with architectural model.

The teams are assigned with compilation of the architectural (in Allplan 2012, Revit Architecture 2012 or ArchiCAD16 2012), structural (Simulation in Dlubal REFM 2012; Sofistik 2012 or Scia 2012, drawings in Tekla Structures 2012; Revit Structure 2012 or Allplan 2012), and ventilation (in Plancal 2102 or Revit MEP 2012) models, as well as the light simulation and energy certificate (Figure 2). For the thermal simulation TAS 9.2 (2012) is used, for light simulation Dialux 4.9 and for energy certificate Archiphysik 10. Planning documentation was handed out, consisting of a functional programme, site-plan with orientation and set origin, layer-structure and colour scheme for latter room-stamps.

The time-schedule of the design-class is strictly organized; the experiment is taking place in the period of one semester. We have organised three presentations, where in the first one the architectural model is presented, in the second presentation the structural and thermal and in the final presentation the optimised, full model containing all the information (Figure 3). Between the presentations the reviews with teachers as well as tutorials provided by software vendors are taking place.

Figure 4 presents the final model as delivered by one of the student-teams (Team 3) at the final presentation, including architectural model with visualization, model of loadbearing structure and maximal slab deformation under load, model of ventilation and energy and HVAC concept (Figure 5).

On the level of technology, the experiment is examining the fitness of various software constellations for data transfer, import and export, documenting the data loss and needed rework if data-loss has occurred. In terms of process, the efficiency and efficacy of multi-disciplinary teams working with BIM: efficiency of the employed BIM methods for data-exchange, communication effort, and work-allocation (work-flows); and on people-level

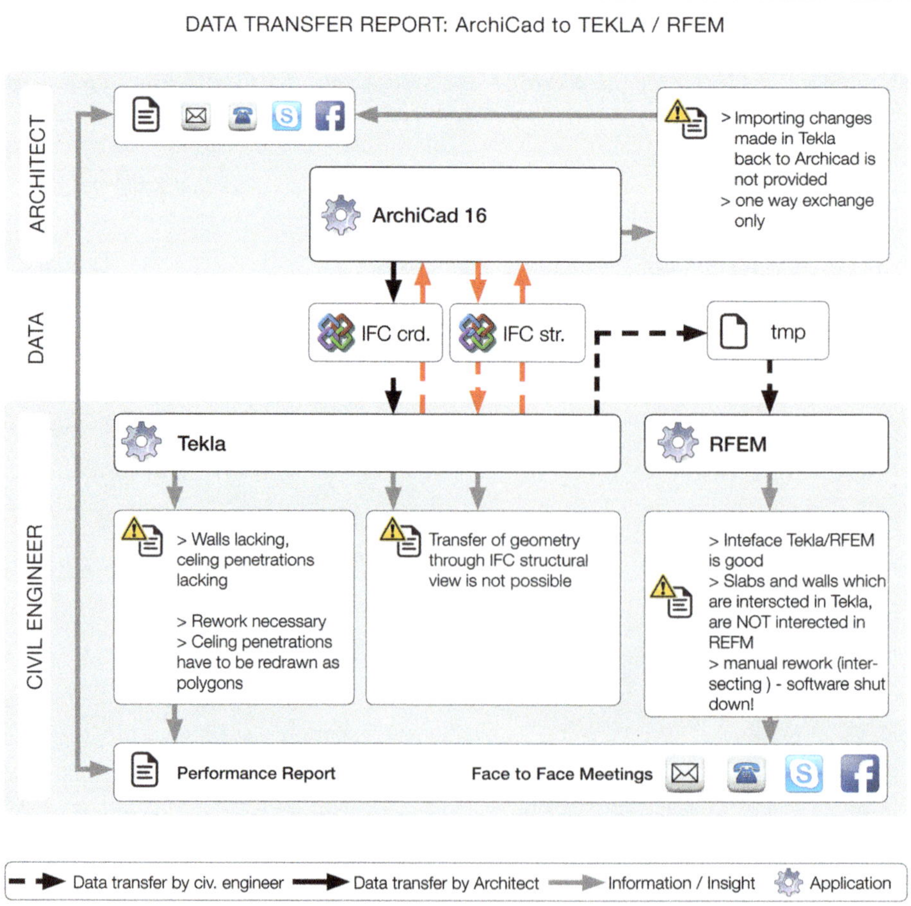

Figure 5 Mistake tree for data exchange from architectural into structural model, based on protocols, Team 3.

satisfaction and conflict levels are assessed. Through the mandatory protocols and time-sheets the problems related to the technology (data-transfer inconsistencies or losses, semantics) but also to the process-people related problems (conflicts, communicational difficulties, lack of work-flow definitions or responsibilities etc.) can be tracked (Figure 4). Additionally, an e-learning platform has been set up, with a forum for tutor feedback as well as for student-communication, scheduling and posting of tasks is taking place.

Results and discussion

People-related problems

The first data on satisfaction was gathered at the point close to the second presentation of structural and thermal models – basically one data exchange step has taken place – export from architectural model towards FEM (Finite Element Method) Software and thermal simulation software. In the student workshop through rough questionnaire answered by 19 students (three architects, two engineers, 14 BS), a) satisfaction with BIM-technology, b) satisfaction with teamwork, c) satisfaction with process (workflows), d) conflict-level and e) stress-levels were questioned on the scale ranging from 0 (low) to 6 (high) (Table 1).

The general BIM-technology dissatisfaction resulted from data transfer problems, as reported especially by the BS students using TAS simulation software, where data exchange uses gbXML standard. Mostly all of the architectural models had to be newly drawn in TAS, due to the data loss or wrong interpretation by TAS. It was reported by the students that the time effort for the adaptation of the imported model was equal to the time effort for creation of the new model (two days).

Technology-related problems

When passing architectural geometry into structural analysis software two types of IFC-Files are used: Coordination View and Structural Analysis View (Building Smart 2013). Software for Finite Element Method (FEM) calculation requires the Structural Analysis View of an IFC-File. But only a few CAD-Programs support the export of this type (Table 2). Additionally, not every FEM-Software supports the import of the physical model (Coordination View –Table 3). As a consequence, an intermediate step is needed to transfer the model from architecture to structural model: an import into a program which can import and export both types of files, a step which goes along with a loss of information (Figure 6). Figure 5 displays a few typical problems when importing from physical (architectural) model into analytic model (structural engineering) – the construction line of intersecting walls is not intersected in the analytic view, which requires remodelling after the import of physical model. This problem originates in the semantics of modelling, namely that architectural models are a set of spaces which require closed elements, whereas structural engineers model a building as a set of loadbearing elements, slabs, columns and plates. Due to this incompatibility it was i.e. not possible to import an Allplan-Model as an IFC-File into the FEM-Software Sofistik, because both programs support the type which cannot be read by the other one.

The IFC 2×3 (structural analysis view standard) still leads to variety of problems: especially more complex geometry such as sloped or rounded walls, roof elements and openings are very likely to cause problems or even disappear when being imported.

For example, problems were the identified with Tekla Structures: stairs become boxes, openings disappear, and round elements become rectangles. Findings from data transfer from Allplan to Scia: when round walls are used, the model takes long period of time to get imported in Scia (hours). To illustrate further problems, we assembled a scene containing several pertinent elements, exported an IFC file and imported this into different structural analysis programs. Figure 7 illustrates the architectural model; Figures 8 and 9 the interpretations by different FEM software - a completely different result when importing the exact same IFC file.

Table 1 Results of the first questionnaire

Question	Mean value	Median value
Satisfaction with BIM	1,89	2
Satisfaction with teamwork	3,84	4
Satisfaction with process	2,37	2
Stress level	4,16	5
Conflict Level	1,37	1

Table 2 Compatibility with IFC – Structural Analysis View

CAD		FEM	
ArchiCAD		Dlubal RFM	⇆
Allplan		Scia Engineering	⇆
Revit	⇆	Sofistik	⇆
Tekla	⇆		

Table 3 Compatibility with IFC – Coordination View

CAD		FEM	
ArchiCAD	⇆	Dlubal RFM	←
Allplan	⇆	Scia Engineering	⇆
Revit	⇆	Sofistik	
Tekla	⇆		

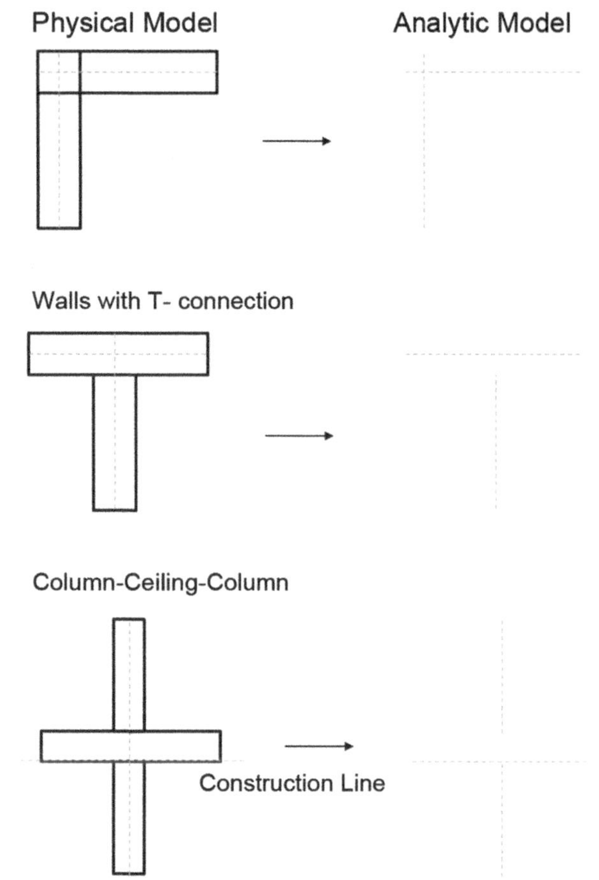

Figure 6 **Results from export from architectural model, import in RFM simulation.**

Considering the data transfer from architectural model to thermal simulation as the main problem appeared the incompatibility of TAS software with the IFC standard. TAS can only import the proprietary gbxml interface, which again can only be produced by Revit. When working with ArchiCad, Encina plug in for creation of gbxml can be used. In any case – using direct transfer via gbxml or via Encina, numerous problems related to the geometry were identified: walls are not transferred correctly, and were reworked in TAS instead of Revit; missing windows, etc.

Conclusion

This paper presented the first results of an experiment: simulation of BIM-supported multi-disciplinary design for energy-efficient office building, using various BIM-tools for architectural, structural, energy and ventilation modelling, and thermal simulation. For both structural engineering and thermal simulation, the data transition becomes difficult as soon as there is complex geometry involved, such as round walls, which have caused problems in all software-combinations. A problem of semantics of building models is a constant issue – architects use different room-stamps than BS, the pillars are drawn from slab to slab whereas structural engineers work with one continuous pillar from top to bottom slab. Further difficulties originate from incompatible software-combinations such as Allplan to Sofistik, or ArchiCAD to Sofistik (see Tables 2 and 3). For building physics this applies to the Allplan to TAS, since TAS does not support IFC standard, and Allplan does not have plug in or possibility of producing a gbxml file. Such constellations can lead to significant problems in the current BIM-supported planning practice, if e.g. an architectural office using ArchiCAD has to work with the structural engineers using Sofistik, since the data transfer will not be possible and purchase of additional software or of additional "BIM-services" will lead to increase of the planning costs.

In terms of comparison of One-Platform BIM versus Open-Platform BIM, it can be concluded that One-Platform BIM constellation, as closed system, does not exist on the market yet. The One-Platform BIM Software (Nemetschek Allplan and Autodesk Revit) both leave the original platform in order to conduct structural calculation and simulation; however offer proprietary interface to these software or even plug-ins (Revit to Sofistik). Even with proprietary interfaces the complex geometry causes problems in transfer through very long

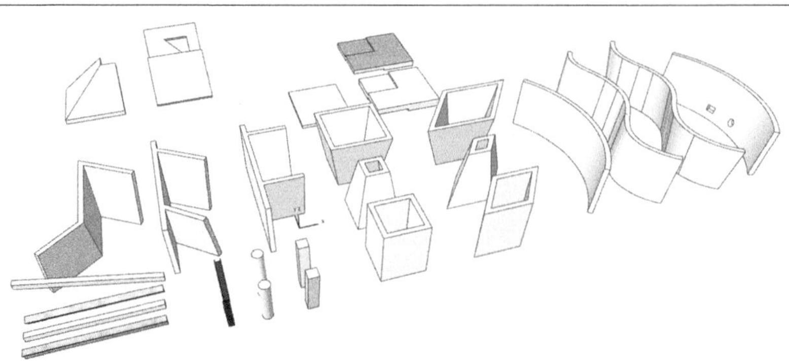

Figure 7 **Architectural model from ArchiCAD16a.**

Figure 8 Import in Dlubal REFM.

transfer-time (Allplan to Scia). The Open-Platform BIM, using IFC interface has proved as time-efficient and exact in transfer, if there is a standardised setting used for IFC transfer from architectural in the structural model and under condition that simple geometry is involved.

Our findings in comparison to the student experiment executed in 2004/05 by Plume and Mitchell (2007) show that ArchiCAD has made significant progress (at that time it was very limited in exporting an IFC file, today it is the software with the best functioning IFC translator); however the question of the building-model semantics for different disciplines and the difference in the grade of required detailing for each model has not been solved yet. Our findings basically confirm the findings of the Rosewood experiment – we experience similar data losses and wrong interpretations at export and import of IFC; Rosewood experiment works with the same version of IFC. Further development of IFC is urgently necessary, since software-side interfaces are still underdeveloped, as well standards IFC 2×3 and 2×4 are still lacking many information (e.g. surface materials (announced for 2×4), real window frame geometry, 3D wall/slab layers.)

The satisfaction with the BIM-technology at intermediate stage of the design class has been reported as low, due to the very difficult data-transfer, inconsistency and data losses, especially so for the thermal simulation, where models had to be redrawn.

Processes-satisfaction has been found as weak: workflows are poorly organised among team-members, there are many problems in allocation of responsibilities. In many teams it is expected from the architect to undertake all of the major adaptions of the architectural model in order to make it fit for the transfer (the consultants are not ready to adopt the imported models). Teams often report that some team members are often not available.

On the level of people-related problems, despite the reported low conflict level team-satisfaction is only average. We were able to observe a lack of team affiliation

Figure 9 Import in Scia.

with most of the teams, often a bonding between two disciplines can be observed and the third one is not playing along. This phenomenon might be referred to the lack of an organised kick off meeting. Some of the problems originate in the lack of professional knowledge (e.g. design of an office building or energy efficient facade) and in general to the lack of experience and knowledge in collaborative planning. In some cases the lacking of the team-building/bonding, and following the aim to „just finish the project" led even to an increase of fragmentation of the design process (architect defines everything, the consultants only optimise in following steps), which is exactly the opposite of the expected BIM-effect. We can confirm the argument by Sturts Dossick and Neff (2011) – BIM-technology is advantageous for exchange and presentation of explicit knowledge, but does not support the tacit knowledge of how the buildings are designed. Our first findings also imply that BIM-technology does not support integrated practice by itself, for the support of the collaboration other means are necessary such as well organised formal (kick-off meeting) and opportunities for informal communication (von Both and Zentner 2004).

Finally, BIM as presently used, hardly changes the work-flow between architects and structural engineers, not only due to technical interface limitations but even more so due to a logical contradiction. FEM models require a far lower level of detail than the architectural model delivers, so any automatically converted model necessarily needs to be post-processed manually by the engineer in order to simplify the model for reasonable meshing and resulting calculation times. When importing a coordination view IFC into calculation software, the discretization of architectural models into FEM-suitable meshes is carried out within the import and is hence forced to accept the model geometry "as it is". For example, a small rounded wall opening, let's say for a drainage pipe, will produce a complex mesh in the FEM model when being imported. For the structural system however, this opening is irrelevant, but still causes enormous effort in calculation and will hence be deleted by the engineer. Once simplified, such a wall element cannot be re-exported into the central model, because otherwise the opening would be missing. Vice versa, once the architectural model changes and is re-imported, the opening is back. The possible solution is either the radical improvement of FEM-software performance concerning calculation time for meshing; or enabling of the FEM software to directly manage the referential architectural model. The problem of bi-directional model-management remains one of the greatest challenges, not only because of the technical issues, but mostly because of the process issues: definition of the rights (who may change what and when?) related to the change management.

As possible solution of the bi-directional model management the model server architecture is proposed (Kiviniemi et al. 2005). Jørgensen et al. (2008) develop different scenarios for the use of separate models, separate models with aggregate model and one shared model, where rights, accessibility and ownership is exactly defined, however with the limitation of the model server using ArchiCad, as the only software properly handling the IFC import and export.

As first future step, detailed statistical analysis of the mandatory protocols and time-sheets will be carried out in order to gain more knowledge on performance of One-Platform BIM versus Open-Platform BIM, as well as of communication effort, work-allocation, satisfaction and conflict levels. The recorded feed back workshop (interviews) will be coded and analysed, to gain qualitative information on process efficacy, not only efficiency.

In September of 2013 the second experiment will be conducted, in the framework of second multi-disciplinary design class, where we will be able to use the first findings and propose a framework for data-exchange procedures as well as for careful design of communication, including a kick-off workshop for team building. Finally we will compare the results of the two experiments, evaluate the benefits and compile the guidelines for the planning practice and standardization bodies.

Competing interests

The authors declare that they have no competing interests.

Authors' contributions

IK organized the experiment, carried out the survey, evaluated the results and drafted the manuscript. LO and CM carried out the experiment, compilled and evaluated the data. CA supported the experiment-design. All authors read and approved the final manuscript.

Acknowledgements

This paper is a result of an on-going research project "BIM-Sustain" funded by FFG (Austrian Research Funding Agency) and following BIM-software vendors: A-Null, Artaker, b.i.m.m. Gasteiger, Construsoft Tekla, Dlubal REFM, Nemetschek Allplan and Plancal. We gratefully acknowledge the work of the academic partners: Institute for Management Sciences, Prof. S. Köszegi and Dr. M. Filzmoser and Institute for Building Physics and Ecology, Prof. A Mahdavi, L. Skoruppa and K. Kiesel, and to the students taking part in this project of Vienna University of Technology.

References

Allplan. (2012). http://www.nemetschek-allplan.at/software/architektur/allplan-architektur.html, last access December 2013.

Arayici, Y, Coates, P, Kiviniemi, A, Koskela, L, & Kagioglou, M. (2011). *BIM implementation and Adoption Process for an Architectural Practice.* USA: FIATECH Conference.

ArchiCAD16. (2012). http://a-null.com/ARCHICAD16.php, last access August 2013.

Azhar, S. (2011). Building Information Modeling (BIM): Trends, Benefits, Risks, and Challenges for the AEC Industry. *Leadership Management Engineering, 11*(3), 241–252.

Barlish, K, & Sullivan, K. (2012). How to measure the benefits of BIM — A case study approach. *Automation in Construction, 24*, 149–159.

Bazjanac, V, & Kiviniemi, A. (2007). *Reduction, Simplification, Translation and Interpretation in Exchange of Model Data.* Maribor: CIB W78.

Becerik-Gerber, B, & Rice, S. (2010). The perceived value of Building Information Modelling in the U.S. Building Industry. *Journal of Information Technology in Construction.* ISSN 1874–4753, at http://www.itcon.org/2010/15.

Building Smart. (2013). http://www.buildingsmart-tech.org/specifications/ifc-overview, last accessed August 2013.

Computer Integrated Construction Research Program. (2012). *BIM Planning Guide for Facility Owners, Version 1.01, May.* University Park, PA, USA: The Pennsylvania State University. http://bim.psu.edu.

Dlubal REFM. (2012). http://www.dlubal.de, last access August 2013.

Dossick, C, Anderson, A, Iorio, J, Neff, G, & Taylor, J. (2012). Messy Talk and Mutual Discovery: Exploring the Necessary Conditions for Synthesis in Virtual Teams. In *Working Paper Proceedings.* The Netherlands: Engineering Project Organizations Conference Rheden.

Fellows, R, & Liu, AMM. (2012). Managing organizational interfaces in engineering construction projects: addressing fragmentation and boundary issues across multiple interfaces. *Construction Management and Economics, 30*(8), 653–671.

Forlati, S, et al. (2006). *Getting Started. Wonderland Association, Platform for Architecture.*

Gilligan, B, & Kunz, J. (2007). *VDC Use in 2007: Significant Value, Dramatic Growth, and Apparent Business Opportunity.* Stanford University: CIFE Technical Report #TR171.

Gu, N, & Kerry London, K. (2010). Understanding and facilitating BIM adoption in the AEC industry. *Automation in Construction, 19*, 988–999.

Jørgensen, KA, Skauge, J, Christiansson, P, Svidt, K, Birch Sørensen, K, & Mitchell, J. (2008). Use of IFC Model Servers Modelling Collaboration Possibilities in Practice. In *Aalborg University, Department of Production, Department of Civil Engineering, Aarhus School of Architecture, Department of Building Design.*

Jung, Y, & Joo, M. (2011). Building information modelling (BIM) framework for practical implementation. *Automation in Construction, 20*, 126–133.

Kiviniemi, A, Fischer, M, & Bazjanac, V. (2005). Integration of Multiple Product Models: IFC Model Servers as a Potential Solution. In *Proc. of the 22nd CIB-W78.*

McGraw-Hill Construction. (2010). *The Business Value of BIM in Europe. Smart Market Report.* McGraw-Hill Construction Research and Analysis Aliances.

Nemetschek Allplan. (2012). http://www.nemetschek-allplan.at, last access December 2013.

Owen, R, et al. (2010). Challenges for integrated Design and Delivery Solutions. *Architectural Engineering and Design Management, 6*(4), 232–240.

Penttilä, H, & Elger, D. (2008). New Professional Profiles for International Collaboration in Design and Construction. In *26th eCAADe Conference Proceedings, Antwerpen (Belgium) 17–20 September 2008* (pp. 333–340).

Peterson, F, Hartmann, T, Fruchter, R, & Fischer, M. (2011). Teaching construction project management with BIM support: Experience and lessons learned. *Automation in Construction, 20*, 115–125.

Plume, J, & Mitchell, J. (2007). Collaborative design using a shared IFC building model—Learning from experience. *Automation in Construction, 16*, 28–36.

Prins, M, & Owen, R. (2010). Integrated Design and Delivery Solutions. *Architectural Engineering and Design Management, 6*(4), 227–231.

Project Delivery For Public and Private Owners. (2010). *NASFA, COAA, APPA, AGC and AIA.* www.aia.org/ipd.

Rekkola, M, Kojima, J, & Mäkeläinen, T. (2010). Integrated Design and Delivery Solutions. *Architectural Engineering and Design Management, 6*(4), 264–278.

Autodesk Revit. (2012). http://www.autodesk.de/products/autodesk-revit-family/overview, last access December 2013.

Revit Architecture. (2012). http://www.autodesk.de/products/autodesk-revit-family/features/architecture, last access December 2013.

Revit Structure. (2012). http://www.autodesk.de/products/autodesk-revit-family/features/structure, last access December 2013.

Revit MEP. (2012). http://www.autodesk.de/products/autodesk-revit-family/features/mep, last access December 2013.

Rizal, S, & van Berlo, L. (2010). Tool for Benchmarking BIM Performance of Design, Engineering and Construction Firms in The Netherlands. *Architectural Engineering and Design Management, 6*(4), 254–263.

Sacks, R, Kaner, I, Eastman, MC, & Jeong, YS. (2010). The Rosewood experiment — Building information modelling and interoperability for architectural precast facades. *Automation in Construction, 19*, 419–432.

Scia. (2012). http://www.scia-software.de, last access August 2013.

Sofistik. (2012). http://www.sofistik.de, last access August 2013.

Sturts Dossick, C, & Neff, G. (2011). Messy talk and clean technology: communication, problem-solving and collaboration using Building Information Modelling. *Engineering Project Organization Journal, 1*(2), 83–93.

Succar, B. (2009). Building information modelling framework: A research and delivery foundation for industry stakeholders. *Automation in Construction, 18*, 357–375.

TAS. (2012). http://www.edsl.net/main/Software.aspx, last access August 2013.

Tekla Structures. (2012). http://www.tekla.com/, last access August 2013.

von Both, P. (2011). Produktdatenmodellierung - aktuelle Entwicklungen und Möglichkeiten der Vernetzung von Produkt- und Prozessebene. *1.* In *Internationaler BBB-Kongress, Dresden 15.09.2011.*

von Both, P, & Zentner, F. (2004). LUZIE Lebenszyklusbezogene Einbindung der Zielplanung und des Zielcontrolling in den Integralen Planungsprozess. In *Technical report, FIA Projekt, Bundesministerium für Wirtschaft und Arbeit, Bundesministerium für Bildung und Forschung.*

6

Review and analysis of augmented reality literature for construction industry

Sara Rankohi* and Lloyd Waugh

Abstract

Research has identified various beneficial capabilities for augmented reality technologies in the AEC industry such as virtual site visits, comparing as-built and as-planned status of projects, pre-empting schedule disputes, enhancing collaboration opportunities, and planning/training for similar projects. This paper provides an expanded foundation for future research by presenting a statistical review of augmented reality technology in the AEC industry. The review is based on articles found within eight well-known journals in architecture, engineering, construction, and facility management (AEC/FM) until the end of the year 2012. The review further narrows the literature within these journals by considering only those 133 articles found through a key word search for "augmented reality." The selected journal articles are classified within the following dimensions: improvement focus, industry sector, target audience, project phase, stage of technology maturity, application area, comparison role, and technology. The number of articles within these dimensions are used to identify maturing and emerging trends in the literature as well as to synthesize the current state-of-the-art of augmented reality research in the AEC industry. In summary, the AR literature has increasingly focused on the demonstration of visualization and simulation applications for comparison of as-planned versus as-built statuses of the project during the construction phase to monitor project progress and address issues faced by field workers. In addition, the future trend is toward using web-based mobile augmented systems for field construction monitoring.

Keywords: Augmented reality; AEC industry; Project phase; Construction management; Literature review

Introduction

The complex nature of the architecture, engineering, construction, and facility management (AEC/FM) industry and its high demand for access to information for evaluation, communication and collaboration, increases the industry's need for information technologies. Recent visualization technologies such as virtual and augmented reality technologies are ideal in this environment.

Overview of augmented reality

Augmented reality gives a view of the real world where elements are superimposed by computer generated files such as graphics, sounds, videos, or digital information. From the first see-through head-mounted AR display developed in the 1960s by Ivan Sutherland at Harvard (Sutherland, 1968), to the enhanced HD^{4}AR and Mobile Augmented Reality System (MARS) developed by Golparvar et al. (Bae et al. 2012), augmented reality technologies have been used in various disciplines and arenas, e.g. engineering, entertainment, aerospace, medicine, military, and automotive industry, as a frontline technology to meet visualization difficulties in their specific domain (Behzadan and Kamat 2011).

Application areas

The AEC industry is also moving to embrace more AR technologies for improving various stages of construction projects. This advanced computer technology provides significant advantages through simulation and visualization of the construction industry, e.g., allowing the observer to interact with both the actual and the virtual objects and to monitor the construction progress by comparing the as-planned and as-built status of the project (Shin and Dunston 2008). AR technologies can benefit the AEC/FM industry in at least three levels: visualization, information retrieval, and interaction (Dong and Kamat 2013). Various applications of AR have been recommended for the AEC/FM industry by different researchers. Dunston and Wang (2005)

* Correspondence: s.rankouhi@unb.ca
The Department of Civil Engineering, University of New Brunswick, Fredericton, New Brunswick E3B 5A3, Canada

proposed AR systems for AEC industry to support all phases of the constructed facility project life cycle. Wang et al. (2007) explore potential AR applications in heavy construction equipment operator training. Golparvar-Fard et al. (2009) developed a 4-dimensional AR model for automating construction progress monitoring, data collection, processing and communication in construction phase of the project. Behzadan et al. (2011) investigate a mobile 3-dimensional AR system for visualizing dynamic site operations during the construction phase. Waugh et al. (2012) developed a web-based augmented panoramic environment to document construction progress. Park et al. (2013) presented a conceptual framework that integrates AR with building information modeling (BIM) to detect construction defects. These applications demonstrate the potential of this technology for future use in this domain.

AR technologies

Although the application of augmented reality technologies in construction projects has tremendously increased in recent years, these technologies are still in the research stage and their full potential is not fully achieved. There are limitations that should be addressed before these technologies will become dominant in the AEC industry such as tracking technologies, and rendering software. Moreover, appropriate application areas for different types of AR will continue to evolve.

Historically, many AR technologies were not applied on construction sites due to tracking and alignment problems, instead they were generally used at the home office for simulation or collaboration during the design phase of a project. However as the technology developed in recent years, the majority of these technologies have been used on construction sites for progress monitoring and defect detection. Moreover, trend analysis shows that web-based and wireless network technologies are becoming more and more popular in recent years, and these types of AR technologies are interesting technologies for further research and application.

A list of various input mechanisms, output mechanisms, and tracking technologies for AR systems can be found in Wang (2009). Portable and mobile AR systems including: radio-frequency based tracking technologies such as GPS, WLAN, indoor GPS; infrastructure-dependent technologies such as fiducial markers; infrastructure-independent tracking technologies such as gyroscopes; and image-based tracking techniques have been studied in many research projects (Bae et al. 2013). Moreover, using cloud computing technologies for web-based and ubiquitous AR systems has been explored in recent years (Wang et al. 2013).

Related studies

Shin et al. (2008) study various application areas for augmented reality technologies in industrial construction based on technology suitability. The research assesses different work tasks from the human factors perspective and presents a comprehensive map, which identifies eight work tasks including layout, excavation, positioning, inspection, coordination, supervision, commenting, and strategizing out of seventeen classified work tasks which could potentially benefit from AR systems.

Wang (2009) gives a detailed review of AR in the AEC industry, and gives a review of several major research efforts prior to 2009, and categorizes various AR technologies with their advantages and disadvantages.

Wang et al. (2013) reviews 120 articles published between 2005 and 2011 in various journal and conferences databases with a focus on augmented reality technologies in the built environment. The paper classifies all available toolkits for augmented reality prototyping in five categories: 2D marker AR-PC and web-cam based, 2D marker AR-mobile, 3D object recognition-mobile, marker-less tools, GPS-compass based AR. In their research, AR literature is classified in three categories: (1) application area; (2) AR system layers: concept and theory (with four sub-layers including: algorithm and modeling, conceptual framework, evaluation framework, and technology adoption), implementation (with two sub-layers: software and hardware), evaluation (with two sub-layers: effectiveness and usability), and industry adoption; (3) other technical criteria. The paper explores state-of-the-art technologies in each category and proposes future research directions.

Chi et al. (2013) discusses trends in AR applications for the AEC/FM with a specific focus on four AR technologies: localization, natural user interface, cloud computing, and mobile devices. The paper reviews 101 articles and outlines future trends and opportunities for applying AR in the AEC/FM industry in six directions: (a) field exploration based on hybrid localization, (b) in-field gesture or kinesthetic control of AR interface, (c) integration with location-specific information, (d) accessing field information using ubiquitous services, (e) portable AR devices in the field, (f) context-aware augmented reality in AEC/FM fields.

Main contributions

To apply AR technologies in AEC projects efficiently and to achieve their full potential in this domain, it is essential to systematically identify application areas in which AR can be used for better performance. This statistical review seeks to answer to these questions: what are key AR application areas in the AEC industry based on suitability of AR technologies? what are the gaps in this area which can potentially benefit from AR technologies? Based on future trends, predict how AR technologies can be further improved for future applications?

This paper presents an in-depth statistical literature review of augmented reality technologies in construction industry over a fourteen-year period (1999–2012). The goals of this review are (1) to synthesize the current state-of-the-art and trends of augmented reality technologies for construction projects, and (2) to identify key application areas which could significantly affect the construction industry. These goals are accomplished by classifying the literature in categories defined by the authors found in the literature.

The scientific contribution of this review is the presentation of a comprehensive multi-dimensional categorization for specifying AR technology and characteristics in the AEC industry. This literature review gives the researcher a broad view of the stage of AR technology maturity in built environment, which can be used to guide new augmented reality system design as well as to help evaluate existing systems for the construction industry.

This paper offers construction practitioners and researchers an assessment of the application areas of augmented reality technologies including the purposes for which these technologies have been applied in different project phases. The paper qualitatively aggregates the results of 133 research studies of AR technologies in construction projects to show researchers and practitioners how augmented reality models have been applied to address project challenges. Based on the trend analysis that is conducted, past research is studied and future research directions are recommended.

Research method

The research methodology used in this paper is illustrated in Figure 1: (A) to select the journals and articles, (B) to review the selected articles, (C) to define relevant categories to classify the articles, (D) to classify the articles in the defined categories. Step (B) and step (C) iterate until the final results are achieved. Step (E) is described in section 6 and section 7 which presents discussion and conclusion.

This paper is an extension of our conference paper (Rankouhi and Waugh 2012); the research methodology is similar to the methodology used previously. Five new journals are added to our database: the ASCE Journal of Computing in Civil Engineering (CCE), the Journals of Advanced Engineering Informatics (AEI), the Journal of Computer Aided Civil and Infrastructure Engineering (CACIE), the Emerald Journal of Engineering, Construction and Architectural Management (ECAM), and the Emerald Journal of Construction Innovation Information, Process, Management (CIIPM). These journals cover a wider range of database in civil engineering research domain, and increase the number of selected articles. In addition, five new dimensions (research methodology, improvement focus, industry sector, comparison role, and location) and various new categories are added to our previous defined dimensions.

Selection of the journals and articles

Eight diverse academic journals (listed in Table 1) were selected within the domain of AEC/FM to record the evolution of AR technology in the AEC industry. Selection of these journals is based on their prominence in the English language field of information technologies in construction engineering and management research.

The articles were selected in two phases. In phase I, a total of 199 articles were found in these eight journals using the search phrase "augmented reality." In phase II, articles that were published in 2013 (due to the lack of a full year at the time when the search was conducted) and articles such as Calendars, Editors Notes, Subject Index, and Content of Volume were excluded. The total number of selected articles was reduced to 133. The number of articles found in each journal is listed in Table 1.

Review and identification of the article characteristics

This section describes statistics based on information provided by the authors, whereas the next section describes our interpreted categories. The number of articles by year and journal is depicted in Table 2.

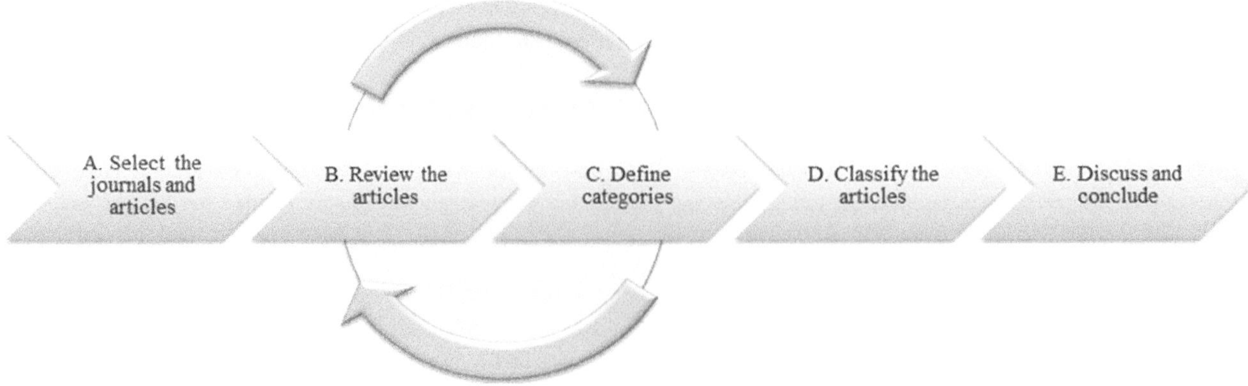

Figure 1 Research methodology.

Table 1 The number of selected journal articles during each phase

Journal	Phase I	Phase II
Journal of Automation in Construction (AIC)	87	43
Advanced Engineering Informatics (AEI)	31	24
ASCE Journal of Computing in Civil Engineering (CCE)	27	20
Journal of Information Technology in Construction (ITCON)	22	19
Journal of Computer Aided Civil and Infrastructure Engineering (CACIE)	14	11
ASCE Journal of Construction Engineering and Management (CEM)	13	11
Emerald Journal of Engineering, Construction and Architectural Management (ECAM)	4	4
Emerald Journal of Construction Innovation: Information, Process, Management (CIIPM)	1	1
Total	**199**	**133**

The distribution of articles by journal and year of publication is depicted in Figure 2. The maximum number of articles in a single year is published in both 2011 and 2012 (26 articles or 20%). The data shows highest numbers of articles for individual years in the AIC (11 articles in 2011), the ITcon (10 articles in 2008), and the CCE (8 articles in 2012) journals. The results show that the increasing trend in the number of articles is dominated by AIC, AEI, and CCE in recent year. Eighty-five percent of the articles were published in the most recent five years.

The final characteristic identified in this section is the percentage of articles based on the first author's country of residence which is shown in Figure 3. With 63 articles (47%), first authors residing in the USA have the highest number of the articles about AR technology in the AEC industry. The remaining counts show that Australia has the second place while both Canada and South Korea are in the third place.

Definition of categories

To better comprehend and further segregate the literature, we defined dimensions and categories to be used in this paper; each article was then compared to these defined dimensions for identification of its principal focus area or to determine the percentage of articles including reference to that classification. Table 3 shows the defined dimensions and the relevant categories. Each dimension is further explained in the following section.

Categorization of the articles

This section discusses the classification of the current state of AR technology literature in the AEC industry. For all but three sub-sections, the articles are classified based on their principal focus and each article is counted once. The exceptions to their approach are sub-sections 5.1, 5.4, and 5.8; in these sub-sections instead of selecting a single "principal focus," we identified the categories to which the article "made reference." In these three sub-sections only we report percentages and do not report counts.

Research methodology

In this section articles are classified based on their research methodology which is divided in five categories: case study, experimental/empirical study, proof of concept (or proof of principle study), questionnaire (survey/interview), and literature review.

A case study is a research method in which detailed consideration is given to the development of a particular case over a period of time. An experimental or empirical study is an empirical scientific method in which an experiment arbitrates between competing models or hypotheses. A proof of concept or a proof of principle study is a research method in which a certain method or model would be recognized to demonstrate its feasibility or to verify that a certain concept, theory, or prototype has the potential of being used. Questionnaires (as well

Table 2 Number of articles by journal and year of publication

	Total	%	2012	2011	2010	2009	2008	2007	2006	2005	2004	2003	2002	2001	2000	1999
AIC	**43**	**32%**	6	11	6	4	7	1	3	1	2				1	1
AEI	**24**	**18%**	7	7	6	3		1								
CCE	**20**	**15%**	8	1	2	5	2	1	1							
ITcon	**19**	**14%**	1	4		2	10	1			1					
CACIE	**11**	**8%**	3		2	2			3							1
CEM	**11**	**8%**	1	3	2	2	1			1					1	
ECAM	**4**	**3%**			1			2	1							
CIIPM	**1**	**1%**			1											
Total	**133**	**100%**	**26**	**26**	**20**	**18**	**20**	**6**	**8**	**2**	**3**	**0**	**0**	**0**	**2**	**2**
Cumulative			26	52	72	90	110	116	124	126	129	129	129	129	131	133
Percent			20%	39%	54%	68%	83%	87%	93%	95%	97%	97%	97%	97%	98%	100%

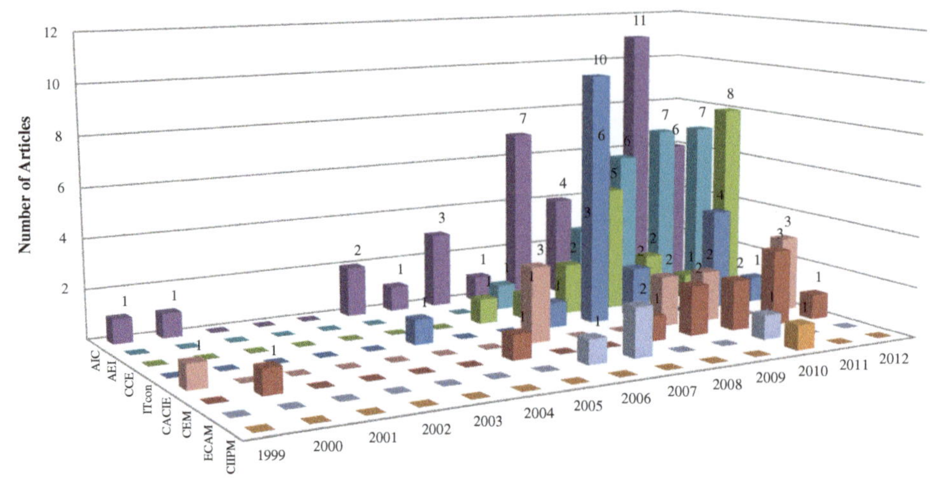

Figure 2 Distribution of articles by journal and year of publication.

as surveys and interviews) are research techniques in which qualitative and qualitative data analysis could be conducted based on the information gathered from research participants. Literature review (historical and documentary research, trend studies), is a research method which considers the critical points of current knowledge including substantive findings as well as theoretical and methodological contributions to a particular topic (Cohen et al. 2007).

Figure 4 depicts the percentages of articles based on their research method. Results show that large number of the articles use case studies to develop their research, while equal number of authors select an experimental method to conduct research in this area. Table 4, presents a list of selected reference articles for each category of research method dimension.

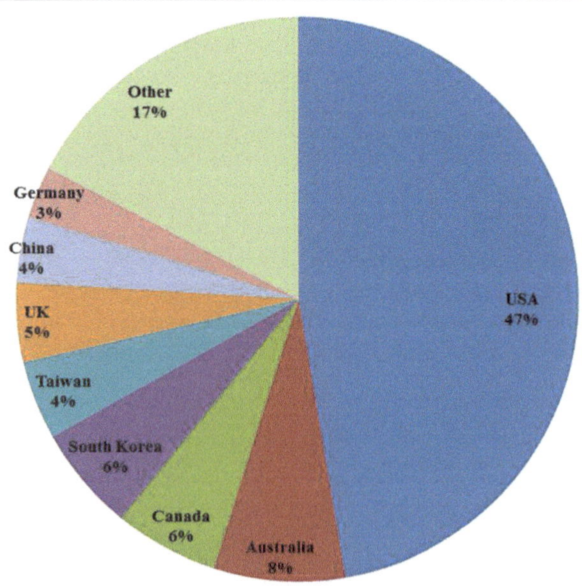

Figure 3 Number of articles by first author's country of residence.

Improvement focus

Articles are classified in four categories based on where the improvement which the article proposes would occur: (1) AEC industry, (2) organization, (3) projects, and (4) individuals. Moreover the organization category

Table 3 Defined dimensions and their categories

Dimensions	Categories
Research methodology improvement focus	Case study, Experimental/empirical study, Proof of concept, Questionnaire, Literature review
	AEC industry, Organization (Facility owner, Contractor, Designer), Projects, Individuals
Industry sector	Building commercial, Municipal/infrastructure, Heavy/highway, Residential, Industrial
Target audience	Design team, Project manager, Worker/technician, Inspector, Project end user, Building systems engineers, Student, Other
Project phase	Initiation, Design, Procurement, Construction, Maintenance
Stage of technology maturity	Theory, Framework, Sub-system technical issues, System development, System application
Application area	Simulation/visualization, Communication/collaboration, Information modeling, Information access/evaluation, Progress monitoring, Education/training, Safety/inspection
Comparison role	Comparison modes (Model vs. model, Model vs. reality, Reality vs. reality),
	Comparison purpose (Progress monitoring, Defect detection, Evaluation the model, Updating the model, Validating the model)
Technology	User perspective (immersive, non-immersive), Device (mobile, non-mobile), Delivery (web-based, standalone)
Location	Home-office, Field

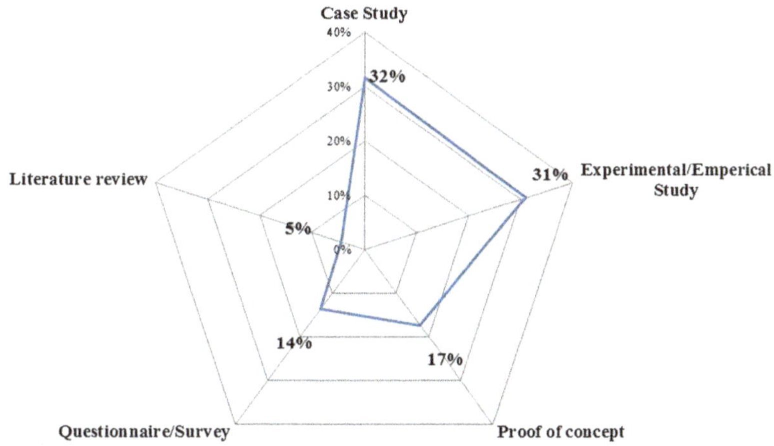

Figure 4 **Percentage of articles based on the research method.**

is divided into three subcategories of organization type including: (a) facility owner, (b) designer, and (c) contractor.

Figure 5 illustrates the number of articles within each improvement focus category. As shown, 69 articles (52%) have a principal focus on projects, while 27 articles (20%) have a principal focus on individuals in the construction industry. In addition, 19 articles (14%) and 12 articles (10%) have a principal focus on the AEC industry and the organization level, respectively.

Industry sector

In the construction industry various project types can benefit from AR technologies including: (1) municipal/infrastructure, e.g., evaluation of dynamic city models and an emission model for transportation (Aschwanden et al. 2012), (2) residential, e.g., virtual and augmented reality for designing and customizing mass housing (Duarte 2005), (3) building/commercial, e.g., visualizing high-rise building construction strategies (Russell et al. 2009), and virtual and augmented reality technologies for maintenance of exterior closures and interior finishes of walls and in the construction of buildings (Sampaio et al. 2012), (4) heavy/highway, e.g., developing virtual reality system for optimized simulation of road design data (Kang L. S. et al. 2010), and segmentation and recognition of highway assets using image-based 3D point clouds and semantic Texton forests (Golparvar-Fard et al. 2012), and (5) industrial, e.g., application areas for augmented reality in industrial construction (Shin and Dunston 2008).

Table 4 Reference article for research methods dimension

Research method	Reference
Case study	Dai et al. (2011), Turkan et al. (2012), Peña-Mora et al. (2012)
Experimental/empirical study	Wang and Dunston 2006, Wang et al. (2008), Behzadan and Kamat (2008)
Proof of concept	Roh et al. (2011), Yabuki et al. (2011), German et al. (2012)
Questionnaire/survey	Wang and Dunston (2006), Kuo et al. (2011), Chi et al. (2012)
Literature review	Malkawi et al. (2004), Wang et al. (2013)

Figure 6 presents the number of articles within each industry type category. As shown, 34 articles (26%) have a principal focus on building/commercial as an industry type for AR technology. Municipal/infrastructure, heavy/highway, industrial, and residential categories have 18 articles (14%), 14 articles (11%), 13 articles (10%) and 8 (6%) articles respectively. Thirty-six articles focus on multiple areas while these categories were not applicable for 10 articles. Table 5, presents a list of selected reference articles for each category of industry sector dimension.

Target audience

Due to the complexity of construction projects and the collaborative nature of the AEC industry, the application of AR systems has a wide range of target audiences. To classify these articles the following audiences were chosen partially based on Muramoto et al. (2008): (1) worker, e.g., machine operators and technicians, (2) design team, e.g., architects, interior and exterior designers, (3) schedule and budget professional, in particular referred to as project manager, (4) building systems engineer, e.g., structural, mechanical, and electrical engineers, (5) inspector, e.g., project safety officers (6) engineering student, (7) project end user, e.g., building occupants, office employees, (8) other stakeholder, e.g., clients, and building owners. If an article proposed a change in, or enhancement of, the work of one of these audiences, it was classified in that category.

As noted above, in this section instead of giving the number of articles with a "principal focus on" a category, we report the percentage of articles "including reference to" that category, since in this section each article may refer to more than one category. Figure 7 presents the

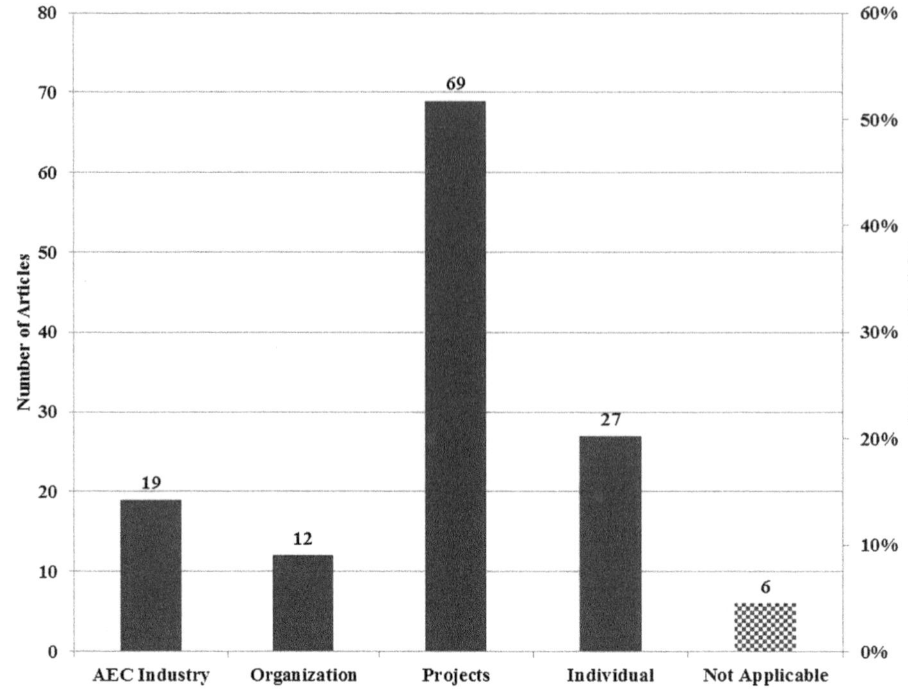

Figure 5 Number of articles by improvement level.

percentage of articles by target audience. The results indicate that the largest number of articles include reference to workers as the target audience. A list of selected reference articles for each category of target audience dimension is shown in Table 6.

Project phase

The life cycle of a construction project consists of a sequence of steps or project phases to be completed in order to reach project goals and objectives. These phases are defined by N. Dawood (2009) as: (1) initiation and

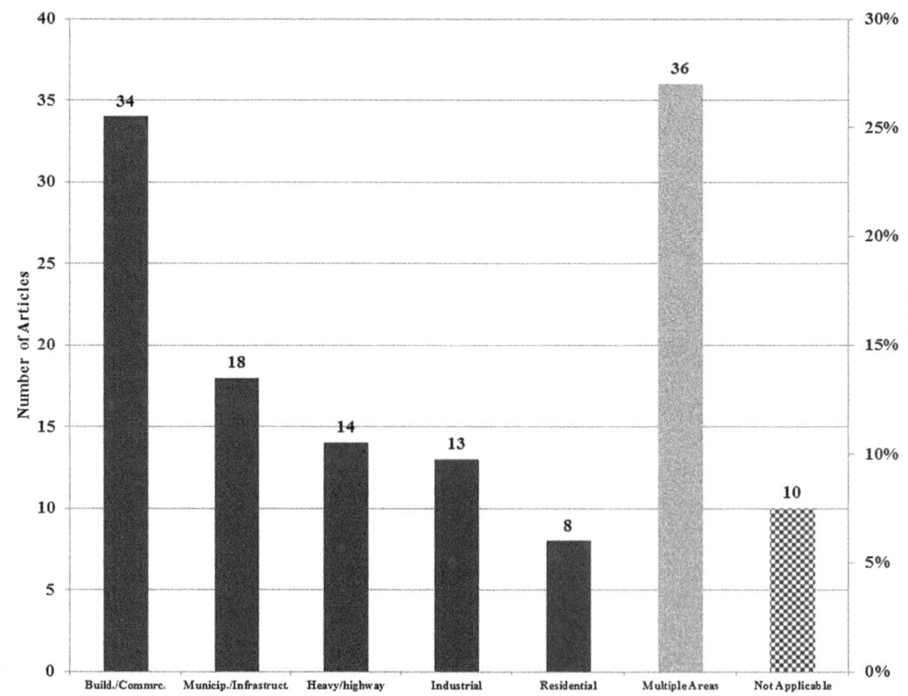

Figure 6 Number of articles by industry sector.

Table 5 Reference article for industry type dimension

Industry type	Reference
Building/commercial	Behzadan and Kamat (2008), Akhavian and Behzadan (2012)
Municipal/infrastructure	Dai et al. (2011), Fathi and Brilakis (2011)
Heavy/highway	Esch et al. (2009), Jordon et al. (2012)
Industrial	Shin and Dunston (2008), Shin and Dunston (2009)
Residential	Wang et al. (2012)

outline design, (2) design development, (3) [procurement], contract and pre-construction, (4) construction, and (5) maintenance. The number of the articles by project phase is depicted in Figure 8. A list of selected reference articles for each category of project phase dimension is shown in Table 7.

Figure 9 illustrates the number of articles for each project phase by year of publication. In this diagram articles with a focus on multiple phases are excluded (reducing the total to 98 articles). The highest number of articles in a single year is for the construction phase in the year 2012. The focus on the design phase of a project reached its highest number (5 articles) in the year 2008. Figures 8 and 9 show that the highest number of articles occur in the construction phase of a project for AR technologies and applications.

Stage of technology maturity

From a stage of technology maturity perspective, the articles are divided into five categories: (1) theory, (2) framework, (3) sub-system technical issues, e.g., investigation of tracking, positioning and orienting issues for AR-based technology for steel column inspection (Shin and Dunston 2010), (4) proposed system development, e.g., development of ARVISCOPE (AR animation scripting language) and ROVER a mobile computing framework for information modeling and simulation of construction operation (Behzadan et al. 2011), and (5) system application demonstration and production, e.g., application of D^{4}AR for construction progress monitoring (Golparvar-Fard et al. 2011a), application of AR Training System for training the operation of heavy construction equipment (Wang et al. 2007).

Figure 10 illustrates the number of articles within each stage-of-technology-maturity category. Results also show that only four articles (3%) have a principal focus on AR theory, while six articles (5%) have a focus on multiple areas (i.e., more than one of the previous stages); these multiple areas are typically a combination of application demonstration and proposed system development. A list of selected reference articles for each category of stage of technology maturity dimension is shown in Table 8.

Application area

Augmented reality technology has many applications in the AEC industry. We classify AR application areas in the AEC industry as follows: (1) visualization or simulation, (2) communication or collaboration, (3) information modeling, (4) information access or evaluation, (5) progress monitoring, (6) education or training, and (7) safety or inspection.

Figure 11 presents application areas for AR technologies in the AEC industry. As shown, 26 articles (20%) have a principal focus on visualization and simulation as an application area for AR technology. Thirteen articles focus on multiple application areas, while these subcategories were not applicable for 8 articles. A list of selected reference articles for each category of AR application area dimension is shown in Table 9.

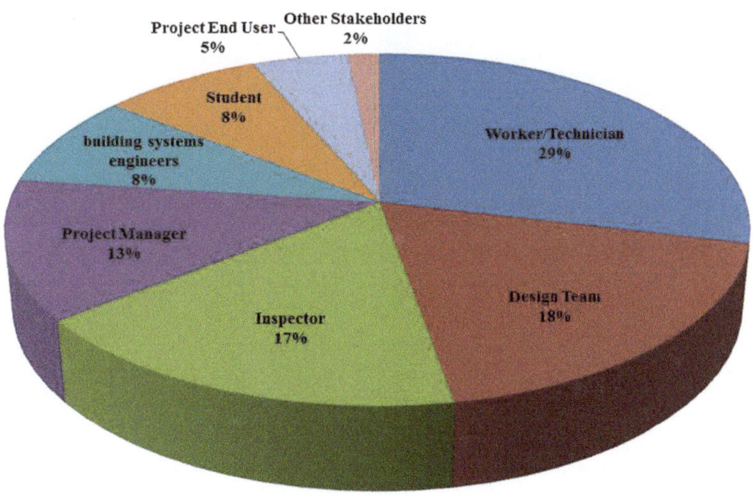

Figure 7 Percentage of articles by target audiences.

Table 6 Reference article for target audience dimension

Target audience	Reference
Worker/technician	Lucas and Thabet (2008), Chi et al. (2012)
Design team	Leicht et al. (2009), Gu et al. (2011)
Inspector	Shin and Dunston (2010), Zhu and Brilakis (2010)
Project manager	Golparvar-Fard et al. (2011b), Roh et al. (2011)
Building system engineers	Lee and Akin (2011), Shin and Dunston (2008)
Student	Wang et al. (2007)
Project end user	Jang et al. (2008), Wang et al. (2012)

Comparison role

Construction participants use augmented reality technologies to compare different statuses of a project. Articles that "make reference" to the comparison role of AR technologies are divided in two categories; (1) comparison modes: (a) reality versus reality, e.g., comparing two construction site 360 degree panoramas at two different times for virtual reality documentation of Inuvik Super School (Waugh et al. 2012), (b) model versus reality, e.g., integrating sequential as-built and as-planned representation with D4AR tools (Golparvar-Fard et al. 2011a), and (c) model versus model, e.g., the application of Experimental Virtual Environment (EVE) (Savioja et al. 2003); and (2) comparison purpose, e.g., progress monitoring, defect detection, validating the model, updating the model, and evaluating the model.

Sixty-two articles (47% of articles) have reference to the comparison role of augmented reality technologies in construction industry. For these 62 articles only, Figure 12 illustrates the number and the percentage of articles within each comparison mode category. A list of selected reference articles for each category of comparison mode dimension is shown in Table 10.

Figure 13 depicts the percentage of articles within each comparison purpose category for these 62 articles. As shown, 32% refer to comparison for progress monitoring, 28% for defect detection, 16% for evaluating the model, 13% for updating the model, and 11% for validating the model. A list of selected reference articles for each category of comparison purpose dimension is shown in Table 11.

Technology

Augmented reality technology, which typically layers virtual information on a real scene, utilizes different hardware (personal computers (PC), laptops, head mounted displays (HMD), GPS, data gloves, smart boards, etc.) and software (AutoCAD, Photoshop, AC3D, 3D Studio, building information model (BIM), etc.). From a technology perspective articles are classified into three categories: (1) user experience: (a) immersive or (b) non-immersive, i.e., desktop-based; (2) device: (a) mobile, (b) stationary or non-mobile; (3) delivery: (a) web-based, (b) standalone.

Devices such as HMD and data-gloves create immersive AR systems, in which users feel immersed in a

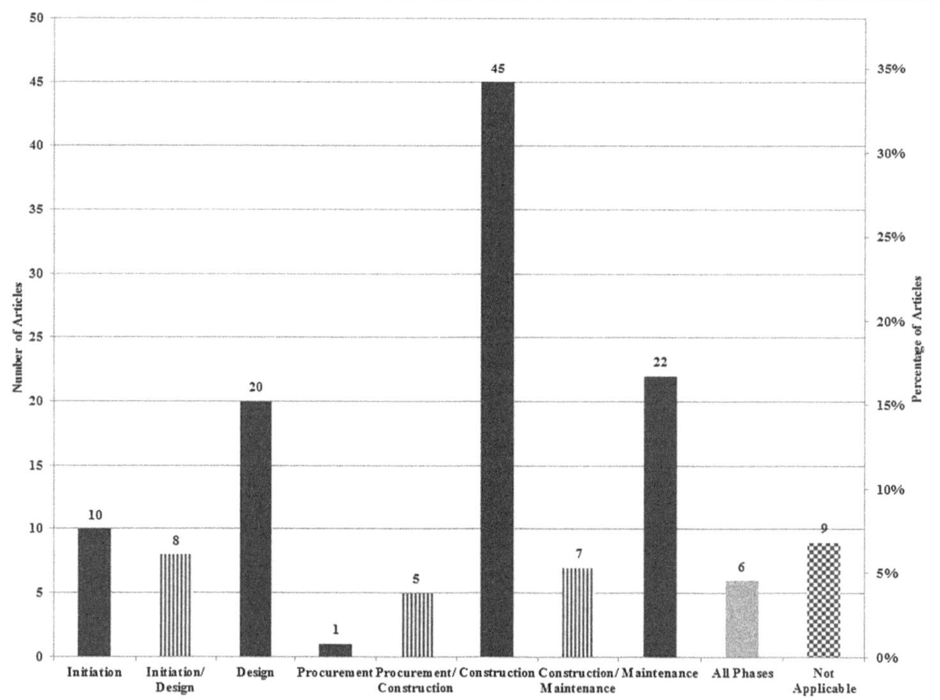

Figure 8 Number of articles by project phase.

Table 7 Reference article for project phase dimension

Project phase	Reference
Initiation	Dunston and Wang (2005), Gu et al. (2011)
Design	Kang et al. (2010), Germani et al. (2012)
Procurement (procure.)	Ajam et al. (2010)
Construction (constr.)	Behzadan and Kamat (2008), Golparvar-Fard et al. (2011b)
Maintenance	Zhu and Brilakis (2010), Victores et al. (2011)

virtual environment just as they usually feel in a real environment. Due to the improving performance of handheld devices and recent solutions to technical difficulties such as tracking, there is an opportunity for augmented reality systems to become portable, as well there is a growing interest in the use of mobile AR applications. Web-based augmented reality technologies can deliver project information to remote locations and are very useful to manage virtual projects around the globe. Finally, due to the wide range of AR applications, these technologies can be used both on-site for progress monitoring and in an office (not-on-site) for design control.

From the user experience perspective, 32 articles have a principal focus on immersive AR technologies, 76 articles (57%) have a principal focus on non-immersive or desktop-based AR technologies, while 25 articles were not applicable. Figure 14 presents the number of articles with immersive and non-immersive technologies as a principal focus by year.

Figure 15 presents the number of articles within the device category that had a principal focus on mobile and non-mobile AR technologies in the AEC industry. The diagram implies an increasing trend in mobile AR technologies in AEC industry. Of the selected articles, one was published in 2000 that discussed mobile AR technology; while 7 articles (5%) focus on mobile AR technologies in both year 2011 and year 2012; 41 articles were not applicable to this category.

Figure 16 depicts the number of articles within the delivery category that had a principal focus on web-based and on standalone AR technologies in the AEC industry. Fifty-two articles (40%) were not applicable to this category. A list of selected reference articles for each category of technology dimension is shown in Table 12.

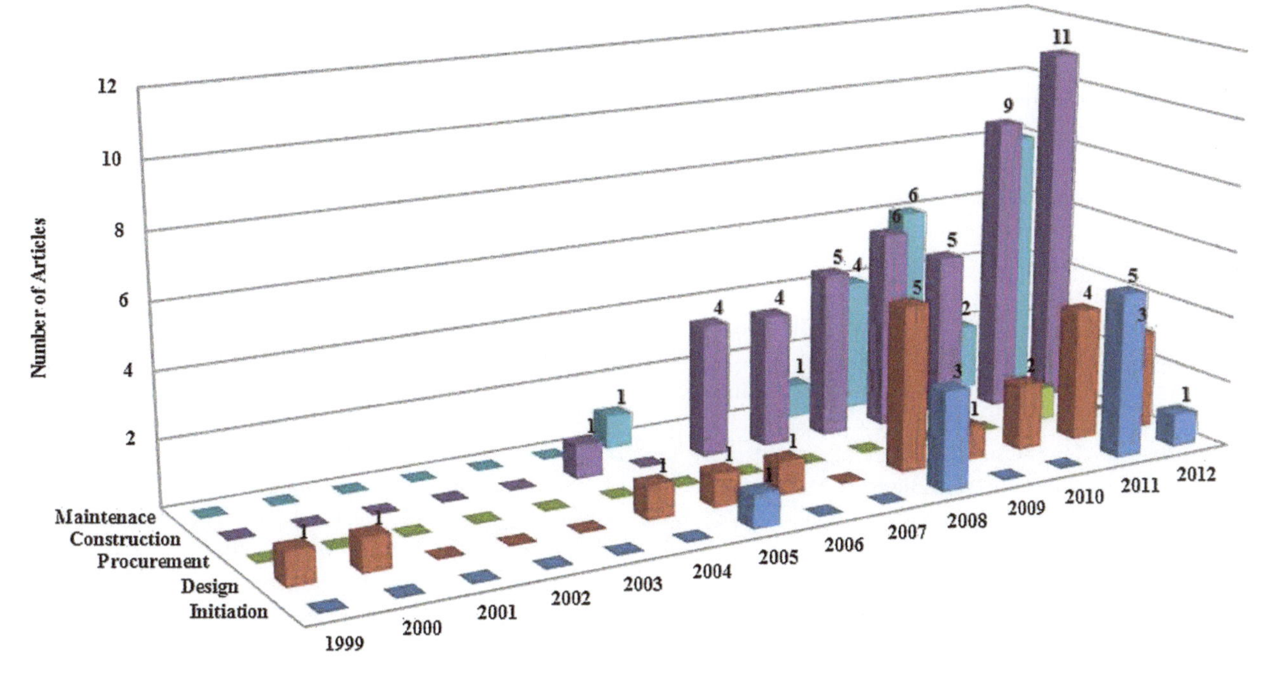

	1999	2000	2001	2002	2003	2004	2005	2006	2007	2008	2009	2010	2011	2012
Initiation							1			3			5	1
Design	1	1				1	1	1		5	1	2	4	3
Procurement													1	
Construction						1		4	4	5	6	5	9	11
Maintenace							1			1	4	6	2	8

Figure 9 Number of articles by project phase and year (articles spanning multiple phases are excluded).

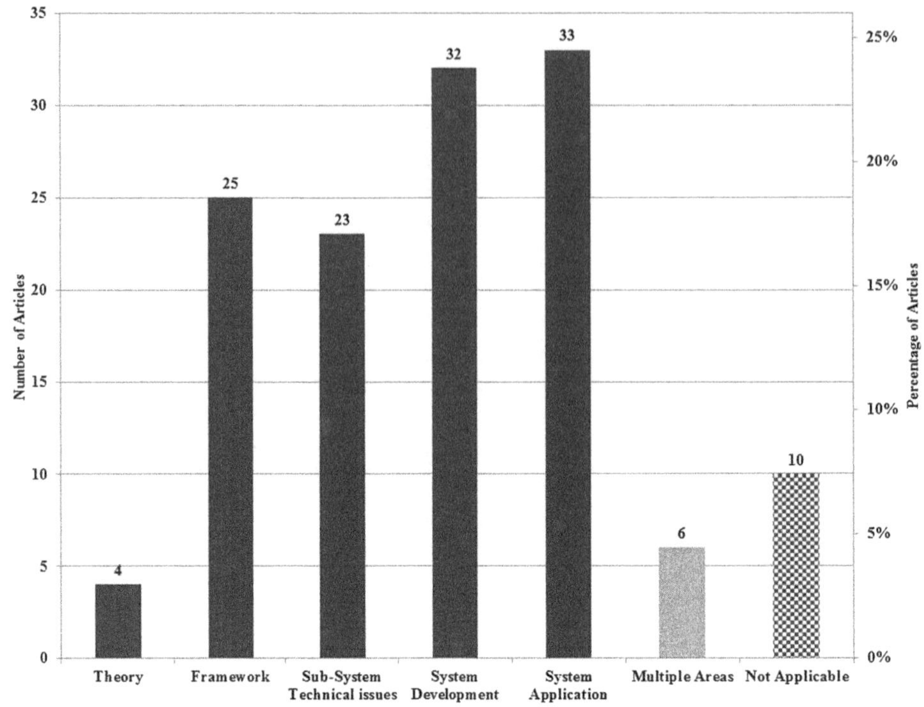

Figure 10 Number of articles by stage of technology maturity.

Location

Augmented reality technologies can be applied in different locations during a construction project. From a location perspective, AR technologies can be classified in two categories: (a) field, e.g., using robot-aided tunnel inspection and maintenance system on construction site (Victores et al. 2011), (b) home office, e.g., virtual environments for synchronous and remote collaborative design (Germani et al. 2012).

Figure 17 depicts the number of articles within the technology category that had a principal focus of on-site and on not-on-site. The diagram implies an increasing trend in the on-site application of AR technologies in construction projects. The highest number of articles in a single year is for the on-site technologies in the year 2012. Twenty-two articles (17%) were not applicable to this category. A list of selected reference articles for each category of location dimension is shown in Table 13.

Table 8 Reference article for stage of technology maturity dimension

Stage of technology maturity	Reference
Theory	Dunston and Wang (2011), Pradhan et al. (2012)
Framework	Dunston and Wang (2011), (Wang et al. 2010)
Sub-system technical issues	Zhu and Brilakis (2010), Huang et al. (2012)
System development	Liang et al. (2011), (Golparvar-Fard et al. 2011a)
System application	Jang et al. (2008), Roh et al. (2011)

Review

Discussion

Figure 18 shows the total number of articles in defined dimensions and categories, in which increasing trends over the period are indicated by an up arrow and dominant categories are indicated by bold font.

The following results are concluded for the categories defined in this research.

- Journals: *Automation in construction* has the highest overall number of articles among the journals. The maximum number of AR technology articles published in these eight journals in a single year, occurred both in 2011 and 2012. Eighty-three percent of the articles were published in most recent five years.
- First authors: *USA*, with more than half of articles, is the dominant residence of the first authors.
- Research methodology: *Case studies* and *experimental studies* with 32% and 31% respectively, are the most frequent research method among selected articles.
- Improvement focus: The majority of the articles focus on *projects* rather than on the AEC industry, organization or individual level.
- Industry sector: *building/commercial* with 35% have the highest number of articles, whereas, *residential* have the least number of articles.

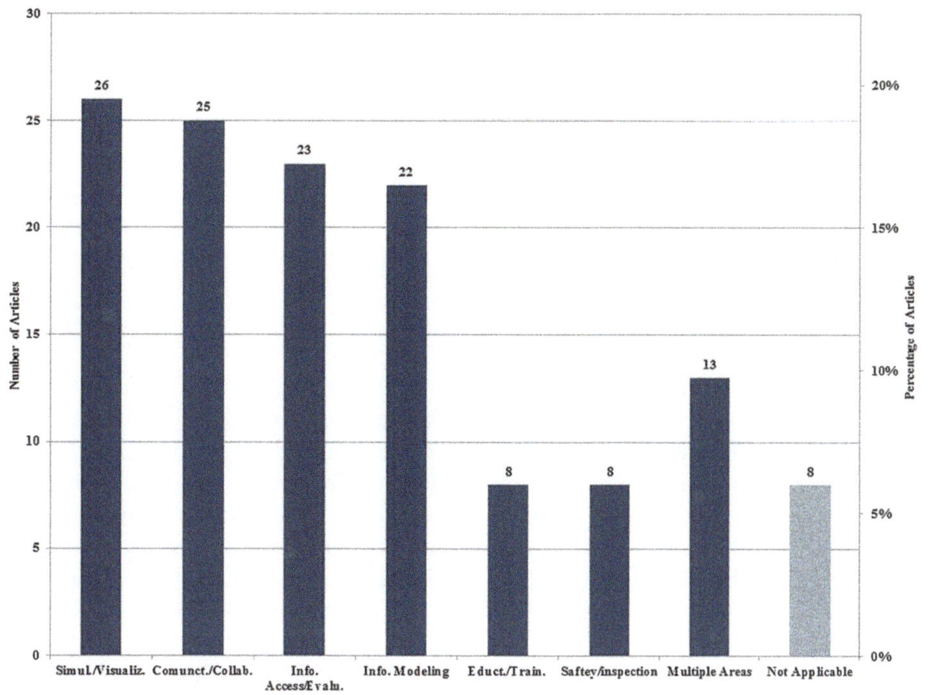

Figure 11 Application areas for AR technology.

- Target audience: The most frequent focus is the *workers* (e.g., machine operators and technicians), whereas the least focus is on project end users.
- Project phase: The most frequent focus is the *construction phase* with the maintenance phase being next with approximately half as many articles. Twenty-six articles cover two phases (e.g., initiation/ design) or all phases. Procurement phase shows a lack of focus in the area of AR systems.
- Stage of technology maturity: The most number of articles focus on AR *system application* rather than system development or sub-system technical issues.
- Application areas: Approximately half of the articles had a principal focus on *visualization/simulation* or *communication/collaboration*, and just a few articles focus on education/training and safety/inspection.

Table 9 Reference article for AR application area dimension

Application area	Reference
Simulation/visualization	Kang et al. (2010), Liang et al. (2011)
Communication/collaboration	Hammad et al. (2009), Gu et al. (2011)
Information access/evaluation	Zhu and Brilakis (2010), Khoury and Kamat (2009)
Information modeling	Behzadan and Kamat (2011), Styliadis (2008)
Education/training	Wang et al. (2007), Jardón et al. (2012)
Safety/inspection	Zhu et al. (2012), Li and Liu (2012)

- Comparison role: 47% of the articles made reference to the comparison role of augmented reality technologies for comparing different statuses of projects. More than half of those articles focused on comparing a *model with a reality* to monitor progress and detect construction defects. Comparing model vs. model captured the least attention in this area. In addition, majority (%60) of the comparison

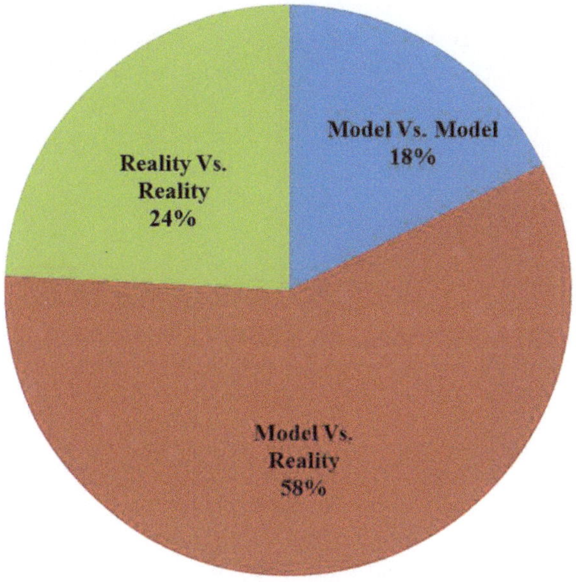

Figure 12 Number and percentage of articles by comparison modes.

Table 10 Reference article for comparison mode dimension

Comparison mode	Reference
Model vs. Model	Wang et al. (2008), Gu et al. (2011)
Model vs. Reality	Shin and Dunston (2009), Behzadan and Kamat (2011)
Reality vs. Reality	Zhu and Brilakis (2010), Brilakis et al. (2011)

articles focus on field audience (for progress monitoring and defect detection), while less than half of the comparison articles focus on home office audience (for model improvement).

- Hardware system: From the user-experience perspective most of the articles discussed *non-immersive technologies*. From the delivery perspective, most of the articles discussed *standalone technologies*, and from device perspective, most of the articles focus on *non-mobile* technologies. A spike in the number of articles focusing on mobile technologies occurred in 2011.
- Location: The most frequent focus is AR technologies which can be applied in the *field* (rather than in the home-office). The field category included almost 80% of those articles which referred to a location.

Future trends

Table 14 provides a list of the categories for which there was a significant and consistently-increasing trend in the most recent five years. There were no categories for which the number of articles was consistently-decreasing over the 14 year period.

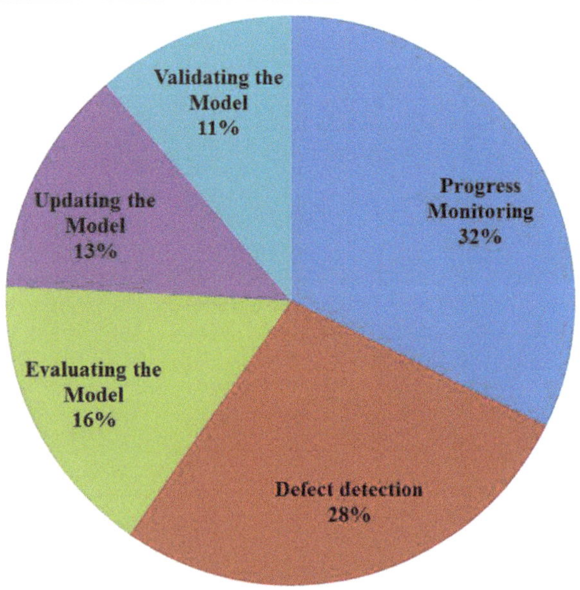

Figure 13 Number and percentage of articles by comparison purpose.

Table 11 Reference article for comparison purpose dimension

Comparison purpose	Reference
Progress monitoring	Bohn and Jochen (2009), Golparvar-Fard et al. (2011a)
Defect detection	Zhu and Brilakis (2010), Shin and Dunston (2010)
Evaluating the model	Gül et al. (2008)
Updating the model	Gu et al. (2011)
Validating the model	Isikdag and Underwood (2010)

AR technologies provide proven benefits especially in the areas of visualization/simulation and communication/collaboration, however these benefits are not yet widely adopted by AEC industry participants nor have they been incorporated into industry-wide workflow processes. As a result, industry participants choose to pilot (i.e., system development and application) AR technologies on a few projects rather than adopting or piloting the technology across their organization.

Building/commercial projects provide a good test bed for visualization and communication of different perspectives of a project, since these projects typically entail more complexity and more need for integration than an infrastructure, heavy/highway, or residential projects. However, the trend in the most recent five years shows that heavy/highway projects are getting an area of focus for new AR technologies. We also predict that use on industrial projects will grow rapidly as technologies are improved and confidence is gained.

We predict expansion of AR technologies from a principal role in the construction and maintenance phases to other phases (especially the design and procurement phases) as the ability to compare virtual models with previous virtual models (and realities with previous realities) to monitor project progress and detect construction defections, rather than the narrow focus of comparing a current construction phase reality with a final design phase model.

The uniform distribution of target audiences among the design team, the project management team, and on-site personnel reflects integration being the essential purpose of AR technologies.

We predict continued growth in the use of internet and web-based devices to enhance integration of perspectives. Collaborative, ubiquitous, and internet-based AR systems enable users to update and synchronize the information from a remote location. Cloud computing technologies could help next generation construction professionals to access massive amount of

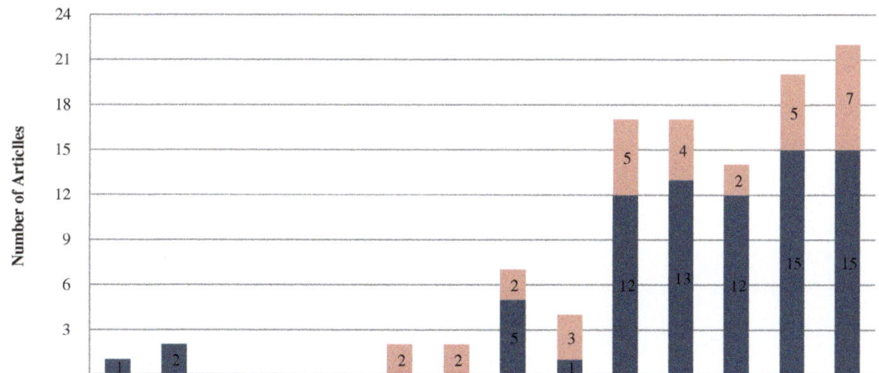

	1999	2000	2001	2002	2003	2004	2005	2006	2007	2008	2009	2010	2011	2012
Immersive						2	2	2	3	5	4	2	5	7
Non-immersive	1	2						5	1	12	13	12	15	15

Figure 14 Immersive and non-immersive AR technology by year.

field information such as BIM rapidly and conveniently (Chi et al. 2013).

We also predict continued growth in the use of mobile and portable multi-user AR devices to display (and capture) models and realities. The next generation mobile technologies are likely to have natural user interface (Wang et al. 2013), which would be controlled by human movement and gestures, and makes it easy for field workers to use AR systems on construction sites. It is also predicted that next generation mobile AR systems would have context-aware and location-aware applications. We also speculate that the cost of immersive hardware is, and will continue to be, an impediment to its widespread use.

Conclusion

A structured methodology was used to identify 133 articles on the topic of augmented reality from eight prominent AEC industry journals. The first article was published in 1999; a significant increase in the number of articles on this topic occurred during the year 2008. In addition to statistics on the counts of articles by year and the first author's country of residence, ten interpreted dimensions were developed for classification of these articles. Literature show field workers and project managers have high interests in using non-immersive and desk-top standalone (individual) AR technologies during construction phase of a project to compare as-planned versus as-built statuses to monitor progress and defect detection. Whereas, it is predicted that future trend, is more toward using collaborative and internet based mobile AR systems which have applications not only in construction phase, but also in procurement and maintenance phases of a project. Due to various benefits of AR technology for construction industry, the application of AR systems for initiation and procurement

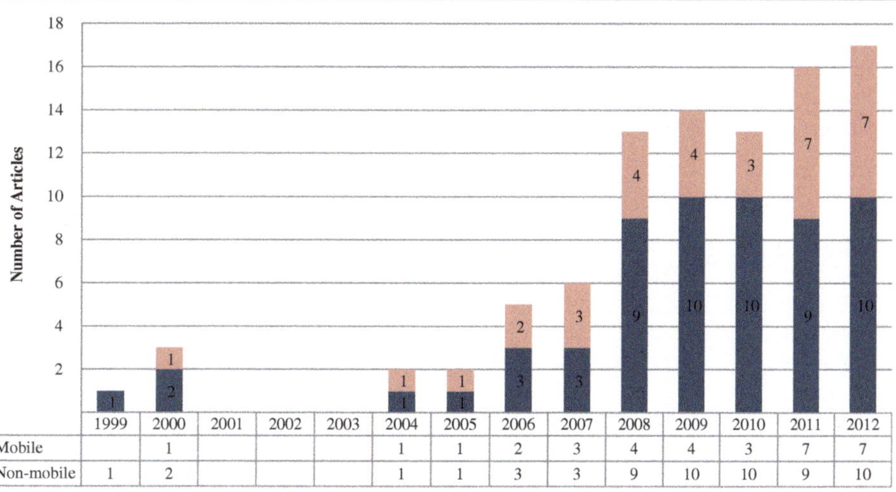

	1999	2000	2001	2002	2003	2004	2005	2006	2007	2008	2009	2010	2011	2012
Mobile		1				1	1	2	3	4	4	3	7	7
Non-mobile	1	2				1	1	3	3	9	10	10	9	10

Figure 15 Number of articles for mobile and non-mobile AR technology by year.

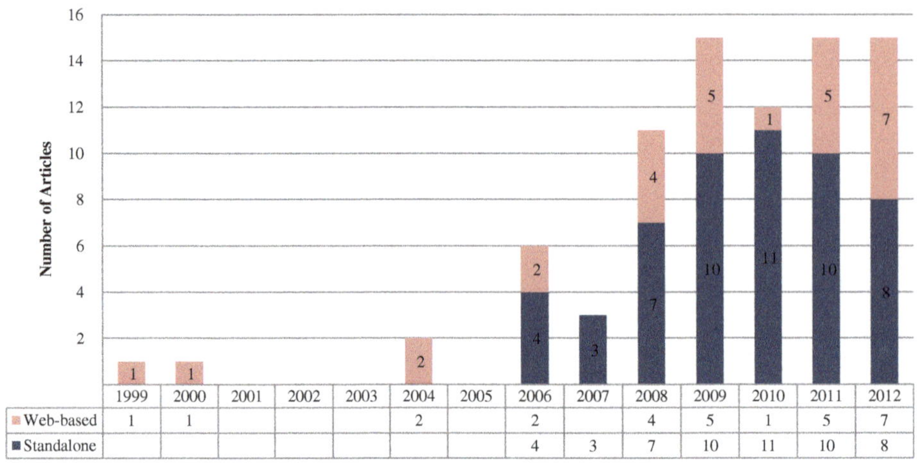

Figure 16 Number of articles for web-based and standalone AR technology by year.

phase of a project to compare model vs. model and reality vs. reality is recommended. Moreover, lightweight mobile and immersive AR systems are also recommended for field personnel due to dynamic environment of construction fields.

Based on this review, our recommendation for researchers in the area of AR technology is that the current trial systems should have narrow applications and there is an essential need for more comprehensive systems. There is an opportunity for more research on the application of AR systems during the procurement phase of construction projects, since literature shows a lack of research in these phases of project. In addition, the assessment of success of developed systems should be validated by researcher and practitioner from academy and industry. From a theory and framework perspective, integration of multiple projects (within an organization) and integration of multiple organizations (within the industry) could be a considered for future research in the area of AR systems. Integration might be easier in less complex types of work, i.e., residential projects. Comparisons are fundamental to AR system within the AEC industry to be able to monitor projects and defect detections, therefore, our testing of five comparison purposes (progress monitoring, defect detection, evaluating the model, updating the model, validating the model) need to be further investigated.

We assume that construction industry practitioners would assess an AR system based on the system contents, features, and value. From the content perspective they would seek current (possibly real-time) information as well as a historical record that may be integrated with traditional project information (e.g., BIM). From a feature perspective, they would seek a user friendly interface (possibly internet-based) that can be integrated into their content workflow process and that facilitate the comparison of project statuses over time. From the value perspective, they would seek an affordable cost (initial and ongoing) for which the payback period is short. We assume the benefits of AR contribute to this payback are virtual site visits, defect detection, pre-empting dispute resolution, photographic as-built, and training of personnel. Currently, most

Table 12 Reference article for technology dimension

Technology	Reference
Immersive	Behzadan and Kamat (2011)
Desktop-based	Golparvar-Fard et al. (2009), Wang and Dunston (2008)
Mobile	Khoury and Kamat (2009), Behzadan and Kamat (2011)
Stationary	Gül et al. (2008), Bohn and Jochen (2009)
Web-based	Muramoto et al. (2008)
Standalone	Styliadis (2008), Lee and Akin (2011)

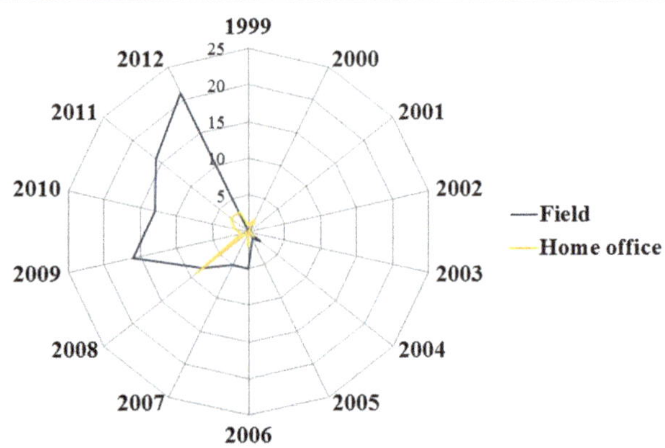

Figure 17 Number of articles for field and home-office AR technology by year.

Table 13 Reference article for location dimension

Location	Reference
Field	Kamat et al. (2011), Golparvar-Fard et al. (2011a), Lee and Akin (2011)
Home-office	Muramoto et al. (2008)

Table 14 Significant trends

Dimension	Category	2008	2012	Factor
Journal	AIE	0	7	∞
Industry sector	Heavy/highway	1	5	5.0
Project phase	Maintenance	1	8	8.0
Delivery	Mobile	4	7	1.8
Location	Field	8	21	2.6
Total number of articles	All categories	20	26	1.3

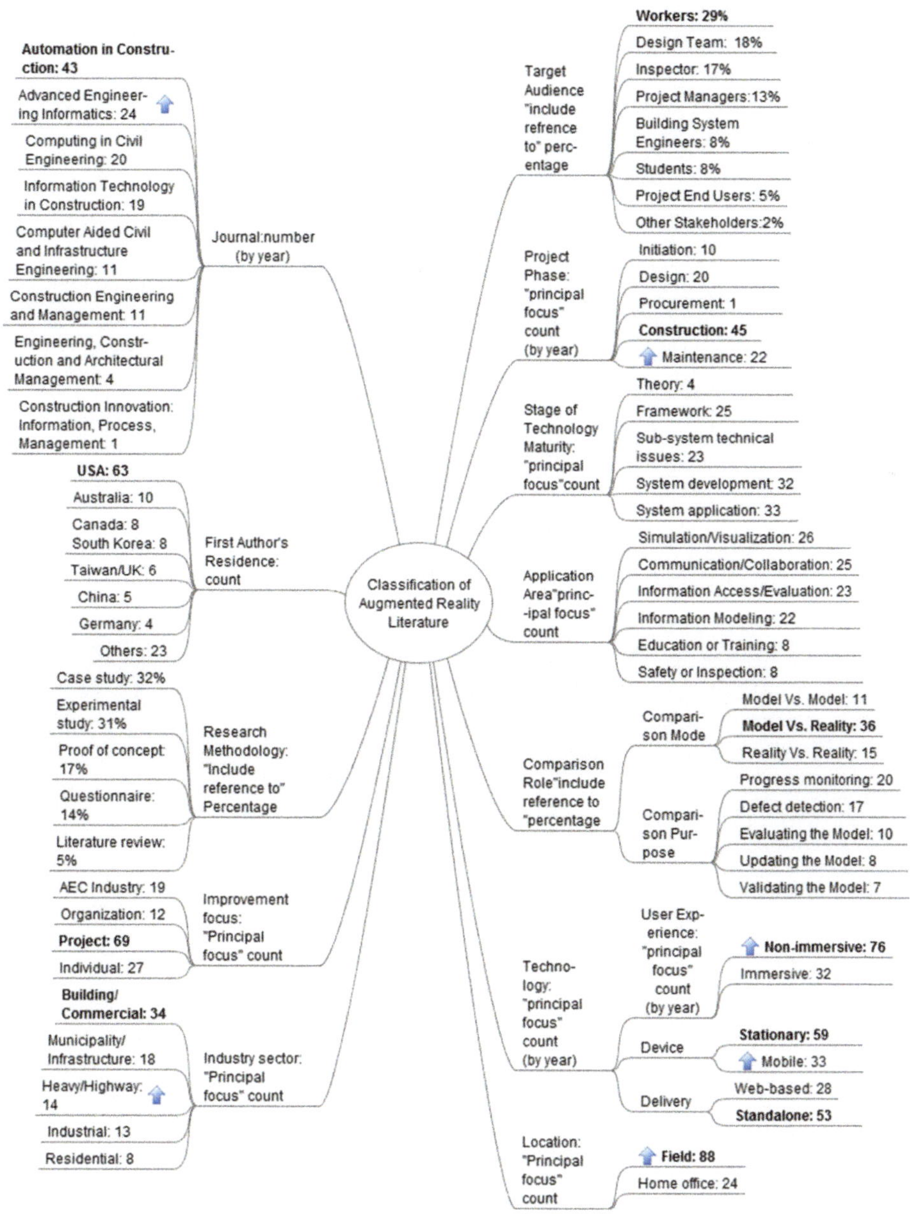

Figure 18 Literature review summary.

of systems found in the literature are trial/demonstration, hence they are developed for specific purposes they do not have all of the above criteria, however some new systems offers some valuable feature and may provide a competitive advantages. As the technology is rapidly evolving, it is recommended to the construction participants to monitor this developing area closely in order to get the latest update.

Competing interests

The authors declare that they have no competing interests.

Authors' contributions

Both authors contributed extensively to the work presented in this paper. SR searched the databases, selected the articles, reviewed and analyzed the literature, and prepared the manuscript. LW defined dimensions and categories, supervised the overall review, and edited the manuscript. Both authors read and approved the final manuscript.

References

Ajam, M, Alshawi, M, & Mezher, T. (2010). Augmented process model for e-tendering: towards integrating object models with document management systems. *Journal of Automation in Construction, Elsevier, 19*, 762–778.

Akhavian, R, & Behzadan, AH. (2012). An integrated data collection and analysis framework for remote monitoring and planning of construction operations. *Journal of Advanced Engineering Informatics, Elsevier, 26*, 749–761.

Aschwanden, GDPA, Wullschleger, T, Müller, H, & Schmitt, G. (2012). Agent based evaluation of dynamic city models: a combination of human decision processes and an emission model for transportation based on acceleration and instantaneous speed. *Journal of Automation in Construction, Elsevier, 22*, 81–89.

Bae, H, Golparvar-Fard, M, & White, J. (2012). *Enhanced HD4AR (Hybrid 4-Dimensional Augmented Reality) for Ubiquitous Context-aware AEC/FM Applications, CONVR 2012, 12th International Conference on Construction Applications of Virtual Reality*. Taipei, Taiwan: National Taiwan University.

Bae, V, Golparvar-Fard, M, & White, J. (2013). High-precision vision-based mobile augmented reality system for context-aware architectural, engineering, construction and facility management (AEC/FM) applications. *Visualization in Engineering*, 1–3.

Behzadan, AH, & Kamat, VR. (2008). Automated generation of operations level construction animations in outdoor augmented reality. *ASCE Journal of Computing in Civil Engineering, ASCE, 23*, 405–417.

Behzadan, AH, & Kamat, VR. (2011). Integrated information modeling and visual simulation of engineering operations using dynamic augmented reality scene graphs. *Journal of Information Technology in Construction, ITcon, 16*, 259–278.

Bohn, JS, & Jochen, T. (2009). Benefits and barriers of construction project monitoring using high-resolution automated cameras. *ASCE Journal of Construction Engineering and Management, ASCE, 25*, 66–74.

Brilakis, I, Fathi, H, & Rashidi, A. (2011). Progressive 3D reconstruction of infrastructure with videogrammetry. *Journal of Automation in Construction, Elsevier, 20*, 884–895.

Chi, HL, Chen, YC, Kang, SC, & Hsieh, SH. (2012). Development of user interface for tele-operated cranes. *Journal of Advanced Engineering Informatics, Elsevier, 26*, 641–652.

Chi, HL, Kang, SC, & Wang, X. (2013). Research trends and opportunities of augmented reality applications in architecture, engineering, and construction. *Journal of Automation in Construction, Elsevier, 33*, 116–122.

Cohen, L, Manion, L, & Morrison, K. (2007). *Research methods in education* (Sixth Editionth ed.). Abingdon, Oxon OX14 4RN: Routledge-Taylor and Francis e-Library. ISBN 13: 978-0-203-02905-3.

Dai, F, Lu, M, & Kamat, VR. (2011). Analytical approach to augmenting site photos with 3D graphics of underground infrastructure in construction engineering applications. *ASCE Journal of Computing in Civil Engineering, ASCE, 25*, 66–74.

Dawood, N. (2009). VR-Roadmap: a vision for 2030 in the built environment. *Journal of Information Technology in Construction, ITcon, 14*, 489–506.

Dong, S, & Kamat, VR. (2013). SMART: scalable and modular augmented reality template for rapid development of engineering visualization applications. *Visualization in Engineering*, 1–1.

Duarte, PJ. (2005). A discursive grammar for customizing mass housing: the case of Siza's houses at Malagueira. *Journal of Automation in Construction, Elsevier, 14*, 265–275.

Dunston, PS, & Wang, X. (2005). Mixed reality-based visualization interfaces for architecture, engineering, and construction industry. *ASCE Journal of Construction Engineering and Management, ASCE, 131*(12), 1301–1309.

Dunston, PS, & Wang, X. (2011). An iterative methodology for mapping mixed reality technologies to AEC operations. *Journal of Information Technology in Construction, ITcon, 16*, 509–528.

Esch, G, Scott, MH, & Zhang, E. (2009). Graphical 3D visualization of highway bridge ratings. *ASCE Journal of Computing in Civil Engineering, ASCE, 23*, 355–362.

Fathi, H, & Brilakis, I. (2011). Automated sparse 3D point cloud generation of infrastructure using its distinctive visual features. *Journal of Advanced Engineering Informatics,.Elsevier, 25*, 760–770.

German, S, Brilakis, I, & DesRoches, R. (2012). Rapid entropy-based detection and properties measurement of concrete spalling with machine vision for post-earthquake safety assessments. *Journal of Advanced Engineering Informatics, Elsevier, 26*, 846–858.

Germani, M, Mengoni, M, & Peruzzini, M. (2012). An approach to assessing virtual environments for synchronous and remote collaborative design. *Journal of Advanced Engineering Informatics, Elsevier, 26*, 793–813.

Golparvar-Fard, M, Peña-Mora, F, & Savarese, S. (2009). D^4AR – A 4-dimensional augmented reality model for automating construction progress monitoring, data collection, processing and communication. *Journal of Information Technology in Construction, ITcon, 14*, 129–153.

Golparvar-Fard, M, Bohnb, J, Teizerb, T, Savaresec, S, & Peña-Mora, F. (2011a). Evaluation of image-based modeling and laser scanning accuracy for emerging automated performance monitoring. *Journal of Automation in Construction, Elsevier, 20*, 1143–1155.

Golparvar-Fard, M, Peña-Mora, F, & Savarese, S. (2011b). Integrated sequential as-built and as-planned representation with D^4AR tools in support of decision-making tasks in the AEC/FM industry. *ASCE Journal of Construction Engineering and Management, ASCE, 137*(12), 1099–1116.

Golparvar-Fard, M, Balali, V, & De la Garza, JM. (2012). Segmentation and recognition of highway assets using image-based 3D point clouds and semantic Texton forests. *ASCE Journal of Computing in Civil Engineering*, 1943-5487. doi:10.1061/(ASCE)CP.1943-5487.0000283.

Gu, N, Kim, MJ, & Maher, ML. (2011). Technological advancements in synchronous collaboration: The effect of 3D virtual worlds and tangible user interfaces on architectural design. *Journal of Automation in Construction, Elsevier, 20*, 270–278.

Gül, LF, Gu, N, & Williams, A. (2008). Virtual worlds as a constructivist learning platform: evaluations of 3d virtual worlds on design teaching and learning. *Journal of Information Technology in Construction, ITcon, 13*, 578–593.

Hammad, A, Wang, H, & Mudu, SP. (2009). Distributed augmented reality for visualizing collaborative construction tasks. *ASCE Journal of Computing in Civil Engineering, ASCE, 23*, 418–427.

Huang, S, Yang, Y, & Chu, C. (2012). Human-centric design personalization of 3D glasses frame in marker less augmented reality. *Journal of Advanced Engineering Informatics, Elsevier, 26*, 35–45.

Isikdag, U, & Underwood, J. (2010). Two design patterns for facilitating building information model-based synchronous collaboration. *Journal of Automation in Construction, Elsevier, 19*, 544–553.

Jang, W, Healy, W, & Skibniewski, M. (2008). Wireless sensor networks as part of a web-based building environmental monitoring system. *Journal of Automation in Construction, Elsevier, 17*, 729–736.

Jardón, A, Victores, JG, Martínez, S, & Balaguer, C. (2012). Experience acquisition simulator for operating microtuneling boring machines. *Journal of Automation in Construction, Elsevier, 23*, 33–46.

Kamat, VR, Martinez, JC, Fischer, M, Golparvar-Fard, M, Peña-Mora, F, & Savarese, S. (2011). Research in visualization techniques for field construction. *ASCE Journal of Construction Engineering and Management, ASCE, 137*(10), 853–862.

Kang, LS, Moon, HS, Dawood, N, & Kang, MS. (2010). Development of methodology and virtual system for optimised simulation of road design data. *Journal of Automation in Construction, Elsevier, 19*, 1000–1015.

Khoury, HM, & Kamat, VR. (2009). Evaluation of position tracking technologies for user localization in indoor construction environments. *Journal of Automation in Construction, Elsevier, 18*, 444–457.

Kuo, CH, Tsai, MT, & Kang, SC. (2011). A framework of information visualization for multi-system construction. *Journal of Automation in Construction, Elsevier, 20*, 247–262.

Lee, S, & Akin, O. (2011). Augmented reality-based computational fieldwork support for equipment operations and maintenance. *Journal of Automation in Construction, Elsevier, 20*, 338–352.

Leicht, RM, Messner, JI, & Anumba, CJ. (2009). A framework for using interactive workspaces for effective collaboration. *Journal of Information Technology in Construction, ITcon, 14*, 180–203.

Li, Y, & Liu, C. (2012). Integrating field data and 3D simulation for tower crane activity monitoring and alarming. *Journal of Automation in Construction, Elsevier, 27*, 111–119.

Liang, X, Lu, M, & Zhang, JP. (2011). On-site visualization of building component erection enabled by integration of four-dimensional modeling and automated surveying. *Journal of Automation in Construction, Elsevier, 20*, 236–246.

Lucas, J, & Thabet, W. (2008). Implementation and evaluation of a VR task-based training tool for conveyor belt safety training. *Journal of Information Technology in Construction, ITcon, 13*, 637–659.

Malkawi, AM. (2004). Developments in environmental performance simulation. *Journal of Automation in Construction, Elsevier, 13*, 437–445.

Muramoto, K, Jemtrud, M, Kumar, S, Balakrishnan, B, & Wiley, D. (2008). Emerging technology in tele-collaborative design studio between Pennsylvania State University and Carleton University. *Journal of Information Technology in Construction, ITcon, 13*, 660–673.

Park, C, Lee, DY, Kwon, OS, & Wang, X. (2013). A framework for proactive construction defect management using BIM, augmented reality and ontology-based data collection template. *Journal of Automation in Construction, Elsevier, 33*, 61–71.

Peña-Mora, F, Thomas, JK, Golparvar-Fard, M, & Aziz, Z. (2012). Supporting civil engineers during disaster response and recovery using a Segway mobile workstation Chariot. *ASCE Journal of Computing in Civil Engineering, ASCE, 26*, 448–455.

Rankouhi, S, & Waugh, LM. (2012). *Augmented reality technologies for AEC projects: a literature review* (CONVR 2012, 12th International Conference on Construction Applications of Virtual Reality). Taipei, Taiwan: National Taiwan University.

Roh, S, Aziz, Z, & Peña-Mora, F. (2011). An object-based 3D walk-through model for interior construction progress monitoring. *Journal of Automation in Construction, Elsevier, 20*, 66–75.

Russell, A, Staub-French, S, Tran, N, & Wong, W. (2009). Visualizing high-rise building construction strategies using linear scheduling and 4D CAD. *Journal of Automation in Construction, Elsevier, 18*, 219–236.

Sampaio, AZ, Gomes, AR, & Santos, JP. (2012). Management of building supported on virtual interactive models: construction planning and preventive maintenance. *Journal of Information Technology in Construction, ITcon, 17*, 121–133.

Savioja, L, Mantere, M, Olli, L, Äyräväinen, S, Gröhn, M, & Iso-Aho, J. (2003). Utilizing virtual environments in construction projects. *Journal of Information Technology in Construction, ITcon, 8*, 85–99.

Shin, DH, & Dunston, PS. (2008). Identification of application areas for Augmented Reality in industrial construction based on technology suitability. *Journal of Automation in Construction, Elsevier, 17*, 882–894.

Shin, DH, & Dunston, PS. (2009). Evaluation of augmented reality in steel column inspection. *Journal of Automation in Construction, Elsevier, 18*, 118–129.

Shin, DH, & Dunston, PS. (2010). Technology development needs for advancing augmented reality-based inspection. *Journal of Automation in Construction, Elsevier, 19*, 69–182.

Styliadis, AD. (2008). Historical photography-based computer-aided architectural design: demolished buildings information modeling with reverse engineering functionality. *Journal of Automation in Construction, Elsevier, 18*, 51–69.

Sutherland, I. (1968). *A Head-Mounted Three Dimensional Display, Proc. Fall Joint Computer Conference 1968* (pp. 757–764). Washington, DC: Thompson Books.

Turkan, Y, Bosche, F, Haas, CT, & Haas, R. (2012). Automated progress tracking using 4D schedule and 3D sensing technologies. *Journal of Automation in Construction, Elsevier, 22*, 414–421.

Victores, JG, Martínez, S, Jardón, A, & Balaguer, C. (2011). Robot-aided tunnel inspection and maintenance system by vision and proximity sensor integration. *Journal of Automation in Construction, Elsevier, 20*, 629–636.

Wang, X. (2009). Augmented reality in architecture and design: potentials and challenges for application. *International Journal of Architectural Computing, 7*(2), 309–326.

Wang, X, & Dunston, PS. (2006). Compatibility issues in augmented reality systems for AEC: an experimental prototype study. *Journal of Automation in Construction, Elsevier, 15*, 314–326.

Wang, X, & Dunston, P. (2007). Design, strategies, and issues towards an augmented reality-based construction training platform. *Journal of Information Technology in Construction, ITcon, 12*, 363–380.

Wang, X, & Dunston, PS. (2008). User perspectives on mixed reality tabletop visualization for face-to-face collaborative design review. *Journal of Automation in Construction, Elsevier, 17*, 399–412.

Wang, X, Gu, N, & Marchant, D. (2008). An empirical study on Designers' perceptions of augmented reality with in an architectural firm. *Journal of Information Technology in Construction, ITcon, 13*, 536–552.

Wang, K, Hou, Z, & Gong, W. (2010). Automated road sign inventory system based on stereo vision and tracking. *Journal of Computer-Aided Civil and Infrastructure Engineering, 25*, 468–477.

Wang, C, Cho, YK, & Gai, M. (2012). As-is 3D thermal modeling for existing building envelopes using a hybrid LIDAR system. *ASCE Journal of Computing in Civil Engineering, ASCE*. doi:10.1061/(ASCE)CP.1943-5487.0000273.

Wang, X, Kim, MJ, Love, PED, & Kang, SC. (2013). Augmented reality in built environment: classification and implications for future research. *Journal of Automation in Construction, Elsevier, 32*, 1–13.

Waugh, LM, Rausch, B, Engram, T, & Aziz, F. (2012). *Inuvik Super School VR Documentation: Mid-Project Status*. Quebec, Canada: CSCE, 15th International Conference on Cold Regions Engineering. August 19–22.

Yabuki, N, Miyashita, K, & Fukuda, T. (2011). An invisible height evaluation system for building height regulation to preserve good landscapes using augmented reality. *Journal of Automation in Construction, Elsevier, 20*, 228–235.

Zhu, Z, & Brilakis, I. (2010). Parameter optimization for automated concrete detection in image data. *Journal of Automation in Construction, Elsevier, 19*, 944–953.

Zhu, Z, German, S, & Brilakis, I. (2012). Visual retrieval of concrete crack properties for automated post-earthquake structural safety evaluation. *Journal of Automation in Construction, Elsevier, 20*, 874–883.

7

Tangible mixed reality for remote design review: a study understanding user perception and acceptance

Xiangyu Wang[1*] and Phillip S Dunston[2]

Abstract

Background: The design review process is often expensive due to the need for face-to-face meetings between the involved parties. Distributed design collaboration is made possible by advances in networking techniques. A tangible Mixed Reality (MR)-based virtual design prototype was created as a distributed virtual environment (DVE) for the purpose of improving remote design review collaboration. This tangible MR system has been developed to a point that experimental evaluation is necessary in order to understand the strengths and weaknesses of various features of the system.

Methods: In this paper, we evaluated the tangible MR system against a commercial 3D distributed design review system in three aspects: the investigation of how users experienced virtual models in the tangible MR system as compared with the commercial system, the measurement of the users' attitude towards the effectiveness of the tangible MR system, and the discoveries of usability issues involved in the tangible MR interface through usability studies.

Results: The findings from user feedback suggest that the tangible MR system may facilitate problem-solving and the quantity of work in a given amount of time and that virtual design displayed in the mixed scene was a useful aid in the design error detection task.

Conclusion: These findings are useful for the improvement of future generations of the MR system. Also the suggestions can be further generalized to become usability guidelines for the MR developers in other applications and domains.

Keywords: Mixed reality; Tangible interface; Usability

Background

Distributed design collaboration is made possible by advances in networking techniques. For instance, there have been several well recognized interactive digital tabletop systems, either for collocated design groups or for networked collaborative design groups such as the TeleDesign interface of Shu and Flowers (1992) and the WebShamn (Tuikka and Salmela, 1999; Merrick et al., 2011; Singh et al., 2011). Interactive digital tabletop systems have been widely applied to many aspects of human life such as education (Khandelwal and Mazalek, 2007), office (Wigdor et al., 2007), military (DiGiovanni et al., 2005) entertainment (Cowell et al., 2004) and games (Nilsen and Looser, 2005). They offered novel and exciting approaches for humans to interact with various scenarios.

Computer-supported collaborative design review has been widely used Mavrikios et al., (2011). The collaborative conceptual design has been a major area of research work, mainly addressing several web- and agent based approaches to support collaboration during the early stages of product development, conceptual design tools and frameworks, conflict resolution, and team/project management for conceptual design (Wang et al., 2002). Extensive research work on collaborative computer-aided design (CAD) has also been reported, addressing issues, such as co-design

* Correspondence: x.wang@arch.usyd.edu.au
[1]Australasian Joint Research Centre for Building Information Modelling, School of Built Environment, Curtin University, Perth, Australia
Full list of author information is available at the end of the article

systems and feature/assembly based representations, web-based visualization, 3D representations for web-based applications, and 3D streaming over networks (Khan et al., 2006). Several synchronous and asynchronous co-design systems, providing collaborative modeling functionality, have been developed. Most of them are based on the client–server model, while recently, some systems providing real-time online collaboration, based on a peer-to-peer network, have also been presented (Chen, 2001). The integration of different commercial client CAD systems into a co-design platform has been demonstrated in some cases (Huang et al., 2010). Virtual reality (VR) based systems for collaborative product modeling were also suggested in the past (Chryssolouris et al., 2009). Shared product visualization and collaborative design review has been another major area of research and development work (Dong and Kamat, 2013). Methods of sharing virtual product representations over the web and a number of CAD-integrated shared workspaces have been presented in the scientific literature for distributed design review (Hou et al., 2008; Hren and Jezernik, 2009). Shared VR-based environments have also been used to support interactive collaboration in product design review (Escorpizo et al., 2011; Liang et al., 2013). Most of the reported research activities focus on providing collaboration tools for specific phases of the product development process. Research works on internet-based product information sharing and product data management (PDM) related applications have also been widely documented (Hu et al., 2010). Apart from product design issues, several researchers have also worked on the development of methods and tools to support real-time collaboration for distributed activities related to manufacturing and product assembly (Zhen et al., 2010; Wu et al., 2012). The development of web-based manufacturing systems has also been extensively investigated (Lan, 2009).

Mixed Reality (MR) is a powerful user interface paradigm enhancing a user's perception by incorporating the computer-generated information into the real world. Augmented Reality (AR), which is a sub-mode of Mixed Reality and sometimes interchangeable with MR in certain cases, can insert digital information into a predominantly real environment (Dunston and Wang, 2005). Verlinden et al. (2009) suggested the concept of augmented prototyping that projects the perspective images of the product on the physical object made by rapid prototyping techniques (Verlinden et al., 2003). Greenberg and Fitchett (2001) presented toolkits called Phigets that allow designers to explore a tangible user interface (TUI) for interactive product design. In addition to the toolkit, powerful tools including stereoscopic display systems, head mounted displays (HMD), data gloves and haptic devices have been introduced (Burdea and Coiffet, 2003) and combined to construct Virtual and Augmented Prototyping systems that provide realistic display of products in a simulated environment and offer various interaction and evaluation means (Wang et al., 2009). For instance, Bochenek et al. (2001) compared the performance of four different VR displays in a design review setting and mentioned that the best approach for design review activities could be a combined technology approach. As for AR adoptions for design review, Moeslund et al. (2003) and Sidharta et al. (2006) both employ optical see-through Head Mounted Displays to inspect designs, but little support is provided to the act of reviewing. Verlinden (2003) attempts to improve design reviews in the domain of Industrial Design Engineering by providing a specific prototyping and annotation device employing physical mockups. The IMPROVE project explicitly addresses design reviews for product design; its main focus lies with hardware and algorithms for photo-realistic rendering (Santos et al., 2007). The annotation facility is minimal.

Although various ways above have been proposed to support collocated Augmented Prototyping and design review of industrial and architectural products, more research is still needed in the following aspects. AR should be made available and accessible to geographically distributed people in product design evaluation. The prototyping environment should be available at low cost without strong restriction of its accessibility to remote users. Applications of AR have been predominantly focused on industrial and architecture product design, however, there is very little noted research on AR adoptions in the area of building services, which this paper is presenting. In this paper, we address these aspects by proposing a tangible Mixed Reality-based distributed virtual environment (DVE) — MRDVE — for distributed concurrent collaboration, which provides a platform which allows groups of people to share the same work and communications space over a network. There have been several noted research efforts that investigated the applications of distributed Mixed and Augmented Reality technologies for collaborative interior design applications (Ahlers et al., 1995), videoconferencing systems (Barakonyi et al., 2004), tutoring sessions in teaching multi-variate calculus (Orozco et al., 2006). Other examples include the collaborative augmented reality system Studierstube (Schmalstieg, 2002), SharedSpace (Billinghurst and Kato, 1999), and mediaBlocks (Ullmer et al., 1998). The development of MR/AR technologies for the architecture, design, and engineering industry should be guided by research aimed at adapting the features of the MR/AR systems to the different demands of the various situations in which it may be used. The user acceptance and usability issue are thus of great importance. There have been several attempts to evaluate the effectiveness of Virtual Reality technology in design and design collaboration. Usability testing has been widely implemented in Virtual

Environments (Hix et al., 1999; Bowman et al., 1999; Gabbard et al., 1999; Hinckley et al., 1997). Although there has been much speculation about the potential of MR/AR, considerable work needs to be done in usability assessment of these mixed environments (Dias et al., 2003; Regenbrecht and Schubert, 2002; Tang et al., 2003). The common feature of these noted research efforts is the attempt to use well-designed systematic usability tests to create detailed knowledge about the usability aspects of the various interaction and information presentation techniques involved in remote MR/AR systems. Very few empirical studies have been done in the area of MR/AR, which creates a knowledge gap. Wang and Dunston (2008) implemented an experimental evaluation that was intentionally geared and implemented to investigate how users experienced virtual models in the MR system and to measure the users' (1) experience, (2) satisfaction, and (3) attitude towards the effectiveness of viewing the 3D models through the MR system as compared with viewing those 3D models on paper.

The most state-of-the-art of commercial design review software provide environments for remote collaborative review. Since industrial construction practitioners are familiar with such remote design review tools, positive results from comparing the MR tool to one such commercial tool may persuade practitioners to move toward the adoption of MR/AR to change the way they work. Based on this motivation, experimentation on the tangible and distributed MR-based virtual space system is presented, a follow-up study to the one by Wang and Dunston (2008) mentioned above. The purpose of this experimentation is to gather and generate detailed knowledge, through well-designed empirical tests, about the usability aspects of the various interaction and information presentation techniques involved in the MR-based virtual space system from a practical perspective. Specifically, the experimentation assessed the effectiveness of the tangible MR system as compared with a representative commercially available remote collaboration system. Although the experimenters chose the mechanical design review as the context for evaluating the MR system, the results and learned lessons could possibly be lent to other similar types of design review activities.

Methods: System Prototyping and Experimental Evaluation

Prototyping the tangible mixed reality-based distributed virtual environments

Figure 1 depicts the generic system architecture of the MR-based virtual space system. This prototype was based on the previous version of an MR prototype introduced by Wang and Dunston (2008) for face-to-face design review collaboration but is a distributed version with extra remote functions and features. Commercial NetMeeting software is used for audio communication only. The hardware has been operated on a Pentium 4 PC with a 1.6 GHz processor (central server) and on different PCs and laptops (local clients) with head-mounted displays (HMDs). The components or devices consist of the central server, network, client computer, HMD, markers and arbitrary real objects. The central server is configured with a CAD application. The central sever can control what information is sent to each client. The network configuration connects server and clients, where an IP network with multicast support was realized on top of the network infrastructure. Client computers are configured with the same MR software, which may receive real-time online data packets for graphics objects. Multiple-marker recognition was used as the tracking technology where a high contrast tracking marker is essential. By manipulating the cube (real objects) to control the virtual tracking ball, users can naturally interact with the virtual object in the virtual graphic world, the users can tell the location of the real cube with reference to the virtual object. Therefore, by using the tracker cube, the user can track his/her hand's position relative to the viewed virtual model by knowledge of the spatial relationship between the virtual model and the tracker ball. In the following sections, the tangible interface and virtual network computing components are elaborated because they are essential for implementing and understanding the user acceptance study presented in the latter part of the paper.

Tangibility

Tangible interfaces enable the integration of convenient arbitrary objects into the real collaboration environment for effective object manipulation. Tangible interaction techniques applied in the system are incorporated via tracking markers for Mixed Reality registration. The users can manipulate (rotate, tilt, move around) the virtual model design by manipulating the physical tracking marker underneath it. Video-based fiducial recognition was used as the tracking technology where a high contrast tracking marker is essential. The computer performs image processing on the video image captured from the camera to find the specially marked tracking marker(s). Once an appropriate marker is recognized, a virtual model is then registered to the position of the marker. The resultant composite image reveals the insertion of the 3D virtual model into the user's real world view. Whenever the spatial relationship between the camera and the tracking marker is changed, the virtual model moves in concert with the marker. The other key tangible interaction technique applied is a tracker cube used for interpreting the spatial relationship of virtual and real entities. The tracker cube is actually a specially marked, small-size real cube with a virtual ball overlaid on it (see Figure 2).

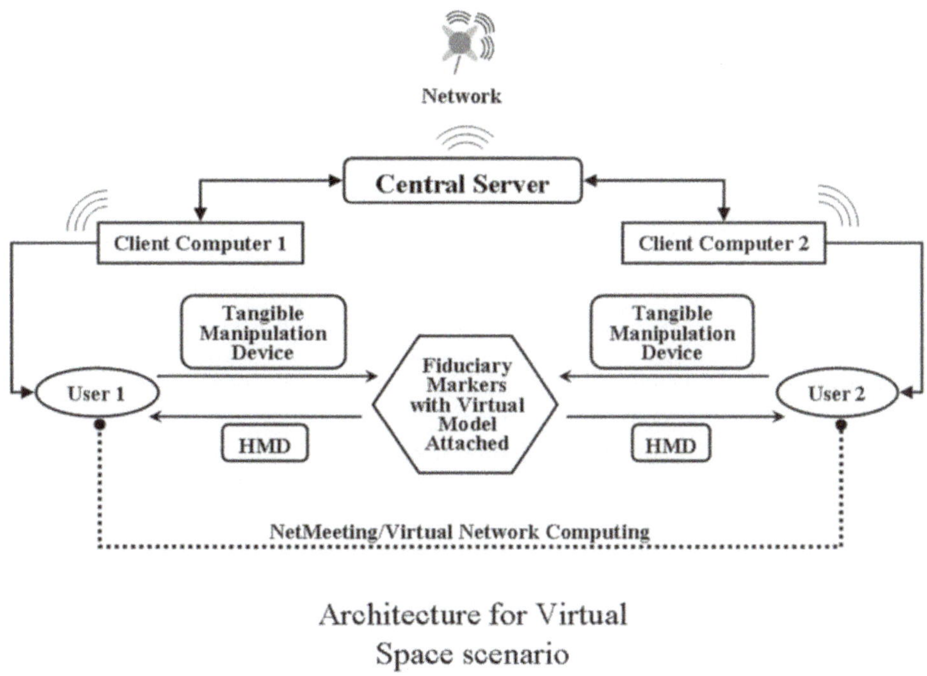

Figure 1 System architecture of the MR-based virtual space.

View windows

The system design is based on a distributed shared Mixed Reality scene window. When communicating the model with others over the network and transferring attention back and forth between discussion with other people and observation of his/her own 3D models, the user must periodically share the perspective and focus with the other collaborators in order to ensure consensus in their decisions. Such consensus can enable collaborator A to keep track of the series of actions taken by collaborator B from the latter's viewpoint. For example, collaborator B asks A to navigate to a target by passing three walls in front and then turning 60 degrees left and then passing another five walls forward in a virtual environment. Collaborator A might be confused by such verbal instructions and might fail to reach the target. An effective collaboration is thus difficult to accomplish. Effective perspective sharing through certain groupware software can reduce such cognitive load that is required otherwise.

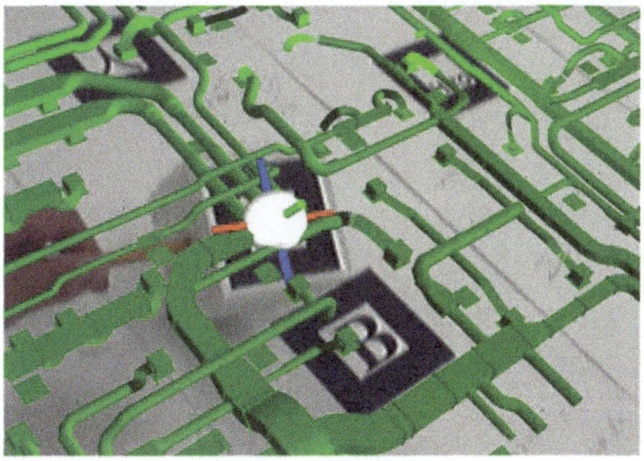

Figure 2 Illustration of the virtual tracking ball for positioning through manipulation of a real cube (Wang and Dunston 2008).

Experiment setup

As stated above in Section 1, the entire experimentation was implemented using the procedure/method that was set by the authors in a study comparing face-to-face collaboration facilitated by paper and MR media (Wang and Dunston, 2008). This procedure was used in that instance to measure the users' perspective on a tabletop Mixed Reality collocated application. The remaining discussion details the evaluation of the MRDVEs using the same method. The experiment aimed to collect data for investigating the subjects' experience with the MR system, gauge the subjects' attitude towards the MR system, and discover usability issues. Three questionnaires were administered to individual participants in an experiment where pairs of subjects worked collaboratively using both the MR-based virtual space system and a leading commercial design review application to detect design errors: the first questionnaire was devised to measure the users' attitude towards the effectiveness of the MR system; the second questionnaire investigated how users experienced virtual models in the MR system as compared with the commercial system; the third questionnaire aimed to reveal usability issues involved in the MR system interface through a usability study.

Comparison benchmark: vommercial3D design review tool

The commercial design review application chosen for comparison, NavisWorks Roamer, is sophisticated commercial 3D design review software which enables geographically distributed users to review 3D designs (see Figure 3). For interaction with models in NavisWorks Roamer, there are nine navigation modes to control how you move around the main navigation view six camera-centric modes (walk, look around, zoom, zoom box, orbit, fly) and three model-centric modes (pan, examine, turnable). In a camera-centric mode, the camera moves within the scene, whereas in a model-centric mode, the model moves inside the scene. NetMeeting software is incorporated into NavisWorks Roamer as a built-in functionality. It was the experimenter's responsibility to install the marker in its precise position and calibrate the MR system. The marker's final relevant position was not subject to be altered afterwards. For example, in order to guarantee that the camera did not lose the tracking image, the marker was set still in the space. The human subjects only needed to familiarize themselves with the MR interface and how to interact with the MR information.

For collaborative design review, NavisWorks Roamer essentially supports asynchronous collaboration, where one collaborator finds the suspicious error, redlines it, saves the current viewpoint, and then sends the information to the other collaborator. The other collaborator then receives the file and loads the saved viewpoint for review. However the MR interface support synchronous collaboration and thus allows real-time view slaving and sharing. Based on the collected data, the time spent on sending views in NavisWorks was quite short, averaging less than half a minute, which almost constitutes synchronous collaboration. This situation is similar to sending and receiving emails between two persons who are waiting at both ends for each exchange. Furthermore, during the experiment, the experimenter intentionally excluded recording the time that was spent on file transfer in order to make its comparison with MR fundamentally fair. A stop watch was used to capture the task duration without file transfer. When the subjects filled in the questionnaires, the experimenter alerted them to not consider the file transfer time as a drawback/effect when rating/evaluating the NavisWorks performance.

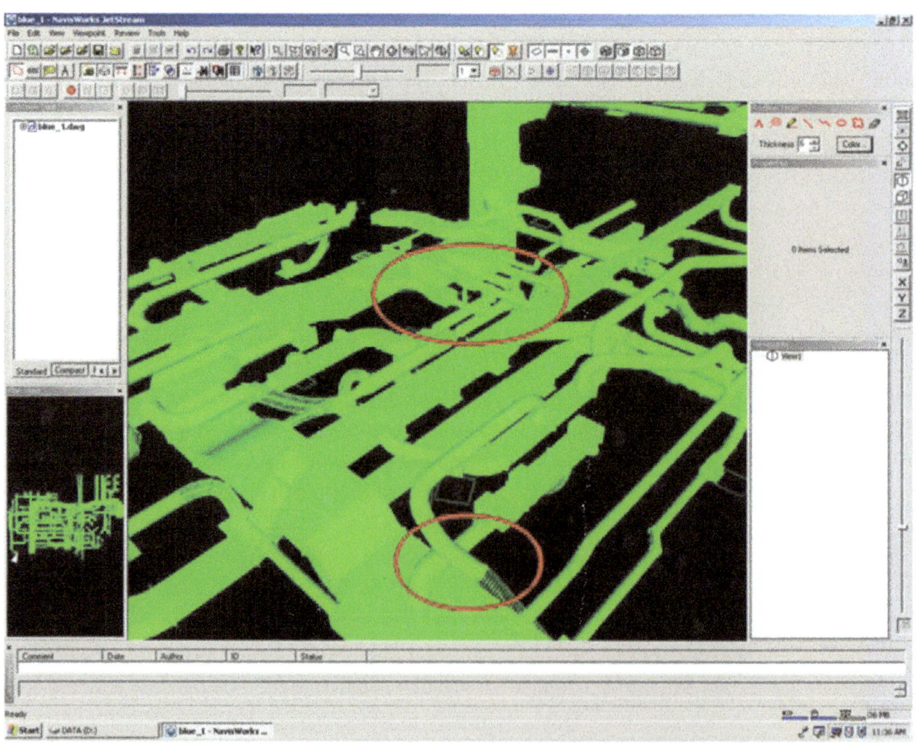

Figure 3 Screenshot of NavisWorks Roamer with redlining and interaction functions demonstrated.

Hypotheses

As noted above, post-experience questionnaires were used to collect data for investigating the subjects' experience with the MR system, gauge the subjects' attitude towards the MR system, and discover usability issues. Our prior experience in testing applications of MR technology has demonstrated that user issues of visual quality, maintaining spatial awareness, and functional utility (i.e., comfort and ease of use) are important to potential users embracing such a system (Wang and Dunston, 2006; Wang and Dunston, 2008). Therefore, the first questionnaire implemented was designed to evaluate the subjects' experiences of using the tangible MR-based virtual space system and NavisWorks Roamer in terms of the following five hypotheses:

- Hypothesis 1 asserts the following: *When compared to NavisWorks Roamer, the MR-based virtual space system will involve significantly better quality of visual presentation;*
- Hypothesis 2 asserts the following: *When compared to NavisWorks Roamer, the MR-based virtual space system will induce significantly higher physical comfort;*
- Hypothesis 3 asserts the following: *When compared to NavisWorks Roamer, the MR-based virtual space system will involve a significantly more natural and intuitive interaction with the model;*
- Hypothesis 4 asserts the following: *When compared to NavisWorks Roamer, the MR-based virtual space system will involve a significantly higher level of immersion;*
- Hypothesis 5 asserts the following: *When compared to NavisWorks Roamer, the MR-based virtual space system will give users significantly more ability to maintain sense of location and orientation.*

Subjects, tasks, materials and method

Sixteen (12 men, 4 women) graduate student participants were recruited for the study (8 groups with 2 in each group). The participants' ages ranged from 25 to 33. Important to the backgrounds of those graduate students was their lack of any 3D computer-aided drawing and animation experience, which practically eliminated experience-based bias. While selecting the subjects, a screening was conducted before the experiment in the format of a simple oral interview. All participants were pre-screened to ensure they had no sophisticated experience with CAD, VR, and MR applications. The modules for design review in the experiments were all taken from real construction industry designs and thus a proper level of complexity in the modules was ensured. These precautions can make the experience factor insignificant to bias the results. Furthermore, none of the participants had any known impairment in spatial cognition and reasoning, any brain damage, or eye (vision) problems that were not corrected by standard eyeglass prescription. It is important to note that there was no cross-over between participants in this experiment and the earlier experiment by Wang and Dunston (2008) comparing the MR approach to the paper-based method. Thus learning curve effects and related biases were avoided as well.

Industry participants in design reviews typically want to view and browse the design independently. Therefore, the experimental task was to review and inspect the design model independently and then figure out the design errors. The error detection task was chosen because it emphasizes design model comprehension and opportunities for collaboration and it involves a high demand of mental interpretation and maintaining a viable spatial mental model. Real 3D CAD models, obtained from an industry partner, were used as designs to be reviewed using the two methods. Small-scale cluttered ductwork models were used for collaboration between two individuals in separate rooms but connected over the network. Before their collaborations, each team of participants was briefed consistently on the types of errors that might be found. Two ductwork model designs were confoundedly used for the two treatments (MR and NavisWorks Roamer) respectively (see Figures 4 and 5). Both designs had a similar spatial layout but different locations of the errors. Three questionnaires were designed to be completed by the subjects with the purpose of investigating the users' experience with the MR system, measuring users' attitude towards the MR system, and implementing usability studies.

Treatments

The Treatment 1 was to perform the review using commercial NavisWorks Roamer where the pair of subjects

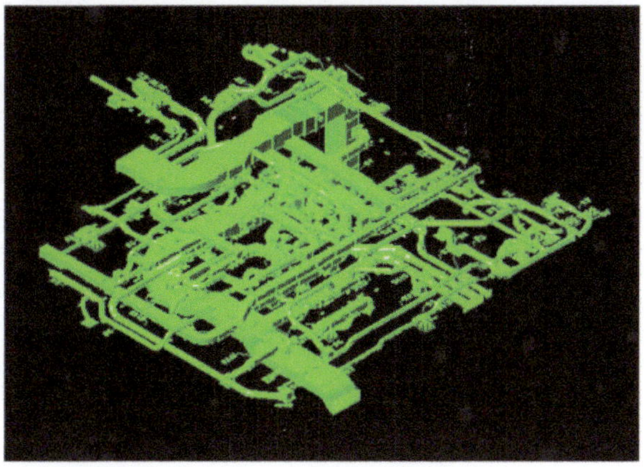

Figure 4 Design Layout of AutoCAD Design I for the two treatments.

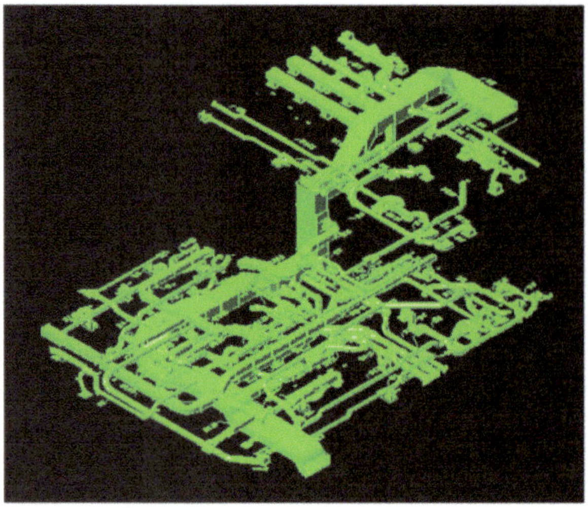

Figure 5 Design layout of AutoCAD design II for the two treatments.

could interact with the 3D ductwork design in a desktop window environment (see Figure 6a) in two different rooms. The Treatment 2 was to use the MR system (user can move his/her marker in hand and view the model from different angles via HMD) where the subjects in two different rooms were able to see their own virtual ductwork model floating above the corresponding real tracking marker via their HMD (see Figure 6b). Subjects can change viewpoints by either moving around or manipulating the real tracking marker underneath the virtual ductwork model.

Experiment process

Each experiment session lasted three hours, including training session, actual experiment session, and post-experiment questionnaire session.

- Training session: The subjects were instructed to use two treatments/methods (MR and NavisWorks Roamer) to review a design. They were assigned enough time to practice how to use the different platforms.

- Actual experiment session: In the real experiment, each team consisted of two subjects (collaborators): one in office A that "owns" Part I of the design and one in office B that "owns" Part II. Each collaborator reviewed the design of the combined models. Each group performed a trial with each method (treatment) to check the two respective sets of the model in a specified sequence. During this process, the

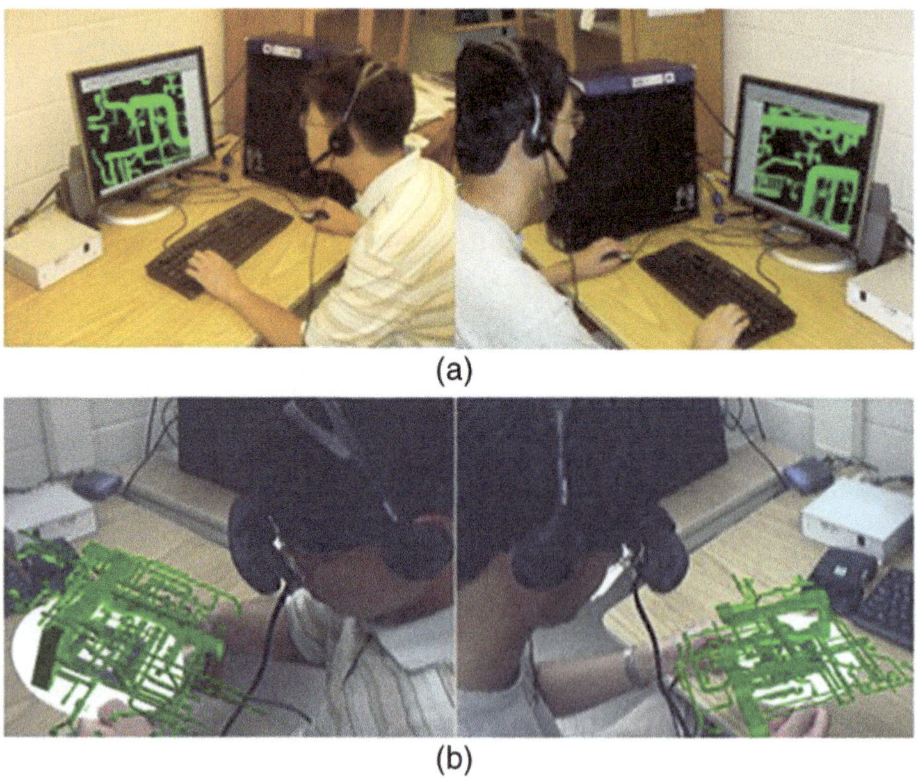

Figure 6 Screen Shots of two Visualized Collaboration Treatments. (a) NavisWorks View and **(b)** MR View.

subjects checked the design errors (see Figure 7 for an example).

- Post-experiment questionnaire session: At the end of the session, each user or group was asked for their comments and observations on the system, as well as presented a set of questionnaires on the comparison of the MR system and NavisWorks Roamer regarding various features

The participants' performances were measured by subjective matrices, such as questionnaires, as well as objective matrices, such as task achievement and duration. It was generally observed that the subjects using the MR system completed the task much faster than the subjects using NavisWorks Roamer. It is believed that the perception and spatial cognition advantages of the MR system resulted in an offloading of some mental processing to the MR system and thus enabled the subjects to complete the task in less time. Further details of the time comparison are described elsewhere by Wang (2005). Furthermore, special useability questionnaires and associated data collection strategies like interviews and observations were developed in order to assess the participatory process and certain features of the MR space. Finally, a statistical model was also developed to arrange experimental sessions and collect data. A statistical analysis parameter (p-value) was used to test inter-factor correlations for reliable results.

Development of questionnaires

The first questionnaire was developed to evaluate the subjects' individual experiences of the virtual model in the MR system as compared with how they experienced the design in the NavisWorks Roamer environment. The questionnaire ratings were on a five-point scale, with a value of 5 representing excellent and 1 representing poor. The questions and a summary of the results are presented in Table 1, including average ratings and standard deviations for each of the methods' features. Aspects covered include quality of visual presentation, physical comfort, naturalness and intuitiveness of interaction with the model, level of immersion, ease of navigation, ability to maintain sense of location and orientation, overall suitability for making decisions and performing tasks on design models.

The second questionnaire was developed to evaluate the subjects' experiences of the virtual model in the MR-based virtual space system for conducting collaborative work compared with their experiences in the NavisWorks Roamer environment. In order to minimize the questionnaire's bias and/or order effects on the results, half of the subjects evaluated the MR system relative to the NavisWorks Roamer, as in Questionnaire I, while the other half evaluated the NavisWorks Roamer relative to the MR system, as in Questionnaire II (see Table 2). The five-scale rating is from "totally agree" to "neutral" to "totally disagree".

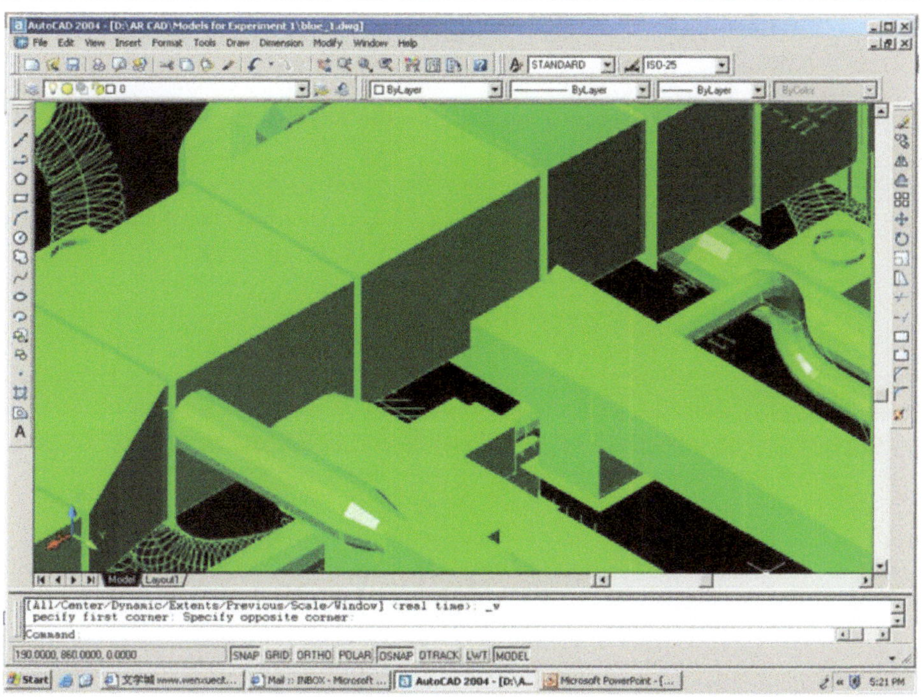

Figure 7 An example of design error pattern: interference between two rectangular ducts highlighted in red.

Table 1 Evaluation of the subjects' experiences of using the MR system and the NavisWorks Roamer, rated from 1 (poor) through 5 (excellent)

No.	Questions	NavisWorks Roamer Mean (S.D.)	MR Mean (S.D.)	F Value	p-value	Significant Difference?
1	Quality of visual presentation	3.1 (0.99)	3.6 (0.74)	2.14	0.1543	No
2	Physical comfort	3.27 (1.22)	2.73 (0.83)	2.09	0.1602	No
3	Naturalness and intuitiveness of interaction with model	3.26 (1.03)	3.87 (0.64)	3.66	0.0661	No
4	Level of immersion	2.73 (0.88)	4.37 (0.81)	27.78	<0.0001	Yes
5	Ease of navigation	3.2 (1.15)	3.87 (0.99)	2.9	0.0994	No
6	Ability to maintain sense of location and orientation	3.2 (1.16)	4.1 (0.7)	5.2	0.0305	Yes
7	Overall suitability for making decisions and performing tasks on design models	3.13 (0.99)	4.0 (0.71)	5.92	0.0216	Yes

The third questionnaire aimed to investigate further some of the questions not fully addressed by the previous questionnaires. The questionnaire was used to explore the usability issues involved in the current version of the MR system and identify the need for any remedies. The questions in the questionnaire were answered using a five-step scale where 1 _represented very little and 5 _represented very much. Table 3 illustrates the mean ratings and standard deviations for the 15 usability questions. The second column lists the usability issues considered; the third gives the mean rating value and standard deviation for each usability issue from the opinions of the subjects.

In Table 3, Questions 1–11 were designed to measure the influence of the interface of the MR system on the effectiveness of the subjects' performances. Questions 12–15 were more related to the comfort issues involved in using the MR system. Comfort was measured in the areas of arm strain, neck strain, dizziness, and nausea.

Results and Discussion

Questionnaires

A one-way analysis of variance (ANOVA) was performed on the individual ratings for both the MR system and NavisWorks Roamer. The results of mean ratings, standard deviation, F-values, and p-values for comparisons are illustrated in the Table 1.

Initially, it is noted that the subjects rated the quality of visual presentation of the model viewed from the MR system (Mean, 3.6; S.D., 0.74), which is marginally higher than the NavisWorks Roamer (Mean, 3.1; S.D., 0.99). The difference is not significant with p-value $(0.1543) > 0.05$ and $F(1, 14) = 2.14$, which does not support Hypothesis 1. This is different from the results obtained from the previous

Table 2 Statements for soliciting measuring the subjects' opinions on the value of the MR system for collaborative work

Questionnaire I	Questionnaire II
1. I felt that 3D interactivity in the MR system aided design comprehension.	1. I felt that 3D interactivity in the NavisWorks Roamer aided design comprehension.
2. Overall, compared with NavisWorks Roamer, the AR system better facilitates design collaboration tasks.	2. Overall, compared with AR, the NavisWorks Roamer better facilitates design collaboration tasks.
3. The MR system better facilitated communication.	3. The NavisWorks Roamer better facilitated communication.
4. The MR system better facilitated creativity.	4. The NavisWorks Roamer better facilitated creativity.
5. The MR system better facilitated problem-solving.	5. The NavisWorks Roamer better facilitated problem-solving.
6. The MR system increased the overall quality of output from the collaboration.	6. The NavisWorks Roamer increased the overall quality of output from the collaboration.
7. The MR system better facilitated the quantity of work I could complete in a given amount of time.	7. The NavisWorks Roamer better facilitated the quantity of work I could complete in a given amount of time.
8. The MR system increased the quality of my contribution to the project.	8. The NavisWorks Roamer increased the quality of my contribution to the project.
9. The MR system increased my satisfaction with the outcome of the collaboration.	9. The NavisWorks Roamer increased my satisfaction with the outcome of the project.
10. The MR system increased understanding between my collaborator and me.	10. The NavisWorks Roamer increased understanding between my collaborator and me.

Table 3 Means (S.D.s) for usability questions

No.	Questions	Mean (S.D.)[a]
1	Did the MR visual display (head-mounted-display) create difficulties for performing?	4.31 (0.96)
2	Did you often feel disoriented?	2.93 (1.18)
3	Did you have a natural perspective that gives you a compelling sense of your hand-motion while manipulating the tracking cube?	3.38 (0.88)
4	With the MR system, are you isolated from and not distracted by outside activities?	3.75 (1.13)
5	Did the surrounding real background help your spatial comprehension of the model?	3.56 (1.15)
6	Was the FOV (field of view) appropriate for supporting this activity?	2.63 (1.2)
7	Does visual output have adequate stability of the image as you move with no perceivable distortions in visual images?	2.81 (1.17)
8	Does visual output have acceptable degree of response delay with no perceivable distortions in visual images?	3.31 (1.14)
9	Is the MR display effective in conveying convincing scenes of models appearing as if in the real world?	3.88 (0.88)
10	Can you predict responses to your actions?	4.0 (0.57)
11	Did you have satisfactory control over the system?	3.38 (0.88)
12	Is the MR system comfortable for long-term use?	2.06 (1.28)
13	Is tracking marker lightweight, portable, non-encumbering, and comfortable thereby avoiding issues of limited your mobility and fatigue?	3.94 (0.85)
14	Did you experience excessive eye fatigue?	3.5 (1.09)
15	Did you experience nausea during your interaction with MR system?	1.75 (1.0)

[a] The question response format was a scale with steps from 1 (very little) to 5 (very much).

experiment by Wang and Dunston (2008) where an MR system was proved to provide more realistic and vivid virtual models than a corresponding 3D color drawing on paper. The virtual model in the MR-based virtual space system possessed a similar level of fidelity as the 3D model in NavisWorks Roamer, however, the current quality of the virtual model in the MR system is somewhat limited (bottlenecked) by the resolution of the head-mounted display. The current virtual models in use were converted from original 3D AutoCAD models through the PolyTrans Converter, where certain rendering information is lost. A more sophisticated converter might preserve more rendering data, therefore obtaining a higher level of fidelity in the MR virtual model. Some subjects commented that more shading and shadow rendering might further improve the visual quality and thus enhance the user's spatial cognition.

The subjects felt more physical comfort using NavisWorks Roamer (Mean, 3.27; S.D., 1.22) than the MR system (Mean, 2.73; S.D. 0.83). However, the significance was not attested with only a p-value $(0.1602) > 0.05$. This does not support Hypothesis 2. The users' neck and back strain (approximately 10–20 minutes of use) in the MR system was induced by the long-time uncomfortable poses for inspecting details of the model and the physical weight of the HMD.

Interaction with the virtual model via the nine functions in the NavisWorks Roamer seemed to be as natural and intuitive as in the MR system according to the ratings (NavisWorks: Mean, 3.26; S.D., 1.03; MR: Mean, 3.87; S.D., 0.64). The difference is insignificant according to the p-value > 0.05, and therefore Hypothesis 3 is not supported. This similarity can be explained as the trade-off between the unnaturalness of MR due to the demand of more physical effort and the intuitiveness of interaction with the virtual model via the virtual tracking ball. The subjects assessed the level of immersion for the MR system at an average rating of 4.37 (S.D., 0.81) compared with 2.73 (S.D., 0.88) for the NavisWorks Roamer with a significant p-value indication, which strongly implies that the MR system provides the users a greater sense of being present with the model. Thus Hypothesis 4 is well supported.

Subjects did not identify much difference in the ease of navigation between these two treatments, rating both of the tools as providing a mediate ease of navigation (MR: Mean, 3.87; S.D., 0.99; NavisWorks Roamer: Mean, 3.2; S.D., 1.15). However, for inspecting the details of the design model, the MR system enabled the users to move more easily to a more advantageous viewpoint (closer and directly in front of the target) to complete the error verification more quickly. This advantage was observed by the experimenter in a large number of trials. The subjects had learned the layout of the 3D design space and expressed more confidence in the error detection task without further effort to interpret the occlusion and depth cues among the objects because they were reviewing the design model in a 3D manner.

Hypothesis 5 is well supported because subjects using the MR system had higher ability to maintain sense of location and orientation as they navigated the model than when using NavisWorks Roamer (MR: Mean, 4.1; S.D., 0.7; NavisWorks Roamer:

Mean, 3.2; S.D., 1.16). The difference is regarded as significant with a p-value (0.0305) < 0.05. Many subjects commented that there was easy loss of spatial orientation in using NavisWorks Roamer. NavisWorks Roamer provides the subjects with the bird's-eye view (exo-centric) and close-view (ego-centric) for review. As the subjects tried to switch between these two views through the viewing functionalities (e.g., zoom out from a small object to the complete view of the design model), they had to mentally transform the view in their mind to accommodate the change of these two reference frames. Therefore it was often observed and reported that the subjects became lost or disorientated, especially after using the interactive viewing techniques to switch between these two views. This disorientation and re-accommodation was an additional mental burden to the cognitive attention switching. Some subjects using the NavisWorks Roamer reported that they felt frustrated because it was hard to even know the previous viewpoint of the design and they had to spent additional time to learn the model again from the new perspective. In contrast, fewer cases were reported from the users of the MR system since MR provided a smooth and natural way to transit back and forth. The subjects using the MR system could offload this centricity transformation task to the intuitive viewing mechanism of the MR system. When the subject leaned back allowing the camera mounted on the HMD to capture the complete design, a bird's-eye view (exo-centric) was created. For inspecting the details of design model, the MR system enabled the users to move their viewpoint to a more advantageous position (closer to the target, with the target directly in front of them) to complete the task more quickly. When the subjects approached a target closely enough, an ego-centric view was created. The extra time spent in using the NavisWorks Roamer to accommodate the switch of reference frames was certainly mostly cognitive overhead. Therefore, the frequent loss of orientation could be interpreted as the inefficiency margin of the NavisWorks Roamer against the MR system in the subjects' spatial orientation activity.

The final point of comparison asked for a general impression of the two methods. Subjects believed that the MR system was more suitable for making decisions and performing tasks on design models than the NavisWorksRoamer system.

Results of questionnaire 2

For the convenience of interpreting the data from the two questionnaires in Table 2, it was rationalized that the respondents who 'totally agree' with a statement of Questionnaire I were regarded to 'totally disagree' with the corresponding statement of Questionnaire II. Therefore, the percentages for these responses were added together. A similar summation of results was applied to the other four levels of the response scale. The data from the two questionnaires was thus tallied with respect to attitudes regarding the MR system and is visually presented in Figure 8.

As indicated in the Figure 8, about 69% of the respondents felt that 3D interactivity in the MR system aided design comprehension in total or partial agreement with the statement. Question 2 asked if the MR system overall better facilitates design collaboration tasks than the NavisWorks Roamer, with which about 69% of respondents agreed. Most of the respondents (63%) believed that the MR system better facilitated communication and creativity and a total of 65% of the respondents totally or partially agreed with the statement that the MR system better facilitated problem-solving. The increase in the overall quality of output from the collaboration is marginally positive (57%), however, the belief that the MR system better facilitated the quantity of work in a given amount of time and increased the quality of the user's contribution to the project is more marked (70% and 70% respectively). It is surprising that these impressions are not stronger in light of the clear time benefits that were observed. The belief that there has been an increase in self-satisfaction from the collaboration as a result of using the MR system is well supported by 63% of the users. Only one-third of respondents, however, believed that the MR system increased understanding between their collaborator and them. One-third held a neutral attitude and one-third disagreed with the statement. A summary of the results convey the conclusion that the MR system both aids individual performance and also supports the collaboration, but there are implications that improvements are yet desirable.

The MR system was rated to offer the advantage of improved collaboration as expected. By sharing the same viewpoint continuously and synchronously between two remote subjects via VNC platform, one subject could maintain the same changing of viewpoints and frame of reference along with the other subject who led the discussion and was in charge of navigating in the MR environment. On the contrary, the subject using the NavisWorks Roamer had to iteratively go through the process of finding the suspicious error, redlining it, saving the current viewpoint, and then sending the information to the other subject. The other subject then received the file and loaded the saved viewpoint for review. Therefore, the receiving subject had to execute additional mental transformation to adapt to the static viewpoint saved by the sending subject for creation of a discussion context. The subjects' task performance using the MR system was therefore improved by reducing the frequency of such electronic exchanges of the static red-lining comments and saved viewpoint between subjects.

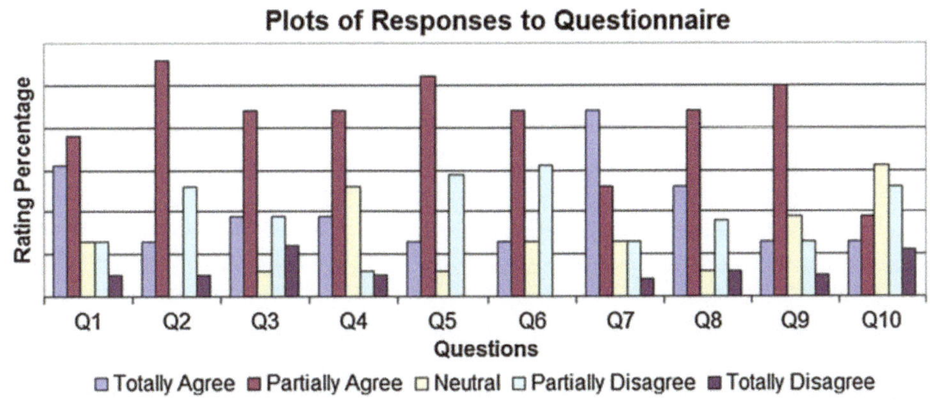

Figure 8 Plot of Responses to Attitude Questionnaire.

Results of questionnaire 3

This section explored some of the usability problems identified from Table 3 and possible redesign recommendations as given for the future generations of MR-based virtual space systems. The following discussion and interpretations are based on the selected questions from Table 3.

Questions 1, 2, 4, 6, 7, 8, 9, 14, and 15 deal with the usability of the HMD. From the ratings in the first question, the respondents reported that the MR visual display did create difficulties for performing (Mean, 4.31; S.D., 0.96). The difficulty mainly resulted from having to wear the cumbersome and uncomfortable HMD. The fact that one hand of the user needed to hold the HMD to maintain a stable view of the mixed scene, increased the arm muscle fatigue. Also, the screen in the HMD was designed to make the user feel that the screen was actually placed several feet away from the eyes. The faraway and low resolution display resulted in a low clarity of the virtual model compared with a standard computer monitor. The fidelity of the HMD is totally dependent on the advance of technology in the industry. More advanced displays might be a promising option for exploration and development. HMDs are well-suited for applications (e.g., architectural and interior design) where complete visual immersion or absence of distractions is required. From the relatively high rating in question 4 (Mean, 3.75; S.D., 1.13), it seemed that the users did not feel much distracted by outside activities (distractions), which implied that the HMD was useful in focusing the users' minds on the task at hand. This result is consistent with the subjects' assessment of the level of immersion for the MR system at an average rating of 4.37 in Table 3. For question 6, the rating for appropriateness of the FOV for supporting this activity was relatively low (Mean, 2.63; S.D., 1.20). The broader the FOV, the better sense the user should have for the environment and communication with other users. This sense is much reliant on the HMD and video camera as the subjects' FOV is constrained to the governing FOV of the display or the camera. The limited field-of-view of the video camera currently used, constrained the user's view to some extent. The visual output had moderate stability of the image as far as allowing the subjects to move with no perceivable distortions in visual images (Mean, 2.81; S.D., 1.17) as indicated by Question 7. The stability of the vision tracker (fiducial recognition) is still insufficient for ensuring stable Augmented Reality applications. Thus improved or alternative methods are required for maintaining a proper registration. The further substantiation of the tracked six-degree-of-freedom data by redundant tracking methods can be a feasible solution. Hybrid technology tracking systems should be examined for applicability and performance regarding stability. In response to Question 8, respondents thought the MR system had a relatively acceptable degree of response delay with no perceivable distortions in visual images (Mean, 3.31; S.D., 1.14). Thus, the system lag was tolerable and did not affect the perception of the visual image by users and therefore did not hinder their performance. The subjects experienced a rather convincing scene of models appearing as if in the real world (Mean, 3.88; S.D., 0.88) as reported for Question 9. The virtual model looked convincingly to be floating in the air of the real environment, which was due to the stereoscopic feature of the MR display with two display panels creating high object presence effects. The relatively high rating implied that the combination of virtual model and real world reached seamlessness to some extent. It is interesting to note regarding Question 14 that the rating (Mean, 3.5; S.D., 1.09) implies that some subjects might have experienced light eye fatigue. The subjects normally wore the HMD for only 20–30 minutes for the entire experiment. If a user has to wear it for a longer time, eye fatigue might emerge more prominently as a negative. On the contrary, regarding the last question, the subjects experienced very little nausea during their interaction with the MR system (Mean, 1.75; S.D., 1.0). These issues need to be further examined and addressed if long-duration tasks are to be performed. It was also revealed from some additional short-answer questions in the questionnaire that about 87.3% of responders believed that they would not be resistant to using MR or similar MR systems in the future. About 81.3% believed that they would embrace the opportunity to use the MR system again in the future. Almost none of the users experienced "blurred vision", "dizziness", "nausea", "difficulty focusing", or "loss of vertical orientation" after exposure to the MR system.

Questions 3 and 13 deal with the naturalness of interaction. In response to the question whether the subjects had a natural perspective that gave them a compelling sense of their hand-motion while manipulating the tracking marker (Question 3), the rating (Mean, 3.38; S.D., 0.88) indicates a mild agreement. From Question 13, the subjects revealed a belief that the tracking marker was lightweight, portable, non-encumbering, and comfortable thereby avoiding issues of limiting their mobility and fatigue (Mean, 3.94; S.D., 0.85). These results are consistent with the rating (Mean, 3.87; S.D., 0.64) from Item 3 in Table 1.

Questions 5, 10, 11, and 12 deal with the issues regarding the overall control of the MR system. The surrounding real background seemed to marginally help the subjects' spatial comprehension of the model (Mean, 3.56; S.D., 1.15) as evidenced by responses to Question 5, an expected result. The real background could have served as a stationary landmark reference for orientation of the virtual model, especially while users were turning the tracking marker. Question 12 responses indicate almost all the subjects had a fairly uncomfortable experience with long-term use of the MR system, with a very low mean rating of 2.06 (S.D., 1.28). This result is in accordance with the high physical discomfort rating in the first questionnaire discussed earlier. The bulky HMD is the most significant usability issue of the MR system. Because the design review collaboration task might require long-duration, intensive work, the MR system would require moderately high levels of ease of use and user comfort. It was found that using the MR system produced neck strain in some users. The tracking and interaction techniques in future generations of the MR system must be easy, comfortable, and intuitive to the user, so that the focus can be on the task and not on the system itself.

Conclusion

Based on the experimental evaluation of the MR-based virtual space system, several conclusions were noted. It was observed that the subjects using the MR system completed the task much faster than the subjects using the NavisWorks Roamer commercial visualization software. It appears the increased 3D visual perception, spatial cognition, and spatial layout of the design offloaded the required mental processing to the MR system and thus enabled the subjects to complete the task in a shorter time. Subjects felt more physical comfort using NavisWorks Roamer than the MR system. The subjects rated the level of immersion for the MR system, on average, much higher than for the NavisWorks Roamer application, which implies that MR definitely provides the users a sense of being present or isolated with the model, more than with the static model in NavisWorks Roamer. The subjects believed that the MR system was more suitable for making decisions and performing tasks on design models than the NavisWorks Roamer. MR seemed to better facilitate communication, creativity and problem-solving. The increase in the overall quality of output from the collaboration is marginally positive, however, the belief that the MR system better facilitated the quantity of work in a given amount of time and increased the quality of the user's contribution to the project is more marked. Perhaps more importantly, there is a belief that there has been an increase in self-satisfaction from the collaboration as a result of using MR.

The usability studies implemented with the third questionnaire revealed usability issues and meanwhile helped to produce possible remedies. These suggestions are useful for the improvement of future generations of the MR system and summarize below Also these suggestions can be further generalized to become usability guidelines for the MR developers in other applications and domains.

- Participants suggested a zoom feature to facilitate inspection of some tiny objects inside dense model structures.
- Participants expected a quicker response in VNC for the real-time collaboration. There was still some latency (lag) in synchronizing the MR scenes between the two collaborators because the VNC conveyed the video stream which occupied too much of the available computing resource. The performance was heavily dependent on the computing resource and the network bandwidth.
- Participants suggested a better user-friendly display. The current HMD was not comfortable enough for long term use. The HMD kept falling and needed one hand to hold it to a correct position, which created extra hand occupation and distracted the users from focusing on the work at hand.
- Participants requested a camera with an automatic focusing function. If the user pulled his or her head back farther, the camera failed to recognize the tracking marker even though it was actually in the FOV of the user.

- Participants suggested adding shadows and shading into the virtual design to help the users obtain more depth cues for better apprehension of the spatial relationships.

Even though these experimental results are preliminary, they proved useful in deriving guidelines and further developments and experimentation. Learning from the preliminary results of the usability study, the authors now have to refine the MR interface to tap into more potentials (Wang et al., 2012). One important factor not addressed specifically in the experiment was the subject's level of expertise. The testing of expert users could produce quite different results. There are very few potential users of most MR applications who could be considered experts, so the current results are useful, but an understanding of how performance is affected by greater experience with conventional tools would have an added value. Future work will hopefully lead to the implementation of the MR system on real construction projects. The real improvements in performance and productivity with MR can be then measured and quantified with site assembly activities in a real project context. The transfer of the animated MR system from laboratory-based applications to real construction applications has higher potential for system flexibility and tracking, e.g. to enable assembly in a limited way on construction sites. Using portable MR devices such as wireless HMD or cameras will enable more stable tracking (images will not be lost when occluding the path between tracking targets and camera). Where the tracking targets (markers) are not able to be posted, markerless tracking techniques such as tracking using the salient geometric features of real spatial objects might be adopted. Since tracking is one of the most essential elements in determining whether the MR system is effective or not in real projects, future work should focus on the integration of the current tracking technologies and develop more robust tracking methods.

Competing interests
Both authors declare that they have no competing interests.

Authors' contributions
XW prototyped the tangible mixed reality system package, designed the experiments and evaluation methods, and carried out the experiments. PD co-authored the system package, and co-designed the experiment. Both authors read and approved the final manuscript.

Acknowledgement
This research has been funded by National Science Foundation grant CMS-0239091. Opinions, findings, conclusions, or recommendations presented in this paper are those of the authors and do not necessarily reflect the views of the National Science Foundation. Models used in the experiment were provided by BMW Constructors, Inc. of Indianapolis, Indiana, U.S.A.

Author details
[1]Australasian Joint Research Centre for Building Information Modelling, School of Built Environment, Curtin University, Perth, Australia. [2]Purdue University, School of Civil Engineering, West Lafayette, India, USA.

References
Ahlers, KH, Kramer, A, Breen, DE, Chevalier, PY, Crampton, C, Rose, E, & Greer, D. (1995). *Distributed augmented reality for collaborative design applications.* Paper presented at the Computer Graphics Forum, © The Eurographics Association and John Wiley & Sons Ltd.

Barakonyi, I, Fahmy, T, & Schmalstieg, D. (2004). *Remote collaboration using augmented reality videoconferencing* (Paper presented at the Proceedings of Graphics interface 2004, pp. 86–89). Waterloo, Ontario, Canada: Canadian Human-Computer Communications Society School of Computer Science, University of Waterloo.

Billinghurst, M, & Kato, H. (1999). *Collaborative Mixed Reality* (Paper presented at 'Mixed Reality – Merging Physical world and Virtual World', proceedings of the International Symposium on Mixed Reality (MR'99)). Yokohama, Japan.

Bochenek, GM, Ragusa, JM, & Malone, LC. (2001). Integrating virtual 3-D display systems into product design reviews: some insights from empirical testing. *International Journal of Technology Management, 21*(3), 340–352. Inderscience Publishers.

Bowman, DA, Johnson, DB, & Hodges, LF. (1999). *Testbed Evaluation of virtual environment interaction techniques* (Paper presented at the Proceedings of the ACM symposium on Virtual reality software and technology, pp. 59–75). New York, NY, USA: ACM.

Burdea, G, & Coiffet, P. (2003). Virtual Reality Technology. Presence: Teleoperators & Virtual Environments, *12*(6), 663–64.

Chen, M. (2001). *Design of a virtual auditorium* (Paper presented at the Proceedings of the ninth ACM international conference on Multimedia, pp. 19–28). New York, NY, USA: ACM.

Chryssolouris, G, Mavrikios, D, Pappas, M, Xanthakis, E, & Smparounis, K. (2009). A web and virtual reality-based platform for collaborative product review and customisation. In *Collaborative design and planning for digital manufacturing* (pp. 137–152). Springer.

Cowell, A, May, R, & Cramer, N. (2004). *The Human-Information Workspace (Hi-Space): Ambient Table Top Entertainment* (pp. 133–61). Entertainment Computing–ICEC 2004.

Dias, M, Jorge, J, Carvalho, J, Santos, P, & Luzio, J. (2003). *Usability Evaluation of Tangible User Interfaces for Augmented Reality. Paper presented at the Augmented Reality Toolkit Workshop, 2003* (pp. 54–61). IEEE International.

DiGiovanni, C, Bowen, N, Ginsberg, M, & Giles, G. (2005). Quarantine Stressing Voluntary Compliance. *Emerging Infectious Diseases, 11*(11), 1778.

Dong, S, & Kamat, VR. (2013). SMART: scalable and modular augmented reality template for rapid development of engineering visualization applications. *Visualization in Engineering, 1*(1), 1.

Dunston, PS, & Wang, X. (2005). Mixed Reality-based Visualization Interfaces for the AEC Industry. *Journal of Construction Engineering and Management, American Society of Civil Engineers (ASCE), 131*(12), 1301–1309.

Escorpizo, R, Reneman, MF, Ekholm, J, Fritz, J, Krupa, T, Marnetoft, S-U, & Tucki, G. (2011). A conceptual definition of vocational rehabilitation based on the ICF: building a shared global model. *Journal of occupational rehabilitation, 21*(2), 126–133.

Gabbard, JL, Hix, D, & Swan, JE. (1999). User-Centered Design and Evaluation of Virtual Environments. *Computer Graphics and Applications, IEEE, 19*(6), 2–10.

Greenberg, S, & Fitchett, C. (2001). *'Phidgets: easy development of physical interfaces through physical widgets'* (Symposium on User Interface Software and Technology, pp. 209–218).

Hinckley, K, Tullio, J, Pausch, R, Proffitt, D, & Kassell, N. (1997). *Usability analysis of 3d rotation techniques* (Paper presented at the Proceedings of the 10th annual ACM symposium on User interface software and technology). New York, NY, USA: ACM.

Hix, D, Swan, JE, Gabbard, JL, McGee, M, Durbin, J, & King, T. (1999). *"User-Centered Design and Evaluation of a Real-Time Battlefield Visualization Virtual Environment." Paper presented at the Virtual Reality, 1999. Proceedings* (pp. 96–103). IEEE.

Hou, J, Su, C, Zhu, L, & Wang, W. (2008). *Integration of the CAD/PDM/ERP system based on collaborative design.* Paper presented at the Computing, Communication, Control, and Management, 2008. CCCM'08. ISECS International Colloquium.

Hren, G, & Jezernik, A. (2009). A framework for collaborative product review. *The International Journal of Advanced Manufacturing Technology, 42*(7-8), 822–830.

Khan, S, Peng, Y, Steinbach, E, Sgroi, M, & Kellerer, W. (2006). Application-driven cross-layer optimization for video streaming over wireless networks. *Communications Magazine, IEEE, 44*(1), 122–130.

Hu, Y, Zhou, X, & Li, C. (2010). Internet-based intelligent service-oriented system architecture for collaborative product development. *International Journal of Computer Integrated Manufacturing, 23*(2), 113–125.

Huang, C-H, Hsiung, P-A, & Shen, J-S. (2010). UML-based hardware/software co-design platform for dynamically partially reconfigurable network security systems. *Journal of Systems Architecture, 56*(2), 88–102.

Khandelwal, M, & Mazalek, A. (2007). *Teaching table: a tangible mentor for pre-k math education* (Paper presented at the Tangible and embedded interaction: Proceedings of the 1 st international conference on Tangible and embedded interaction, pp. 191–194). New York, NY, USA: ACM.

Lan, H. (2009). Web-based rapid prototyping and manufacturing systems: a review. *Computers in Industry, 60*(9), 643–656.

Liang, JS, Chao, K-M, & Ivey, P. (2013). VR-based wheeled mobile robot in application of remote real-time assembly. *The International Journal of Advanced Manufacturing Technology,* , 1–15.

Mavrikios, D, Alexopoulos, K, Xanthakis, V, Pappas, M, Smparounis, K, & Chryssolouris, G. (2011). *A Web-Based Platform for Collaborative Product Design*. Review and Evaluation Springer, City.

Merrick, K, Gu, N, & Wang, X. (2011). Case Studies Using Multiuser Virtual Worlds as an Innovative Platform for Collaborative Design. *Journal of Information Technology in Construction, 16*, 165–188. International Council for Research and Innovation in Building and Construction (CIB), Rotterdam, Netherlands.

Moeslund, TB, Stoerring, M, Broll, W, Aish, F, & Liu, Y. (2003). The ARTHUR system: an augmented round table. *Computer, 1*(1), 277–282.

Nilsen, T, & Looser, J. (2005). *Tankwar-Tabletop war gaming in augmented reality* (Paper presented at the 2nd International Workshop on Pervasive Gaming Applications, PerGames).

Orozco, C, Esteban, P, & Trefftz, H. (2006). *"Collaborative and Distributed Augmented Reality in Teaching Multi-Variate Calculus." Paper presented at the IASTED International Conference on Web based Education (WBE 2006).* Puerto Vallarta, Mexico.

Regenbrecht, H, & Schubert, T. (2002). *Measuring presence in augmented reality environments: design and a first test of a questionnaire* (Paper presented at the Proceedings of the Fifth Annual International Workshop Presence 2002, pp. 9–11).

Santos, P, Stork, A, Gierlinger, T, Pagani, A, Araújo, B, Jota, R, Bruno, L, et al. (2007). Improve: Collaborative Design Review in Mobile Mixed Reality. *Virtual Reality*, 543–53.

Schmalstieg, D, & Hesina, G. (2002). *Distributed applications for collaborative augmented reality* (Paper presented at the Virtual Reality, pp. 59–66). Virtual Reality, 2002. Proceedings. IEEE.

Shu, L, & Flowers, W. (1992). *Groupware experiences in three-dimensional computer-aided design* (Paper presented at the Proceedings of the 1992 ACM conference on Computer-supported cooperative work, pp. 179–186). New York, NY, USA: ACM.

Sidharta, R, Oliver, J, & Sannier, A. (2006). *Augmented reality tangible interface for distributed design review* (Paper presented at the Computer Graphics, Imaging and Visualisation, pp. 494–470). Sydney, Qld.

Singh, V, Gu, N, & Wang, X. (2011). A Theoretical Framework of a BIM-based Multi-disciplinary Collaboration Platform. *Automation in Construction, 20*(2), 134–144.

Tang, A, Owen, C, Biocca, F, & Mou, W. (2003). *Comparative effectiveness of augmented reality in object assembly* (Paper presented at the Proceedings of the SIGCHI conference on Human factors in computing systems, pp. 73–80). New York, NY, USA: ACM.

Tuikka, T, & Salmela, M. (1999). Webshaman: Shared Virtual Prototypes for Product Designers in the World Wide Web. *ACM SIGGROUP Bulletin, 20*(1), 51–55.

Ullmer, B, Ishii, H, & Glas, D. (1998). *mediaBlocks: physical containers, transports, and controls for online media* (Paper presented at the Proceedings of the 25th annual conference on Computer graphics and interactive techniques, pp. 379–38). New York, NY, USA: ACM.

Verlinden, JC, De Smit, A, Peeters, AWJ, & Van Gelderen, MH. (2003). Development of a Flexible Augmented Prototyping System. *Journal of WSCG, 11*(1), 496–503.

Verlinden, J, & Horváth, I. (2009). "Analyzing Opportunities for Using Interactive Augmented Prototyping in Design Practice. *AI EDAM (Artificial Intelligence for Engineering Design, Analysis and Manufacturing), 23*(3), 289.

Wang, L, Shen, W, Xie, H, Neelamkavil, J, & Pardasani, A. (2002). Collaborative conceptual design—state of the art and future trends. *Computer-Aided Design, 34*(13), 981–996.

Wang, X. (2005). *Applications of Mixed Reality in Architecture, Engineering, and Construction| Specification, Prototype, and Evaluation.* PURDUE UNIVERSITY.

Wang, X, & Dunston, PS. (2006). Compatibility Issues in Augmented Reality Systems for Aec: An Experimental Prototype Study. *Automation in construction, 15*(3), 314–326.

Wang, X, & Dunston, PS. (2008). User Perspectives on Mixed Reality Tabletop Visualization for Face-to-Face Collaborative Design Review. *Automation in construction, 17*(4), 399–412.

Wang, X, Chen, J, & Zhou, H. (2009). Experimentally Exploring the Benefits of Augmented Reality Technology in Design Conceptualisation. *International Journal of Design Engineering, 1*(3), 278–299. Inderscience Publisher.

Wang, X, Love, P, Klinc, R, Kim, M, & Davis, P. (2012). Integration of E – learning 2.0 with Web 2.0. *Journal of Information Technology in Construction, 17*, 387–396. International Council for Research and Innovation in Building and Construction (CIB), Rotterdam, Netherlands.

Wigdor, D, Perm, G, Ryall, K, Esenther, A, & Shen, C. (2007). *Living with a tabletop: Analysis and observations of long term office use of a multi-touch table* (Paper presented at the Horizontal Interactive Human-Computer Systems, 2007. TABLETOP'07, pp. 60–67). Newport, RI.

Wu, D, Zhen, X, Fan, X, Hu, Y, & Zhu, H. (2012). A virtual environment for complex products collaborative assembly operation simulation. *Journal of Intelligent Manufacturing, 23*(3), 821–833.

Zhen, X-J, Wu, D-L, Hu, Y, & Fan, X-M. (2010). A real-time simulation grid for collaborative virtual assembly of complex products. *International Journal of Computer Integrated Manufacturing, 23*(6), 500–514.

Empirical assessment of a RGB-D sensor on motion capture and action recognition for construction worker monitoring

SangUk Han[1], Madhav Achar[2], SangHyun Lee[3*] and Feniosky Peña-Mora[4]

Abstract

Background: For construction management, data collection is a critical process for gathering and measuring information for the evaluation and control of ongoing project performances. Taking into account that construction involves a significant amount of manual work, worker monitoring can play a key role in analyzing operations and improving productivity and safety. However, time-consuming tasks involved in field observation have brought up the issue of implementing worker observation in daily management practice.

Methods: In an effort to address the issue, this paper investigates the performances of a cost-effective and portable RGB-D sensor, based on recent research efforts extended from our previous study. The performance of an RGB-D sensor is evaluated in terms of (1) the 3D positions of the body parts tracked by the sensor, (2) the 3D rotation angles at joints, and (3) the impact of the RGB-D sensor's accuracy on motion analysis. For the assessment, experimental studies were undertaken to collect motion capture datasets using an RGB-D sensor and a marker-based motion capture system, VICON, and to analyze errors as compared with the VICON used as the ground truth. As a test case, 25 trials of ascending and descending during ladder climbing were recorded simultaneously with both systems, and the resulting motion capture datasets (i.e., 3D skeleton models) were temporally and spatially synchronized for their comparison.

Results: Through the comparative assessment, we found a discrepancy of 10.7 cm in the tracked locations of body parts, and a difference of 16.2 degrees in rotation angles. However, motion detection results show that the inaccuracy of an RGB-D sensor does not have a considerable effect on action recognition in the experiment.

Conclusions: This paper thus provides insight into the accuracy of an RGB-D sensor on motion capture in various measures and directions of further research for the improvement of accuracy.

Keywords: Motion capture; Action recognition; Motion classification; RGB-D sensor; Machine learning

Introduction

During a construction project, data collection is critical to the evaluation and control of ongoing project performances. The complexity of construction environments and the dynamics of moving equipment and human resources, however, often pose a challenge in undertaking such tasks on a jobsite. Particularly, the time-consuming tasks required for worker monitoring can give rise to the issue of implementing field observation in a daily management practice (Johnson and Sackett 1998). For efficient field data acquisition, research efforts have thus been made to investigate and propose available sensing devices—such as cameras, laser scanners, and the combination of sensors (e.g., ultra wide-band and physiological status monitoring devices)—for the tracking of human movements and the analysis of construction activities (Cheng et al. 2013; Gong and Caldas 2011; Peddi et al. 2009; Gonsalves and Teizer 2009). The previous studies provide valuable insight into the analysis of human postures and actions, but further research is still needed for the capture of an articulated motion and the modeling of its kinematics.

* Correspondence: shdpm@umich.edu
[3]Department of Civil & Environmental Engineering, University of Michigan, Ann Arbor, MI 48109, USA
Full list of author information is available at the end of the article

Along this line, an RGB-D sensor—such as the Microsoft Kinect sensor—has gained great attention as a cost-effective and readily available device for motion capture.

Since it was released in 2010, the Kinect has been actively studied as a motion capture device to record the movement of human subjects. In this regard, action recognition techniques—in particular—have been explored for the detection of specific actions using the motion capture data for use with operation and safety analysis in construction. For example, Weerasinghe et al. (2012) propose a Kinect-based tracking framework for the localization of workers and the analysis of their movement patterns, which could potentially be used for productivity measurement. For operation analysis, Escorcia et al. (2012) also present an action recognition technique to classify construction workers' actions based on the color and depth information from a Kinect. On the other hand, Ray and Teizer (2012) utilize a Kinect for the pose analysis of construction workers to classify awkward postures based on ergonomic rules during safety and health training, and Han et al. (2013) study the unsafe action detection of workers for safety behavior monitoring with motion capture data from a Kinect. These studies have thus demonstrated the great potential of the Kinect to gather motion information from a jobsite, as well as the great potential of its applications to construction management. To validate the proposed approach, however, the prior work has mainly focused on the performances of motion classification and detection rather than the accuracy of estimated postures and actions (e.g., 3D human skeleton models). The results in the studies suggest that pose estimation is computationally verified to a certain extent, but the accuracy of the Kinect solely when used for motion capture still remains unexplored. Taking into account that one of the main uses of the Kinect is to estimate 3D body skeletons of humans and track their movements over time, the thorough assessment of a Kinect-based motion capture system will thus help elucidate: (1) up to what degree of accuracy a Kinect sensor can detect and track the 3D positions of body parts; (2) to what research areas the Kinect can potentially be applied, depending on the accuracy; and (3) which processes of motion analysis cause computational errors for the debugging of action recognition systems.

This paper evaluates the performance of the Kinect sensor on motion capture and action recognition for construction worker monitoring. An experimental study is undertaken to compare the accuracy of a Kinect with a commercial marker-based motion capture system, VICON, which has been used as the ground truth in prior work (Dutta 2012; Stone and Skubic 2011; Fernández-Baena et al. 2012). A VICON tracks the 3D locations of reflective markers attached to body parts with multiple cameras (e.g., 6 or 8 cameras), thereby minimizing occlusions and producing accurate tracking results. Extended from our previous work (Han et al. 2012), this paper performs the error analysis based on: (1) the estimated 3D positions of body joints, (2) the recomputed 3D rotation angles at particular joints, and (3) the effect of the motion capture accuracy on motion detection. The rest of this paper is organized as follows. Background section provides a background on the Kinect sensor and its performance evaluation. Methods section demonstrates a research methodology used to compute and analyze the three types of errors for the comparative study. Experiment section describes the experimental process for the collection of motion capture datasets with both a Kinect and a VICON. Results, including the error analysis, are presented and discussed in Results and discussion section. Finally, Conclusion section summarizes the findings of this study and suggests the direction of future research.

Background

This section summarizes the pros and cons of an RGB-D sensor (i.e., Kinect) for motion capture, and reviews previous work on the performance evaluation of a Kinect motion capture system. Based on the literature review, further research efforts required in this domain are identified.

An RGB-D sensor for motion tracking and analysis

The Kinect sensor was initially developed as a motion-sensing device for video gaming. A Kinect consists of two main components—one is a RGB camera that produces images at a 640 × 480 resolution, while the other is a depth sensor that measures the depth information of the image (Rafibakhsh et al. 2012). In addition, the depth sensor is comprised of both a projector and an infrared (IR) camera, all of which projects a structured IR light onto the scene and measures the depth by analyzing the distortion of the IR light (Weerasinghe et al. 2012; Khoshelhan 2011). Accordingly, the Kinect allows not only for the 3D reconstruction of a scene with point clouds but also for the 3D skeleton extraction of a human subject as combined with the motion capture solutions (e.g., OpenNI, Microsoft Kinect for Windows SDK, iPi Soft Motion Capture). In terms of the image processing for motion capture, the measured depth can be used for the building of 3D human models through 2D pose estimation (i.e., 2D skeletons with depth), as well as for the direct inference of 3D poses by integrating the depth into the pose estimation process. On the other hand, the use of IR light brings about constraints in the practical application of a Kinect to a field setting. For example, the Kinect's sensitivity of IR light to sunlight may cause unreliable motion capture outcomes in an outdoor environment, and its operating ranges for motion capture are also known to be limited (e.g., 0.8–4 m) (Weerasinghe et al. 2012; Han et al. 2013). Nevertheless, previous studies report that the operating distance for object tracking can be extended up to 10 m (Rafibakhsh et al. 2012) and 7.5 m (Ray and Teizer 2012) from a camera; hence, further investigation is required to clarify the range issue. Though limited to indoor applications, the Kinect still has the following notable advantages for motion sensing: (1) it requires no additional body attachment (e.g., markers, a special suit), which allows for worker observation without the interference of ongoing work; (2) the cost of a sensor (e.g., approximately 150–250 USD) is quite competitive, compared with other motion capture systems (e.g., approximately 96–120K USD for a marker-based VICON system) (Han et al. 2013); (3) the minimum number of sensors for motion tracking is only one Kinect; and (4) it provides an easy-to-use and easy-to-carry means for data collection in a field setting.

Previous work on the performance evaluation of an RGB-D sensor

For motion capture, performances of the Kinect can broadly be evaluated in terms of the functionalities such as the depth measured by a sensor and body part positions estimated by motion capture solutions. This section summarizes the previous work on depth measurement and discusses issues in the pose estimation assessment.

A principal function of the Kinect sensor is to compute the depth (i.e., the distance from a sensor) as a laser scanner does. Due to its low cost compared with that of a laser scanner (e.g., 10–130K USD) (Golparvar-Fard et al. 2011), previous studies have investigated the accuracy and resolution of Kinect depth data for the 3D

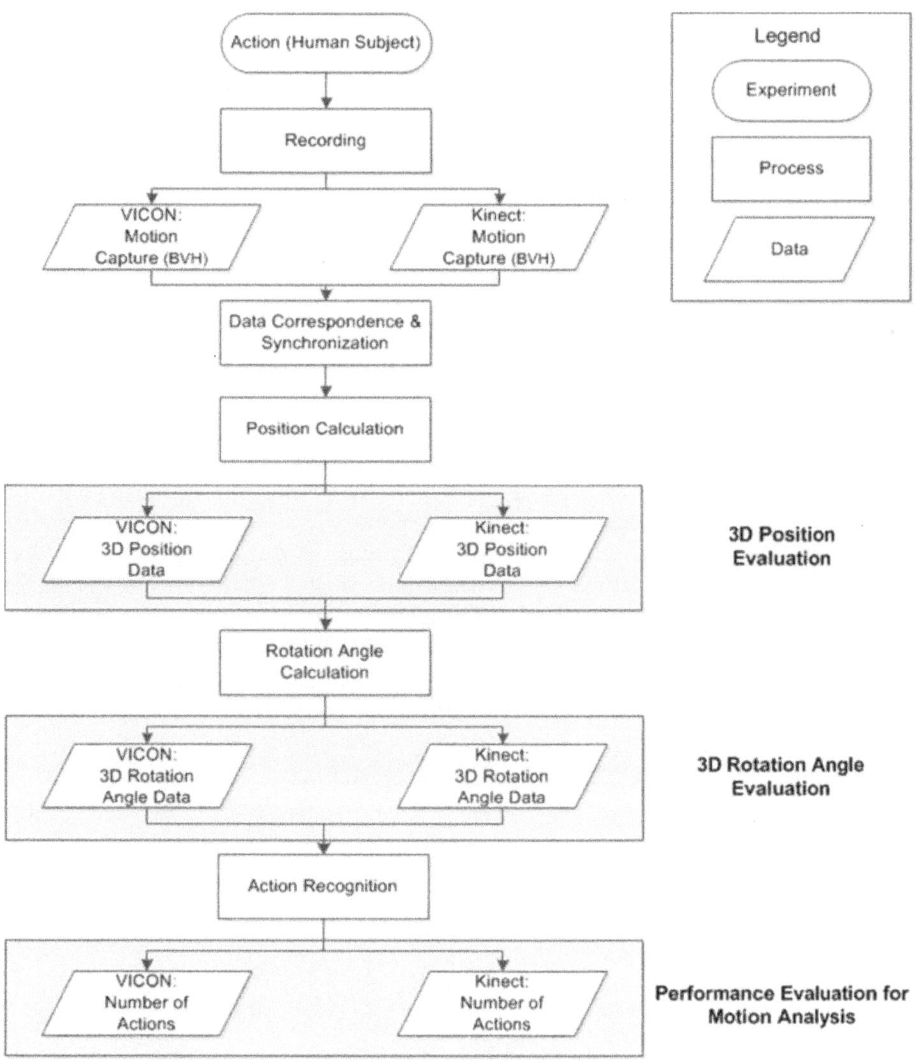

Figure 1 Overview of evaluation processes.

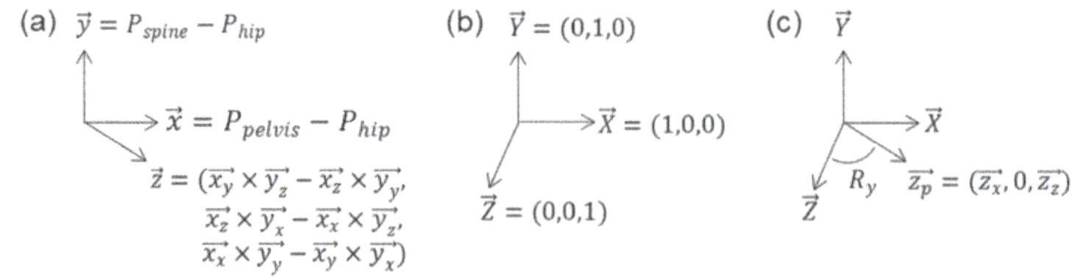

Figure 2 Y-axis rotation for data correspondences; (a) a local coordinate system of motion capture data, (b) a global coordinate system, and (c) Y-axis rotation between (a) and (b).

modeling of indoor environments, as well as for motion sensing. Khoshelham and Elberink (2012) report that the depth discrepancies between pairs of point clouds generated by a Kinect and a high-end laser scanner (i.e., Faro LS 880) are less than 3 cm for 84% of the point pairs, and that the point spacing in the depth direction (i.e., resolution) is about 2 mm, 2.5 cm, and 7 cm at the 1-, 3-, and 5-m distance. Rafibakhsh et al. (2012) also compare the accuracy and resolution of a Kinect with a laser scanner (i.e.. a Faro Focus3D scanner) and reveal that the average distance error between the point pairs is 3.49 cm, and that the resolution of the Kinect is about 4 times less than that of a laser scanner at 1.7- to 3.4-m distances from a sensor. Dutta (2012) measures the differences in distances between a Kinect and a VICON for a 0.1-m cube over a range of 1–3 m from a sensor, and the Root-Mean-Square Errors (RMSEs) are 6.5 mm in a horizontal direction, 5.7 mm in a vertical direction, and 10.9 mm in depth. On the other hand, Stoyanov et al. (2011) evaluate the accuracy of a Kinect in comparison with a laser scanner using the Three-Dimensional Normal Distributions Transform (3DNDT), which is a spatial representation accuracy evaluation technique, and conclude that the Kinect sensor performs well within 3.5-m distances. In Chow et al. (2012), a 3D reconstruction model of a mannequin is computed and compared with a laser scanner, and an RMSE of 11 mm is observed. In sum, previous studies reviewed herein conclude that the depth measurement and resolution of the Kinect are promising within a short range (e.g., 3 m), though not as accurate as those of a laser scanner, particularly in longer ranges.

The accuracy of motion capture data obtained with the Kinect has also been investigated. In Livingston et al. (2012), human skeletons tracked by a Microsoft software development kit are evaluated based on the positions of body joints (e.g., arms and hands) along a meter stick, and the average error and standard deviation in this experiment are 5.6 mm and 8.1 mm, respectively. Fernandez-Baena et al. (2012) conduct an experiment associated with rehabilitation treatments to compare the accuracy between a Kinect—combined with Natural Interaction Technology for End-user (NITE)—and a VICON in terms of the rotation angles of knee, hip, and shoulder joints, defined as angles

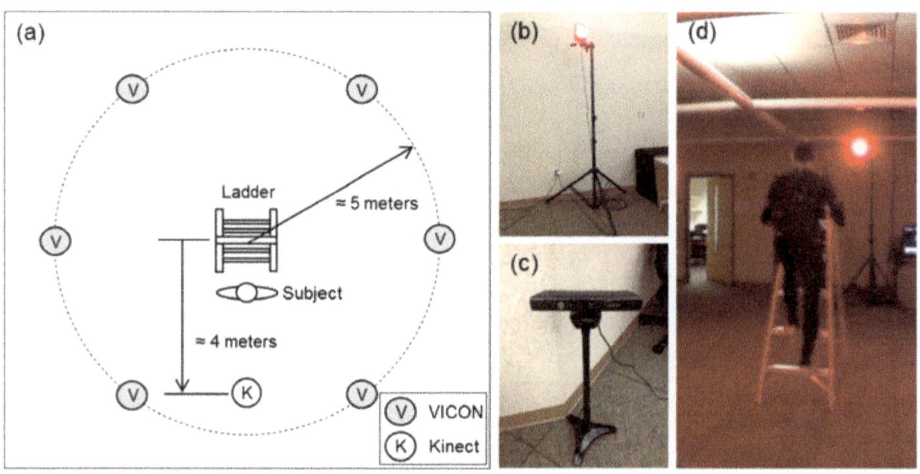

Figure 3 Experimental settings; (a) configurations of Kinect and VICON sensors, (b) a VICON sensor, (c) a Kinect, and (d) a human subject wearing a black suit and attaching reflective markers.

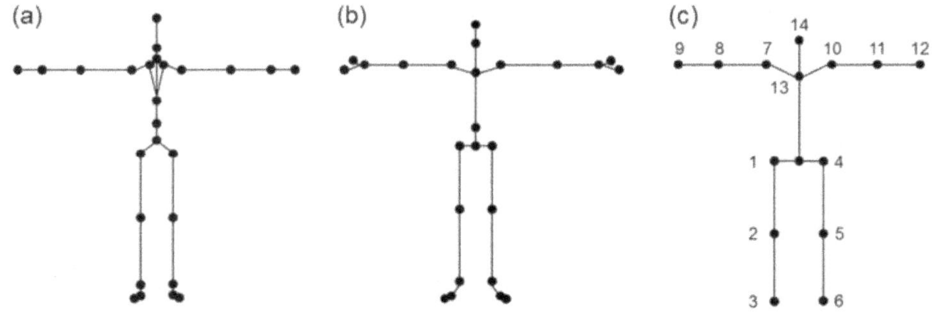

Figure 4 Human skeleton models; (a) a Kinect, (b) a VICON, and (c) a converted skeleton model (number: body joint ID).

between two vectors of body parts (e.g., one from knee to foot); the results show that the differences in rotation angles range from 6.78 to 8.98 degrees for a knee, from 5.53 to 9.92 degrees for a hip, and from 7 to 13 degrees for a shoulder. In the study of physical rehabilitation by Chang et al. (2012), the trajectories of the right hand, right elbow, and right shoulder that are tracked by a Kinect with OpenNI/NITE middleware are visually compared with those of marker-based OptiTrack motion capture system; the trajectories of a hand and an elbow are matched between two systems, while a shoulder is not accurately tracked by a Kinect system. To apply the Kinect to construction, however, further research efforts are required to address the following issues on the assessment of its motion capture performances: (1) the motions involved in construction activities need to be investigated, (2) the tracking results of full body joints need to be evaluated due to the characteristics of construction activities (i.e., manual work), and (3) the impact of the Kinect system's performances on action recognition needs to be studied for the analysis of construction worker monitoring and operation.

Methods

The objective of this paper is to assess the accuracy of Kinect motion capture data for the motion analysis of construction operations; Figure 1 illustrates an overview of evaluation processes comparing the outputs of VICON and Kinect motion capture systems. The evaluations are based on the error analysis of tracked 3D positions of full body joints, the 3D rotation angles at body joints used as a feature for motion classification, and the effect of the accuracy on action recognition. To compute the tracking errors, a VICON is used as the ground truth for motion tracking, and the iPi Motion Capture solution (http://ipisoft.com) is used with Kinect sensors to track the 3D positions of a human subject and extract 3D skeletons; the iPi Motion capture system estimates human poses mainly based on the depth measurements of a human body, and is thus less affected by a performer's appearance (e.g., special black suit and markers required by a VICON). In the experiment, human motions are thus simultaneously recorded with a Kinect and a VICON, and corresponding body joints of both systems—synchronized in time and space domains—are compared to compute the errors of Kinect outcomes. In addition, the ethics of this study including human subjects has been approved by the University of Michigan Institutional Review Board and the reference number is HUM00061888.

Data correspondence and synchronization

To compare the pose estimation results of a Kinect and a VICON, coordinate systems and data frames of both systems are matched through the rotation of coordinate systems and the synchronization of frames. For the spatial correspondence, local coordinate systems of both (i.e., coordinate systems defined by each system—an x-axis defined by the pelvis and a y-axis defined by the spine) are rotated about the y-axis into a global coordinate system (i.e., an absolute coordinate system newly defined for the coordinate system matching—a subject always faces the front) (Figure 2). In this experiment, a local coordinate system is defined based on the positions of a hip (i.e., P_{hip}), a spine (i.e., P_{spine}), and a pelvis (i.e.,

Table 1 Description of body parts and their joint IDs in Figure 4c

Body part ID	1	2	3	4	5	6	7
Body part	Left upper leg	Left leg	Left foot	Right upper leg	Right leg	Right foot	Left arm
Body part ID	8	9	10	11	12	13	14
Body part	Left forearm	Left hand	Right arm	Right forearm	Right hand	Neck	Head

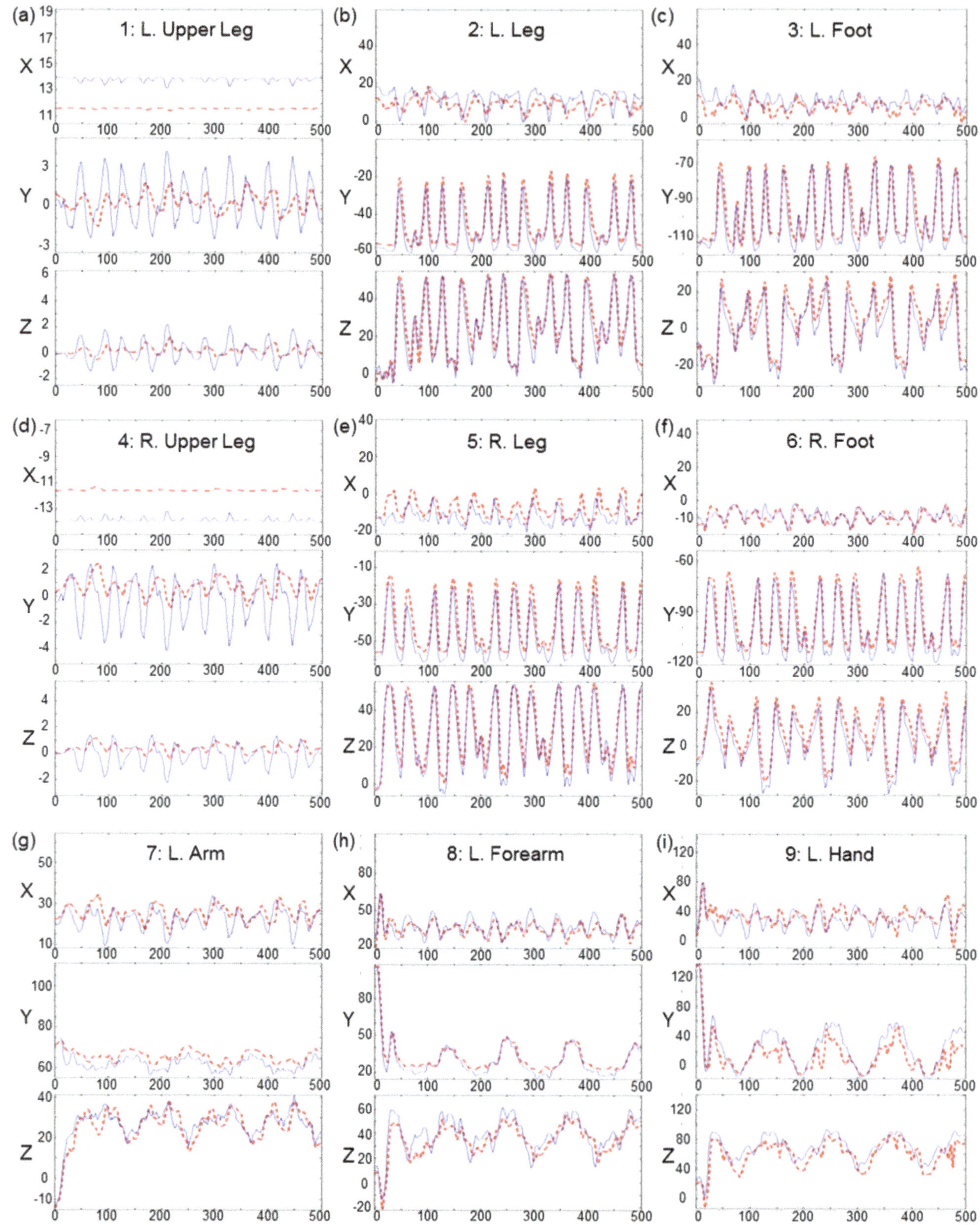

Figure 5 3D position trajectories of a Kinect and a VICON in x-, y-, and z-directions over the first 500 frames; (a) left upper leg, (b) left leg, (c) left foot, (d) right upper leg, (e) right leg, (f) right foot, (g) left arm, (h) left forearm, and (i) left hand.

P_{pelvis}) tracked by motion capture systems. The y-axis rotation angle, R_y, is calculated using Eq. (1):

$$R_y = \cos^{-1}\left(\overrightarrow{z_{p,x}} \times \overrightarrow{Z_x} + \overrightarrow{z_{p,y}} \times \overrightarrow{Z_y} + \overrightarrow{z_{p,z}} \times \overrightarrow{Z_z}\right) \quad (1)$$

where $\overrightarrow{z_{p,x}}, \overrightarrow{z_{p,y}}, \textit{ and } \overrightarrow{z_{p,z}}$ denote x, y, and z components of $\overrightarrow{z_p}$ in Figure 2c, and $\overrightarrow{Z_x}, \overrightarrow{Z_y}, \textit{ and } \overrightarrow{Z_z}$ denote x, y, and z components of $\overrightarrow{Z}$ in Figure 2b. Then, entire datasets of both systems are rotated using a rotation matrix obtained from R_y. In this manner, skeleton models of both systems face the front (i.e., z-axis), thus allowing for the comparison of skeletons in the same coordinate system regardless of viewpoints.

In the experiment, the synchronization of a pair of datasets is manually performed by identifying the same frame. For instance, we observe the frame in which a performer contacts a ladder's rung with a foot, and then we search for the exact frame among adjacent frames (e.g., 2 frames before and after the frame) by selecting the moment minimizing the distance between two datasets. In addition, the frame rates of the two systems are different (e.g., 120 frames per second for a VICON, and 30 frames per second for a Kinect). In the case of a VICON, thus 1 frame for every 4 is selected for the performance comparison. The accuracy is evaluated using RMSE in Eq. (2):

$$\mathrm{RMSE} = \sqrt{\sum_{i=1}^{n} \frac{\left(x_{v,i}-x_{k,i}\right)^2}{n}} \quad (2)$$

where x_v denotes a VICON data value, x_k denotes a Kinect data value at each frame (i), and n is the total number of frames.

Action recognition

To evaluate the impact of motion tracking accuracy on action recognition, this paper adopts the action detection framework presented in our previous work (Han et al. 2013). The framework consists of the dimension reduction of high-dimensional motion data, similarity measurements between a pair of motion data, and motion classification based on the measured similarity. First, dimension reduction is needed due to the high dimensions in motion data (e.g., 78), which hinder efficient and accurate action detection. Thus, we use Kernel Principal Component Analysis (Kernel PCA) (Schölkopf et al. 1998) to map motion data onto a 3D space, and then we compare the trajectories of datasets in the low-dimensional coordinate. In this space, a trajectory represents a sequential movement of postures (i.e., actions), and actions can be recognized by comparing the temporal patterns of transformed datasets. For the pattern recognition, temporal-spatial similarity between a pair of datasets is quantitatively measured using Dynamic Time Warping (DTW) (Okada and Hasegawa 2008). In this study, DTW measures Euclidean distances between datasets by warping the datasets in a time domain so as to compare datasets, even the sizes (i.e., durations) of which are different. For the performance evaluation, thus the similarity between a motion template (i.e., one trial of action datasets) and the entirety of the data is computed over all of the frames, and the behavior (e.g., fluctuation) of measured similarities is compared to investigate the effect of motion capture systems on the detection accuracy. Eventually, we perform the action detection that recognizes actions based on similarities by observing the ones with less similarity than a threshold (i.e., a classifier learned through classification); the detection results of Kinect and VICON datasets are compared in terms of accuracy (i.e., the fraction of correctly classified actions among all sample actions), precision (i.e., the fraction of correctly detected actions among detected ones), and recall (i.e., the fraction of correctly detected actions among ones that should be detected).

Experiment

To collect motion capture data, a lab experiment was conducted in the University of Michigan 3D Lab (Han et al. 2012); experimental configuration and scenes are illustrated in Figure 3. In this experiment, actions during

Table 2 3D position comparison (cm) of body joints between a Kinect and a VICON

ID	1	2	3	4	5	6	7	8	9	10	11	12	13	14	Mean
3D	2.7	8.3	8.8	3.1	10.0	9.5	6.8	9.1	24.3	6.8	12.4	21.7	19.0	7.7	10.7
(Std.)	(0.4)	(3.2)	(3.0)	(0.7)	(3.7)	(3.6)	(2.3)	(3.5)	(12.0)	(2.7)	(4.9)	(12.2)	(1.2)	(2.3)	(5.3)
X	2.3	5.2	4.5	2.3	5.9	2.9	4.5	4.4	11.3	4.1	8.7	14.7	4.1	4.8	5.7
(Std.)	(0.2)	(3.9)	(3.0)	(0.2)	(3.7)	(2.7)	(3.2)	(4.0)	(10.6)	(3.3)	(5.1)	(10.9)	(3.2)	(3.9)	(5.1)
Y	1.3	4.4	4.5	1.8	6.0	5.8	3.6	3.5	17.4	3.1	4.6	10.9	17.3	4.7	6.4
(Std.)	(1.3)	(3.0)	(3.7)	(1.3)	(3.1)	(4.1)	(1.5)	(3.1)	(13.7)	(1.5)	(2.5)	(10.4)	(1.0)	(1.0)	(5.1)
Z	0.6	4.8	6.0	0.9	5.4	7.0	3.7	7.1	12.6	4.4	7.6	11.7	6.6	3.8	5.9
(Std.)	(0.6)	(4.2)	(3.6)	(0.6)	(4.5)	(3.4)	(3.6)	(6.3)	(9.0)	(3.7)	(6.1)	(11.6)	(3.6)	(3.0)	(5.4)

(Unit: cm).

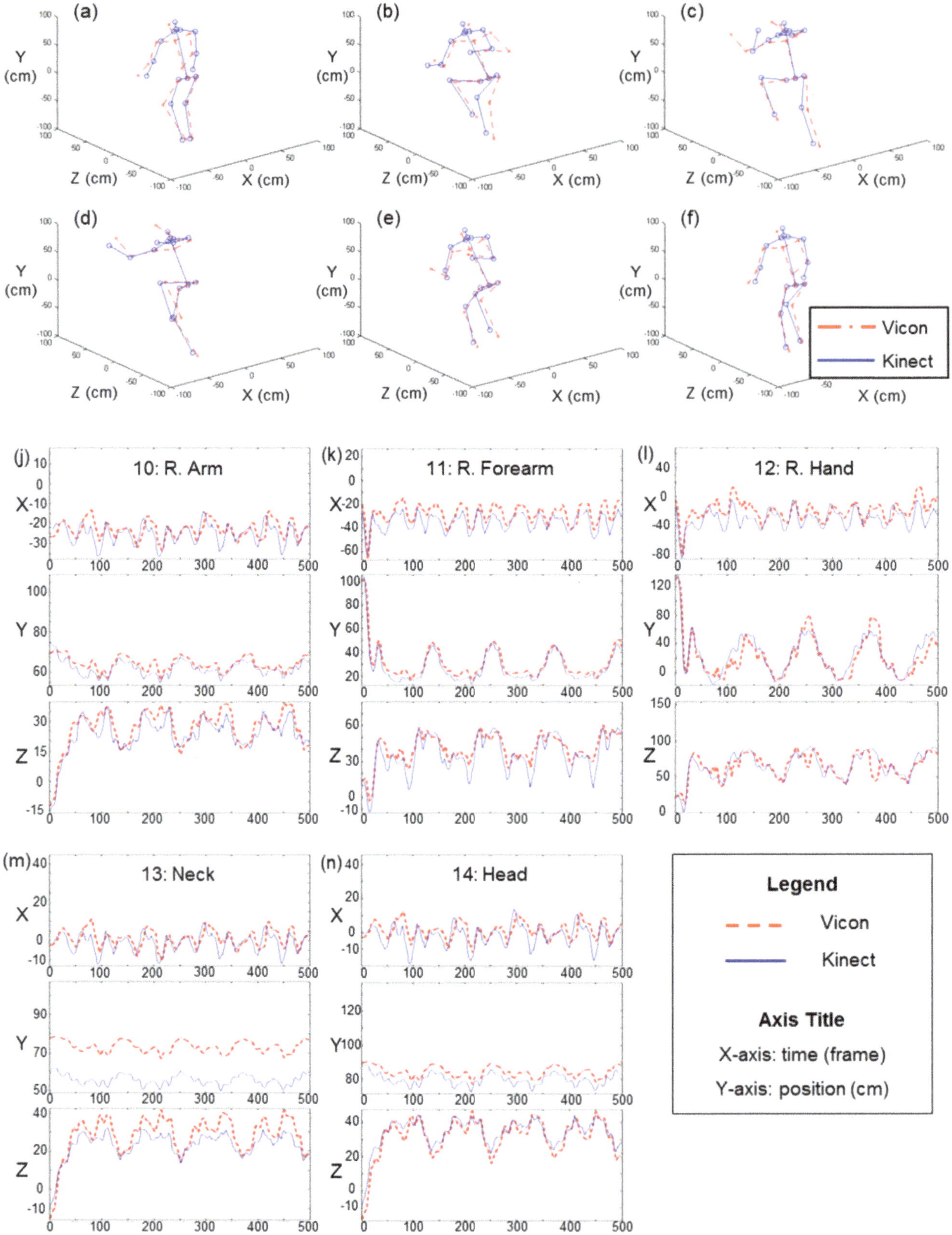

Figure 6 Comparisons of 3D position estimation between a Kinect and a VICON for a trial of ladder climbing; frames (a) 310, (b) 335, (c) 360, (d) 385, (e) 410, and (f) 435. 3D position trajectories of a Kinect and a VICON in the x-, y-, and z-directions over the first 500 frames; **(j)** right arm, **(k)** right forearm, **(l)** right hand, **(m)** neck, and **(n)** head (Note: graphs are scaled for each body part).

ladder climbing were recorded and analyzed; in construction, 16% of fatalities and 24.2% of injuries were caused by falls from a ladder in 2005 (CPWR 2008). 25 trials of each action (i.e., ascending and descending) taken by 1 subject were recorded with six 4-mega-pixel VICON sensors and a Kinect sensor. In total, 3,136 and 12,544 frames were collected with the Kinect and the VICON, respectively; and the datasets were synchronized for each system to have 3,136 frames for the comparison.

In this experiment, human skeleton models of the VICON and Kinect systems were slightly different in terms of the hierarchical structures of a human body; graphical illustrations of skeleton models extracted from each system are presented in Figure 4. Thus, for the comparison, corresponding body joints between the two systems are selected to convert the two models into the same form of a skeletal model (Figure 4c), and positions of such joints, as well as their rotation angles, are computed from motion capture data. For instance, motion capture data used in this study was in the Biovision Hierarchy (BVH) format (Meredith and Maddock 2001), in which a human posture at each frame is represented only with 3D Euler rotation angles. The BVH format also defines the 3D positions of body joints (i.e., translations) in an initial pose (e.g., T-pose as shown in Figure 4). This rotation and translation information forms a transformation matrix allowing for the computation of the 3D positions of all body joints in a global coordinate system (Meredith and Maddock 2001). To re-calculate Euler rotation angles (e.g., rotations in an order of x-, y-, and z-axes in this study) with respect to the converted skeleton model, an axis-angle between two body parts is first computed, a quaternion is defined with the axis-angle and axis vector, this quaternion forms a rotation matrix, and lastly a rotation angle is computed based on the rotation matrix (Han et al. 2012). Consequently, the 3D positions and rotation angles of each body part (Figure 4c) are compared to evaluate the tracking performances of the two systems; Table 1 describes body joint IDs corresponding to body parts in Figure 4c.

Results and discussion

To assess the performance of the Kinect as a motion capture system, we compare it with the VICON in terms of the results of: (1) 3D positions of body joints, (2) 3D rotation angles, and (3) motion detection for the datasets simultaneously collected through a lab experiment. Based on the error analysis, the applicability of the Kinect to the motion analysis of construction workers is discussed.

3D Position evaluation

To compare the 3D positions of body joints tracked by both systems, postures at each frame were iteratively rotated about the y-axis in a global coordinate system (Figure 2) over all of the temporally synchronized frames. Figure 5 visualizes skeleton models extracted from both systems at selected frames in the coordinate where two datasets are mapped. In this manner, the inspection of entire frames (i.e., animations) visually confirmed that the data correspondence and synchronization were successfully carried out for the two datasets. Through the visual investigation, we found that overall a Kinect model was closely matched with a VICON model, while hands and feet in particular were not exactly located in the same place.

For the quantitative assessment, RMSEs of body parts are computed over the entire frame using Eq. (2). Table 2 summarizes the RMSEs and standard deviations on distance differences in x-, y-, and z-directions, as well as in a 3D space; body part IDs refer to Figure 4 and Table 1. The temporal trajectories of the 3D positions of both systems in the first 500 frames are also presented in Figure 6. Compared with a VICON, a Kinect produces the discrepancy of 10.7 cm in a 3D coordinate, and no significant disparity in each direction was identified. The results show that the largest RMSEs are caused by the tracking of both hands (i.e., IDs 9 and 12) among body parts, and the large standard deviations of hands also indicate that the locations of such body parts are inconsistently estimated over the frames. Yet, Figures 6i and 6l imply that the patterns of a Kinect at large are still similar with those of a VICON. In addition, a large RMSE—the third greatest after that of the two hands—is found in a neck (i.e., ID 13). However, the standard deviation is relatively small, and most

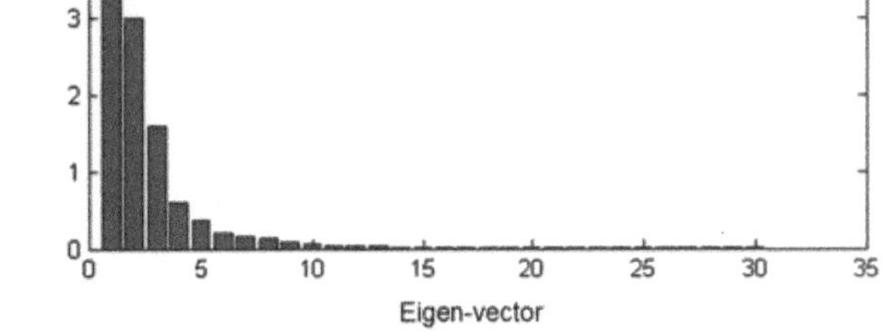

Figure 7 Eigen-decomposition for the internal dimensionality estimation.

Table 3 Rotation angle comparison (degree) at body joints between a Kinect and a VICON

ID	1	2	4	5	7	8	10	11	13	Mean
3D	5.1	5.6	6.2	8.2	13.9	34.2	18.9	49.0	4.4	16.2
(Std.)	(5.2)	(5.1)	(4.9)	(6.2)	(7.8)	(29.6)	(15.4)	(40.0)	(3.8)	(18.0)
X	6.6	7.3	6.7	8.1	12.1	31.2	18.7	38.6	6.0	15.1
(Std.)	(6.2)	(7.3)	(6.1)	(8.1)	(6.3)	(27.7)	(15.6)	(34.5)	(2.0)	(16.5)
Y	3.3	3.5	3.9	7.2	6.3	21.9	5.2	48.3	0.2	11.1
(Std.)	(3.2)	(2.9)	(2.7)	(5.1)	(5.5)	(19.6)	(5.2)	(33.8)	(0.2)	(13.5)
Z	5.6	5.9	7.9	9.2	23.4	49.5	32.9	60.0	7.0	22.4
(Std.)	(5.6)	(4.1)	(5.1)	(5.1)	(10.6)	(38.4)	(21.0)	(49.7)	(6.3)	(22.7)

(Unit: degree).

errors result from differences in a y-direction; this suggests that the tracking positions of a neck by the two systems are slightly different, as shown in Figure 6m. Next, relatively large RMSEs are caused by forearms (i.e., IDs 8 and 11), legs (i.e., IDs 2 and 5), and feet (i.e., IDs 3 and 6). As observed with hands, the trajectories of those body parts also similarly fluctuate over time with both a Kinect and a VICON (Figures 6h, 6k, 6b, 6e, 6c, and 6f).

The results show that the discrepancy between the Kinect motion capture system and a marker-based system is 10.7 cm on average in 3D positions. However, the estimated trajectories reveal that the Kinect sensor can still capture patterns of movements well, even with only one sensor. On the other hand, the use of one sensor may introduce the issue of occlusions. In the experiment, a Kinect sensor was positioned at the rear of a performer (Figure 3), and hence the performer's hands were frequently occluded by the performer, himself/herself. Also, forearms and legs, which were sometimes occluded as a performer climbed up and down a ladder, caused larger errors than other body parts. This implies that occlusions may have been a major source of errors in this experiment.

3D Rotation angle evaluation

In this experiment, rotation angles were the outcomes of both motion capture systems. However, other types (e.g., joint angles) of motion data—which can efficiently characterize human postures—can be obtained from motion capture systems and used as a feature for motion analysis. Taking into account that the selection of discriminating features significantly affects the classification performances (Mangai et al. 2010), we compared three data types in our previous study: rotation angles, joint angles (i.e., horizontal and vertical joint angles between a body part and x-y and x-z planes in a global coordinate system), and position vectors (i.e., normalized vectors of body parts) (Han, Lee, and Peña-Mora: Comparative study of motion features for similarity-based modeling and classification of unsafe actions in construction, submitted). The result reveals that, in the experiment, rotation angles outperformed the other two data types in applying the motion detection framework, which is also adopted in this paper. In this respect, 3D rotation angles used as inputs for motion analysis are compared to evaluate the accuracy of the Kinect and its impact on action recognition.

For the assessment, rotation angles were computed according to the converted skeleton model in Figure 4c. A rotation angle at a particular joint is defined as the angle rotating a vector of the joint (i.e., a vector from the joint to its child joint) from a corresponding vector in an initial pose. Thus, end-joints such as body part IDs 3, 6, 9, 12, and 14 are excluded from the comparison as not defined. Thus at the available joints, RMSEs of rotation angles in the x-, y-, and z-directions, as well as mean RMSEs of the three directions, were computed

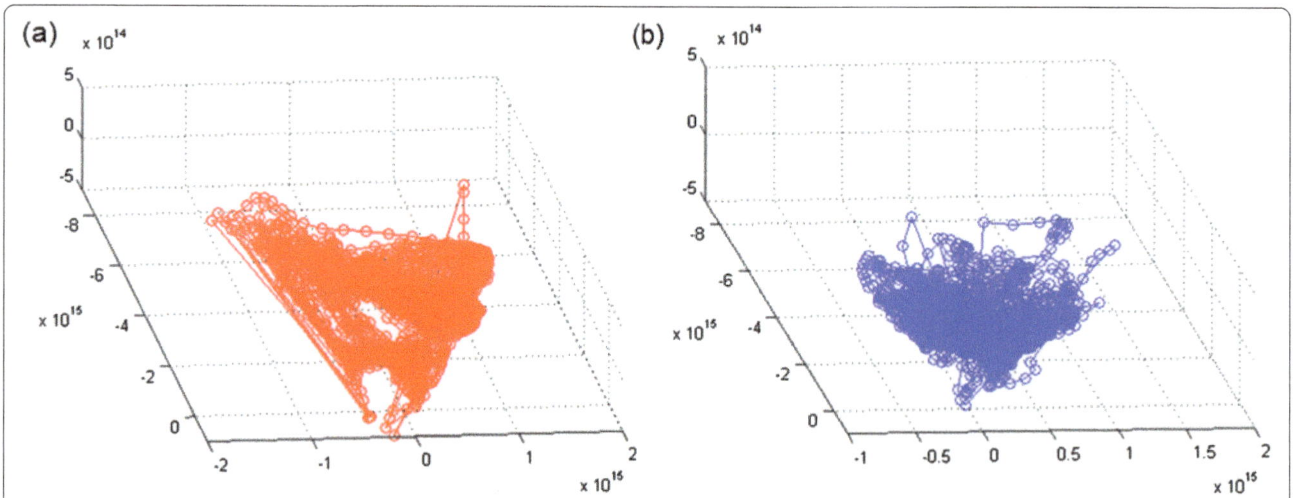

Figure 8 Dimension reduction results of (a) a VICON and (b) a Kinect (axis: eigen-vector; and unit: eigen-value).

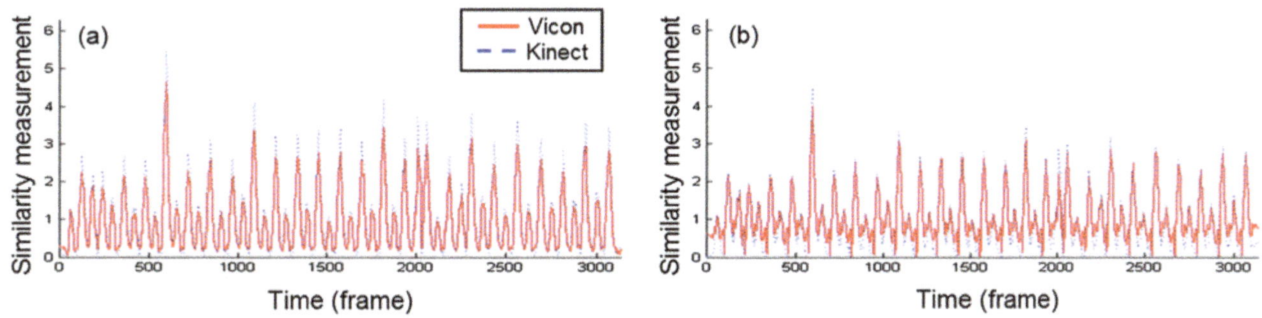

Figure 9 Comparison of similarity measurements between a VICON and a Kinect; (a) ascending, and (b) descending actions.

through all of the frames, and the results are presented in Table 3 (extended from Han et al. 2012). Compared to a VICON, overall the mean difference of 16.2 degrees and the standard deviation of 18.0 degrees were observed for the average of the three directions. For each direction, large errors occur in an order of z-, x-, and y-axis rotations (the z-axis has the largest error). In particular, the largest errors of up to 49 degrees were caused by forearms (i.e., IDs 8 and 11), which define the hand position. This implies that position errors at forearms and hands, which determine the rotation angles at forearms, can heavily magnify errors of rotation angles as combined. This phenomenon also explains the large errors of arms (i.e., IDs 7 and 10). The position errors of arms and forearms were not relatively large, but the combination of errors produces the second largest errors of rotation angles at the arms. Except for forearms and arms, the rotation angle errors of other body parts were less than 10 degrees.

Performance evaluation for motion analysis

To evaluate the performance of the Kinect for motion analysis, we applied a motion detection method (Han et al. 2013) to motion capture datasets from a Kinect and a VICON, and compared the results of detection based on conventional measures of classification performances (i.e., accuracy, precision, and recall). For motion analysis, dimensionalities of motion datasets were first reduced using kernel PCA. To determine the dimension to be reduced, eigen-decomposition was performed for the estimation of internal dimensionality in the datasets. As shown in Figure 7, the first three eigen-vectors have large eigen-values, which means that most information can be represented with three dimensions. In this regard, motion datasets were transformed onto a 3-dimensional coordinate; Figure 8 illustrates the distributions of each dataset in the low-dimensional space. In this space, a data point represents posture information at one frame, and hence the trajectories describe actions as changing postures over time. In Figure 8, the minimum and maximum values of each system are slightly different; for example, the ranges of x, y, and z values of Vicon are $[-8.5*10^{15}, 1.1*10^{15}]$, $[-1.0*10^{15}, 1.6*10^{15}]$, and $[-6.0*10^{14}, 4.3*10^{14}]$, while the ranges of x, y, and z values of Kinect are $[-4.9*10^{15}, 0.9*10^{15}]$, $[-0.4*10^{15}, 1.5*10^{15}]$, and $[-3.5*10^{14}, 3.2*10^{14}]$, respectively. However, the results indicate that the motion data captured by both systems could be mapped onto the same space. More importantly, despite large errors associated with body parts (e.g., arms and forearms) in rotation angles, the result of mapping (i.e., the transformation of high-dimensional data onto a low-dimensional space) reveals that the patterns of motion data can be preserved though dimension reduction; action detection is based on the comparison of patterns (i.e., trajectories) in a 3D space.

To compare the trajectories between actions, temporal-spatial similarities are measured using the DTW. In this experiment, one trial of datasets among 25 for each ascending and descending action was used as a motion template to compare its similarity with testing data and detect similar actions when the similarity is higher—or the distance is smaller—than a threshold. To avoid a

Table 4 Detection error comparison

Data source	Action type	# of actions in data	# of correctly detected actions		# of incorrectly detected actions		Acc. (%)	Prec. (%)	Rec. (%)
			Template (TP)	Other action (TN)	Not detected (FN)	Mis-detected (FP)			
Vicon	Ascending	25	25	25	0	0	100	100	100
	Descending	25	25	25	0	0	100	100	100
Kinect	Ascending	25	24	25	1	0	98	100	96
	Descending	25	25	25	0	0	100	100	100

biased assessment, a motion template from a Kinect and a VICON were compared with the same testing dataset (i.e., an entire frame of VICON data) for the detection; for instance, consistent errors (e.g., constantly estimating locations of a hand at a wrong but similar place) caused by a Kinect over the frames can positively affect the detection accuracy. The similarities measured over all of the frames are illustrated in Figure 9. Notwithstanding errors in Kinect data, the fluctuations of both datasets behave similarly over time. This result suggests that the errors of a Kinect system have not significantly affected the motion analysis in this experiment. Detection results (Table 4) also show that the accuracy, precision, and recall of a Kinect system are 98%, 100%, and 96%, respectively; only one trial among 25 was not detected.

Conclusions

This paper evaluates the performance of an RGB-D sensor (e.g., Kinect sensor) as a motion capture system based on the accuracy in estimated 3D positions and computed rotation angles, and the sensor's impact on action recognition. We conducted an experiment to collect motion capture data for 25 trials of ladder climbing actions, and we analyzed the accuracy on the datasets to identify the sources of errors. In the experiment, a 3D position RMSE and standard deviation were 10.7 cm and 5.3 cm, compared with a VICON. In the case of rotation angles, the RMSE and standard deviation were 16.2 degrees and 18.0 degrees, respectively. The rotation angles were used for motion detection, and the results show that among 25 trials, only 1 case of an ascending action was incorrectly detected (i.e., accuracies of 98% and 100% for ascending and descending actions, respectively). The experimental study implies that the inaccuracy of the Kinect motion capture system, particularly on occluded body parts, did not have a considerable effect on action recognition. However, the Kinect system produces large errors in estimating the positions of body parts, which can even increase errors as converted into rotation angles. The relatively lower accuracy of the Kinect system than that of marker-based systems can thus limit its application to construction; for example, the Kinect system may not be suitable for applications requiring high accuracy such as hand-related ergonomic analysis. Moreover, further investigation of Kinect performance evaluation on various actions (e.g., walking, running, lifting and carrying an object, and slipping) in construction operations is required for the thorough review of the feasibility of a Kinect for construction applications (e.g., productivity and safety). In addition, occlusions by a performer or other moving objects might be common in construction; thus a single Kinect motion capture system may potentially produce noise in a field setting. In this respect, further investigation on the use of multiple Kinect sensors is required to collect reliable motion information on a jobsite.

Competing interests

The authors declare that they have no competing interests.

Authors' contribution

SH carried out the motion analysis studies, participated the sequence alignment and drafted the manuscript. MA undertook the experiments to collect data using a motion capture system. SL and FP directed the entire processes of this study, provided suggestions on each procedure in the data collection and analysis, and reviewed the manuscript. All authors read and approved the final manuscript.

Author details

[1]Department of Civil and Environmental Engineering, University of Illinois at Urbana-Champaign, Urbana, IL 61801, USA. [2]Department of Computer Science, University of Michigan, Ann Arbor, MI 48109, USA. [3]Department of Civil & Environmental Engineering, University of Michigan, Ann Arbor, MI 48109, USA. [4]Department of Civil Engineering and Engineering Mechanics, Earth and Environmental Engineering, and Computer Science, Columbia University, New York, NY 10027, USA.

References

Chang, CY, Lange, B, Zhang, M, Koenig, S, Requejo, P, Somboon, N, Sawchuk, AA, & Rizzo, AA. (2012). *Towards pervasive physical rehabilitation using Microsoft Kinect* (pp. 159–162). San Diego, CA: 2012 6th international conference on pervasive computing technologies for healthcare (pervasiveHealth). May 21–24, 2012.

Cheng, T, Migliaccio, GC, Teizer, J, & Gatti, UC. (2013). Data fusion of real-time location sensing and physiological status monitoring for ergonomics analysis of construction workers. *Journal of Computing in Civil Engineering, 27*(3), 320–335.

Chow, J, Ang, K, Lichti, D, & Teskey, W. (2012). *Performance analysis of a low-cost triangulation-based 3D camera: Microsoft Kinect system.* Melbourne, Austrailia: The XXII Congress of the International Society for Photogrammetry and Remote Sensing.

CPWP – The Center for Construction Research and Training. (2008). *The construction chart book: the U.S. construction industry and its workers.* Washington, D.C: CPWP.

Dutta, T. (2012). Evaluation of the Kinect sensor for 3-D kinematic measurement in the workplace. *Applied Ergonomics, 43,* 645–649.

Escorcia, V, Dávila, MA, Golparvar-Fard, M, & Niebles, JC. (2012). *Automated vision-based recognition of construction worker actions for building interior construction operations using RGBD cameras.* West Lafayette, Indiana: Proceeding of 2012 Construction Research Congress (CRC). May 21–23, 2012.

Fernández-Baena, A, Susín, A, & Lligadas, X. (2012). *Biomechanical validation of upper-body and lower-body joint movements of Kinect motion capture data for rehabilitation treatments* (pp. 656–661). Bucharest: 2012 4th international conference on intelligent networking and collaborative systems. Sep.19–21, 2012.

Golparvar-Fard, M, Bohn, J, Teizer, J, Savarese, S, & Peña-Mora, F. (2011). Evaluation of image-based modeling and laser scanning accuracy for emerging automated performance monitoring techniques. *Automation in Construction, 20*(8), 1143–1155.

Gong, J, & Caldas, CH. (2011). *Learning and classifying motions of construction workers and equipment using bag of video feature words and Bayesian learning methods* (pp. 274–281). Miami, FL: Proceeding of 2011 ASCE International Workshop on Computing in Civil Engineering. June 19–22, 2011.

Gonsalves, R, & Teizer, J. (2009). *Human motion analysis using 3D range imaging technology.* Austin, Texas: 26th International Symposium on Automation and Robotics in Construction (ISARC). June 24–27, 2009.

Han, S, Achar, M, Lee, S, & Peña-Mora, F. (2012). *Automated 3D human skeleton extraction using range cameras for safety action sampling* (12th international conference on construction applications of virtual reality (CONVR)). Taipei, Taiwan: National Taiwan University.

Han, S, Lee, S, & Peña-Mora, F. (2013). Vision-based detection of unsafe actions of a construction worker: a case study of ladder climbing. *ASCE Journal of Computing in Civil Engineering*. in press.

Johnson, A, & Sackett, R. (1998). Direct systematic observation of behavior. In HR Bernard (Ed.), *A handbook of methods in cultural anthropology* (pp. 301–331). Walnut Creek, California: AltaMira Press.

Khoshelhan, K. (2011). Accuracy analysis of Kinect depth data. In DD Lichti & AF Habib (Eds.), *International Archives of the Photogrammetry, Remote Sensing and Spatial Information Sciences* (pp. 133–138). Calgary, Canada: Proceeding of ISPRS workshop laser scanning 2011 (Volume XXXVIII-5/W12th ed.). Aug. 29–31, 2011.

Khoshelham, K, & Elberink, SO. (2012). Accuracy and resolution of Kinect depth data for indoor mapping applications. *Sensors, 12*(2), 1437–1454.

Livingston, MA, Sebastian, J, Ai, Z, & Decker, JW. (2012). *Performance measurements for the Microsoft Kinect skeleton* (pp. 119–120). Costa Mesa, CA: 2012 IEEE in Virtual Reality Workshops (VR). March 4–8, 2012.

Mangai, UG, Samanta, S, Das, S, & Chowdhury, PR. (2010). A survey of decision fusion and feature fusion strategies for pattern classification. *The Institution of Electronics and Telecommunication Engineers (IETE) Technical Review, 27*(4), 293–307.

Meredith, M, & Maddock, S. (2001). *Motion capture file formats explained.* University of Sheffielf: Department of Computer Science. http://www.dcs.shef.ac.uk/intranet/research/public/resmes/CS0111.pdf. Accessed 8 May 2013.

Okada, S, & Hasegawa, O. (2008). *Motion recognition based on dynamic-time warping method with self-organizing incremental neural network.* Tampa, FL: Proceeding of 19th International Conference on Pattern Recognition (ICPR).

Peddi, A, Huan, L, Bai, Y, & Kim, S. (2009). *Development of human pose analyzing algorithms for the determination of construction productivity in real-time* (pp. 11–20). Seattle, WA: Proceedings of the 2009 Construction Research Congress, Building a Sustainable Future. April 5–7, 2009.

Rafibakhsh, N, Gong, J, Siddiqui, MK, Gordon, C, & Lee, HF. (2012). *Analysis of XBOX Kinect sensor data for use on construction sites: depth accuracy and sensor inference assessment* (pp. 848–857). West Lafayette, IN: Proceeding of 2012 Construction Research Congress (CRC).

Ray, SJ, & Teizer, J. (2012). Real-time construction worker posture analysis for ergonomics training. *Advanced Engineering Informatics, 26*, 439–455.

Schölkopf, B, Smola, A, & Müller, K. (1998). Nonlinear component analysis as a kernel eigenvalue problem. *Neural Computation, 10*, 1299–1319.

Stone, E, & Skubic, M. (2011). Evaluation of an inexpensive depth camera for in-home gait assessment. *Journal of Ambient Intelligence and Smart Environments, 3*, 349–361.

Stoyanov, T, Louloudi, A, Andreasson, H, & Lilienthal, AJ. (2011). *Comparative evaluation of range sensor accuracy in indoor environments.* Orebro, Sweden: European Conference on Mobile Robots (ECMR). Sep. 7–10, 2011.

Weerasinghe, IPT, Ruwanpura, JY, Boyd, JE, & Habib, AF. (2012). *Application of Microsoft kinect sensor for tracking construction workers* (pp. 858–867). West Lafayette, IN: Proceeding of 2012 Construction Research Congress (CRC).

The potentials of and barriers to the utilization of advanced computer systems in remote construction projects: case of the Kingdom of Saudi Arabia

Bhzad Sidawi* and Abdulsalam Alsudairi

Abstract

Background: An extensive body of knowledge indicated the positive impact of the Advanced Computer based Management Systems (ACMS) on various aspects of project management, while highlighting barriers that hinder adoption, diffusion, and utilization of the ACMS by the construction industries around the world. Remote projects have their unique management problems and these are caused mainly by the remoteness of the project. Little research was undertaken concerning this issue, particularly in the Persian Gulf region, and it has highlighted few unique communications and management problems such as the loose control, lack of human resources, infrastructure and experience.

Methods: This research investigated the use of ACMS by large companies in the Eastern province, Kingdom of Saudi Arabia (KSA), and how it would help these companies sorting out a number of present projects' management problems. Subsequently, a field study i.e. a questionnaire survey and interviews was carried out.

Result: The field study revealed significant association between frequent management problems with little use of ACMS and the domination of use of traditional communications and management systems. This paper argues that the use of traditional systems and the traditional way of sorting out construction problems limit the applicability of ACMS.

Conclusion: The present researchers recommend the use of customized ACMS associated with the application of lean and sustainable management principals as these would help overcoming barriers and providing intelligent solution for the strategic, technical, and social issues of the remote construction sites.

Keywords: Mobile systems; Web-based project management systems; Remote construction projects

Background

Remote construction projects exist in many regions throughout the world such as the Sahara desert, Antarctic regions, the Arabian Peninsula desert, the Australian desert, the Empty Quarter etc. The dilemma in managing remote projects is highlighted by Deng et al. (2001), Kestle and London (2002, 2003), Kestle (2009), McAnulty and Baroudi (2010), and Thorpe (2000). These authors have pointed out that the remoteness thus the loose control is major cause of the management problems. They suggested possible causes such as the lack of human resources, infrastructure and experience of how to manage these remote projects. In the Kingdom of Saudi Arabia, remote construction sites have unique problems and these have been highlighted by Justanyah and Sidawi (2011), Sidawi 2010a,b, 2012a,b.

The literature review suggested that some of the project management and communications problems are caused by the use of inappropriate tools and systems for communication, coordination, and management. For example, Yang et al. (2007) suggested that intense need for project information and effective communications by the project team cannot be met by traditional communications and information management systems since these systems have shortcomings and are incapable of fulfilling project

* Correspondence: Bsidawi@ud.edu.sa
Architecture Department, College of Architecture and Planning, University of Dammam, 31451 Dammam, P O Box 2397, Kingdom of Saudi Arabia

duties and objectives. One of these shortcomings is that traditional systems provide limited access to information, which is considered one of the key barriers to successful project management practices (Vadhavkar and Pena-Mora 2002; Pena-mora et al. 2009). Many of these project failures are caused by inadequate organization and management of the construction process (e.g., a weak coordination of processes and uncertainty about available information) (Wamelink et al. 2002).

These management problems can be sorted out by the use of ACMS such as mobile, Web-based Project Management Systems (WPMS), augmented reality and BIM (Alshawi and Ingirige 2003 and Stewart and Mohamed 2004, Nitithamyong and Skibniewski 2004, 2006, Charoenngam et al. 2004, Arayici and Aouad 2010, Arayici et al. 2012). These systems use wireless, satellite, Internet-based, or mobile tools and networks and it helps - to a certain degree - construction industry firms manage the increasing complexity of normal construction projects. They have also helped fulfill project objectives such as quality, scope, time, and cost. Sustainable practices and measures would also help in sustaining the quality of projects, eliminate waste, and minimize the cost (Arayici et al. 2012). Previous research however, highlighted the barriers that affect the adoption, diffusion, and full application of the technology (Alshawi and Ingirige 2003, Nuria 2005; Leskinen 2006, 2008, Nitithamyong and Skibniewski 2007).

In the KSA, some large companies such as Aramco and Royal commission of Jubail (RCJ) have a number of remote construction projects. These projects are of different sizes i.e. medium, large to very large. They are in remote locations and some operate in undeveloped and environmentally sensitive regions. They are far from the supervision team office, the contractor's office, and major urban concentrations. During construction, all project parties experience countless difficulties and cumbersome management problems. These potential problems negatively affect project quality and cause substantial delays and increases in costs. This paper investigates the potential use of ACMS by large companies in the Eastern province in managing remote projects and barriers that hinder the potential use, and how ACMS would help these companies in sorting out project's problems and improving their management practices.

Literature review

Remote construction sites problems

The dilemma of managing remote projects is highlighted by Deng et al. (2001), who mentions that the extensive physical distance between project participants, sometimes extending over national boundaries, is the primary cause of delays in decision making. The project team has to not only tackle traditional management problems but those that specifically occur as a result of the remote locations of these often environmentally sensitive sites (Kestle 2009; Kestle and London 2002, 2003). These sites are often far from logistic support and suffer a continuous shortage of materials and specialized labor (Kestle and London 2002, 2003). Kestle (2009) investigated the management problems of remote projects and reports lack of project pre-planning, certainty, and/or clarity concerning project process integration. There were also misinterpretations and miscommunications of project results and needs issues. A centralized decision-making process and lack of delegated authority to field personnel often hindered progress and communications at critical emergency response and recovery stages. Kestle and London (2002) suggested a framework for the design management of remote sites. The framework emphasizes the following management functions: serving, controlling, organizing and economizing.

McAnulty and Baroudi (2010) conducted a survey of top and mid-tier construction contractors with experience in remote construction projects in Australia. They find that contractors experience difficulty attracting and retaining skilled workers; working in remote locations has a negative impact on an employee's family life. It is difficult to procure and access materials and equipment in remote areas and severe climatic factors in remote areas have a negative impact on productivity. There is lack of infrastructure and communications. The researchers suggest a number of possible solutions such as the need for appropriate material management systems and design cost information specifically for remote construction works. They recommended that unique types of costing issues should be included in the project's cost estimation at the pre-construction stages of project; these include: mobilization/demobilization, accommodation, inclement weather downtime, site allowances, delivery, and productivity.

Justanyah and Sidawi (2011), and Sidawi 2010a,b, 2012a,b examined the case of Saudi Electric Company (SEC). SEC has a number of remote projects and it experiences difficulties in running these projects. These problems can be summarized as the following:

A. Procurement and risk Management: There is a frequent shortage of materials. The delivery of materials and equipment is constrained by road/highway regulations and bad conditions of some remote roads. Projects have much higher risk margins than ordinary projects. This is caused by the ad hoc approach and both sides i.e. the owner and the contractor do not accurately plan projects.
B. Cost, time, scope, and quality management: There is serious delay in decision-making, loose control, and infrequent visits to the remote site result in wasted

time, excessive costs, unfocused scope, and poor construction quality.

C. Human Resources: SEC has a staff shortage so employees are incapable of doing all required site visits. There is a lack of security and shortage in skilled workers. The impact of the harsh working conditions has negative impact on the productivity of SEC's supervisors.

D. Infrastructure and communications: Land ownership in some remote areas is not definite or known; legal disputes are likely to occur. There is a lack of or no infrastructure. Contractors and SEC supervisors still use traditional communications and management tools. Decisions are made autocratically and SEC's project managers are not able to control and coordinate integration of a project's aspects and the typical management style is non-standard, fragmented, and loose.

The potentials and barriers to the use of ACMS

The potentials of ACMS

Some of a remote project management's problems can be avoided if ACMS were used. So, it is important to know how far these systems would be capable in sorting remote construction problems. These systems include Web based Project Management Systems WPMS and mobile systems that feature mobile tools, personal digital assistants (PDA), wearable computers, wireless tools, four dimensional augmented reality (4D), virtual reality and other technologies. These systems possess the capability to improve communications between project team members and enable teams to share information and quickly solve problems (Charoenngam et al. 2004). They improve team members' ability to manage time and costs (ibid). Davidson and Moshini (1990) and Bowden (2005) stated that construction costs can be reduced by 25% through efficient transfer of information between the construction teams; that transfer can be achieved through ACMS. Ahuja et al. (2010) suggested that adoption of Information Communication Technology (ICT) enables effective communication between dispersed project team members but argues that strategic adoption of ICT. This enables them to grasp effectively the IT benefits (Ahuja et al. 2009).

Alshawi and Ingirige (2003) and Stewart and Mohamed (2004) identified the following benefits of using WPMS: productivity enhancement of communication between project participants, reduction in project delays, heightened awareness of project issues among all parties, and ease of access to and retrieval of project information. Other advantages include: avoiding delays due to the arrival of updated drawings and documents, reducing visits to sites and travelling time to meetings, avoiding drawing mistakes, reducing time and money spent on disputes, sharing and exchanging project information, automating repetitive routine processes, and eliminating paper reports. Thomas et al. (2003) discussed how WPMS - from the point of view of selecting contractors - helps project managers boost contractor performance and confidence by minimizing subjectivity and eliminating the potential for corrupt practices. This improves competitiveness through increased awareness of competitors' strengths and weaknesses and nurtures mutual trust in the exchange of sensitive information such as performance data. Nitithamyong and Skibniewski (2004, 2006) suggested that benefits of using WPMS can be categorized into four main areas. These categories include cost reduction and time saving, enhancement of communications and collaboration, improvement of productivity and partnership, and support of e-commerce and the customer. A number of researchers anticipate that WPMS will replace traditional project management methods (Becerik 2005; Zou and Roslan 2005) and these methods are drivers of WPMS adoption. Several aspects support this claim including increased competitive pressures, expectations of revenue growth, the ability to compete globally, and the desire to reengineer the business to respond to market challenges (Nitithamyong and Skibniewski 2006).

Leskinen (2006, 2008) argued that it is difficult to make direct assessment of which mobile systems would benefit the construction industry. The most important intangible benefits include improving customer service, gaining a competitive advantage, acquiring more timely management information, supporting core business functions, avoiding competitive disadvantages, improving management information, improving product quality, improving internal communication, implementing changes through innovation, improving external communication, and enhancing the jobs of employees. In recent years, the development of laser scanning and video and image-based 3D reconstruction system is enabling remote and virtual walk through on actual construction sites. These systems have the ability to minimize the travel times of supervisors and may increase the frequency of progress, quality and safety inspections by providing project supervisors with systems that are easily applicable (Golparvar-Fard et al. 2011, Jaselskis et al. 2011). The utilization of Building Information Modeling (BIM) by the project team would provide a more streamlined business process, associated project and site management methodologies including complete facilitation of construction knowledge during the full lifecycle of a building project (Arayici and Aouad 2010). Arayici et al. (2012) argued that BIM implementation serves as a useful alternative to addressing key construction sector issues, and offer solutions to these in order to increase productivity, efficiency, quality; reduce costs, lead times and duplications, via effective collaboration and communication of stakeholders in remote construction projects. These key findings, in terms of challenges, also lend support to the classification of the key challenges for construction project management in remote construction projects, such as human resources, cost, time, scope, and quality management; procurement and risk management, infrastructure and communication.

Justanyah and Sidawi (2011) and Sidawi 2010a,b, 2012a,b have studied the potentiality of Communications and Project Management Systems (CPMS) and whether it would help SEC in avoiding some construction problems, sorting out efficiently the site queries and improving the management of these projects. They pointed out that the CPMS would support SEC's management activities, particularly during the construction stages that witness long delays in providing feedback to the construction enquiries. CPMS can be used to provide fast feedback to the site personnel regarding the most problematic construction problems such as these related to the monitoring of the construction process, quality of work, procurement of materials and productivity levels.

The section above highlighted the problems and capabilities of ACMS regarding each of the project management aspects (see also Table 1). The previous research indicated that remote construction sites have unique problems that need unique and customized IT solutions. It also indicated that remote construction sites have some common problems such as these related to time, potential risk, pre-planning, cost, and logistic support despite that they are in different geographical locations.

Barriers to the use of ACMS

Despite fast developments in IT and the creation of many IT applications for the construction industry, some issues still hinder the applicability of these systems to construction project management. There is a problem with regard to the diffusion of IT in the construction industry and the absorption of IT into work practices. This includes the level of strategic IT investment by construction industry firms (Alshawi et al. 2009). Other barriers include IT technical shortages, deployment of the system on an ad hoc basis, isolated project management practices, and costly systems (Alshawi and Ingirige, 2003, Nuria, 2005, Leskinen, 2006, 2008). To minimize such barriers and enable ICT adoption, the following issues should be investigated (Margherita and Petti, 2010):

- *Strategy: the action plan deriving from an integrated view of organization's goals and priorities, people expectations, and potential benefits;*

Table 1 The management problems of remote construction projects and potential ACMS solutions

Management problems	Potential ACMS solutions
The extensive physical distance between project participants, sometimes extending over national boundaries, is the primary cause of delays in decision making (Deng et al., 2001)	ACMS can improve team members' ability to communicate and manage time and costs (Charoenngam et al., 2004)
Lack of project pre-planning, certainty, and/or clarity concerning project process integration (Kestle, 2009)	ACMS use would reduce cost and save time, enhance communications and collaboration, improve productivity and partnership, and support the e-commerce and the customer (Nitithamyong and Skibniewski, 2004, 2006)
These sites are often far from logistic support and suffer a continuous shortage of materials and specialized labor (Kestle and London, 2002, 2003). Contractors experience difficulty attracting and retaining skilled workers; and it is difficult to procure and access materials and equipment and site's conditions has a negative impact on productivity (McAnulty and Baroudi, 2010)	Construction costs can be reduced by 25% through efficient transfer of information by ACMS between the construction teams (Davidson and Moshini, 1990, Bowden, 2005)
Misinterpretations and miscommunications of project results and needs issues; and a centralized decision-making process and lack of delegated authority to field personnel often hindered progress and communications (Kestle, 2009)	ACMS use enables productivity enhancement of communication between project participants, reduction in project delays, alleviation of awareness of project issues among all parties, and easy access to and retrieval of project information (Alshawi and Ingirige 2003, and Stewart and Mohamed, 2004)
There is frequent shortage of materials, higher risk margins, serious delay in sorting out a number of project queries and Delays in decision-making and decisions are made autocratically, excessive costs, unfocused scope, and poor construction quality, unskilled workers, a lack of or no infrastructure (Justanyah and Sidawi, 2011 and Sidawi, 2010a, b, 2012a, b)	ACMS use minimizes delays due to the arrival of updated drawings and documents, reduces visits to sites and traveling time to meetings, reduces time and money spent on disputes, automates repetitive routine processes, and eliminates paper reports (Alshawi and Ingirige 2003, and Stewart and Mohamed, 2004)
	ASMC have the ability to minimize the travel times of supervisors and may increase the frequency of progress, quality and safety inspections by providing project supervisors with systems that are easily applicable (Golparvar-Fard et al., 2011, Jaselskis et al. 2011).
	ACMS would help SEC's supervisors to monitor the Procurement, supply and consumption of construction materials. It would would help SEC eliminating some causes of potential delays. It helps in monitoring closely and frequently the contractors/ sub-contractors and site personnel activities and construction process, performance and outcomes (Justanyah and Sidawi 2011 and Sidawi 2010a, b, 2012a, b)

- *People: the single individuals' attitude and the overall organizational context which impact on the level of willingness and readiness to change;*
- *Process: the real unit of analysis and trigger of change in terms of alternative redesign scenarios and associated impact; and*
- *Enablers: the potential facilitators of implementation at technological and organizational level*

Nitithamyong and Skibniewski (2007) investigated the impact of project success/ failure factors on performance of WPMS. This is regarding the following aspects: project characteristics, project team characteristics, service provider characteristics and system characteristics. In respect to the project and project team characteristics, they found that strategic, time, cost, quality, Risk, and communication improvement are significantly linked: owner type, project type, project cost, project duration, and starting stage of PM-ASP development. In regards to project team characteristics, the above project factors are significantly linked with the project Parties deciding for the use of PM-ASP.

Anumba et al. (2006) found that the following structural and social impediments to the integration of ICT system: commitment from top management, continuing support for users, highly reliable systems, involvement of users, easy to use systems, effective project management, clear cost justification, well proven systems, full training, precisely defined objectives, facilities to meet specific needs, use of pilots or prototypes and organizational change (see also Grimshaw and Kemp 1989).

Methods

The literature review have highlighted the potentials of the ACMS in a number of geographical locations around the world so it would be useful to test the useful of these systems for large companies within the context of KSA. This study examines the management of remote construction sites by large companies and how the use of ACMS would help these companies improving its' management of remote projects. The objectives of the research are:

- To find out possible management and communications problems of remote construction sites;
- To explore how these problems would be sorted out using ACMS and how ACMS would help in improving management practices; and
- To identify possible barriers to the full application of ACMS

To achieve the research objectives, it is argued that a combination of quantitative (i.e. a questionnaire) and qualitative (i.e. interviews) research tools are used. This combination was chosen because it would enable the researchers in creating and studying the visual representation of data. Also, the findings that relate to each method will be used to complement one another and, at the end of the study, to enhance theoretical or substantive completeness (Ausubel 1968). The research tools i.e. the questionnaire and interviews, were designed in accordance to the literature review, the findings of the pilot study that carried out in 2009 and the previous research on remote construction sites that has been conducted in the KSA in 2010.

The targeted population consists of contractors and companies' supervisors/engineers of Aramco (Saudi Arabian Oil Company), Royal commission of Jubail (RCJ) (petrochemical company), SABIC (Saudi Basic Industries Corporation and it a petrochemicals manufacturer), compendium of construction and consultancy companies which are working on remote sites of University of Dammam. These companies are located or have branches in the Eastern province, KSA and they have a number of remote construction sites of different sizes. This targeting method would provide feedback from the two major project's parties i.e. the owner and the contractor, which would increase the applicability of the proposed solutions regarding management and tools to both of them.

In mid-2012, phone calls were made to the managers and contractors of above mentioned companies asking them whether they would be happy to participate in the field survey. Seventy questionnaire forms were sent out to individuals who expressed their willingness to take part in the field survey. The progress of filling up the questionnaire forms was checked up by phone. These forms were filled in and returned back. However, after examining the returned forms, it was found that 23 of these are invalid as many parts have not been completed. Eventually, the total number of valid questionnaires is 47. The face to face interviews were conducted in January 2013 and these were with one consultant that represents the University of Dammam in respect of supervision of a number of construction projects for the university's, two engineers who represent a contracting company which is working on one of the University's projects, and seven engineers of various specialists i.e. architectural, structural engineering, and HVAC, who are working in the private projects department of the Royal Commission of Jubail. The total number of interviewees was ten. The researchers were not able to conduct any interview with Aramco engineers and contractors as none was willing to participate in the interviews stage.

The analysis of the results was undertaken using SPSS16 for quantitative analysis and a number of data visualization techniques were used. Mean, standard deviation, and skewness values were calculated, and the researchers inspected links between potential factors (i.e. independent variables) and dependent variables.

Cross-referencing (i.e. similarities and non similarities) technique was used to analyze the qualitative data. This technique enabled the researchers to classify the data into categories and make comparisons, and allowed rational interpretation and judgment. The following sections discuss remote construction site challenges and the potentials and barriers to the application of ACMS.

Results and discussion

Remote construction projects problems

On a daily basis, the most requested information by the construction team on the remote construction site from the regional supervision office include the followings: clarification of design information, reporting QA/QC problems, contract drawings, initiation site inspections, reporting the site inspection outcomes, as-built records, and delay records. There is however, substantial delay in dealing with the following site's queries: revisions to submittals, meeting minutes, updated Schedule, clarification of Design information, as-built records, contract drawings and reporting QA/QC problems (Figure 1). Respondents said that there is a strong negative impact of the delay in providing feedback for the following remote project's queries on the project's process and performance: clarification of design information, contract drawings, updated Schedule, contract specifications, reporting QA/QC problems, and budget, and as-built records. Respondents were asked about remote construction sites' problems.

They said that the most frequent problems of these sites are: delay in project time table, unavailability of supervision engineers on the remote project site due to their heavy workload, the company's tendering system that obligate the choice of the lowest bidding value, the shortage of skilled labour on the site, and delay in the approval of contractor's submissions by the supervision engineers (Figure 2).

The ANOVA results showed significant links between the delay in providing information and the remote construction site problems. It showed that the more delay in providing information on the budget is associated with higher occurrence of the following problematic issues on the site: mistakes in construction works, shortage in site equipment, delay in the conduction of field survey by the contractor, unavailability of the supervision engineers during sample testing stages, improper construction methods implemented by the contractor and inadequate equipment used. The more delay in providing the products catalogue is associated with higher occurrence of ineffective planning and scheduling of the project by the contractor (Table 2). The more delay in providing the visitors log is associated with more delay in project time table. The more delay in providing revisions to submittals information is associated with higher occurrence of the use of inadequate equipment. The more delay in initiating site inspections is associated with more delay in project time table. The more delay in reporting QA problems are associated with more delay in project time table.

The interviews revealed the following facts:

- Very few Saudi nationals are working on construction projects in KSA and there is a problem of employing non Saudis who are in KSA or recruiting HR from aboard because of the high recruitment fees on foreign contracts;
- The remoteness of remote projects has an impact of on social life of workers. The higher rank staff would

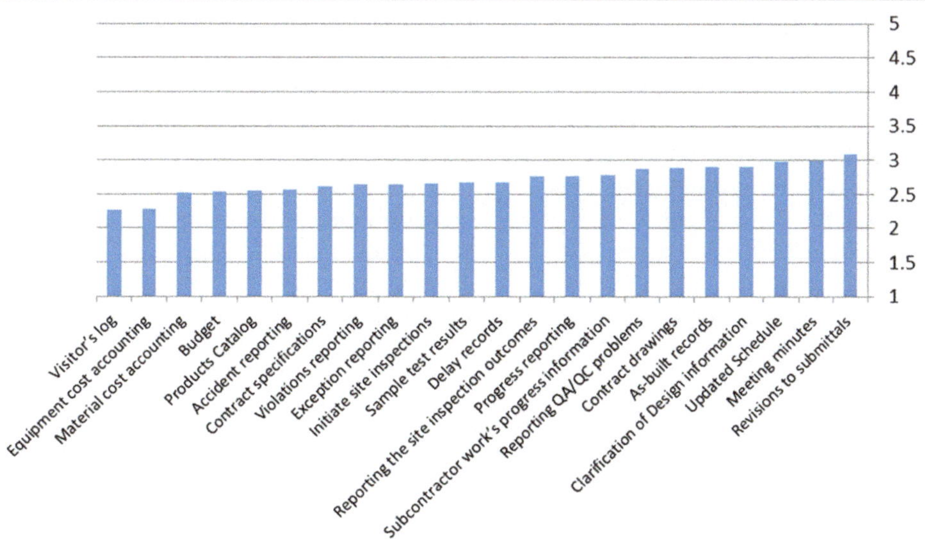

Figure 1 How frequent the respondent companies have experienced a delay in dealing with the remote construction site queries ***(vertical column represents the Mean value, scale: 1 = not at all to 5 = always).***

Figure 2 How frequent the following issues happen during the remote construction project's period (i.e. from start to completion). *(horizontal column represents the Mean value, scale: 1 = does not happen at all to 5 = always happens).*

find working in a remote project very difficult as they will be away from their families and the social life. The absence of electronic social communications would affect negatively the workers' productivity who are away from their families;

- Limited authority is given to the site managers and this sometimes causes substantial delay;
- Due to the cultural differences, social conflicts between workers take place and in one case, these are not reported to the project's owner or to the authorities. Also, one should be careful and choose his words when talking with staff from different backgrounds as some would understand certain words as an abuse or a threat;
- It is noticed that there are differences in productivity between workers from different backgrounds so this should be considered as it would delay the work in some parts of the project;
- all interviewed staff have QA measures but they do not apply sustainability measures;
- in one company, the contractor has weak commitment to providing frequent progress report to the supervision office as this has not been specified in the contract;
- risk management and materials' procurement is programmed. However, each remote project has its' merits, estimation of insurance on labour, equipment, third party, risk level etc. and unexpected issues such as absence of road net or bad roads status or so. Differences in estimating the risk level exist between the contractor, the owner and other concerned parties. This affects how the project would be run by each party and may create conflict of interests between these parties;
- the government bidding and tendering system for design or/and construction of a project takes into account the lowest bidding price regardless of the quality which affect the quality of projects;
- materials were originally specified according to the specification but sometimes the materials that brought to the site did not comply with the specifications;
- one of the main causes of construction problems is the conflict between design documents that are produced by different specialists;
- the contractor should thoroughly survey the site. He should check the utilities i.e. sewage, electricity, water on the site or if there anything that would affect the project's kick-off such as the availability of road

Table 2 ANOVA results showing correlation between the factor; *delay experienced in dealing with the budget query* with a number of dependent variables namely; *the frequency of problematic issues during the remote construction project's period,* (level of significance <0.05)

Dependent variable: *the frequency of problematic issues during the remote construction project's period (i.e. from start to completion*		**Independent variable (factor):** *the delay experienced in dealing with the budget query*				
		Sum of Squares	**df**	**Mean Square**	**F**	**Sig.**
Mistakes in construction works	Between Groups	6.105	4	1.526	2.758	0.043
	Within Groups	18.818	34	0.553		
Shortage in site equipment	Between Groups	15.838	4	3.959	3.934	0.01
	Within Groups	33.215	33	1.007		
Delay in the conduction of field survey by the contractor	Between Groups	15.123	4	3.781	5.469	0.002
	Within Groups	22.121	32	0.691		
Unavailability of the supervision engineers during sample testing stages	Between Groups	19.735	4	4.934	5.869	0.001
	Within Groups	27.739	33	0.841		
Improper construction methods implemented by the contractor	Between Groups	5.554	4	1.388	3.325	0.022
	Within Groups	13.365	32	0.418		
Inadequate equipment used	Between Groups	10.849	4	2.712	4.082	0.008
	Within Groups	22.587	34	0.664		

networks. SEC – for instance- did not update information of some sites so when these sites are possessed by RCJ, they may find electric cables, foundations, SEC temporary station, and septic tanks under the ground;

- changes would happen to the original design scheme. In one occasion, the granite panels for the facades were replaced with precast concrete panels to speed up the construction, and in another incident, solar panels were included in the original design scheme, but later on these were abandoned, and the project was connected to the national electric grid. The reason was that the owner of the project was scared of trying something for the first time and considered it of high risk. In another incident, the desalination station was designed with a certain capacity. Later on, it was discovered that the capacity is not enough and should be extended to cover the present and future needs; and
- some contractors do not have the background to manage certain types of projects and remote projects. They are incapable to manage their financial affairs, materials' procurement and how to provide qualified human resources.

The present and potential use of the ACMS

The systems and tools that are mostly used by respondents for management of the projects and communications between the remote construction site and the supervision office are: e-mail through internet/intranet/wireless/satellite, mobile phone, laptop, site visit, weekly or monthly reports and weekly meetings. Advanced computer tools such as Web-based Project Management system (WPMS), personal digital assistants, virtual private network, and construction Robots are of little use.

Respondents said that the use of the following systems would help in efficiently manage the remote construction projects: e-mail through internet/intranet/wireless/Satellite, intelligent mobile phones, tablet computer, and Web-based Project Management System. On the other hand, they considered the following systems would help less in managing remote sites: the Walkie-talkie, web monitoring cameras and construction robots. Respondents said the ACMS would highly accelerate the feedback of the following site queries: reporting QA/QC problems, contract drawings, reporting the site inspection outcomes, clarification of design information, accident reporting and violations reporting. On the other hand, it was considered of lesser help in providing fast feedback regarding the following sites queries: material cost accounting, equipment cost accounting, and visitor's log (Table 3).

Respondents said that ACMS would be very helpful in achieving the following project lean and sustainable values: pursuing perfection by continuous improvement, specifying precisely value from the perspective of the ultimate customer/owner, identifying clearly the process that delivers what the customer/owner values (the value stream) and eliminating construction works' waste.

On the other hand, it would provide less support to managers who aim to achieve the following values: protecting the KSA people health and reducing the use of natural resources. The interviewees said that ACMS would help:

- Facilitating the procurement of materials whereas requests for materials are sent electronically to the head office for approval. However, as some materials are imported from aboard, there is a need to see the material or a sample before purchasing it.

Table 3 How fast the ACMS would accelerate the feedback of the remote construction site' queries, *(Mean value scale: 1 = do not accelerate to 5 = highly accelerate)*

Type of the remote construction site's query	Mean	Std. deviation	Skewness
Reporting QA/QC problems	4.13	0.992	−0.825
Contract drawings	4.11	0.841	−0.468
Reporting the site inspection outcomes	4.06	0.942	−0.62
Clarification of Design information	4.02	0.944	−1.016
Accident reporting	3.96	1.083	−0.879
Violations reporting	3.94	1.131	−1.001
Progress reporting	3.91	1.08	−1.014
Revisions to submittals	3.89	0.983	−0.209
Initiate site inspections	3.89	1.005	−0.45
Contract specifications	3.83	1.018	−0.693
Updated Schedule	3.83	1.253	−0.72
Sample test results	3.77	1.165	−0.812
Exception reporting	3.74	1.206	−0.879
Meeting minutes	3.68	1.163	−0.72
As-built records	3.64	1.293	−0.73
Delay records	3.63	1.388	−0.752
Subcontractor work's progress information	3.48	1.11	−0.452
Products Catalog	3.36	1.342	−0.368
Budget	3.33	1.266	−0.242
Material cost accounting	3.26	1.307	−0.447
Equipment cost accounting	3.2	1.222	−0.316
Visitor's log	3.13	1.377	−0.135

- sending requests for specialist HR to recruitment agents who recruit HR from overseas. It also helps in mobilization of HR from a project to another as needed.
- overcoming cultural differences and understanding the other and, for instance, including religious festivals holidays in the project schedule
- providing fast social communications between the worker on site and his family at home or abroad is essential and this would bring relief to the worker. Fast communications channels are required.
- delegating certain level of authority to the sites' managers. This would help in mobilization of HR and equipment between sites; and
- providing updated benchmarks/standards/measures on the net so staff would refer to it, while examining certain case. Thus they would take the decision without getting an approval from the higher authorities

However, the management system should be transparent, so any investment opportunity by the headquarters should be known by all concerned departments.

Barriers to the use of ACMS

This study suggests that the above mentioned companies experience a number of unique construction management performance problems. The study found significant links between these performance issues and how far the ACMS would help in providing fast feedback to the site's queries and in achieving project's values. Therefore, these issues would affect the adoption and utilization of the ACMS; and would be considered as barriers to the use of ACMS (see also Nitithamyong and Skibniewski 2007).

The field survey show frequent construction problems and delays in responding to some site queries (see also Deng et al. 2001). Also it showed some unexpected problems i.e. infrastructure, social, psychological as mentioned above (see McAnulty and Baroudi 2010). This would raise the risk level and by turn the cost of remote projects in comparison with normal projects. The interviews show problems related to materials' procurement, transportation and installation, differences in estimating the risk level between project parties (Kestle and London 2002, 2003).

Differences in estimating cost, time, scope, and quality management between parties and the traditional management practices can be seen as obstacles to ACMS full utilization. There is lack of skilled human Resources, and there are cultural differences and a language barrier between workers and communication problems between the senior staff and labor. The cultural differences occasionally create tensions between teams (ibid).

ACMS systems are of little use by the respondents and site personnel (i.e. the contractor, sub-contractor or the owner's employees). The ACMS was considered of little help regarding the financial matters such as Budget, material cost accounting, and equipment cost accounting. Also, the ACMS would provide lesser support to managers who aim to achieve the following values: protecting Saudi Arabia's peoples' health and reducing the use of natural resources. The interviews indicated that in some companies, fragmented and standalone software was used for managing projects. The IT system of the project owner at the company headquarters is sometimes not connected with its departments and mostly not linked with the contractor or other project parties' systems. Also, problems regarding the weak IT infrastructure (e.g. slow servers) are experienced. Staff finds some difficulty in dealing with updated developments of software. In one of the companies, IT staff have good IT skills but without good construction background. None of the interviewees use Building Information Modelling (BIM) programs or receives BIM documents from consultants or design firms. All documents received are AutoCAD ones or based on AutoCAD drawings.

Conclusions: How to maximize the potentials

To maximize the potentials of ACMS, it should be designed to provide innovative and intelligent solutions to the unique remote construction problems. The recommendations for the application of ACMS below can be generalized to other remote construction sites aboard taking into account the location and project's parameters. The design of future ACMS should consider the following issues:

A. **The management side:** The present researchers that design of ACMS would have the following management levels:
   I. **Strategic planning level:** the company's strategy should be programmed into the ACMS system and this includes the strategic management of all its' remote projects in terms of how to coordinate activities, mobilization/demobilization of HR, equipment, materials, transportation, and other logistics issues.
   II. **Project planning and process:** automated design and pre-planning of site activities is essential as this should consider the risky variables' of the environment, the remote site and project as mentioned above. There are differences in risk estimation between project parties, and this should be discussed during the design stage and the level of risk should be agreed by all parties thus embedded in the system and all contract documents.
   The ACMS should provide flexible decision-making mechanisms and consider short and long-term partnering, and sharing information and management and communications tools with contractors and other project parties. This would enhance knowledge integration and help to foster innovative ideas that dramatically improve projects (Barlow 2000). During the construction stage, ACMS should provide precise daily control and follow-up procedures regarding issues such as remote examination of work quality, monitoring productivity of site workers, and calculation of material consumption rates. Some problems seem to be generated during the early stages of the project. These should be programmed as well. ACMS should enable the creation of new hypothesized cases, and the input of existing cases of emergency scenarios and the generation of possible intelligent solutions. ACMS would have an educational/training part that helps various project parties to learn about unique previous problems of the construction of remote projects and possible solutions. This would help staff overcoming present remote site's problems. The system would also have an assessment section wherein it checks whether the contractor has the ability and knowledge to run the designated remote project. ACMS should provide transparency and it should consider how to reduce the negative impacts of the project on the environment and environmental parameters on project performance and processes.
   It is important to provide all the necessary information i.e. benchmarks/standards/measures on the system so it would be used by staff without the need to get the approval from the higher authority. However, financial, sustainable, and environmental aspects of the project should be considered in the design of the ACMS system.
   III. **Human resources level:** the future ACMS would help all parties to respond efficiently to unexpected issues, to quickly sort queries, to lead multinational teams, to sort out potential social conflicts and find out possible solutions, to coordinate with other remote sites' managers and overcome communications problems. Religious festivals and holidays should be considered and frequent social events should be programmed. The system should give the managers the required level of authority and this helps in mobilization and demobilization of HR and equipment between distant projects.
   Staff should be informed about the benefits and advantages of new ACMS systems. Managers should be trained on site and how to manage remote sites virtually. The design of the new ACMS in-house or adoption of new ACMS should be discussed with the contractors and supervisors/engineers to find out their views, perceptions and expectations.

B. **the social and cultural side:** ACMS should help bridging cultural differences and overcoming language barriers between workers and communication problems between the senior staff and labor.

ACMS should help facilitating the procurement of materials whereas requests for materials are sent electronically to the headquarters for approval. ACMS should help in sending requests for specialist HR to recruitment agents who recruit HR from overseas. Through this system, fast social communications should be provided between the worker on site and his family at home or abroad.

Competing interests
The authors declare that they have no competing interests.

Authors' contributions
BS and AA together carried out the filed study. Subsequently, the gathered data was analyzed. All authors contributed to the writing of the manuscript. Both authors read and approved the final manuscript.

References

Ahuja, V, Yang, J, & Shankar, R. (2010). IT-enhanced communication protocols for building project management. *Engineering, Construction and Architectural Management Journal, 17*(2), 159–179.

Ahuja, V, Yang, J, & Shankar, R. (2009). Benefits of collaborative ICT adoption for building project management. *Construction Innovation: Information, Process, Management, 9*(3), 323–340.

Alshawi, M, & Ingirige, B. (2003). Web-enabled project management: an emerging paradigm in construction. *Automation in Construction, 12*, 349–364.

Alshawi, M, Goulding, J, Khosrowshahi, F, Lou, E, & Underwood, J. (2009). *How strategic is IT investment in the construction industry? A UK Perspective, Modern Built Environment –Knowledge Transfer Networks, Intelligent Buildings Index* (pp. 1–3).

Anumba, CEH, Dainty, A, Ison, S, & Sergeant, A. (2006). Understanding structural and cultural impediments to ICT system integration: A GIS-based case study. *Engineering Construction and Architectural Management, 13*(6), 616–633.

Arayici, Y, & Aouad, G. (2010). *Building information modelling (BIM) for construction lifecycle management, in: Construction and Building: Design, Materials, and Techniques* (pp. 99–118). NY, USA: Nova Science Publishers.

Arayici, Y, Egbu, C, & Coates, P. (2012). *Building information modelling (BIM) implementation and remote construction projects: Issues, Challenges and Critiques, ITcon Vol. 17, Special Issue Management of remote construction sites and the role of IT Systems* (pp. 75–92). http://www.itcon.org/2012/5. Accessed 10 July 2013.

Ausubel, DP. (1968). *Educational Psychology: A Cognitive View*. New York, NY: Holt, Rinehart & Winston.

Barlow, J. (2000). Innovation and learning in complex construction projects. *Research Policy, 29*, 973–989.

Becerik, B. (2005). *Critical Enhancements for Improving Adoption of OPM Technologies*. Harvard Graduate School of Design, Barrie Award Winning Reports, PMI Educational Foundation Funded. http://i-lab.usc.edu/documents/Enhancements%20for%20IT%20Adoption_PMI_Sept%202005.pdf. Accessed 10 December 2012.

Bowden, S. (2005). *Application of mobile IT in construction* (PhD Dissertation): University of Loughborough, Department of Civil & Building Engineering.

Charoenngam, C, Ogunlana, SO, Ning-Fu, K, & Dey, PK. (2004). Re-engineering construction communication in distance management framework. *Business process management Journal, 10*(6), 645–672.

Davidson, CH, & Moshini, R. (1990). Effects of organisational variables upon task organisations' performance in the building industry. *Building Economics and Construction Management, 4*, 17–22.

Deng, ZM, Li, H, Tam, CM, Shen, QP, & Love, PED. (2001). An application of the Internet-based project management system. *Automation in Construction, 10*(Elsevier), 239–246.

Golparvar-Fard, M, Peña-Mora, F, & Savarese, S. (2011). Integrated sequential as-built and as-planned representation - with D4AR – 4 dimensional augmented reality - tools in support of decision-enabling tasks in the AEC/FM industry. *ASCE Journal of Construction Engineering and Management*. http://dx.doi.org/10.1061/(ASCE)CO.1943-7862.0000371. Accessed 10 April 2012.

Grimshaw, DJ, & Kemp, B. (1989). Office automation in local government in the UK. *Local Government Studies, 15*(2), 7–15.

Jaselskis, E, Ruwanpura, J, Becker, T, Silva, L, Jewell, P, & Floyd, E. (2011). Innovation in Construction Engineering Education Using Two Applications of Internet-Based Information Technology to Provide Real-Time Project Observations. *ASCE Journal of Construction Engineering and Management*. http://dx.doi.org/10.1061/(ASCE)CO.1943-7862.0000297. Accessed 12 July 2012.

Justanyah, N, & Sidawi, B. (2011). *The dilemma of communications and management of remote construction projects in the kingdom of Saudi Arabia, Sixth International Conference on Construction in the 21st Century (CITC-VI)* (pp. 395–406). Kuala Lumpur, Malaysia: Construction Challenges in the New Decade. ISBN 1-884342-02-5.

Kestle, L, & London, K. (2002). Towards the development of a conceptual design management model for remote sites. In C Formoso & G Ballard (Eds.), *10th Annual Conference on 'Lean Construction (IGLC-10) (Vol. 1, pp. 309–322)*. Gramado: Federal University of Rio Grande Do Sul.

Kestle, L, & London, K. (2003). Remote site design management –the application of case study methodology. In *Paper presented at the Postgraduate construction research conference*. Melbourne: RMIT.

Kestle, L. (2009). *Remote Site Design Management* (PhD thesis). NZ: University of Canterbury.

Leskinen, S. (2006). *Mobile Solutions and the Construction Industry Is it a working combination?* VTT publications. http://www.vtt.fi/inf/pdf/publications/2006/P617.pdf. Accessed 18 February 2011.

Leskinen, S. (2008). *Mobile technology in the Finnish construction industry – present problems and future challenges. 21st Bled eConference eCollaboration: Overcoming Boundaries through Multi-Channel Interaction June 15–18, 2008*. Slovenia: Bled.

Margherita, A, & Petti, C. (2010). ICT-enabled and process-based change: an integrative roadmap. *Business Process Management Journal, 16*(3), 473–491.

McAnulty, S, & Baroudi, B. (2010). *Construction Challenges in Remote Australian Locations*. Leeds, United Kingdom: Association of Researchers in Construction Management (ARCOM) Conference.

Nitithamyong, P, & Skibniewski, MJ. (2004). Web-based construction project management systems: how to make them successful? *Automation in Construction, 13*, 491–506.

Nitithamyong, P, & Skibniewski, JM. (2006). Success/Failure factors and performance measures of web-based construction project management systems: professionals' viewpoint. *Journal of Construction Engineering and Management ASCE, 132*(1), 80–87.

Nitithamyong, P, & Skibniewski, MJ. (2007). *Key success/failure factors and their impacts on system performance of web-based project management systems in construction, ITcon Vol. 12* (pp. 39–59). http://www.itcon.org/2007/3. Accessed 14 January 2011.

Nuria, FM. (2005). *Life cycle document management system for construction. PhD thesis*. Universitat Politechica De Catalunya. http://www.tdx.cat/bitstream/handle/10803/6160/01Nfm01de12.pdf?sequence=1. Accessed 12 October 2011.

Pena-mora, F, Vadhavkar, S, & Aziz, Z. (2009). Technology strategies for globally dispersed construction teams. *Journal of Information Technology in Construction*. http://www.itcon.org/2009/08. Accessed 5 August 2011.

Sidawi, B. (2010a). *The sustainable management of remote construction projects*. Fes, Morocco: Arab Society of Computer Aided Architectural Design AsCAAD. ISBN 978-1-907349-02-7.

Sidawi, B. (2010b). *The use of advanced electronic management systems to manage remote projects in the Kingdom of Saudi Arabia* (pp. 633–642). Leeds, UK: Association of researchers in Construction management conference (ARCOM). ISBN 978-0-9552390-4-5.

Sidawi, B. (2012a). *Remote construction projects' problems and solutions: the case of SEC. ASC 48th International Conference held in conjunction with the CIB Workgroup 89*. UK: Birmingham City University.

Sidawi, B. (2012b). Potential use of communications and project management systems in remote construction projects: the case of Saudi Electric Company. *Journal of Engineering, Project, and Production Management, 2*(1), ISSN 2221-6529 (Print), ISSN 2223-8379 (Online), 14–22, 2012.

Stewart, RA, & Mohamed, S. (2004). Evaluating web-based project information management in construction: capturing the long-term value creation process. *Journal of Automation in Construction, Elsevier Science, 13*(4), 469–479.

Thomas, S, Palaneeswaran, E, & Kumaraswamy, MM. (2003). Web-based centralized multiclient cooperative contractor registration system. *Journal of Computing in Civil Engineering, 17*(1), 28–37.

Thorpe, D. (2000). *E-projects in action – the online remote construction management research project* (CIIA Fifth Annual Conference: Innovation in Construction). Australia: Construction Industry Institute.

Vadhavkar, S, & Pena-Mora, F. (2002). *Empirical Studies of the Team Interaction Space: Designing and Managing the Environments for Globally Dispersed Teams* (International Workshop on the Role of Empirical Studies in Understanding and Supporting Engineering Design Work). Gaithersburg, MD: USA: NIST.

Wamelink, JWF, Stoffelem, M, & der Aalst, V. (2002). *W.M.P. International Council for Research and Innovation in Building and Construction. CIB w78 conference, Aarhus School of Architecture, 12–14 June 2002*. Denmark: CIB.

Yang, J, Ahuja, V, & Shankar, R. (2007). *Managing Building Projects through Enhanced Communication – An ICT Based Strategy for Small and Medium Enterprises, CIB World Building Congress 2007* (pp. 2334–2356). South Africa: CIB.

Zou, P, & Roslan, B. (2005). Different perspectives towards using web-based project management systems in construction: large enterprises versus small- and medium-sized enterprises. *Architectural Engineering and Design Management, 1*(2), 127–143.

Accessibility evaluation system for site layout planning – a tractor trailer example

Jacob Je-Chian Lin, Chi-En Yang, Wei-Han Hung and Shih-Chung Kang*

Abstract

Background: Accessibility is a critical issue in site planning. A good accessibility plan can avoid possible conflicts involving equipment and allows for smooth transportation during the entire project. It can also ensure that sufficient space is available so that all onsite equipment, such as trucks, cranes and excavators can be safely operated. In current practice, the evaluation of accessibility is done manually, relying heavily on the experience of the construction planners. However, the evaluation needs to deal with temporal and spatial information simultaneously, which make the manual evaluations very challenging.

Methods: This research develops a sandbox environment for construction planners. As tractor-trailers are usually critical in accessibility evaluations, the simulation specifically focuses on these two-section vehicles. Four major steps were involved: Step 1: deriving a generic tractor-trailer mathematical model allowing real-time, physics-based computer simulation; Step 2: digitalizing a 2D site plan to the 3D virtual environment; Step 3: developing evaluation methods for accessibility, considering both safety and operability; Step 4: visualizing the evaluation results. We implemented these four steps using Microsoft XNA (a game platform) and Nvidia PhysX (a game engine). An example case is presented which uses the sandbox environment to evaluate the accessibility of a construction site.

Results: From the case study, we found that realistic visualizations and simulations provide solid references for construction planners. They are able to identify potential accessibility problems and unsafe situations in the sandbox environment and avoid them early on in the design and planning stages.

Conclusion: This paper proposed an innovative approach for evaluating accessibility during pre-construction site layout planning. It can present an intuitive and easily understood visualization result which clearly indicated the unsafe parts of the site layout plan.

Keywords: Accessibility; Site layout; Trailer truck; Physics-based simulation; Game engine

Background

Accessibility is one of the most important considerations in site layout planning. The routes of a construction site need to be designed to ensure smooth delivery of materials, safe machine operation and relatively easy relocation of large objects on site. Site planning directly influences the construction progress and often impacts cost if there are any unforeseen conflicts (Su et al. 2012).

Although accessibility appears to be an intuitively important issue in construction site layout planning, it has not been well-researched thus far. In practice, the evaluation of accessibility is still done manually, relying heavily on the experience of the construction planners (Tam & Tong 2003; Elbeltagi et al. 2004; Li et al. 2009). Daily meetings are held to deal with accessibility problems in which the engineers discuss transportation paths and the location of heavy equipment and temporary facilities to ensure that none of the various processes will be in conflict with one another. Without an intuitive tool to evaluate the accessibility of the site, engineers evaluate the space they need by considering qualitative positions (Akinci et al. 2002), which could easily lead to accessibility problems. Conflicts are usually caused by an incorrect assessment of the space occupied by equipment and facilities. Sometimes newly constructed elements/ parts may need to be demolished to create a passage for

* Correspondence: sckang@ntu.edu.tw
Computer-Aided Engineering Group, Department of Civil Engineering, National Taiwan University, 188, Sec. 3, HsinHai Rd, Taipei 10668, Taiwan

other equipment, for example a concrete conveyor truck, to get to its designated area.

Traditional site layout planning is primarily concerned with the positioning, existence and timing of the temporary facilities that are used throughout the construction project (Mawdesley & Al-Jibouri 2003), and the site layout planning problem can be dissected into the following sub-problems: (1) identifying the shape and size of the facilities to be laid out; (2) identifying the constraints between facilities; and (3) determining the relative positions of these facilities (Zouein et al. 2002). In order to study the space utilization of construction sites, research of site layout planning has been carried out and many studies have attempted to solve site layout planning problems. The research within this scope can be classified according to two different criteria: (1) optimization algorithms and (2) visualization.

Under the first criterion, researchers have used different kinds of optimization algorithms to solve site layout planning problems. Zouein et al. (2002) tried to resolve transportation cost problems in construction site layout planning by using genetic algorithm analysis. Mawdesley and Al-Jibouri (2003) used a genetic algorithm to reduce the costs associated with transportation paths, and both Lam et al. (2009) and Ning et al. (2010) combined the genetic algorithm and the Max-min Ant system (MMAS) to provide a computing method for reducing cost and transportation time. Wong et al. (2010) utilized genetic algorithms to optimize a layout for precast facilities on a construction site. One of the limitations of these algorithms is that some of them assume that all the facilities in the construction site can be represented by box shapes. Researchers have tried other approaches of developing automated optimization of site layout planning (Elbeltagi et al. 2001; Lam et al. 2007; Easa & Hossain 2008).

Under the second criterion, research involves the use of visualization for site layout planning. Sadeghpour et al. (2006) utilized computer-aided tools to analyze the appropriate location of the facilities, and simplified 2D visualization tools to analyze their geometric relationships. In terms of developing a site representation, the computer-aided tools approach has some similarity with the approach implemented in this research. Dawood and Marasini (2003) developed a simulation tool that visualizes the construction site in 2D to help improve the plan. Ma *et al.* (2005) developed a 4D system which integrates schedules, 3D models, resources, and site spaces to provide a visualization of the plan. Sadeghpour (2006) used computer-aided tools to offer an intuitive understanding of the site layout by displaying shapes of temporary facilities through direct visualization of their positions. Benjaoran and Bhokha (2010) used a 4D computer-aided design (CAD) model to simulate the construction process to ensure safety. In another research study, a pre-construction planning method was implemented by creating a virtual construction work package. This package can virtually rearrange 3D models to review the design for constructability (Waly & Thabet 2003).

In short, the majority of previous studies have focused on how to reduce the transportation costs of site layout planning. They utilized mathematical models and algorithms to compute the least distance between facilities. Thus, in current site layout planning methods, the location of facilities will be largely determined by considerations such as the distance to materials, the transportation time and the transportation cost (Hegazy & Elbeltagi 1999; Mawdesley et al. 2002; Tam et al. 2002; Zouein et al. 2002). Although these methods can generate a plan with paths of the least distance or lowest transportation costs, the accessibility issue is neglected. In a real site situation, the accessibility of transportation vehicles can directly affect the progress of work. If the transportation path is difficult to navigate due to obstacles or convoluted routes in a narrow working area, even if the distance is shorter, it can lower the efficiency of the operation. Furthermore, it can occasionally cause additional damage and costs (Soltani et al. 2002).

On the other hand, drawbacks can also be found in research under the second category. This research may incorporate 3D modeling of construction elements but still not perform realistic simulations of mobile machinery, a factor that profoundly influences site layout performance. Since accidents involving transportation are one of the most frequently occurring types of accidents at a construction site (IOS & H 2012), safety and accessibility should be given more importance during the planning process and the designing of transportation paths, rather than restricting the considerations only to distance and cost (Thomas et al. 2005; Andayesh & Sadeghpour 2013).

In summary, a new method that focuses on the accessibility issue by realistically simulating mobile units needs to be developed. Winch and North (2006) interviewed 18 experienced site planners and concluded that a construction space planning system should be a decision supporting system instead of a decision making system. Therefore, we aimed to develop a computation-based decision supporting sandbox system to deal with this problem. In this research, we propose a tool that can easily evaluate transportation access and also indicate potential problems of site layout plans. The system is composed of four modules: (1) transportation, (2) site importing, (3) accessibility evaluation, and (4) visualization. Using this system, we can easily simulate transportation processes in a construction site and test the safety and accessibility of the site layout plan. The system can also be used to visualize the test results.

Engineers can therefore easily identify the unsafe parts of a site and mitigate potential accessibility problems.

Research goals

In this research, we developed an accessibility evaluation system for site layout planning. Because tractor-trailers are usually critical in accessibility evaluations, we have specifically focused the simulation on these two-section vehicles. The system can facilitate the achievement of the following goals:

1. The simulation of transportation in an interactive, realistic virtual environment. From this, users can directly simulate transportation process to identify potential safety and accessibility problems under various situations and different site layout plans.
2. The development of an evaluation method for site accessibility. Via the proposed system, users can test their planned site layout and evaluate its accessibility and safety. After evaluation, the proposed system can support a visualized output of the evaluated result to help engineers understand the unsafe areas of the site layout, and therefore being able to work out a lower cost, safer, and improved plan.
3. The importing of the map of the construction site and the building up of the virtual environment of the construction site in a fast and efficient way.

Methods

Accessibility evaluation system

We developed an accessibility evaluation system, which consists of four modules (Figure 1). The transportation module simulates the dynamic motion of a tractor-trailer by using multi-body dynamics, a method commonly used for game physics. The site importing module enables users to generate the construction site environment precisely and efficiently from a 2D blueprint or image file of a construction site. The safety evaluation module is used to evaluate the site's accessibility by adding a multi-layer collision boundary to the transport unit, which detects collisions between the boundary of each layer and obstacles at the site. A formula was then developed for assessing the accessibility of each divided area at the site. The visualization module is used to render the virtual construction site and simulate the transportation activities (i.e. detailed motions of the trailer) taking place there.

Transportation module

The transportation module aims to provide engineers with a real-time interactive system for operating the tractor-trailer in a virtual environment so that problems in the transport process can be discovered efficiently. We used a tractor-trailer, a commonly used transport vehicle in construction, as the transport unit. It has been observed that the most commonly occurring accessibility problems involve tractor-trailers. The tractor-trailer is hard to control because the driver needs to take care of the trailer, and it also has more blind spots than a single unit vehicle. If an entrance or road is not well designed, the tractor-trailer can very easily become stuck. This kind of situation can be prevented by improved accessibility simulation.

The tractor-trailer is the most complicated machine to simulate as it has the most degrees of freedom of all mobile construction machinery. Mobile construction machinery can be classified into three different types: wheel, track, and chain. The wheel type includes typical construction machinery such as a truck. The track type

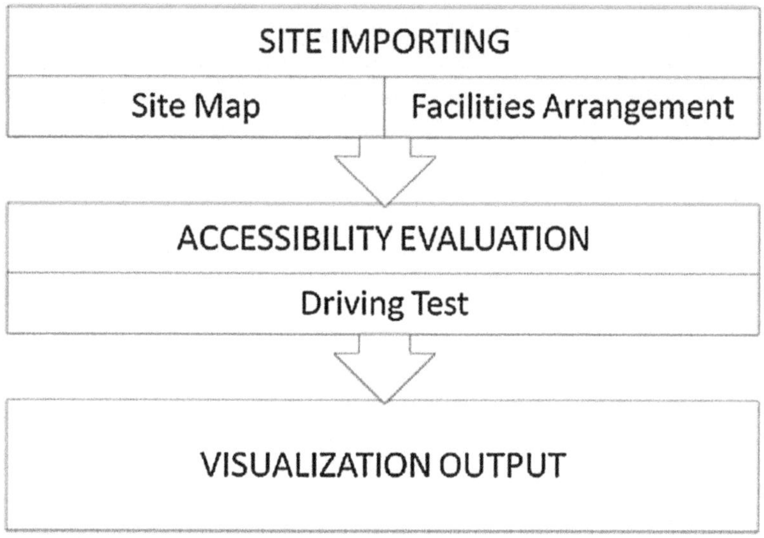

Figure 1 System framework and information flow.

refers more to heavy construction machinery such as an excavator. The chain type consists of multiple rigid bodies and connections such as a tractor-trailer. The chain type is the most complicated because its motion has the most degrees of freedom. Once the accessibility method for simulating the tractor-trailer is established, the user can easily apply it to other kinds of mobile construction machinery. A tractor-trailer is composed of multiple rigid bodies and connections. Mathematical equations can be used to represent the relationships between the connections of rigid bodies, which are controlled by the rules of physics. We used the concept of multi-body dynamics, which are commonly used for simulating and modeling the dynamic motions of articulated mechanisms or equipment (Hung & Kang 2009). We also provided a mathematical model to compare with the developed model to verify its feasibility.

Tractor-trailer model using multi-body dynamics

Multi-body dynamics is mainly used for computing physical feedback between multiple bodies in mutual contact with each other or connected to each other by joints (Erleben 2005). The simulation of multi-body dynamics is generally composed of rigid body dynamics and constraints. By solving the equations of motion (which are used to describe the dynamic behavior of multi-body dynamics), the simulation can calculate the behavior of a multi-body during each time integration (Hung & Kang 2009).

In this study, the rigid bodies in the tractor-trailer are called *actors* (for example, the bodies that steer the wheels of a car). The type of *Joint* determines the motion between a pair of rigid bodies within the entire multi-body. The basic joint used in this paper is the *revolute joint,* which attaches two actors by a hinge-like structure. It only has one degree of freedom of rotation so that the two actors can only rotate on one axis (Hung & Kang 2009).

A tractor-trailer can be separated into two main parts: the tractor and the trailer. The trailer connects to a fill-up tank. To build the tractor-trailer model, the locations of the wheels, tractor, trailer, and fill-up tank must be analyzed and connected correctly. In this paper, the tractor-trailer is simulated as an eight-wheel truck, with four wheels on the tractor and four wheels on the trailer. The construction of the tractor and trailer is similar: revolute joints connect the wheels to the steers and also the steers to the main body of the truck. The only difference between the tractor and trailer is that the trailer is connected to the fill-up tank by another revolute joint. Figure 2 shows the developed model represented by symbols as defined in previous research.

Tractor-trailer mathematical model

The mathematical model of the tractor-trailer is constructed based on a previous study by Rouchon and Fliess (1993) and is governed by the following equations:

$$\dot{x}_0 = \cos(\theta_0)u_1 \tag{1}$$

$$\dot{y}_0 = \sin(\theta_0)u_1 \tag{2}$$

$$\dot{\varphi} = u_2 \tag{3}$$

$$\dot{\theta}_0 = \frac{1}{d_0}\tan(\varphi)u_1, \text{for } i = 1, \ldots\ldots, n \tag{4}$$

$$\dot{\theta}_i = \frac{1}{d_i}\left(\Pi_{j=1}^{i-1}\cos\left(\theta_{j-1}-\theta_j\right)\right)\sin(\theta_{i-1}-\theta_i)u_1 \tag{5}$$

In these equations, $(x_0, y_0, \varphi, \theta_0, \theta_1 \ldots \theta_n) \in R^2 \times (S^1)^{n+2}$ represents the tractor-trailer state, where (x_0, y_0) are the coordinates of the head and φ is the steering angle. The heading of the truck is Θ_0 and the angle of the trailer is Θ_1, both with respect to the *x*-axis. 'u_1' and 'u_2' are the velocity control factors, where 'u_1' is the driving speed and 'u_2' is the angular velocity of the steering angle. 'd' represents the distance between trailers and is a positive constant (Rouchon and Fliess 1993). Figure 3 shows a representation of these variables. Using these equations, we can derive the mathematical relationship between the tractor and trailer and program it into the simulation

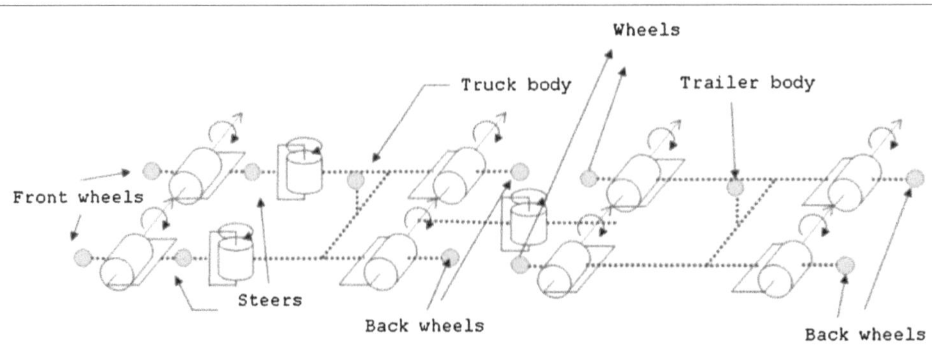

Figure 2 The tractor-trailer model using multi-body dynamics.

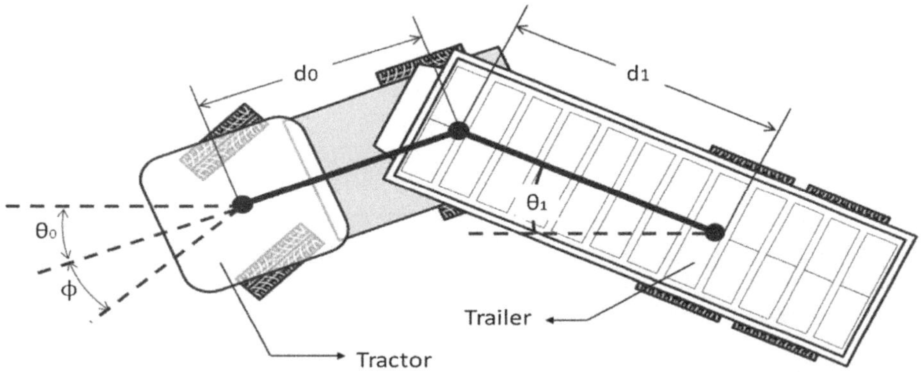

Figure 3 The Mathematical tractor-trailer model is developed based on the research of (Rouchon & Fliess 1993).

engine. We used this approach to simulate the construction tractor-trailer for comparison with our developed model.

Comparison between the two models

The result of the mathematical model simulation is used to verify the model developed in this research. Each simulation used the same degree of turning and velocity so that the trajectory of the tractor-trailer would be a circle. The differences between the two models under different degrees of turning and velocities were then compared.

Motion trajectory

In Figure 4, the velocity and turning degrees are set to the same value for the two models. One turning degree is set at 6°, and the other at 12°. We discovered that the lower the turning degree, the greater the motion trajectory radius, with the difference and error between the two models becoming greater; in other words, the higher the turning degree, the lower the motion trajectory and the smaller the error between the two models.

Motion trajectory with time

In this section, we describe the impact of the addition of the time factor. The comparison results are shown in Figure 5 and they demonstrate that error accumulates as time increases. The possible reasons for this error are discussed in the following section.

Sources of error

The mathematical model is constructed in a very simplified manner. It ignores the friction between the wheels and the ground (Figure 6). On the other hand, the physics engine enhances the contribution of the behavior between the tractor-trailer and the environment to its ultimate trajectory. Hence, there will necessarily be some differences between the two models. However, since the extent of error is acceptable we can conclude that the

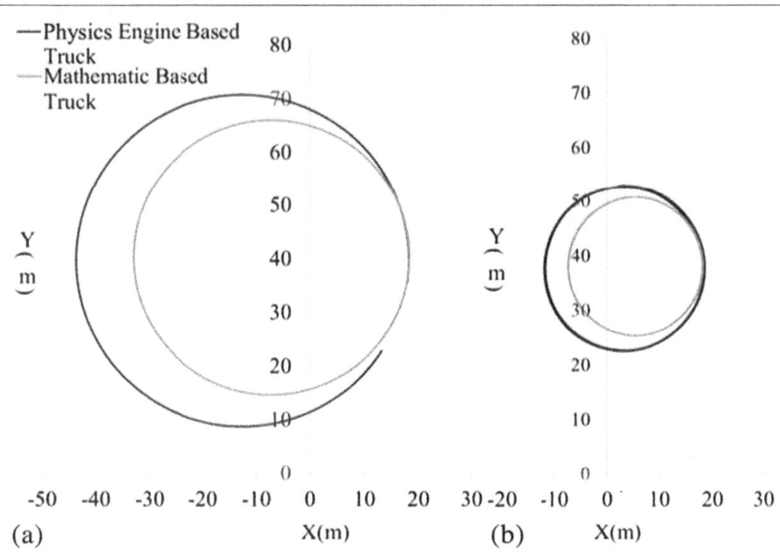

Figure 4 The motion trajectory coordinates of the two models. (a) With 6 degrees of turning and **(b)** 12 degrees of turning.

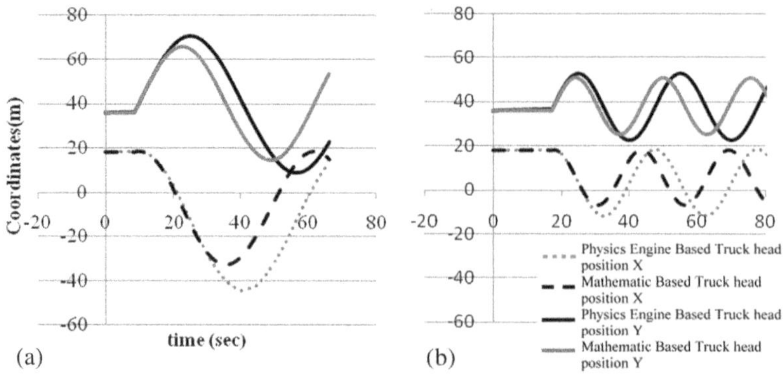

Figure 5 The motion trajectory co-ordinates versus time for the two models. (a) With 6 degrees of turning and **(b)** 12 degrees of turning.

physics model is acting similarly to the mathematics model, and, considering the greater number of factors it considers, it can be conjectured that it models the behavior even more accurately.

Summary

Our aim was to create a tractor-trailer model that approximated real-world behavior. The mathematical model was built based on a previous study, whereas the physics-based model was built using "multi-body dynamics", an approach not yet taken in any other research. The comparison of the physics model and the mathematical model leads to the following conclusions: (1) the mathematical model generates the ideal behavior of the tractor-trailer and the physics-based model behaves in a similar manner; (2) The mathematical model is easier to build and consumes less computational resources; (3) Accumulated error related to friction is present in the physics model and therefore when the proper friction coefficient is applied, the physics model can generate more realistic behavior than the mathematical model; and (4) Collision boundaries can be directly incorporated into the physics model whereas they can only be appended onto the mathematic model, and as a result the physics model is the more desirable model for use in site plan evaluations.

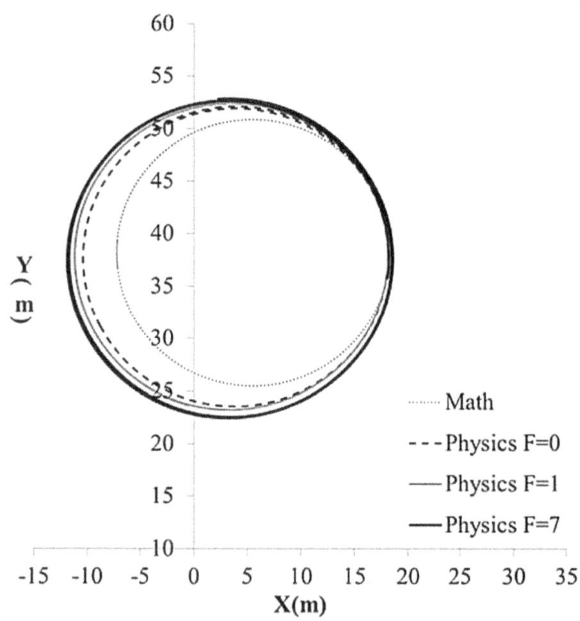

Figure 6 The difference in motion trajectory with different friction settings of the physics model. F represents the friction coefficient setting between the wheels and the ground.

Site importing module

To begin the evaluation the virtual environment of the construction site has to be generated and it must be precise in its dimensions and include all possible obstacles along the transportation routes to ensure that the safety test will be realistic. Since the facilities at the site are changeable during construction, the procedure of site environment generation must also be efficient and automatic. This can then allow engineers to test several layout plans in a fast manner.

To achieve this goal we developed a site importing module which uses the 2D image of the site plan as the input data for generating the virtual construction site, which includes collision checking. Users should manually modify the plan image to a specific scale and then highlight every obstacle with a specific color, which will then be recognized by the system. The system will then render an obstacle at a default size at each highlighted location. Figure 7 shows an example of modifying a 2D site plan. In this case, the site's dimensions are approximately 400 meters in width and 300 meters in height. The resolution was set so that one pixel represented one square meter, and so the 2D plan was converted to 400 pixels in width and 300 pixels in height, being a rather small scale for the modification process. During this process, only necessary boundaries are kept while others are deleted. In the case of Figure 7(b), there are permanent

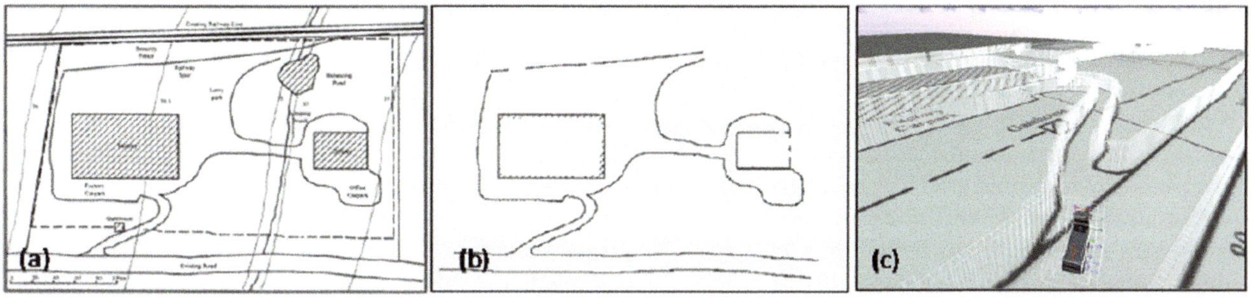

Figure 7 Import necessary obstacles by modifying a 2D plan. (a) Original site image; **(b)** processed image; and **(c)** virtual construction site environment built by the processed image.

facilities such as the factory and office, an existing road, and the walls or fences of the entire site.

The site importing module described takes 2D images of the site plan as input data. The scope of this research did not include an in-depth look at the impact of differing object heights in order to simplify the input process and facilitate computation. The system simply generated columns with a default height to represent the obstacles and boundaries appearing on the site.

Accessibility evaluation module

After the virtual construction environment is built, the accessibility evaluation can commence. In this module, the safety level of the transportation route is numerically assessed. An index, *Safety Factor* (SF), is developed to evaluate the site layout by integrating three different safety parameters. Detailed concepts relating to each safety parameter will be discussed in the following sections.

Safe driving range

The safety range of the driving paths for transportation was evaluated by adding a multi-layer collision boundary (*Safety Bound*) to the transport unit, which detected any collision between the boundary of each layer and obstacles on the site. The safety range score, l_i, was developed to quantify this factor. There are three layers of safety bounds as shown in Figure 8(a). Whenever an obstacle runs into a safety bound, it will gain a safety range score according to the level it touches and the maximum value will be kept. If no bound is touched, then the score will be zero.

Narrow level of route

If a route is narrow, there will be obstacles close to the transport unit on both the left and right hand sides. By separating the multi-layer collision boundary into two parts, we can evaluate the degree of the narrowness of the route. Narrow level, n_i, quantifies this factor. There are four collision boundaries as shown in Figure 8(b). Whenever an obstacle runs into a collision boundary, it will gain a narrow level score according to the level it touches and the maximum value will be kept. Otherwise, the score will be zero.

Curvature of route

If a route is tortuous, more driving skills are required than for driving on a straight route. Therefore, we take the curvature of the route into consideration. The *Instant curvature* of each segment of the route is evaluated while the transport unit is in motion. The curvature score, c_i, which is computed at every time step of the test, is also recorded whenever an obstacle touches one of the safety bounds.

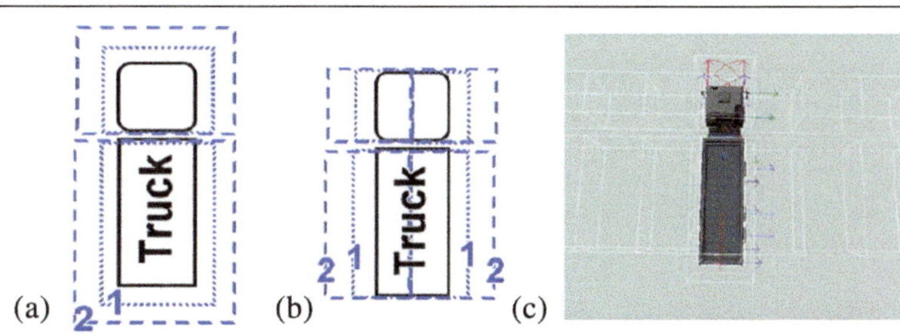

Figure 8 The multi-layer collision boundary. (a) The safety bound; **(b)** the modified safety bound for narrow routes; and **(c)** the safety bound and modified safety bound visualized in the virtual environment.

The safety factor

To execute a safety test for a route, the user needs to drive the tractor-trailer through a desired route, and can repeat this several times in order to test several possible driving conditions. After that, each unit obstacle gains a safety score, s_i, which integrates the three safety parameters (curvature score c_i, narrow level n_i, and safety range score l_i) according to the formula shown in Equation 6. We have adjusted these factors to be between 10^{-2} and 10^{-1} by multiplying them by a weight. In this case, the weight value for C was ten because the value of curvature lay mostly between 10^{-2} and 10^{-1} and would otherwise be too small to be of significance. For N and L, 1/4 and 1/3 were used respectively in order to diminish the value to between 10^{-2} and 10^{-1}. The value of α, β and γ was determined by the relative importance of user cognition and the project type.

In this research α=12, β= 4 and γ=7. These values have been tested by the authors and represented the most significant safety factor value of the construction site. This value is most suitable for construction sites that are more congested and have narrower corners. The value of the weights for each factor depends on the environment of the construction site and the type of transportation vehicle, and so it should be modified as required by engineers.

To turn safety scores into a practical guide for safety evaluation, users can select a certain resolution for the summing up of safety scores in an area. For example, if the chosen resolution is 10 m × 10 m, each divided area will gain a SF by summing up every safety score available within its 100 m^2 scope as shown in Equation 7.

$$s_i = (\alpha \times 10 \times c_i) + \left(\beta \times \frac{n_i}{4}\right) + \left(\gamma \times \frac{l_i}{3}\right) \quad (6)$$

$$SF = \Sigma_i s_i \quad (7)$$

Visualization module

The *visualization module* was designed for visualizing the evaluation results. This module provides a direct and brief method for engineers to quickly check through the evaluation result using visualization rather than having to interpret numerical data. After the evaluation, SF will be presented by the height of a rectangular column situated on each divided area, as shown in Figure 9(a). Engineers can thus gain a broad understanding of the results from a quick glance. In addition, the complete trajectory of the tractor-trailer can be stored and rendered again if required, as shown in Figure 9(b).

Implementation

To implement the proposed system, we used the physics-based simulation engine developed by Hung and Kang (Hung & Kang 2009), which integrates Microsoft XNA and NVIDIA PhysX. XNA is a rendering engine used to generate consecutive images from 3D models in real-time. It also provides a basic game development environment such as handling user inputs and audio playing. PhysX is a physics engine, which calculates the behavior of objects in the virtual environment. PhysX utilizes position-based dynamics to simplify and approximate multi-body dynamics, which is a stable and efficient method allowing the behavior to be simulated in real-time. It is also used to compute the collision detection in the accessibility evaluation module.

The procedure for using this system is as follows. Before executing the system, the user should load the 2D plan of the target site into the "site importing module" so that the system can generate the virtual obstacles. In addition, the user should set the initial position of the transportation unit according to the evaluation scenery. Next, the user launches the system and directs the model along the desired routes; the system will automatically record the safety factors. We used the Xbox360 gamepad as the default device for users to control the transport unit in the virtual environment. After reaching the end of the route, the user can switch to different evaluation modes provided by the "visualization module" and also export evaluation data for further use.

Results and discussion

Example case

A preliminary example case was constructed to test the proposed evaluation system. We utilized the actual site

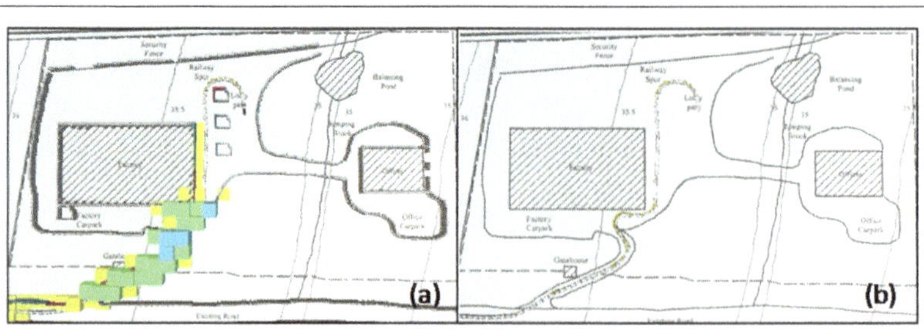

Figure 9 Visualization of the safety test result. (a) SF columns and **(b)** trajectory of the tractor-trailer.

layout data of a previous site planning study by Mawdesley et al. (2002) as shown in Figure 7(a). There are permanent facilities in this construction site, such as the factory, lorry, park, and office. The goal of site planning is to determine the location of temporary facilities, including the reinforcement store, temporary office and batching plant. First, we modified the construction site plan by adding possible locations of temporary facilities according to results obtained by Mawdesley et al. (2002), as shown in Figure 10(a) and (b). Then, the 2D site layout image was sent into the site importing module to generate the collision boundary of the virtual environment. The feasibility of the proposed site layout plan was then evaluated. When the evaluation process was finished, the result was then visualized by the visualization module as shown in Figure 10(c). It clearly indicates the relative safety levels of each area and engineers can then re-plan or modify the unsafe areas accordingly. If the site is small, truck operators can use the simulation system to rehearse the operation around those critical parts, which can reduce unexpected accidents or costs caused by accessibility problems.

The advantages of using this method can be summarized as follows: (1) the construction site is visualized so that during the evaluation process the user can clearly see accessibility problems; (2) the evaluation result is visualized so that the user can easily tell which part of the construction site needs to be modified; (3) the computational method is parametric and the parameters can be changed according to the circumstances of a specific site. The disadvantages of this method are as follows: (1) the user needs to decide the route first, and then evaluate; (2) the construction of the virtual construction site takes time.

Figure 11 shows snapshots of the evaluation process using the developed system: (a) shows the tractor-trailer as it prepares to enter the site through the entrance; (b) is the *physics mode*, which visualizes the physically effective elements in the virtual environment; (c) is the combined view of the physics mode and normal mode, in this mode, we can clearly see that the obstacle changes color when it comes into contact with different levels of the safety bound; (d) shows the visualization of the curvature score using yellow columns (it can be seen that the scores are relatively large around sharp turns); (e) shows a wide area resulting in a narrow level of zero; (f) is a return trip of the tested route; (g) is a top view of the visualization result, whereby the color and height of the columns indicate the value of the Safety Factor at each location; (h) is the *curvature mode* which shows the curvature score columns; and (i) is *trajectory mode* which enables visualization of the trajectory of the tractor-trailer during the entire test.

Benefits

The developed accessibility evaluation system is expected to greatly improve the planning of construction site layouts. The benefits are summarized as follows:

- Construction sector workers all over the world are very vulnerable to accidents and physical injury. The system helps to calculate the use of safety zones around the boundaries of moving vehicles in construction areas which makes incorporating safety into site layout planning much more realistic.
- The system vividly demonstrates the collision detection process of a moving vehicle, which makes the conflicts clearly visible. This can help poorly educated or illiterate operators be aware of safe driving/operating systems without having to read manuals or operation specifications.
- The system can facilitate the accommodation of changeable issues regarding Health and Safety of the workforce, transport management, power distribution, water supply and utility services, material storage, deployment and consumption at variously scheduled intervals allowing operational requirements from a variety of perspectives to be evaluated. Thus, it can provide managers with IT support and sophisticated information for making valuable decisions.
- The system can facilitate the use of spaces for temporary facilities during a designated time

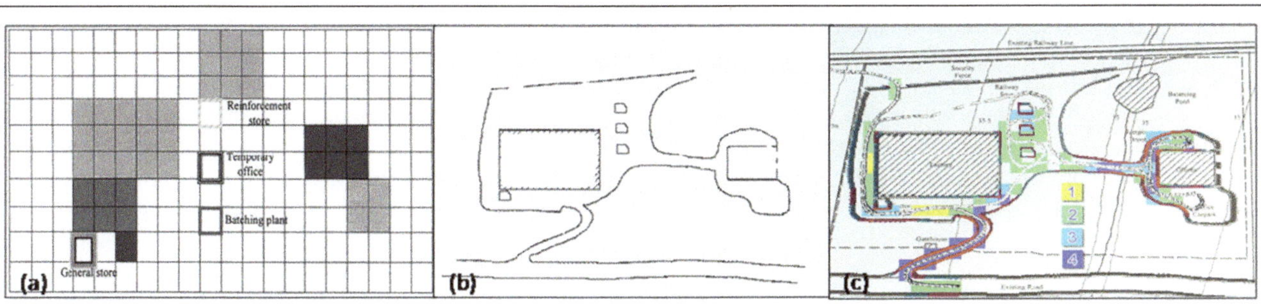

Figure 10 Example case of the evaluation process. (a) Site plan map; **(b)** modified 2D image; and **(c)** visualized evaluation result.

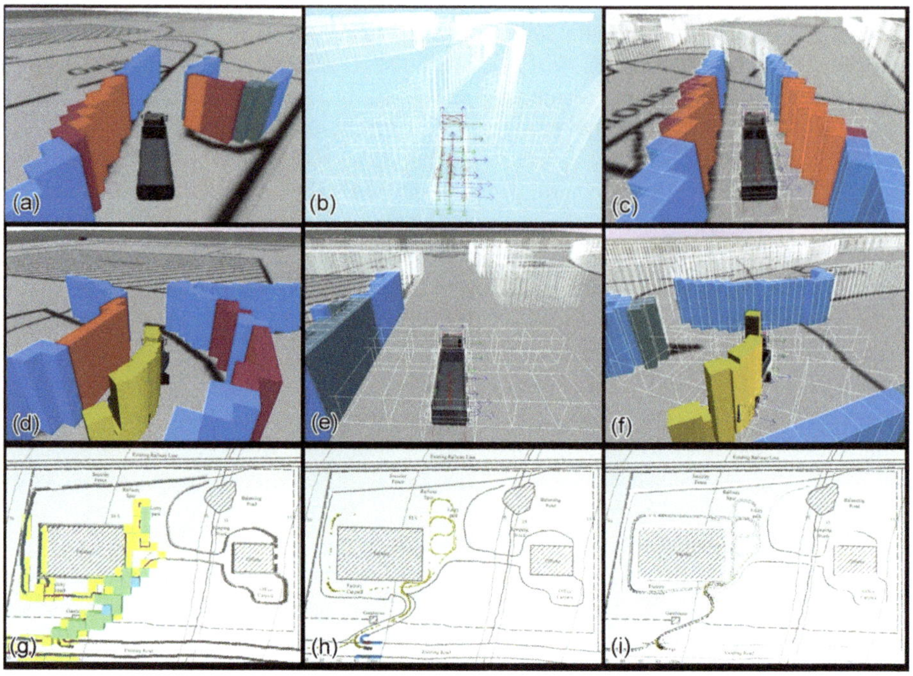

Figure 11 Snapshots of the evaluation process in the developed system. (a)-(c) The basic three viewing modes of the system; **(d)-(f)** visualized narrow level, safety range score and curvature score; and **(g)-(i)** three visualization modes of the evaluation result.

window to help relieve congestion of operations at site.

- The optimum level of exposure for a moving truck can be recognized through color detection. Therefore, the optimal position of the object can be established for strategic areas.
- Analyzing the closeness of objects, it helps in optimizing the positioning and the duration of the activities of certain expensive machine operations. The closeness of the objects can be analyzed by an evaluation of the *Safe Driving Range* safety parameter.
- The system can help to identify problems regarding the changing of site layouts at different phases of construction. Thus, it can help analyze and control the sequence and priority of different phases of work to obtain a high cost-time productivity. During different phases, the user can try to evaluate the accessibility of the site layout.
- It helps to identify the minimum space required to support specific construction operations in restricted space or congested areas, thus improving the welfare of workers and pedestrians. Because the accessibility of the tractor-trailer can affect the space of other operations, the user can understand the required space of operations by the accessibility evaluation of the site.
- The modeled method helps engineers progress from an "experience-based" to a "computation–based" accessibility evaluation, which will make it easier to identify potential problems.

Conclusions

This paper proposed an innovative approach for evaluating accessibility during pre-construction site layout planning. A 3D simulation platform was developed for the verification of the feasibility of a site layout plan. In this system, a physics engine was used to capture the motion and reaction of the objects in the environment in order to approximate real world conditions. To allow work on different planning cases to be carried out efficiently, a site importing module was designed. The visualization module features the advantage of virtual reality, providing engineers with an easy way to check evaluation results and improve current site planning proposals. This paper described an example case using the proposed system to evaluate the safety and accessibility of an actual site layout plan. The evaluation result presented an intuitive and easily understood visualization which clearly indicated the unsafe parts of the site layout plan. The example case results show that the proposed system has the potential to reduce the costs caused by transportation accessibility problems.

The same modeling method used for the tractor-trailer can be applied to different kinds of construction machinery and future research will be directed towards this end so that the developed accessibility evaluation method can be used. The major limitation of this research is that the method has only been tested by one example case, and although it is a relatively complicated scenario, the method must be applied to more construction sites to validate its effectiveness more conclusively. Furthermore, by applying the method to more cases, the user can also identify more safety factors that can be included in the safety score. The system provides a realistic interactive environment for users in which they may simulate any possible route of the tractor-trailer. In future work, the route could be decided automatically by applying a motion planning algorithm.

Competing interest

The authors declare that they have no competing interests.

Authors' contribution

JCL and CEY co-developed the accessibility evaluation system and contributed evenly to the whole study and manuscript writing. WHH developed the simulation environment implementing multi-body dynamics in his previous research, and was included as the basis of this study. SCK was the adviser and proof-read the article. All authors read and approved the final manuscript.

References

Akinci, B, Fischer, M, et al. (2002). Automated generation of work spaces required by construction activities. *Journal of Construction Engineering and Management-Asce, 128*(4), 306–315.

Andayesh, M, & Sadeghpour, F. (2013). Dynamic site layout planning through minimization of total potential energy. *Automation in Construction, 31*, 92–102.

Benjaoran, V, & Bhokha, S. (2010). An integrated safety management with construction management using 4D CAD model. *Safety Science, 48*(3), 395–403.

Dawood, N, & Marasini, R. (2003). Visualisation of a stockyard layout simulator "SimStock": a case study in precast concrete products industry. *Automation in Construction, 12*(2), 113–122.

Easa, S, & Hossain, K. (2008). New Mathematical Optimization Model for Construction Site Layout. *Journal of Construction Engineering and Management, 134*(8), 653–662.

Elbeltagi, E, Hegazy, T, et al. (2004). Dynamic Layout of Construction Temporary Facilities Considering Safety. *Journal of Construction Engineering and Management, 130*(4), 534–541.

Elbeltagi, E, Hegazy, T, et al. (2001). Schedule-dependent evolution of site layout planning. *Construction Management and Economics, 19*(7), 689–697.

Erleben, K. (2005). *Physics-based animation*. Hingham: Mass., Charles River Media.

Hegazy, T, & Elbeltagi, E. (1999). EvoSite: Evolution-based model for site layout planning. *Journal of Computing in Civil Engineering, 13*(3), 198–206.

Hung, WH, & Kang, SC. (2009). *Physics-Based Crane Model for the Simulation of Cooperative Erections* (p. 11). Sydney: 9th International Conference on Construction Applications of Virtual Reality.

IOS&H. (2012). *Construction Industry Accidents Knowledge Platform*. http://www.iosh.gov.tw/CIAKP/Statistics2.aspx.

Lam, KC, Ning, X, et al. (2009). Conjoining MMAS to GA to Solve Construction Site Layout Planning Problem. *Journal of Construction Engineering and Management-Asce, 135*(10), 1049–1057.

Lam, KC, Ning, X, et al. (2007). The application of the ant colony optimization algorithm to the construction site layout planning problem. *Construction Management and Economics, 25*(4), 359–374.

Li, H, Chan, N, et al. (2009). Optimizing construction planning schedules by virtual prototyping enabled resource analysis. *Automation in Construction, 18*(7), 912–918.

Ma, Z, Shen, Q, et al. (2005). Application of 4D for dynamic site layout and management of construction projects. *Automation in Construction, 14*(3), 369–381.

Mawdesley, MJ, & Al-Jibouri, SH. (2003). Proposed genetic algorithms for construction site layout. *Engineering Applications of Artificial Intelligence, 16*(5–6), 501–509.

Mawdesley, MJ, Al-jibouri, SH, et al. (2002). Genetic algorithms for construction site layout in project planning. *Journal of Construction Engineering and Management, 128*(5), 418–426.

Ning, X, Lam, KC, et al. (2010). Dynamic construction site layout planning using max-min ant system. *Automation in Construction, 19*(1), 55–65.

Rouchon, P, & Fliess, M. (1993). *Flatness, motion planning and trailer systems* (pp. 2700–2705). San Antonio, Texas, U.S.A: Proceedings of the 32nd IEEE Conference on Decision and Control.

Sadeghpour, F, Moselhi, O, et al. (2006). Computer-aided site layout planning. *Journal of Construction Engineering and Management-Asce, 132*(2), 143–151.

Soltani, AR, Tawfik, H, et al. (2002). Path planning in construction sites: performance evaluation of the Dijkstra, A*, and GA search algorithms. *Advanced Engineering Informatics, 16*(4), 291–303.

Su, X, Andoh, AR, et al. (2012). GIS-based dynamic construction site material layout evaluation for building renovation projects. *Automation in Construction, 27*, 40–49.

Tam, CM, & Tong, TKL. (2003). GA-ANN model for optimizing the locations of tower crane and supply points for high-rise public housing construction. *Construction Management and Economics, 21*(3), 257–266.

Tam, CM, Tong, TKL, et al. (2002). Site layout planning using nonstructural fuzzy decision support system. *Journal of Construction Engineering and Management, 128*(3), 220–231.

Thomas, HR, Riley, DR, et al. (2005). Fundamental principles of site material management. *Journal of Construction Engineering and Management, 131*(7), 808–815.

Waly, AF, & Thabet, WY. (2003). A Virtual Construction Environment for preconstruction planning. *Automation in Construction, 12*(2), 139–154.

Winch, G, & North, S. (2006). Critical Space Analysis. *Journal of Construction Engineering and Management, 132*(5), 473–481.

Wong, CK, Fung, IWH, et al. (2010). Comparison of Using Mixed-Integer Programming and Genetic Algorithms for Construction Site Facility Layout Planning. *Journal of Construction Engineering and Management-Asce, 136*(10), 1116–1128.

Zouein, PP, Harmanani, H, et al. (2002). Genetic algorithm for solving site layout problem with unequal-size and constrained facilities. *Journal of Computing in Civil Engineering, 16*(2), 143–151.

11

Visualization of CCTV coverage in public building space using BIM technology

Huan-Ting Chen, Si-Wei Wu and Shang-Hsien Hsieh*

Abstract

Background: Nowadays, the use of Closed Circuit Television (CCTV) systems is effective for monitoring traffic, preventing crime, and ensuring safety in many public spaces. However, the effectiveness of CCTV coverage is often achieved through design experience and trial-and-error, instead of being evaluated and visualized using a robust approach.

Methods: Firstly, a method for simulating varifocal CCTV lenses in order to attain different fields of view was developed, allowing real CCTV views to be approximated by adjusting the parametric properties of simulated CCTV cameras in the 3D BIM model. Secondly, an API (Application Programming Interface) plug-in program for a commercially available BIM tool was developed to facilitate the parametric modeling of CCTV systems and the evaluation of the CCTV coverage.

Results: A complete BIM model of an MRT (Mass Rapid Transit) station was chosen as a case study to apply the developed approach to the examination of CCTV coverage. Finally, the overall coverage of the CCTV systems for the MRT station were demonstrated visually and studied in the station's BIM model.

Conclusions: This research has developed a robust visualization approach for evaluating the coverage of CCTV systems in public building spaces. The developed approach is based on Building Information Modeling (BIM) technology and is capable of simulating CCTV systems in a 3D virtual environment in order to evaluate the CCTV coverage.

Keywords: Building Information Modeling (BIM), Closed Circuit Television (CCTV) coverage, Visualization approach, Parametric modeling

Background

In the 1990s, Closed Circuit Television (CCTV) systems were used for public security and crime prevention in urban cities in England. The British government and the police were both in favor of the use of CCTV systems. On the other hand, there was a huge controversy in relation to its cost, benefits, legitimacy, and privacy issues despite its effectiveness in crime prevention (Harris et al., 1998). Nowadays, CCTV is widely used in many public spaces in order to prevent crime, monitor traffic events, and ensure public safety (e.g. Aguado et al., 2009; Teague et al., 2010).

Once a CCTV camera is installed in the field, its relocation is expensive. Therefore, a thorough design of CCTV layouts should be conducted in advance of positioning in order to ensure proper siting and mounting (Lee et al., 1995). Currently, the positions and orientations of most CCTV systems are designed based on 2D diagrams such as floor plans. This may cause many spatial design conflicts with other systems due to a lack of the third dimension in the design space, and lead to unnecessary overlapping of CCTV coverage. Additionally, both CCTV screen views and CCTV coverage are usually unknown before construction is completely finished.

Building Information Modeling (BIM) has been a rapidly-developing technology in the Architecture, Engineering and Construction (AEC) industry in recent years. Using BIM technology, an accurate building model with precise geometry and other AEC-relevant data can be constructed in a virtual environment throughout the lifecycle of the building. BIM models are thus referred to as computer-generated models that utilize parameters of model elements to support the construction, fabrication and procurement activities required to realize the building (Eastman et al., 2011: 1–2). The 3D modeling and visualization provided by the BIM software can effectively discover and solve hidden problems that previously could not be detected until construction was completed. By using different

* Correspondence: shhsieh@ntu.edu.tw
Department of Civil Engineering, National Taiwan University, Taipei, Taiwan

viewpoints, reviewers are able to conduct interference checking between systems and thus reduce unnecessary design conflicts. Also, the computational representation of a BIM model can offer a simultaneous visual representation, which allows discussion of design ideas and analytical tests during the BIM-based design process (Wang et al., 2010).

One of the major differences between BIM and conventional 3D CAD models is the linkages among model views. A conventional 3D CAD model describes a building by independent projected views such as plans, sections and elevations. If one of these views is modified, all other related views must be revised and updated manually; such a process is prone to human error. In addition, these 3D drawings contain only graphical entities, such as lines, arcs and circles. In contrast a BIM model consists of parametric objects that are defined in terms of building model elements and systems, like spaces, walls, beams and columns. BIM software is able to automatically generate conventional building plans, sections, and elevations directly from a 3D BIM model as well as photo-realistic 3D renderings, which up to a few years ago could only be produced by special visualization packages but can now be produced with little additional effort (Sah and Cory, 2008; Azhar, Hein and Sketo 2008). Furthermore, a BIM model is equipped with other non-geometric information needed for tasks in the lifecycle of the corresponding building or facility, including physical properties, functional characteristics, spatial relationships, etc. Therefore, it allows designers, engineers, and other project participants to visually examine the planning and design outcomes in a more intuitive way with richer information support. By the same token, the use of an existing BIM model to examine the coverage of CCTV systems can be a highly efficient and low cost auxiliary approach to aid CCTV layout design.

The objective of this research is to develop a robust visualization approach for evaluating the coverage of CCTV systems in public building space. The expected benefits can be summarized as follows: (1) to ensure intended CCTV coverage and reduce design conflicts with other systems; (2) to shorten the time required to setup CCTV cameras during construction; (3) to provide a better communication medium among CCTV system designers, project owners and facility operators.

In the following sections, the methods and process of modeling a CCTV system in a BIM environment are presented first. Then, an MRT (Mass Rapid Transit) station was chosen as a case study for demonstrating the results of this research. Using the Autodesk Revit Architecture (Autodesk, Inc 2010a) (hereafter shortened to Revit) along with the BIM model of an MRT station, this research employs BIM technology to integrate CCTV systems into the virtual building space of an MRT station. Finally, virtual CCTV screen views were simulated so that the CCTV coverage could be clearly visualized and studied in the virtual environment.

Methods

This section explains the process of modeling a CCTV system in a BIM model. This research uses Autodesk Revit as the development platform because the BIM model of the MRT station used in the case study was originally modeled by Revit. Revit provides a user-friendly and powerful 3D modeling design environment and can import 2D AutoCAD files for assisting with the model construction. It also provides a collaboration platform with a centralized database so that all changes can be synchronized to the cloud server (Azhar, Nadeem et al., 2008). Moreover, Autodesk supports an Application Programming Interface (API) for users to extend the core functionality of Revit through plug-in programs. In this research, a complete solution in Revit for simulating CCTV systems in BIM models is developed.

Using Revit cameras as CCTV cameras

A complete CCTV system is composed of video cameras and TV monitors. This section introduces how the Revit camera works, as well as how it can serve as a CCTV camera in order to generate virtual screen views on TV monitors. Revit supports several kinds of 3D views to allow users document and present projects. Among these different kinds of 3D views, the perspective 3D view can show the building model in a 3D view where components that are further away appear smaller, and components that are closer appear larger, as illustrated in Figure 1. Since both Revit cameras and CCTV cameras can produce perspective views on a screen, it is an appropriate and convenient option in the simulation of CCTV systems. Virtual screen views can also be automatically generated by setting up Revit cameras with several significant parameters applied.

Simulating the varifocal lenses of CCTV cameras

Different CCTV cameras in a building have different fields of view (FOV). Based on the optical properties of camera lenses, different FOVs can be achieved by changing the focal lengths of the lenses. Thus, the first step in the simulation of CCTV systems in BIM models is to come up with an approach to simulate the varifocal lenses of CCTV cameras.

Revit cameras lack parameters concerning focal length (i.e. it is not possible to change the focal length of a certain Revit camera by setting the element parameter). One possible solution is to apply a pre-calculated crop region size (see Figure 2) to the 3D view of a Revit camera, which corresponds to the desired FOV. Hence, it is necessary to figure out the relationship between the focal length (i.e. FOV) and the crop region size. According to the geometric

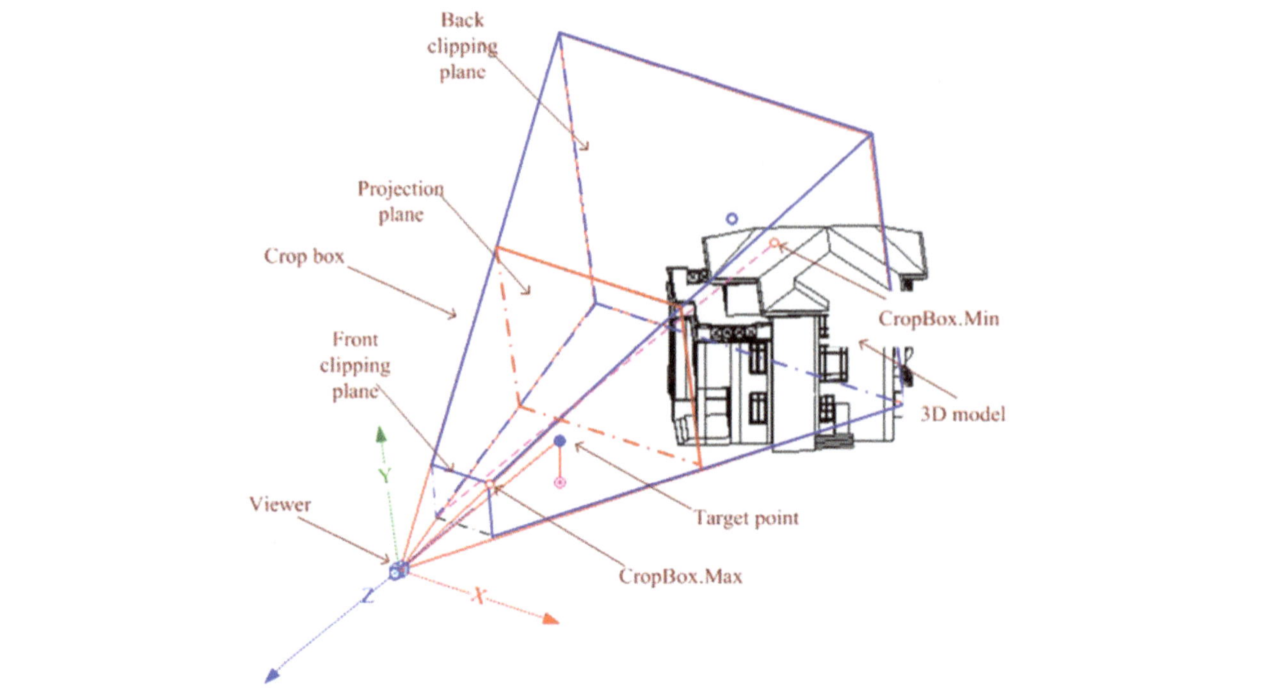

Figure 1 The perspective 3D view in Revit (Autodesk, Inc 2010b).

relationship, the focal length can be obtained by inputting the FOV into the following equation:

$$\textbf{Focal length} = \frac{\frac{1}{2} \times Film\ \ dimension}{\textbf{tan}\frac{FOV}{2}} \tag{1}$$

while the default film dimension is fixed (i.e. the larger the focal length, the smaller the FOV, and vice versa, as shown in Figure 3). The width of the crop region size can now be calculated by using the principle of similar figures (see Figure 4 below). Thus, the crop region size is inversely proportional to the focal length. The height of the crop region can also be obtained by multiplying by the screen aspect ratio of 3/4.

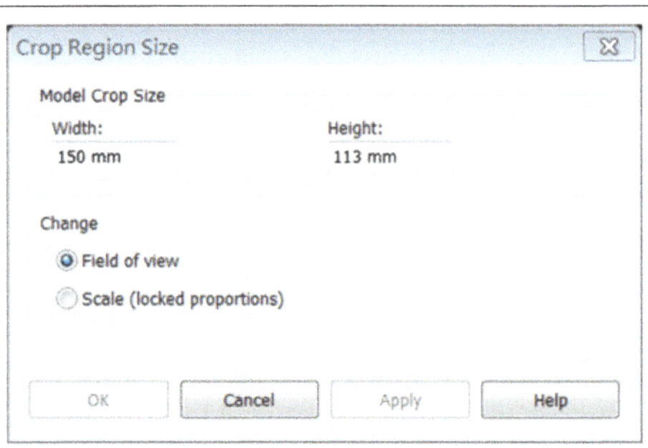

Figure 2 The menu page for setting the crop region size of the Revit camera.

The default configuration of the Revit camera indicates the default field of view, the default focal length, and the default film dimension, which are 50 degrees, 38.6 mm, and 36 mm, respectively. In addition, the default crop region size is 150 mm * 113 mm, as shown in Figure 2. When reconfiguration of the Revit camera for a desired FOV (for example, 60 degrees) is required, the corresponding crop region size can be calculated as follows:

(1) According to Eq. (1) and the default film dimension, 36 mm, the focal length of 60 degrees FOV is calculated to be 31.2 mm.
(2) According to the principle of similar figures, the width of the crop region size corresponding to 60 degrees FOV can be calculated by multiplying 150 mm (i.e., the width of the default crop region size) by a ratio of (38.6 mm/31.2 mm), in which 38.6 mm is the default focal length, and the result is 185.6 mm.
(3) According to the screen aspect ratio, the height of the crop region size can be calculated by multiplying 185.6 mm by 3/4 and the result is 139.2 mm.

Configuring revit cameras with parameter analysis

Virtual CCTV screen views can be simulated using Revit camera views if appropriate parameters are applied. There are several significant parameters to be configured in the CCTV system and it is necessary to relate these

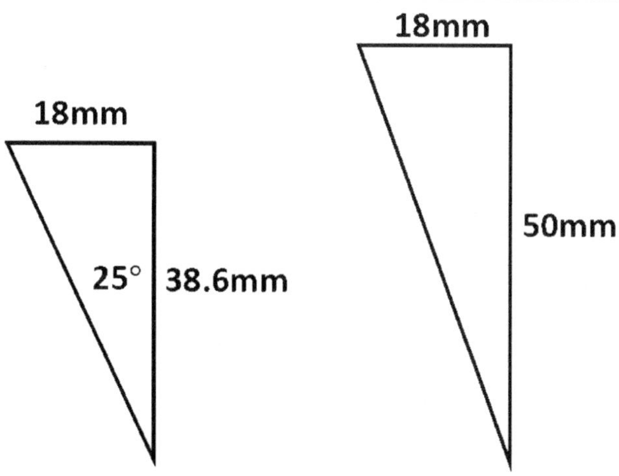

Figure 3 The larger the focal length, the smaller the FOV.

parameters to the parameters of the Revit camera. The most relevant parameters of the CCTV system to this research include FOV, camera height, target height, subject height, subject distance, and the real angle of depression. After analyzing these parameters and determining their reasonable ranges, an optimized real angle of depression (θ) for each CCTV camera can be found based on the spatial relationship shown in Figure 5. By setting up a CCTV camera with a predetermined optimized real angle of depression, construction efficiency is enhanced as the set up time is reduced. The following paragraphs explain each parameter although its reasonable range will vary from case to case. The detailed configuration data will be discussed later in the *Process of modeling CCTV systems with CCTV parameter advisor* section.

1. Field of view (FOV):
   As mentioned in the previous section, a CCTV camera uses a varifocal lens to attain different FOVs, which are determined according to its requirements, location, etc. Since this research aims to examine the existing design of CCTV systems, all the FOVs of the CCTV cameras should be predefined by CCTV designers.
2. Camera height (H):
   The camera height relates to the ceiling height. This dimension is measured from the floor elevation to the ceiling elevation. The corresponding parameter of the Revit camera is "eye elevation".
3. Target height (h):
   Normally, the target of CCTV cameras are people, therefore, the average height of humans is adopted as the target height. The corresponding parameter of the Revit camera is "target elevation".
4. Subject height (%) and Subject distance (D):
   Rotakin, developed by the Home Office in the UK, is the only device specified in BSI EN 50132–7:1996 for testing CCTV camera performance. The Home Office also offers an operation manual (Cohen et al., 2009: 8–9) that defines the concepts of subject height and subject distance. Depending on the nature of the human activity to be observed, five general observation categories have been defined for different purposes, including monitoring and control, detection, observation, recognition, and identification. These five categories are based on the relative size of a person that appears on a screen (see Figure 6). This is defined as the subject height, a percentage of the height of a person divided by the height of the screen. For a specified subject height, the maximum distance from which a person can be seen on the screen is defined as the subject distance (i.e. a person would be too small to be seen on the screen beyond the subject distance for the specified subject height). An independent consultancy company provides an online lens calculator based on the standards of the Home Office, LensCalc (CCTV Advisory Service, 2008), to help determine the subject distance required to achieve specified subject heights for different combinations of sensor size, subject height, and lens focal length (see Table 1). Please note that the subject height is used to specify

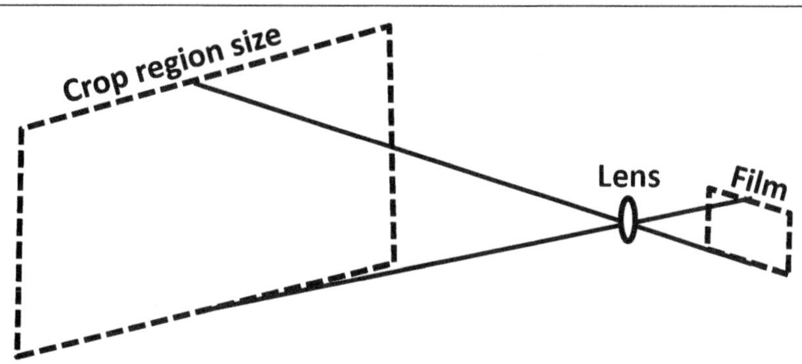

Figure 4 The principle of similar figures between the crop region size and the film dimension.

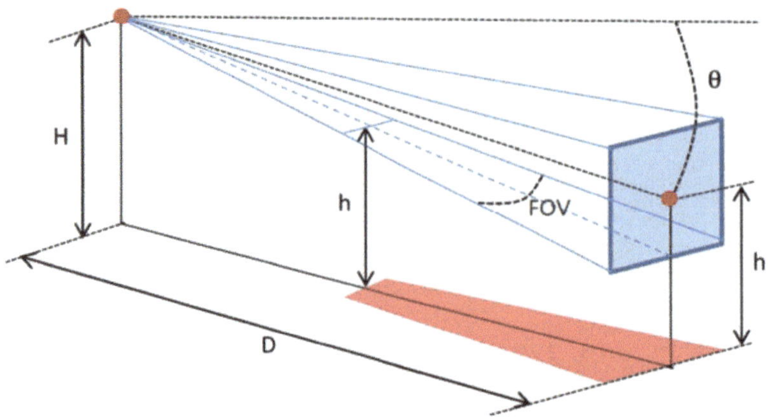

Figure 5 The spatial relationship of CCTV camera parameters.

a particular system requirement rather than to define a general standard. Additionally, the related guidelines were first developed in the PAL (Phase Alternating Line) standard, which is a color encoding standard for analogue television and has been commonly used in the design of CCTV systems in public areas. Nowadays, a smaller subject height may be accepted through the use of new cameras with higher resolutions. Please refer to the *CCTV Operational requirements manual 2009* (Cohen et al., 2009: 9–10) for more information.

5. Real angle of depression (θ):
   The Real angle of depression can be referred to as the end result once the camera height, target height, and subject distance are specified. An optimized real angle of depression can be determined for each CCTV camera based on spatial relationships (see Figure 5). The following formula shows the geometric expression of an optimized real angle of depression (θ):

$$\tan\theta = \frac{H-h}{D} \tag{2}$$

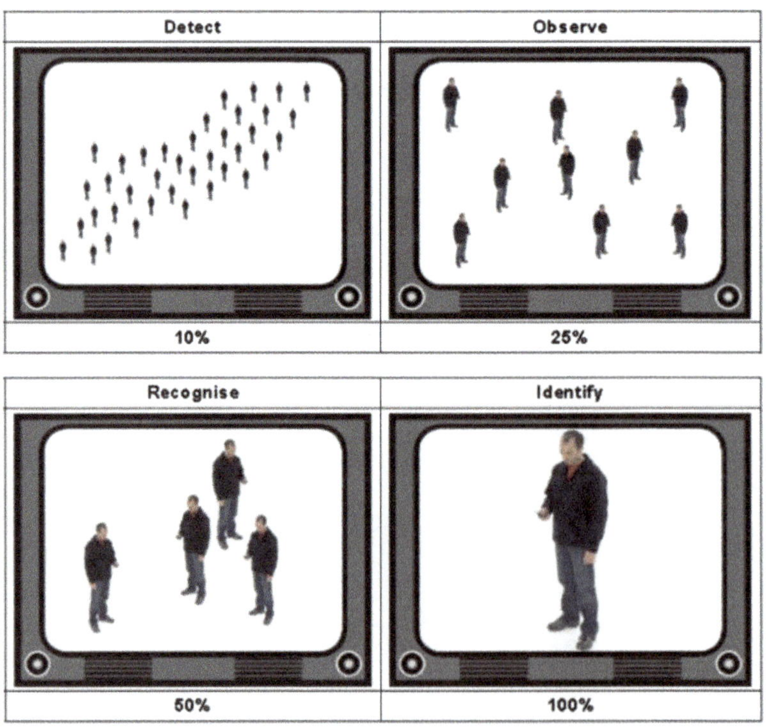

Figure 6 Four out of the five general observation categories in accordance with different subject heights (Cohen et al., 2009: 8–9).

Table 1 Subject distances for different combinations of the subject height and the FOV

FOV	Subject height			
	10%	25%	50%	100%
10°	77.72	38.86	19.43	9.715
35°	21.56	10.78	5.39	2.695
50°	14.6	7.3	3.65	1.825
60°	11.76	5.88	2.94	1.47
95°	6.24	3.12	1.56	0.78
99°	5.8	2.9	1.45	0.725

(Unit: meter).

In addition to the parameters above, the CCTV coverage is also shown in Figure 5. The red trapezoidal area reveals the coverage area of a certain CCTV camera that covers the area where persons standing can be seen by the CCTV for a specified subject height. The measure of that red trapezoidal area can also be calculated and displayed in BIM models using Revit as discussed later in the *Evaluation and visual representation of CCTV coverage* section.

Simulating virtual CCTV screen views

As previously mentioned in the *Using Revit cameras as CCTV cameras* section, a Revit camera object that provides a 3D perspective view can serve as the CCTV camera in the Revit BIM environment. That is, virtual CCTV screen views can be simulated by using Revit camera views if appropriate parameters are applied.

In order to evaluate the existing design of CCTV systems, it is essential to accurately position all CCTV cameras within the BIM model according to the plan views of the CCTV design in 2D CAD files. After importing the 2D CAD files of CCTV plan views and aligning them with the floor plan views of the BIM model (see Figure 7), Revit cameras can be properly positioned with reference to the layout of the two overlapped plan views. Virtual CCTV screen views can also be automatically generated after inputting all of the required parameters (see Figure 8), including eye elevation, target elevation, etc. Figure 9 shows an example of a virtual CCTV screen view simulated by the Revit camera. It is important to carefully determine the values for the parameters of Revit cameras based on those of their corresponding CCTV systems.

Because the MRT station used in the case study is still under construction, the capture of a real view image from a CCTV camera in the station for a validation study is not possible. However, to validate the proposed approach for simulating the CCTV cameras by the Revit cameras, a comparison was made between a real photo

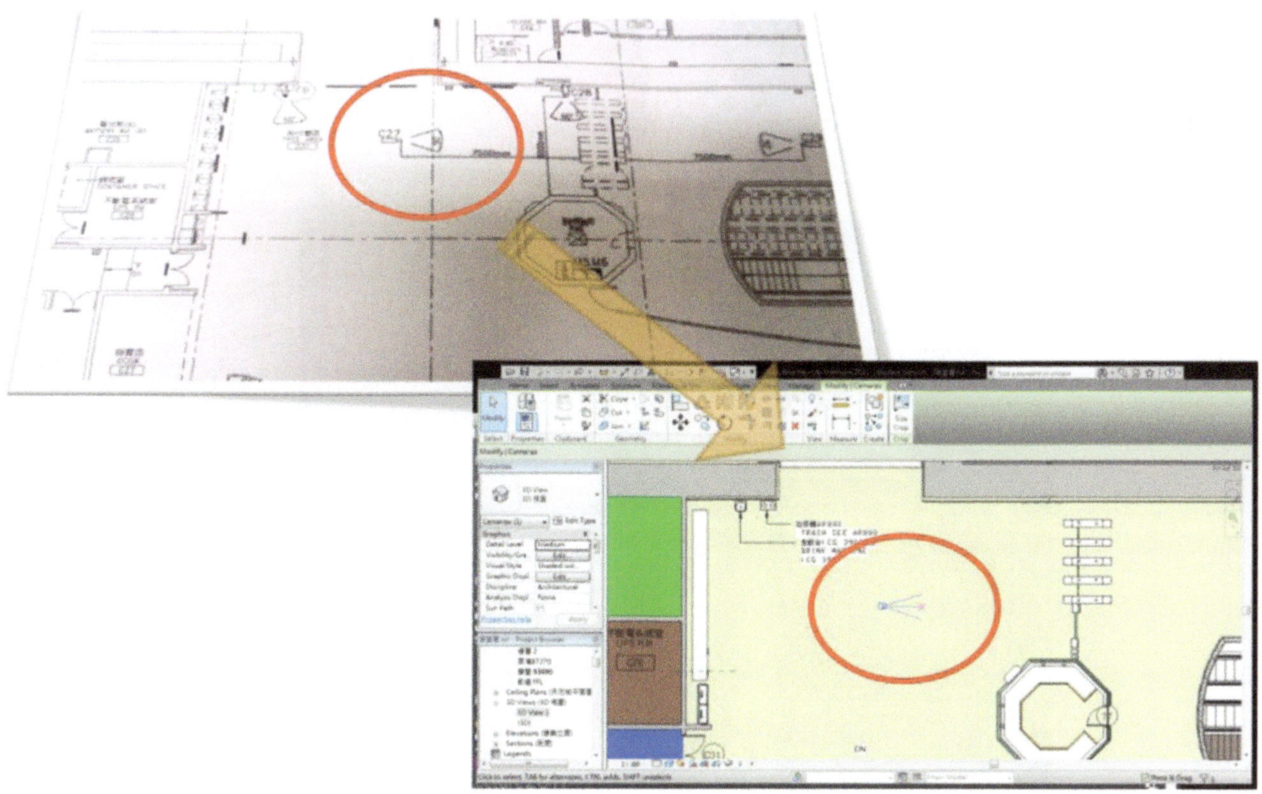

Figure 7 Floor plan views aligned with CCTV plan views.

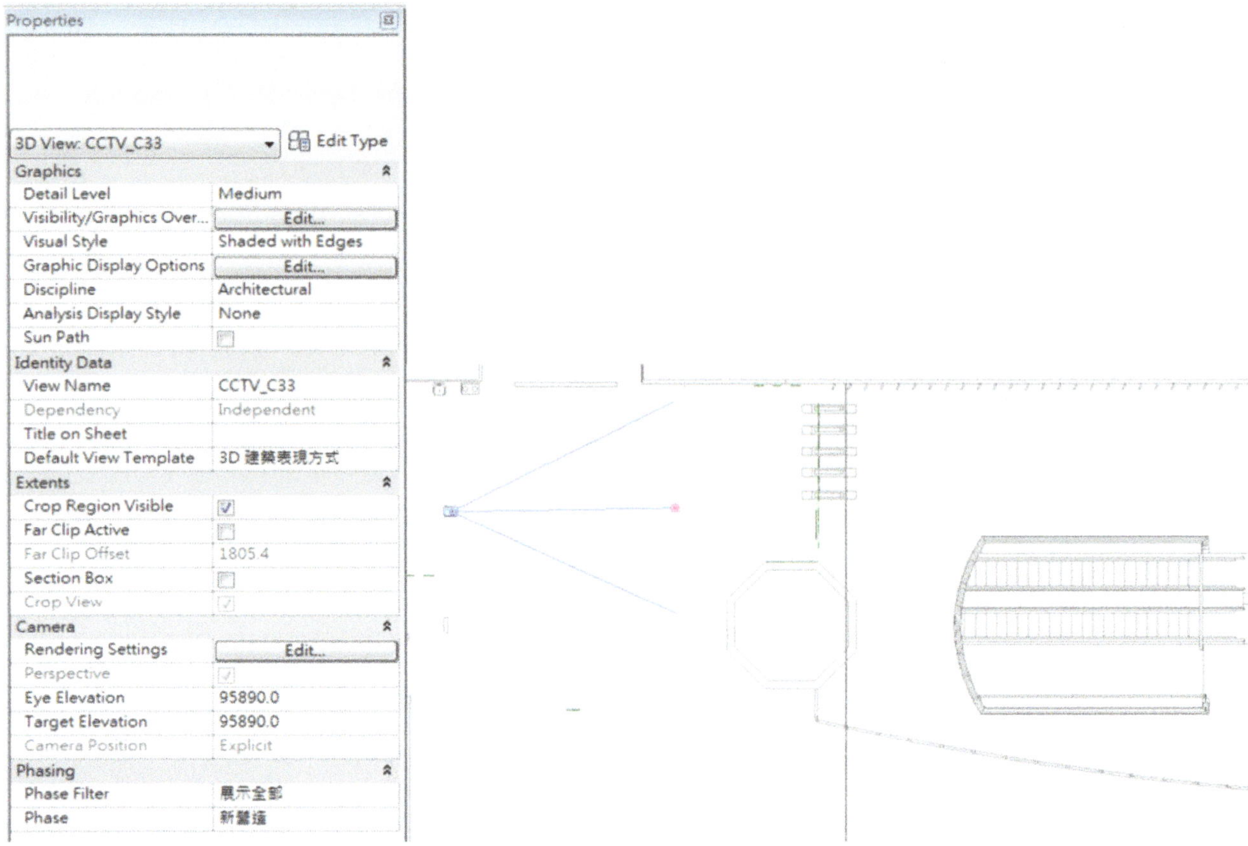

Figure 8 Configuring the parameters of the Revit camera.

Figure 9 A real photo taken with the digital camera (SONY SLT-A77V).

Figure 10 A virtual CCTV screen view simulated by the Revit camera for comparison with the photo in Figure 9.

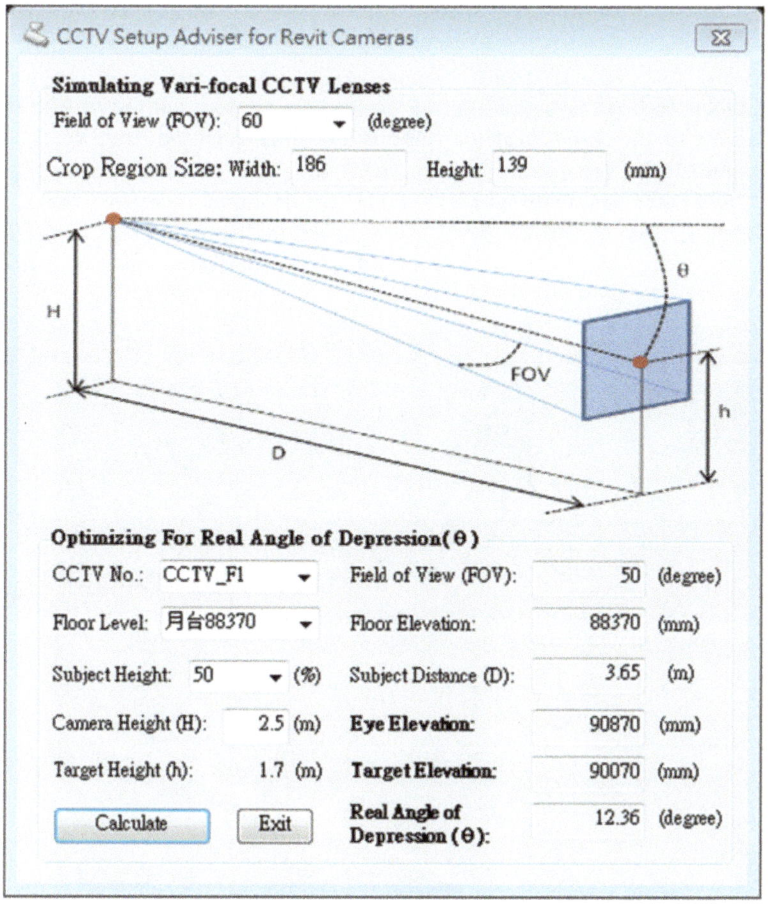

Figure 11 The user interface of the CCTV setup advisor for Revit cameras.

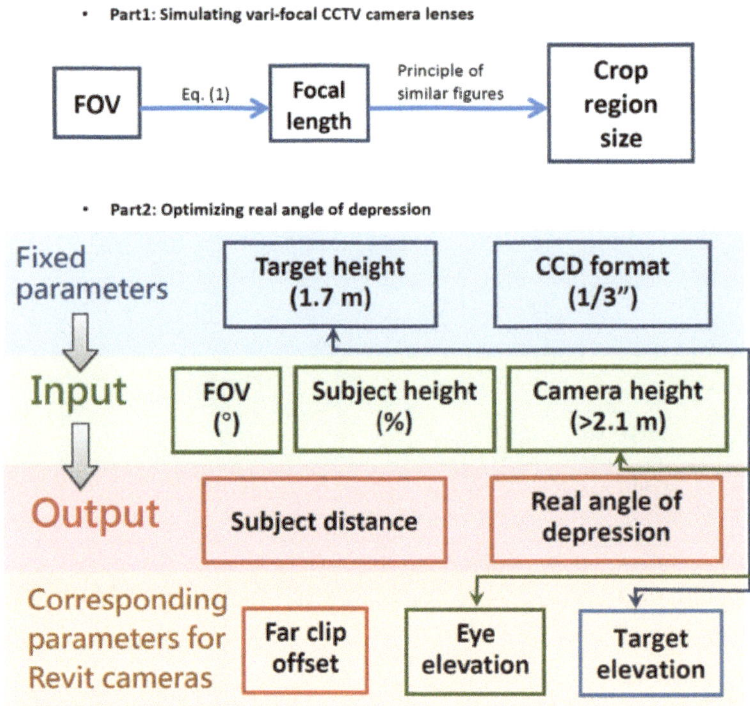

Figure 12 The logical process of the CCTV setup advisor.

taken by a digital camera (SONY SLT-A77V) and an image simulated by a Revit camera at the same location on the 6th floor of the Civil Engineering Research building at the National Taiwan University (see Figures 10 and 9, respectively). It can be seen that a satisfactory simulation result is obtained.

Plug-in development with revit API

Autodesk has released the Revit API (Autodesk, Inc 2010b) to allow users to develop plug-in programs in order to automate repetitive tasks, thus extending the core functionality of Revit. With Revit API, it is also possible to perform customizations based on the

Figure 13 The heights of CCTV cameras are lower than normal due to the influence of the smoke curtain.

Figure 14 Smoke curtains on the concourse level (CCTV No. C29, the location is shown in Figure 24).

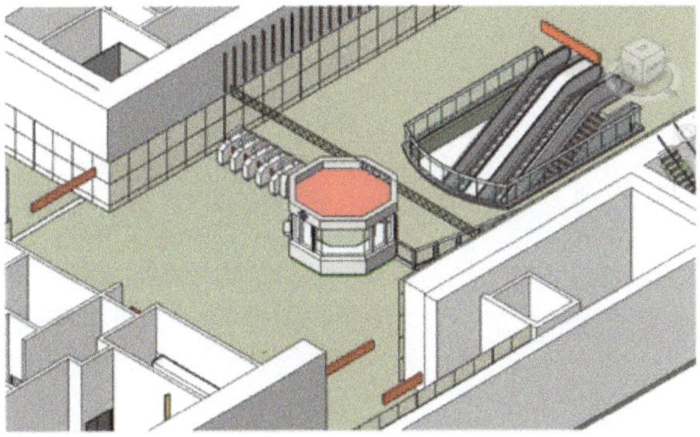

Figure 16 A bird's eye view of smoke curtains and ceiling signboards on the concourse level.

requirements of users. The program developed with Revit API is called Revit add-in that supports all .NET compliant languages for programming.

This research uses Autodesk Revit Architecture 2010 as the BIM software. It also provides an API to enable advanced users to develop external applications in order to incorporate BIM-based parametric design methods (Wang et al., 2010). The API offers access to the active Revit document and its corresponding BIM database. Information can be retrieved from the database, and any External Commands available in the Revit API can perform basic database operations (Yan et al., 2011).

To enhance the performance of Revit in the evaluation of CCTV coverage, a plug-in API program, called CCTV setup advisor for Revit cameras (hereafter shortened to CCTV setup advisor) is developed in this research with the following two considerations. Firstly, it is more succinct to integrate all related functions and steps in Revit to a single interface because modeling CCTV systems in BIM models requires a sequence of procedures. Secondly, it is more time efficient for users to configure many parameters in a single interface due to the fact that there may be many CCTV cameras required even in a single building space.

Figure 15 Ceiling signboards on the platform level. (CCTV No. C2, the location is shown in Figure 25).

CCTV setup advisor, therefore, has two main functions developed with BIM technology: (1) simulating varifocal CCTV camera lenses by Revit cameras and (2) configuring the parameters of Revit cameras to obtain virtual CCTV screen views and optimized real angles of depression. Revit cameras in the Revit API are read-only; therefore, no parameter of Revit cameras can be modified through external API commands. Thus, it is advisable for users to manually input the values of the parameters suggested by the plug-in program.

Figure 11 shows the user interface of the plug-in program. The upper section simulates varifocal CCTV camera lenses, which helps to determine the crop region size depending on different FOVs. The lower section optimizes the real angle of depression, which suggests parameters for Revit cameras based on user inputs and then calculates the real angle of depression. Figure 12 presents the logical process of this plug-in program, including input of parameters, returned outputs, and the relationship between parameters. An additional movie file is provided to show the CCTV setup advisor in more detail (see Additional file 1).

Results and discussion

After successfully modeling a CCTV system in a BIM-based virtual building space, this research conducted a case study to examine CCTV coverage. To make the BIM-based virtual environment correspond better to a real situation, a BIM design model of an MRT station under construction in Taipei city was adopted for the case study.

The application of CCTV systems in public spaces, especially mass transit facilities, for surveillance and protection purposes has grown at a considerable rate (Sanderson et al., 2007; March Networks News, 2008). Some research, focused on long-term experiments of advanced Intelligent CCTV (ICCTV) technologies in some

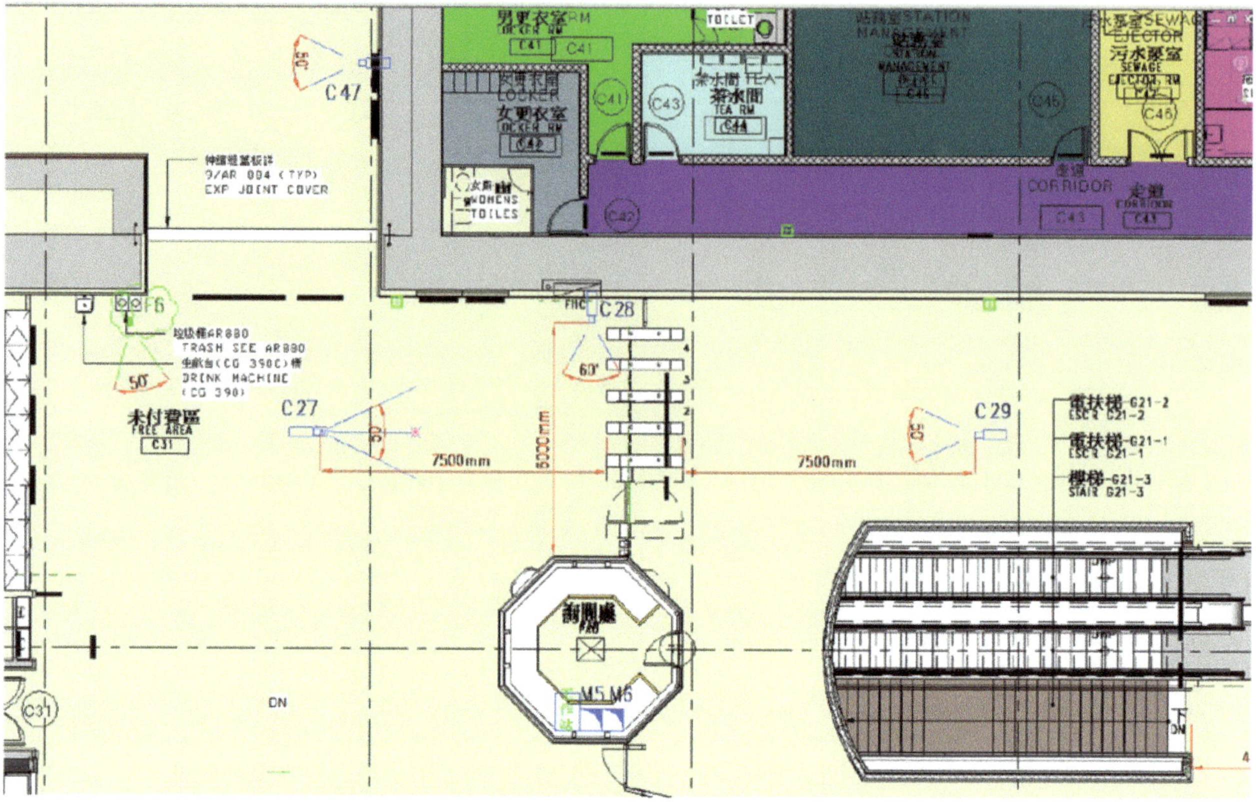

Figure 17 A floor plan view overlapped with the imported CCTV plan view.

sensitive public spaces, like major ports and railway stations, has also been published (Bigdeli et al., 2007). In Taipei city's MRT system, the CCTV systems are mainly used at platforms and in the lobbies of an MRT station for the following functions (Sun, 2005):

1. When the MRT train stops at a station, the train driver can monitor the movement of passengers getting on and off the train via CCTV systems. This helps to ensure the safety of passengers.
2. Station staff can monitor the activities in the station through the CCTV systems so that proper reactions can be made promptly in the case of an emergency.
3. The traffic control center manages the operations of the entire MRT system with the help of the CCTV systems in MRT stations. The operators can visualize all situations in MRT stations and respond correctly to ensure the safety of the trains in service.
4. Station staff can monitor, through CCTV systems, the traffic flows of passengers at all entrances and exits of the station and ensure that entry/egress is smooth.

The purposes of application discussed above also define the design requirements of the CCTV systems in the MRT system. For example, CCTV cameras are needed in certain locations on the platform to ensure full coverage of the movement of passengers getting on and off the trains. They are also needed at all entrances and exits of stations for good coverage of passenger traffic flows.

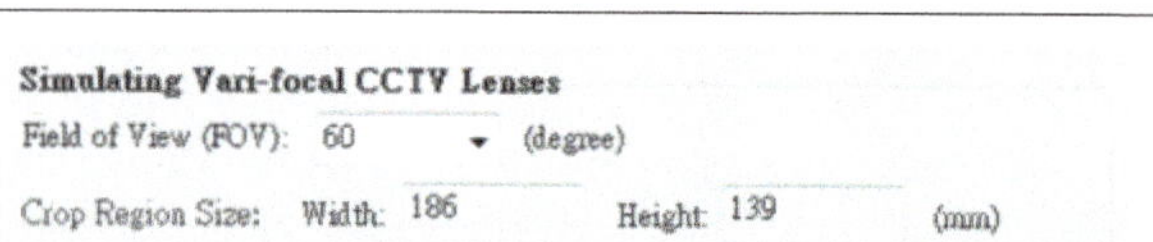

Figure 18 Crop region size advised by the CCTV setup advisor.

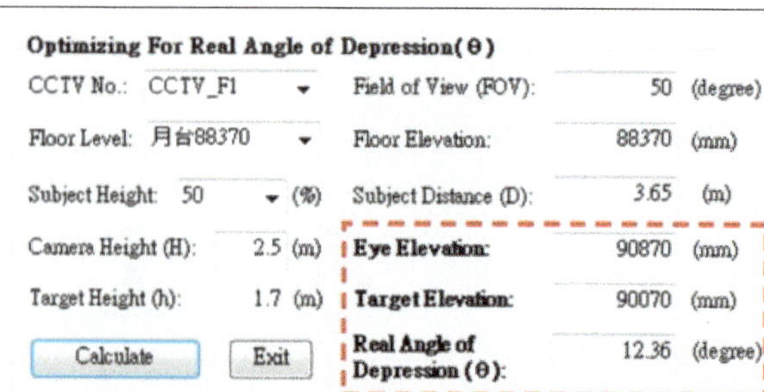

Figure 19 Parameters suggested by the CCTV setup advisor.

Figure 20 A virtual CCTV screen view simulated by the Revit camera (CCTV No. C18, the location is shown in Figure 25).

Modeling of smoke curtains and ceiling signboards

In order to examine any design conflicts between CCTV systems and other systems, several architectural components in an MRT station that may block the views of the CCTV cameras need to be modeled as parametric elements in the BIM model. These include smoke curtains and ceiling signboards.

Traditionally, the placement of a CCTV camera is subject to the conditions of the on-site environment, such as irregular ceiling heights, suspended signboards under the ceiling, smoke curtains, and so on. For example, in the real MRT station shown in Figure 13, the CCTV cameras are located lower than the normal height due to the influence of the smoke curtain. A lower location for a CCTV camera can increase the possibility of being intentionally destroyed and decrease the range of the CCTV coverage. If a better location can be found in advance using the BIM model, a suitable height for the CCTV camera can also be determined. However, the consideration of these architectural components has relatively low priority during the life cycle of the construction project and may not be modeled in the BIM model. This is also the case for the BIM model the authors obtained from the design company.

Figure 21 A simulated CCTV screen view before the checking of design conflicts.

Figure 22 A simulated CCTV screen view after the checking of design conflicts (CCTV No. C10, the location is shown in Figure 25).

Thus, this research models the smoke curtains and ceiling signboards as parametric elements in the Revit model. Among predefined system family types in Revit, the Wall element is most appropriate for modeling parametric smoke curtains and ceiling signboards. By means of modifying several Wall properties, such as base constraint, base offset, top constraint, top offset, and thickness, the appearance of Wall elements can resemble smoke curtains and ceiling signboards. The elements with different locations, heights and sizes can be easily and parametrically modeled in the BIM model.

In order to distinguish smoke curtains from ceiling signboards, the two kinds of elements were assigned different materials, colors, and appearances. Smoke curtains are set to be transparent meshed acrylic sheets, as shown in Figure 14, while ceiling signboards are set to be red opaque acrylic sheets, as shown in Figure 15. In addition, Figure 16 presents a bird's eye view of both smoke curtains and ceiling signboards on the concourse level of the MRT station.

Process of modeling CCTV systems with CCTV parameter advisor

The first step of modeling CCTV systems is to correctly position Revit cameras according to the existing design drawings. Thanks to the assistance from the Department of Rapid Transit Systems, Taipei city government, 2D plan views of the CCTV system in the MRT station were provided in CAD format and therefore could be imported into the BIM model directly. With reference to the imported CCTV plan view, Revit

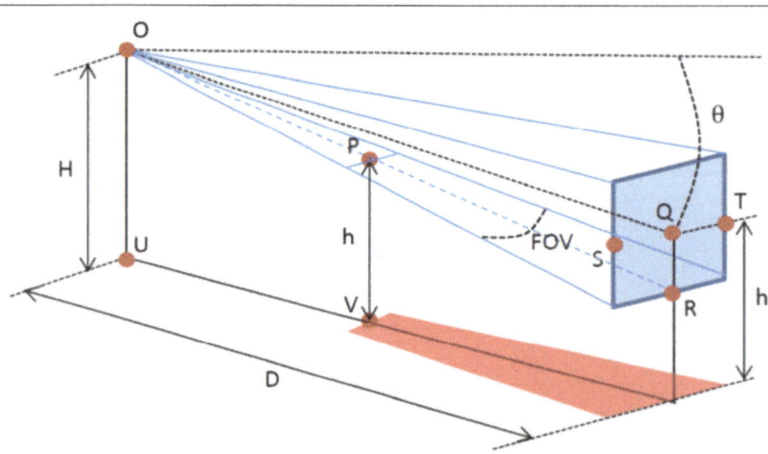

Figure 23 The calculation of the trapezoidal coverage area.

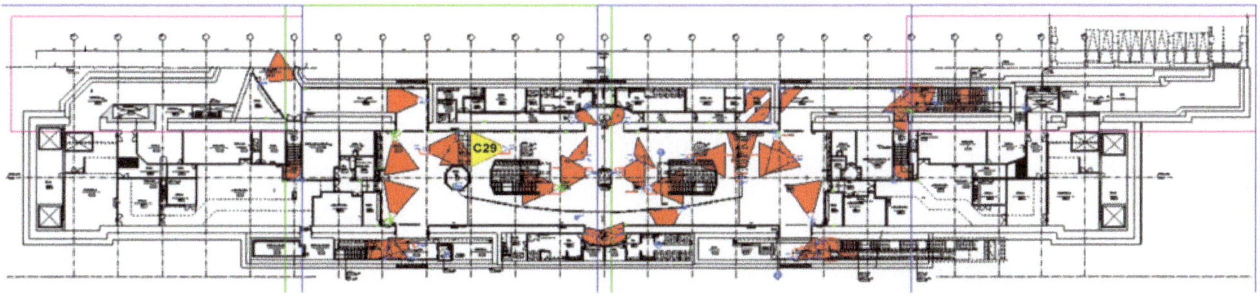

Figure 24 The CCTV coverage of the concourse level shown in a 2D floor plan view (Coverage ratio = 25.73%).

cameras can be placed in accurate positions on the corresponding floor plan view of the BIM model (see Figure 17). For modeling CCTV cameras with different FOVs, the developed plug-in program, CCTV setup advisor, can be used for calculating the corresponding crop region size (see Figure 18).

The second step is to determine the other two necessary parameters for the Revit cameras within reasonable ranges: the eye elevation and the target elevation. As previously mentioned in the *Configuring Revit cameras with parameter analysis* section, the eye elevation depends on the camera height, while the target elevation depends on the target height. Normally, the reasonable range of the camera height may be between 2.3 and 2.5 meters in the environment of an MRT station. Also, it is appropriate to apply the average height of the Taiwanese people to the target elevation, that is, approximately 1.7 meters.

After creating all necessary Revit cameras in the first step, the CCTV setup advisor is able to read in all Revit camera instances as well as the elevations of all floor levels, as shown in Figure 19. Subsequently, users need to key in the correct eye elevation and target elevation values for the Revit cameras (see Figure 8), which can be advised by the CCTV setup advisor after the selection of the desired Revit camera, the floor level, and the subject height, and the input of desired camera height. As for determining the subject height, it should be chosen based on the monitoring requirements of users (see the discussions in the *Configuring Revit cameras with parameter analysis* section and Figure 6). It should be noted that the selection of the subject height does not affect the simulation of the CCTV view but the effective coverage of the simulated CCTV. As shown in Figure 20, a CCTV screen view can be simulated by the corresponding Revit camera view. The simulated views are accessible when users choose a certain Revit camera view from the list of all 3D views via the Revit project browser.

More importantly, the CCTV setup advisor can suggest an optimized real angle of depression for the specified CCTV camera (see Figure 19). This may help to reduce the time for installation and adjustment of CCTVs because the best angle of depression for a CCTV camera has been determined in advance under user-desired conditions (i.e. an optimized real angle of depression exactly corresponds to a virtual CCTV screen view simulated by the Revit camera).

Checking for design conflicts

Prior to the use of 3D technology, detecting interferences between systems was both difficult and time-consuming. Nowadays, 3D BIM technology helps detect clashes in advance through more sensible visual presentations and

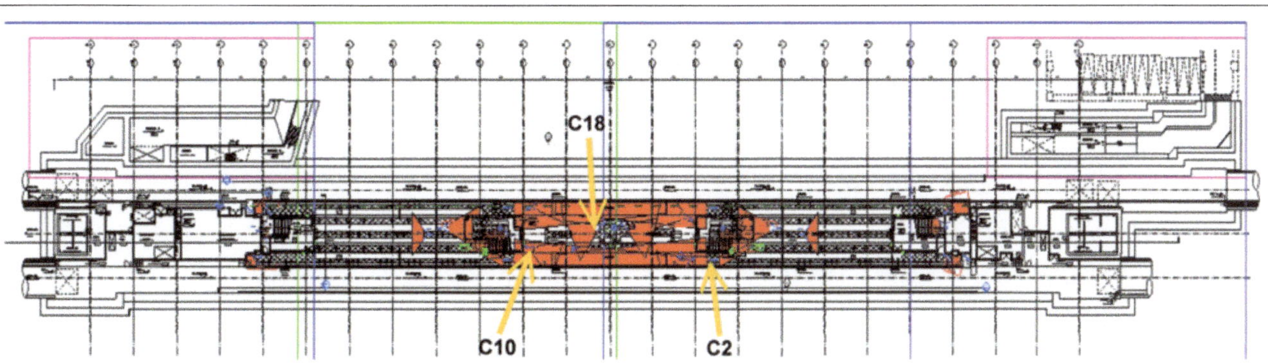

Figure 25 The CCTV coverage of the platform level shown in a 2D floor plan view (Coverage ratio = 42.00%).

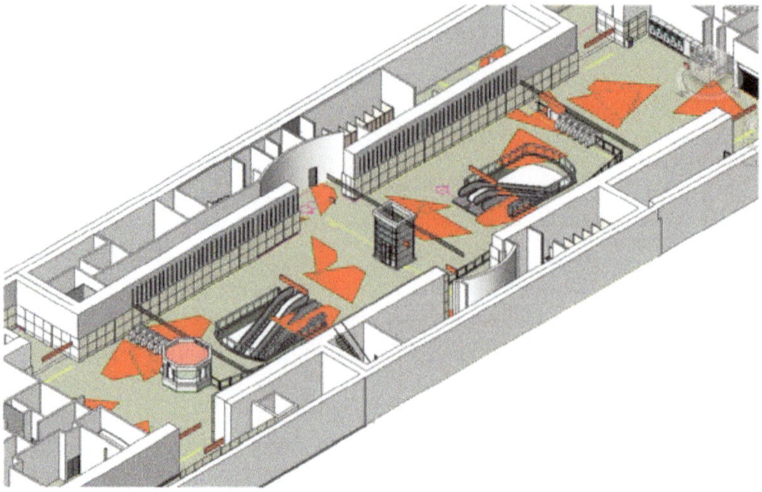

Figure 26 The CCTV coverage of the concourse level shown in a 3D model view (Coverage ratio = 25.73%).

more seamless collaboration platforms (Eastman et al., 2011: 272–273).

One of the aims of this research is to take advantage of BIM technology for inspecting in the design phase whether a desired location for CCTV camera installation is suitable in the future construction phase. Once CCTV screen views can be simulated via Revit cameras, it is simple to check if anything blocks the view of the CCTV camera in the BIM model. Examples of these are smoke curtains and ceiling signboards, which can be modeled as parametric elements in the BIM model of the MRT station. Such design conflicts must be discovered in advance and eliminated to reduce the installation and adjustment time required for the setup of CCTV cameras.

However, it is difficult to give a precise definition of design conflicts due to the fact that how seriously something blocks the CCTV camera's view may depend on subjective assessment by the operators. A more objective way to evaluate whether there is a clash is to use the optimized real angle of depression suggested by the CCTV setup advisor. Once an optimized real angle of depression is applied, most interference is supposed to be avoided. If there is still an obstacle blocking the CCTV camera's view, this implies that the parameters of the CCTV camera should be reset, including the camera height, target height, FOV, or even its original location.

Figure 21 shows a simulated CCTV screen view where almost half of the view is blocked by smoke curtains and ceilings. After adjusting several parameters of the Revit camera with the assistance of the CCTV setup advisor, a clearer view can be obtained as shown in Figure 22. The setup of a CCTV camera must be based on the

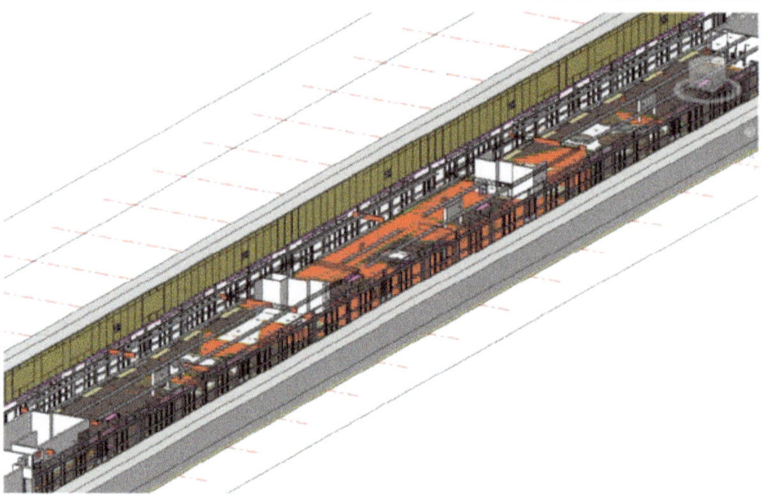

Figure 27 The CCTV coverage of the platform level shown in a 3D model view (Coverage ratio = 42.00%).

monitoring requirements of users. The simulation of CCTV screen views offers an approach for CCTV designers to conduct both detailed modifications and subtle adjustments prior to real construction.

Evaluation and visual representation of CCTV coverage

Even with the highly developed VR technology used today in the construction industry, CCTV systems are more often designed in a 2D environment. As CCTV systems have been introduced in 3D BIM models in this research, it is convenient to display the overall 3D environment and the CCTV coverage area through both 2D and 3D VR approaches.

In order to display visual representations of CCTV coverage, it is essential to define the effective area covered by CCTV cameras. According to Figure 23, the red trapezoidal area reveals the coverage area of a certain CCTV camera. For a known FOV, the area can be determined by the subject distance (D) and the length $\overline{UV}$. Because the subject distance (D) and other parameters, such as h, H, and FOV, have been defined in the *Configuring Revit cameras with parameter analysis* section, the length of $\overline{UV}$ can be calculated using the following steps:

(1) Assuming that the coordinates of the point O and the point Q are (0, 0, H) and (D, 0, h), respectively.
(2) The length of $\overline{OQ} = \sqrt{(\mathrm{H}-\mathrm{h})^2 + \mathrm{D}^2}$.
(3) The length of $\overline{ST} = \overline{2OQ} \times \tan\frac{FOV}{2}$, where the point Q is the midpoint of the line segment, $\overline{ST}$.
(4) Based on the screen aspect ratio (i.e. 3/4) as mentioned in *Simulating varifocal lenses of CCTV cameras* section, the coordinate of the point R is $\left(\mathrm{D}, 0, \mathrm{h} - \frac{1}{2} \times \frac{3}{4}\overline{ST}\right)$.
(5) The coordinate of the point P is $\left(\overline{UV}, O, h\right)$ and the points of O, P and R are in the same straight line. Therefore, the length $\overline{UV} = D(H{-}h)/\left(H = h + \frac{3}{8}\overline{ST}\right)$.

Although the red trapezoidal area can be calculated through the mathematical approach, it is more convenient and efficient to use the "Filled region" function in Revit. Once an area formed by a closed loop of lines is specified, the corresponding filled region can be shown on the floor plan view, and the measure of that area can be calculated automatically.

After all "filled regions" are created for all CCTV cameras in the MRT station, it is crucial to check if there are any overlapping areas among those filled regions. If so, the filled regions with overlapping areas must be redrawn as a single closed loop area to obtain accurate calculations of the areas. Although this may take a lot of time to do, especially when there are many CCTV cameras in an MRT station, it is a necessary step.

Finally, the coverage ratio can be determined by dividing the total area of all filled regions by the overall floor area in the MRT station. The overall effective coverage of the CCTV cameras in the MRT station can be displayed on both 2D floor plan views and 3D model views. Figure 24 and Figure 25 show the CCTV coverage in 2D floor plan views of the concourse level and the platform level, respectively. Figure 26 and Figure 27 show the effective CCTV coverage in 3D model views of the concourse level and the platform level, respectively. An additional movie file is provided to show the CCTV coverage in the MRT station in more detail (see Additional file 2).

Conclusions

This research has developed a robust visualization approach for evaluating the coverage of CCTV systems in public building spaces. Firstly, a method for modeling CCTV systems in virtual building spaces is presented. The emphasis is placed on offering a visual representation of the CCTV coverage in a BIM-based virtual environment. By simulating varifocal lenses and configuring the parameters of Revit cameras, the developed approach simulates the CCTV screen views to provide a better visual demonstration of the working of the CCTV systems. This is advantageous in the checking of design conflicts and effective communication between owners and contractors. The filled regions displayed in the 3D environment are also apparent, allowing accurate visual evaluation of CCTV coverage. Additionally, the plug-in program developed using Revit API, (i.e. CCTV setup advisor) is very helpful for processing the repetitive task of setting up the values of Revit camera parameters. Finally, in the case study of an MRT station, the developed approach is shown to be effective and can be widely applied to other building spaces under similar conditions.

Additional files

Additional file 1: The CCTV setup advisor.

Additional file 2: The CCTV coverage in the MRT station.

Competing interests

The authors declare that they have no competing interests.

Authors' contributions

HTC and SWW together developed the robust visualization approach for evaluating the CCTV coverage under the supervision of SHH. The plug-in API program, called the CCTV setup advisor, was developed by HTC and the case study was conducted by both SWW and HTC. All authors contribute to the writing of the manuscript in the way that HTC and SWW drafted the manuscript and SHH reviewed and revised it. All authors read and approved the final manuscript.

Acknowledgements

The authors would like to thank National Science Council, Taiwan for sponsoring the College Student Research Training Fellowship on this research, Department of Rapid Transit Systems, Taipei City Government for supporting all related data, drawings, and materials needed by the research, Sinotech Engineering Consultants, Ltd. for sharing a complete BIM model of an MRT station, and ST Electronics (Taiwan) Ltd. for supporting technical assistance of CCTV systems and its practical experiences. Also, the authors appreciate all the comments and suggestions received when an earlier version of this paper was first presented in the 12th International Conference on Construction Applications of Virtual Reality (CONVR 2012), Taipei, Taiwan (Chen et. al, 2012).

References

Aguado, M, Jacob, E, Matias, J, Conde, C, & Berbineau, M (2009). *Deploying CCTV as an Ethenet service over the WiMAX mobile network in the public transport scenario* (Proceedings of IEEE International Conference on Communications Workshops (ICC Workshops 2009), pp. 1–5).

Autodesk, Inc. (2010a). *Revit Architecture 2011 user's guide. Autodesk, Inc.* http://images.autodesk.com/adsk/files/revit_architecture_2011_user_guide_en.pdf. Accessed 6 Feb 2013.

Autodesk, Inc. (2010b). *Revit 2011 API develper's guide. Autodesk, Inc.* http://images.autodesk.com/adsk/files/revit2011sdk0.exe. Accessed 6 Feb 2013.

Azhar, S, Hein, M, & Sketo, B (2008). *Building Information Modeling (BIM): benefits, risks and challenges* (Proceedings of the 44th ASC Annual Conference). Auburn, Alabama.

Azhar, S, Nadeem, A, Mok, JYN, & Leung, BHY (2008). *Building Information Modeling (BIM): A new paradigm for visual interactive modeling and simulation for construction projects* (Proceedings of the First International Conference on Construction in Developing Countries (ICCIDC-I)). Karachi, Pakistan.

Bigdeli, A, Lovell, BC, Sanderson, C, Shan, T, & Chen, S (2007). *Vision processing in intelligent CCTV for mass transport security* (Proceedings of the IEEE Workshop on Signal Processing Applications for Public Security and Forensics, pp. 1–4).

CCTV Advisory Service. (2008). *LensCalc - online lens calculator.* http://www.cctv-information.co.uk/lenscalc/. Accessed 6 Feb 2013.

Chen, HT, Wu, SW, & Hsieh, SH (2012). *Studying CCTV coverage in an MRT station using BIM-based VR approach* (Proceedings of the 12th International Conference on Construction Applications of Virtual Reality (CONVR 2012), pp. 90–98). Taipei, Taiwan.

Cohen, N, Gattuso, J, & MacLennan-Brown, K (2009). *CCTV Operational requirements manual 2009. St Albans: Home Office Scientific Development Branch.* http://nactso-dev.co.uk/system/cms/files/127/files/original/28_09_CCTV_OR_Manual2835.pdf. Accessed 6 Feb 2013.

Eastman, C, Liston, K, Sacks, R, & Teicholz, P (2011). *BIM Handbook: A guide to building information modeling for owners, managers, designers, engineers & contractors* (2nd ed.). New Jersey: John Wiley & Sons.

Harris, C, Jones, P, Hillier, D, Turner, D (1998). CCTV surveillance systems in town and city centre management. *Property Management, 16*(3), 160-165.

Lee, LK, Zachariah, M, & Everett, P (1995). *CCTV camera site selection: a field experience* (Proceedings of the Vehicle Navigation and Information Systems Conference (VNIS), pp. 441–446). Washington, D.C.

March Networks News. (2008). *Singapore MRT moves ahead on CCTV system expansion. March Networks.* http://www.marchnetworks.com/Documents/Singapore_MRT_Moves_Ahead_on_CCTV_System_Expansion. Accessed 6 Feb 2013.

Sah, V, & Cory, C (2008). *Building Information Modeling: An academic perspective* (Proceedings of the 2008 IAJC-IJME International Conference). Nashville, Tennessee.

Sanderson, C, Bigdeli, A, Shan, T, Chen, S, Berglund, E, & Lovell, BC (2007). Intelligent CCTV for mass transport security: challenges and opportunities for video and face processing. *Electronic Letters on Computer Vision and Image Analysis, 6*(3), 30–41.

Sun, DS (2005). *Closed Circuit Television System (CCTV System) of Taipei Metro. Taipei MRT Newsletter, No. 211.* http://www2.dorts.gov.tw/news/newsletter/ns211/rp211_06.htm. Accessed 6 Feb 2013.

Teague, C, Green, L, & Leith, D (2010). *Watching me watching you: the use of CCTV to support safer work places for public transport transit officers* (Proceedings of Australian and New Zealand Communication Association Conference 2010). Canberra, Australia: Australian and New Zealand Communication Association.

Wang, J, Li, J, Chen, X, & Lv, Z (2010). *Developing indoor air quality through healthcare and sustainable parametric method* (Proceedings of the 4th International Conference on Bioinformatics and Biomedical Engineering (iCBBE 2010), pp. 1–4). Chengdu, China: IEEE.

Yan, W, Culp, C, & Graf, R (2011). Integrating BIM and gaming for real-time interactive architectural visualization. *Automation in Construction, 20*(4), 446–458.

12

Application of BIM coordination technology to HSR Changhua station

Weilun Lee[1*], Szumin Kang[1], Richard Moh[1], Robert Wu[2], Hungchang Hsieh[3] and Zhehua Shu[3]

Abstract

The arising trend of building information modeling technology application with innovative and revolutionary approach has changed the workflow and deliberation in A/E/C industry. By considering the condensed schedule of the project, operation safety, and information correctness during design and construction stages, Taiwan High Speed Rail Corporation (THSRC) brings BIM technology into Changhua Station construction project. The accumulated massive information helps engineers to coordinate design information among different expertise (architecture, structure, MEP, etc.) into BIM model at earlier phases. The effectiveness of BIM coordination technology increases the design quality and reduces significantly the conflicts of various interfaces. Three-dimension visualized interface increases the problem solving efficiency. During construction stage, construction information is integrated by the model to enhance fluency of construction process among different construction sectors. Application of BIM coordination technology reduces information gap, enhances construction quality and meets the schedule.

Keywords: Building information modeling; Integration; Coordination; Visualization

Background

The HSR Changhua Station (shown as Figure 1) is a planned station of the Taiwan High Speed Rail located in Tian Zhong Township (shown as Figure 2), Changhua County. The elevated station will have two side platforms and is planned to open in 2015. The new station will significantly enhance the availability of inter-regional traffic and promote the local industrial development.

Case description

Design concept of Changhua station

1. Tian Zhong means "the center of the fields" in Chinese, Changhua is famous for its yearly flora exposition and is known as the county of flora. Therefore flora is the main design concept of Changhua high speed rail station.
2. The columns in the lobby combines the flower image, opening and stretching like flower petals. The skylight lets in natural light and the columns are placed in a way to reflect indirect light to illuminate the lobby. Light, structure and flower are presented in harmony.
3. Inside the station, there is a greenhouse surrounding the station. Passengers can enjoy the green view in the station. This is one of the features of Changhua station.

Building information modeling

The revolution of BIM technology has overturned the traditional civil construction workflow and process. Among traditional and innovative thinking, BIM has brought the spirit of enhanced quality and efficiency into civil and architecture industries. The application of BIM technology has provided different meanings to each phase of the whole building life cycle (shown as Figure 3).

The THSR Changhua station project incorporates BIM integration technology at the early design phase in consideration of stability and safety of the railroad operation, high standard of precision and quality, the tight schedule of construction, and other factors. The objectives of the project are (shown as Figure 4):

1. Integrate the curvatures of featured petal-like columns.

* Correspondence: weilun.lee@maaconsultants.com
[1]Moh and Associates, Inc., BIM Mgmt. & Engr. Integration Center, Taipei, Taiwan
Full list of author information is available at the end of the article

Figure 1 New HSR Changhua station.

2. Complete the project under a tight schedule. The construction period is only 31 months.
3. Incorporate BIM integration technology at the early planning phase to reduce the errors of traditional two dimensional CAD drawings so as to reduce the chances of design changes at the construction phase and to increase the overall quality of the project.
4. Provide the platform of 3D visualization to detect piping clash and thus to optimize work efficiency and to reduce jobsite problems.
5. Incorporate the information of the architecture, electrical, mechanical facilities such as facility brands, life and maintenance schedule through BIM technology to increase the operation and management efficiency.

The introduction of BIM technology has changed the workflow of traditional architecture construction. In this project, BIM technology was introduced during the detailed design phase. A great amount of data was integrated to improve the efficiency and to reduce the corrections and errors during the construction (shown as Figure 5).

Taiwan High Speed Rail Corporation found the importance of integrating design outcomes of various disciplines and took it as the priority during the planning stage of the project; and, thus, BIM technology was applied for this purpose. BIM Management and Integrating Center (BIM center) of MOH and Associates was invited to provide the BIM service. Architecture, structural and MEP BIM models were created based on the design of each discipline to clarify problems and to review clashes of each discipline and between different disciplines at the early stage.

Discussion and evaluation

Application of BIM technology in design drawing integration in design phase

Previously, two-dimensional architecture, structural and MEP drawings were combined and organized at the construction site. Frequently errors and information gap occurred while integrating those drawings of different revisions, resulting in waste of resource and decrease of productivity during late construction phase. This project adopted BIM technology (shown as Figure 6) and cloud computing interface (shown as Figure 7) at detailed design stage to assist THSRC to incorporate different types of drawings.

This interface enabled the integration of traditional 2D drawings and made it possible to manage 3D BIM models, 3D parameters, and BIM collaboration. Clash detection reports and clarification reports can be generated with the integrated BIM model. The 3D visualized communication model provides a useful tool for the examination and discussion with THSRC (shown as Figure 8).

To perform the design concept of the graceful posture of flower blossom, it is critical to integrate the curved columns with architecture curvature and MEP. The curved structural column is the featured design of this project. The steel manufacturer pre-fabricated the steel

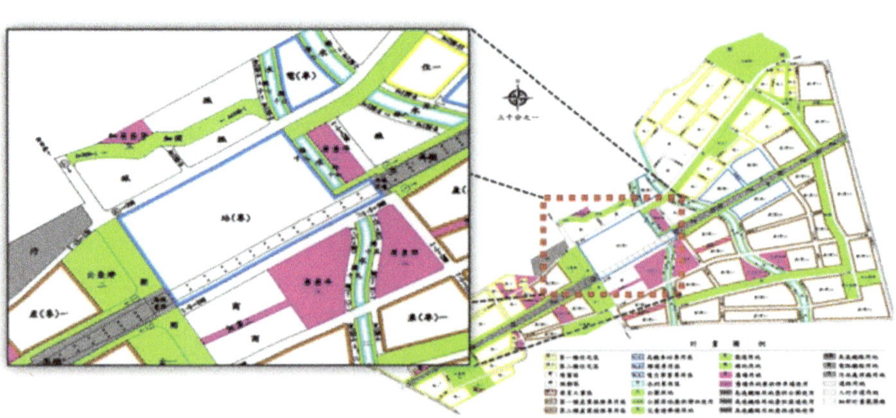

Figure 2 City planning.

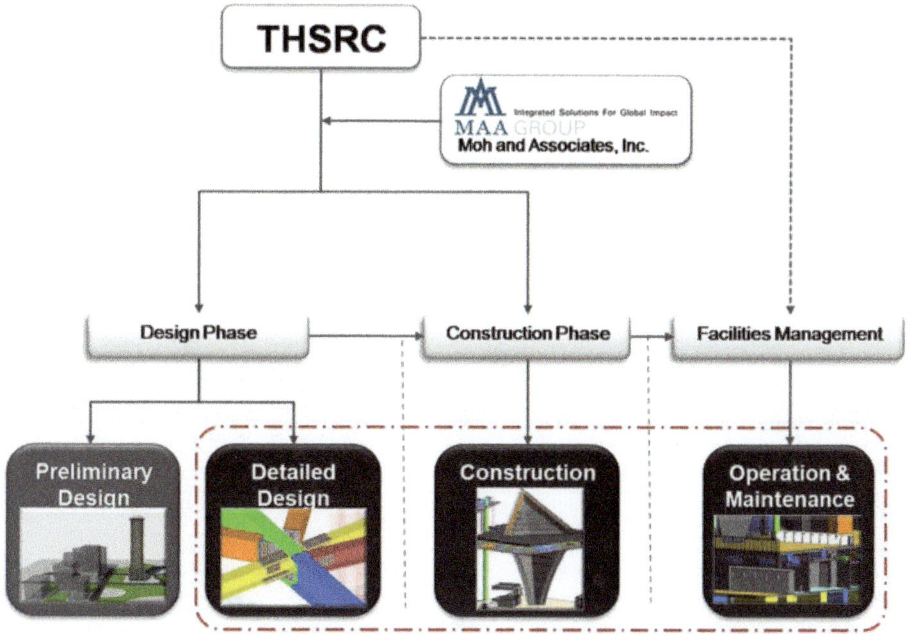

Figure 3 Organization chart.

structure of petal-like columns. BIM center provided the sections and perspective drawings generated from BIM model for the steel manufacturer to produce the curvature.

The production and application of integrated database in construction phase

The application of BIM technology for information synchronizing during the construction

Prior to the construction, BIM model was created according to the contract drawings approved by THSRC. The BIM model is used as a tool to coordinate various professional disciplines, to distribute integrated comments, to report errors and discrepancies in the drawings, and to generate clash detection report. Based on the clash report, THSRC reviewed the drawings submitted by the contractor, subcontractors and manufacturers in order to coordinate and correct the mistakes in the contract drawings before construction starts. Such a process allows for fewer field problems, enhanced construction quality, and a better control in the project procedure (shown as Figure 9).

The approved final version of BIM model was uploaded to the collaboration platform for THSRC and various parties to increase the efficiency of discussion. With the application of this platform, different sub-

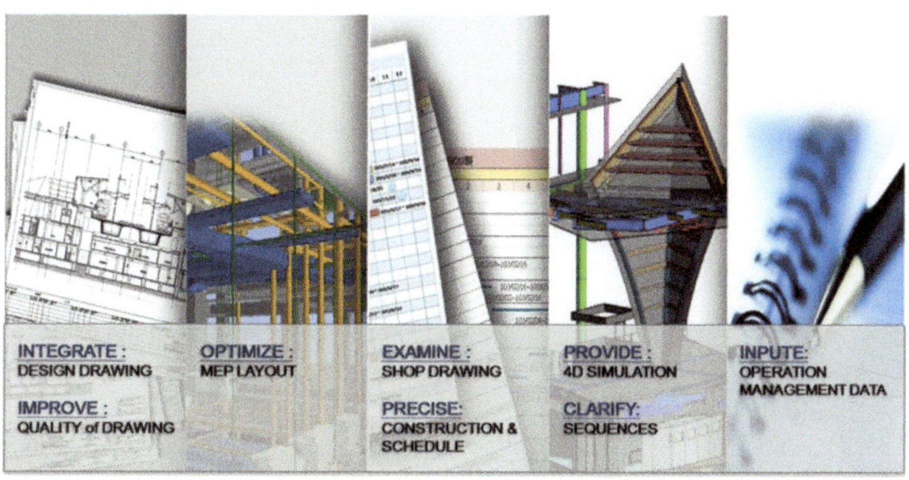

Figure 4 Project objectives.

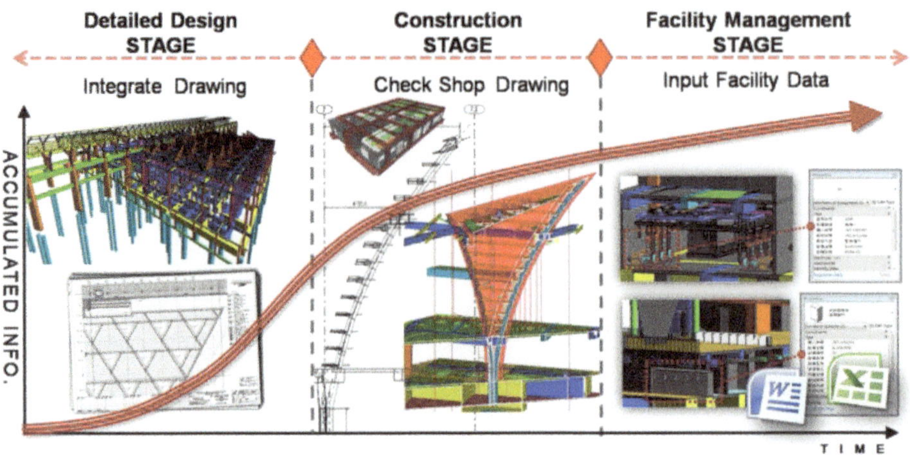

Figure 5 Building life cycle and BIM service flow chart.

contractors and manufacturers were able to discuss and resolve potential issues consistently.

To minimize the field problem and develop solutions, it is necessary to find out potential construction issues; for example, whether the elevation of piping is too high or too low, the clear space between drainage pipelines and surrounding objects, and the relationship between the positions of drain pipe and the elevation of catch basins (Figure 10).

Eighteen petal-like columns were designed to symbolize the characteristic of HSR Changhua station, twelve of them were located at the high ceiling lobby, and the other six were located at the maintenance area of the concourse floor slab. The design of petal-like columns creates the rhythm in the interior space and brings the image of local flower industry into the station. At early planning stage, the petal-like steel columns were integrated with interior decoration detailing, mechanical/electrical/piping and other design elements such as penetration of six petal-like steel columns through the floor slab, connection details of petal-like columns and beams supporting the floor, and the rebar arrangement between the steel column and concrete beam. Preventive discussions which reduce the frequency of design modification during construction are critical for early stage integration (shown as Figure 11).

During construction, drawings of petal-like columns and the ring girder from steel manufacturer were integrated with the sections generated from BIM model to examine the curvature of each petal-like columns and to corporate with architectural and MEP models for clash detection. Feedback loop was created between steel manufacturer and THSRC.

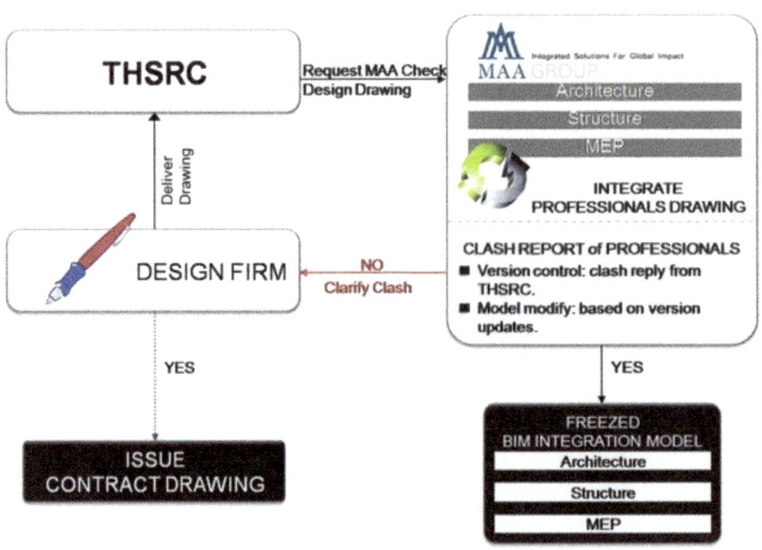

Figure 6 Design integration flowchart.

Figure 7 Exchanges platform.

The application of BIM technology for delivering clash report during construction

Frozen BIM model provided a virtual platform to generate different perspective views for the steel manufacturer. The elevation of the ring girder was incorporated to the perspective views to improve the communication for resolving welding and on-site assembling issues. The ring girder elevation drawings from the manufacturer could also be included into the frozen structural BIM model and steel construction drawings for finding discrepancies. For THSRC, the project outcomes have improved greatly because of the capability of on-site checking and execution. Petal-like steel column checklist is shown as follows:

The clash report and construction report generated by the frozen BIM model were provided to THSRC as a reference for reviewing the construction drawing. The clash report list is shown as follows:

Facility management and other benefits

The BIM model was modified and corrected according to as-built drawings. To enhance the operation smoothness and management efficiency, it is crucial to understand owner's needs such as management method and to study the inspection and maintenance list during operation. MEP and operation facility information such as ticket gates and ticket vending machines were included in the as-built BIM model for different operation and maintenance sectors (shown as Figure 12). The statistic report can be derived from as-built BIM model by the customized parametric chart which includes information such as facility expiration date and maintenance schedule (shown as Figure 13).

Quantity report is another important task of the project. BIM model was tailored to generate the required outputs. Structural BIM model generates steel and concrete quantities; MEP BIM model generates pipe sizes and quantities. With these reports, THSRC was able to perform cross-check with the quantity report from design team, and thus to utilize the construction resource effectively, to reduce the waste, and to better control the budget (shown as Figure 14).

With the demand of animation film production, the ability of data interoperability in BIM technology helped to shorten the time of high quality animation production and to increase the accuracy of animation model by providing the corresponding data.

The 4D construction sequencing simulation is one of the other benefits of BIM model. It was made by integrating the construction information such as project schedule, work flow, site access, resource, and time to simulate the construction sequence. Visualized platform enabled the owner to discuss construction issues with contractors.

Conclusions

BIM technology was adopted throughout the planning stage to the construction phase for Taiwan High Speed Rail, Changhua Station construction project. This platform was used to improve communication, to integrate

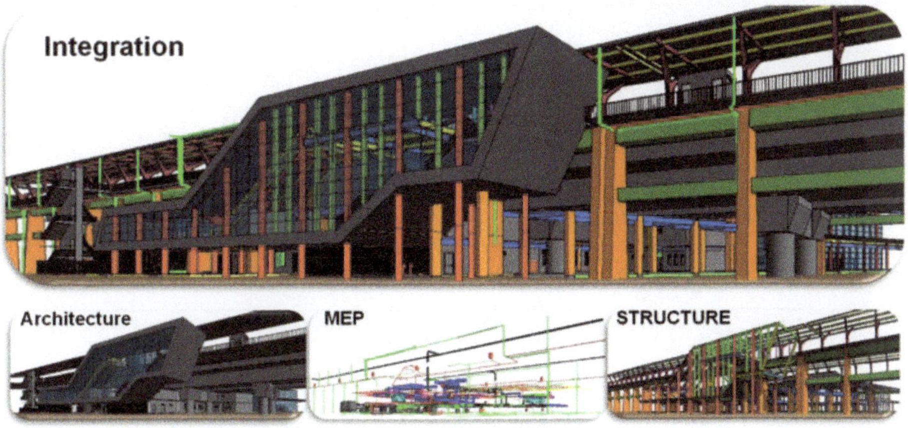

Figure 8 Integration BIM Models of Different Expertise.

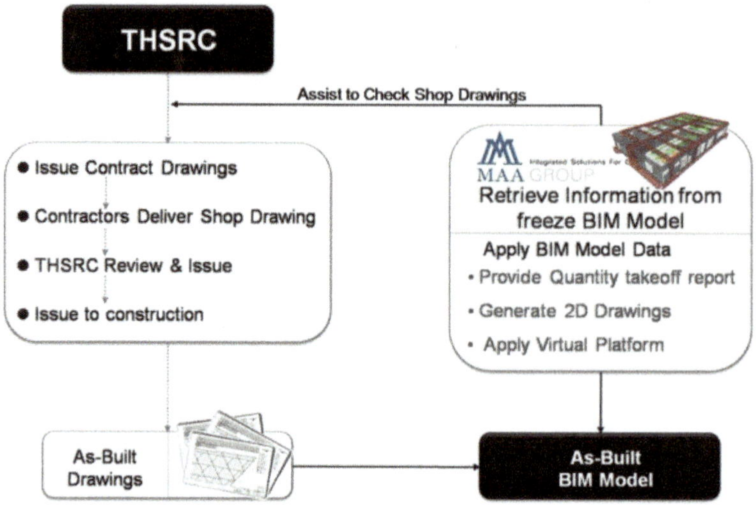

Figure 9 Construction integration flowchart.

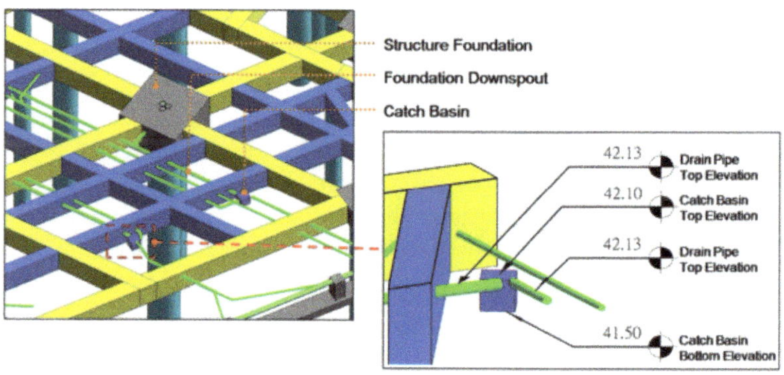

Figure 10 Foundation drainage elevation check point.

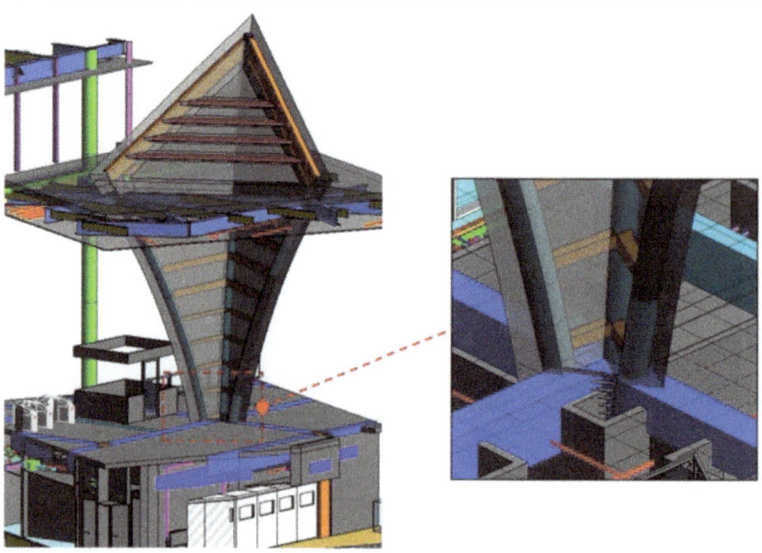

Figure 11 Integration of petal-like columns.

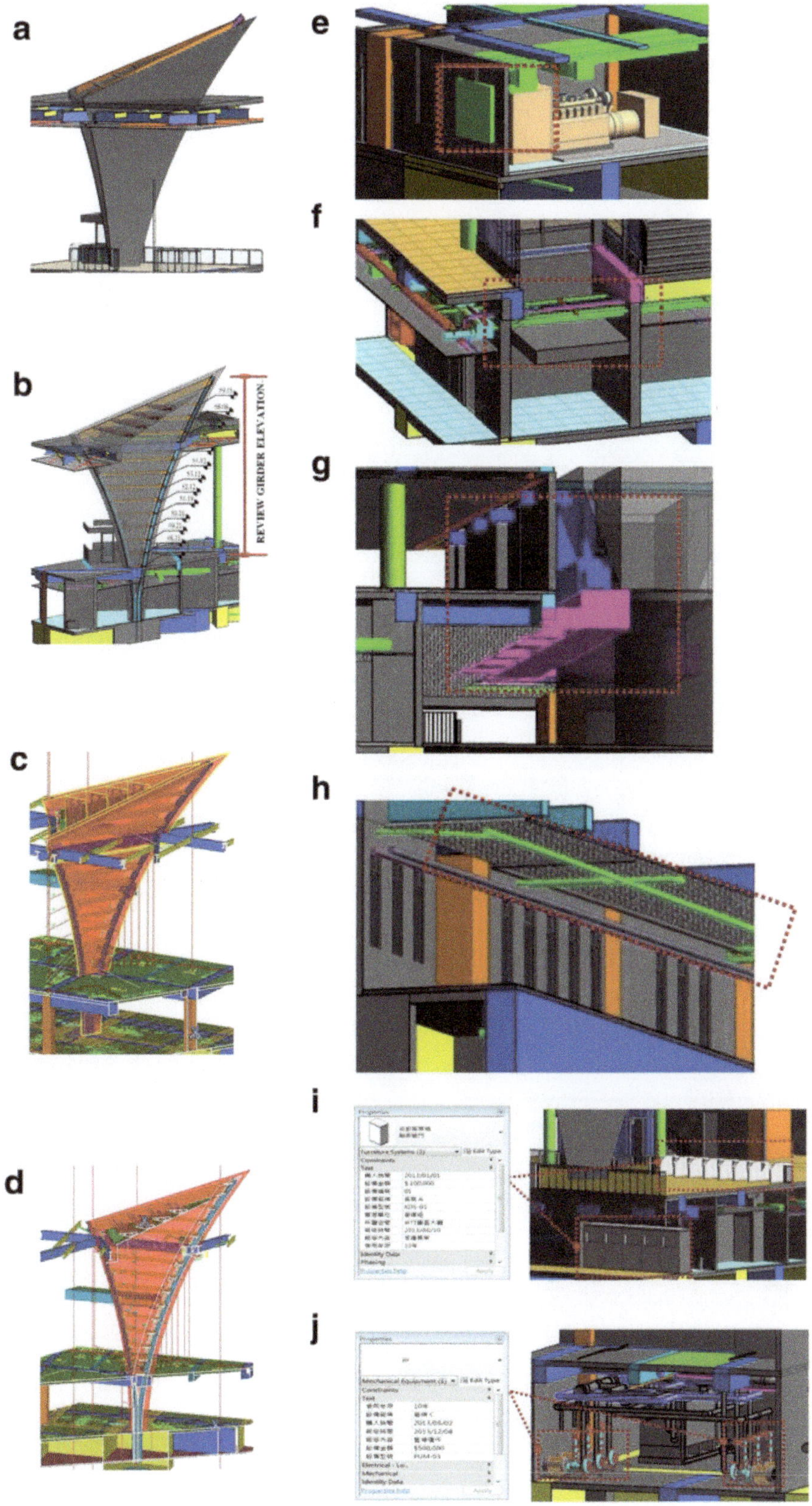

Figure 12 (See legend on next page.)

(See figure on previous page.)
Figure 12 Facility management equipment. a. Integration of disciplines: clash detection of column A between structure and architecture. **b**. Assistance of reviewing girder elevation. **c**. Import 2D CAD into BIM structure model. **d**. Compare section of column A with 2D shop drawing. **e**. RM. No. 177 Generator RM. Incorrect position of exhaust vent and the louver. **f**. RM. No. 124 Station storage RM. MEP piping conflicts with the elevator pit. **g**. LINE 7.2-8.2 & LINE B-E Ground level, duct and vent conflict with exsiting bridge. **h**. Water Piping and duct elevation are below ceiling height. **i**. Operation Equipment maintenance data input. **j**. Mechanical Equipment maintenance data input.

Facilities Management Schedule							
物件名稱	設備廠牌	設備型號	購入時間	使用年限	設備金額	維修時間	維修內容
無障礙驗票閘門	廠商 A	KOS-01	2013/01/01	10年	$ 100,000		
無障礙驗票閘門	廠商 A	KOS-01	2013/01/01	10年	$ 100,000		
無障礙驗票閘門	廠商 A	KOS-01	2013/01/01	10年	$ 100,000		
無障礙驗票閘門	廠商 A	KOS-01	2013/01/01	10年	$ 100,000	2013/04/12	無法運作
無障礙驗票閘門	廠商 A	KOS-01	2013/01/01	10年	$ 100,000		
無障礙驗票閘門	廠商 A	KOS-01	2013/01/01	10年	$ 100,000		
無障礙驗票閘門	廠商 A	KOS-01	2013/01/01	10年	$ 100,000		
無障礙驗票閘門	廠商 A	KOS-01	2013/01/01	10年	$ 100,000	2013/06/10	無法運作
8							
驗票閘門	廠商 B	GAT-02	2013/01/01	8年	$100,000		
驗票閘門	廠商 B	GAT-02	2013/01/01	8年	$100,000		
驗票閘門	廠商 B	GAT-02	2013/01/01	8年	$100,000	2013/07/15	感應異常
驗票閘門	廠商 B	GAT-02	2013/01/01	8年	$100,000		
驗票閘門	廠商 B	GAT-02	2013/01/01	8年	$100,000		
驗票閘門	廠商 B	GAT-02	2013/01/01	8年	$100,000		
驗票閘門	廠商 B	GAT-02	2013/01/01	8年	$100,000		
7							

Figure 13 Facility operation management report.

管道配件明細表				
系統類型	大小	面積	整體大小	數量
SAD	1200x700-1200x700		1200 mmx700 mm-1200 mmx700 mm	1
SAD	700x700-700x700		700 mmx700 mm-700 mmx700 mm	1
SAD	700x700-700x700		700 mmx700 mm-700 mmx700 mm	1
SAD	900x700-900x700-900x700		900 mmx700 mm-900 mmx700 mm-900 mmx700 mm	1
SAD	700x700-700x700		700 mmx700 mm-700 mmx700 mm	1
SAD	700x700-700x700		700 mmx700 mm-700 mmx700 mm	1
SAD	700x700-700x700		700 mmx700 mm-700 mmx700 mm	1
SAD	900x700-900x700		900 mmx700 mm-900 mmx700 mm	1
SAD	1000x700-1000x700		1000 mmx700 mm-1000 mmx700 mm	1
SAD	870x400-870x400		870 mmx400 mm-870 mmx400 mm	1
SAD	870x400-870x400		870 mmx400 mm-870 mmx400 mm	1
SAD	1250x450-1250x450		1250 mmx450 mm-1250 mmx450 mm	1
SAD	1250x450-1250x450		1250 mmx450 mm-1250 mmx450 mm	1
SAD	1250x450-1250x450		1250 mmx450 mm-1250 mmx450 mm	1
SAD	1000x700-1000x700		1000 mmx700 mm-1000 mmx700 mm	1
SAD	1000x700-1000x700		1000 mmx700 mm-1000 mmx700 mm	1
SAD	1000x700-1000x700		1000 mmx700 mm-1000 mmx700 mm	1
SAD	1000x700-1000x700		1000 mmx700 mm-1000 mmx700 mm	1
SAD	1000x700-1000x700		1000 mmx700 mm-1000 mmx700 mm	1
SAD	1000x700-1000x700		1000 mmx700 mm-1000 mmx700 mm	1

Figure 14 Quantity takeoff for MEP.

different versions of drawings and database, and to have each team member work effectively on the same pace. Each team member's design was coordinated to resolve interface issues and to implement clarifications on the drawings. The information gap and labeling errors were reduced significantly at the construction stage. The design delivery, construction quality, and project management were improved significantly.

The functions of BIM technology applied in this project at planning and construction stages are (shown as Figure 15):

1. Clarify the confusions between drawings (of architectural, structural and piping).
2. Provide solutions of conflicts of different disciplines (architecture, structure and MEP).
3. Create multi-discipline BIM model and provide 3D visual inspection for specific areas. Point out the problem area in advance. Reduce the blind spots of integrated 2D construction drawings.
4. Provide quantity reports of different disciplines for checking the quantity reports from the design team.
5. Provide collaboration platform for filing, database sharing and preview. Assist the management of various versions of drawings.
6. Help clarify the clash and difference between architectural drawing and construction drawing.
7. Provide the contractor relevant construction drawings and information.
8. Use 3D BIM model to resolve interface issues during construction.
9. Improve control over project progress, material control, and on-site preparation.
10. Provide as-built BIM model for facility management.

After completion of the project, the as-built BIM model was used for the operation and maintenance of HSR Changhua station by maintaining the useful and comprehensive information. The model was submitted according to owner and operator's requirements in a desired format to keep high performance management and to upgrade efficiency of maintenance work. In addition to the engineering applications throughout the complete lifecycle of the building, BIM technology provides additional benefits such as assistance to the production of high-quality animation, simulation of construction progress; and providing the visualized communication platform for different sectors to understand the project portfolio in an effective way. All these benefits lead the project to a greater business value. The application of BIM technology for HSR Changhua station improves the

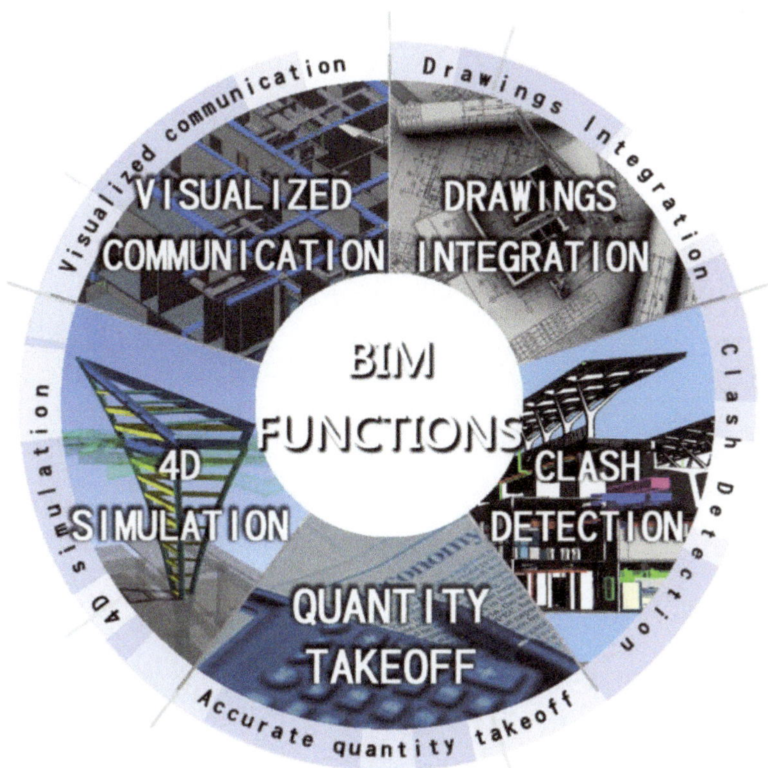

Figure 15 Functions of BIM technology.

efficiency of the project management and enhances the quality of the construction that brings the project to complete on time.

Competing interests
The authors declare that they have no competing interests.

Authors' contributions
WL: BIM modelling, clash detection,4D simulation. SK: Project Manager. RM: Senior Vice President, Project Supervisor. RW: Site Resident Engineer. HH: Project Engineer. ZS: Project Engineer. All authors read and approved the final manuscript.

Acknowledgement
Taiwan High Speed Rail Corporation (THSRC) has commissioned Moh and Associates Inc.(MAA) to provide BIM service in the planning and consulting phases. Foremost, MAA would like to express our sincere gratitude to the directors of THSRC for their continuous support of the project. Besides, the authors would like to thank DANIEL CHU, WEN-JEN FANG, and ROGER LIN of THSRC and WEN-SHAN LIU, BEN LIN, S.C HUANG, and HSIANG-HSIANG YU of MAA for their assistance to develop this paper.

Author details
[1]Moh and Associates, Inc., BIM Mgmt. & Engr. Integration Center, Taipei, Taiwan. [2]Moh and Associates, Inc., Construction Supervision Dep., Taipei, Taiwan. [3]Taiwan High Speed Rail Corporation, Taipei, Taiwan.

13

Automatic design algorithms for securing the ground contact stability of mobile cranes

Donghoon Lee[1], Jun Young Park[1], Jongkwan Ho[2] and Sunkuk Kim[1*]

Abstract

Background: Mobile cranes overturn due to excessive lifting, heavy wind, or when appropriate ground contact pressure is not secured. Such accidents can be prevented by reviewing the force conveyed from the outrigger to the ground upon lifting, as well as the soil bearing force. The stability review of mobile crane overturn is classified into the following stages: reviewing lifting conditions, mobile crane selection and reviewing ground contact stability. However, a precise review of ground contact stability requires expertise and is extremely time-consuming. Thus, this study develops automatic design algorithms for securing the ground stability of mobile cranes for easy and quick stability reviews.

Methods: The stability review method and the algorithm models are created in connection with the proposed conceptual process. The algorithms proposed in this study conduct simulations for all possible candidate cases, save the results in a database, review practical alternatives, and select the solution that minimizes cost. To verify the effectiveness of the automatic design model, a case project was selected and the algorithm was applied assuming several conditions.

Results: Developed algorithms can select candidate cranes according to lifting conditions and design plates through simulations easily and quickly, and design components required for ground contact plate design are collected to build the database that allows easier and quicker stability review.

Conclusion: The algorithms proposed in this study conduct simulations for all possible candidate cases, save the results in a database, review practical alternatives, and select the solution that minimizes cost. Developed processes are useful for analyzing correlations of mobile crane performance, ground contact and soil bearing force reinforcement designs and so forth, and for systematically selecting optimal mobile cranes.

Keywords: Mobile crane; Stability; Lifting plan; Ground contact pressure; Automatic design model

Background

Many crane accidents occur globally, including the overturn of mobile cranes in Phuket, Thailand and England (The Vertikal Press 2014). As the use of cranes increases in South Korea, the number of overturn accidents is gradually increasing, and accidents related to mobile cranes account for approximately 61.4% of all crane accidents (Kim and Lee 2007). Overturn of mobile cranes occurs due to excessive lifting, heavy wind, or when appropriate ground contact pressure is not secured. Such accidents can be minimized by reviewing the force that is conveyed from the outrigger to the ground upon lifting and the soil bearing force. The stability review of mobile crane overturn is classified into the following stages: reviewing lifting conditions, mobile crane selection, and reviewing ground contact stability. For safe operation, stage-by-stage reviews are needed.

Stability should be pre-reviewed, and it is reasonable to select mobile cranes among the candidate cranes that have completed reviews in consideration of cost or other factors. However, engineers or technicians do not set lifting plans that suit the construction conditions for stability analyses when implementing construction projects; instead, equipment is operated based on the information provided by an equipment supplier. Operation relying on such limited information and experience is the fundamental cause of crane disasters (Ho, Han and Kim 2007). In

* Correspondence: kimskuk@khu.ac.kr
[1]Department of Architectural Engineering, Kyung Hee University, 1732 Deogyeong-dearo, Giheung-gu, Yongin-si, Gyeonggi-do, Korea
Full list of author information is available at the end of the article

addition, since such an approach makes precise reviews of mobile cranes and working conditions impossible, it is difficult to select the optimal mobile crane.

There are two factors that prevent engineers or technicians from setting detailed mobile plans. First, it is difficult and extremely time-consuming to consider the ground contact stability applied to mobile cranes. Second, there is not enough of a systematic technique for analyzing a lot of data for stability review of candidates. There have been previous studies that investigated the selection of optimal cranes, taking into account the lifting capacity of mobile cranes and transporting routes (Reddy and Koshy 2002; Al-Hussein et al. 2001; Tantisevi and Burcu 2007). However, those previous studies did not include detailed stability reviews such as the soil bearing force and ground reinforcement. Therefore, this study intends to move beyond the limits of previous studies and develop automatic design algorithms for securing the ground contact stability of mobile cranes. The algorithms proposed in this study focus on the ground stability review process of mobile cranes.

For stability review of cranes, this study limits scope to improve the efficiency of the study and the reliability of its results. Above all, the study subject is limited to the hydraulic crane (or tire crane) that uses an outrigger among mobile cranes. The lifting method can be divided into single lifting and double lifting, and the study is conducted assuming single lifting. In the case of double lifting, more than two working radii and the soil bearing force should be reviewed, and the distance from the crane should be considered. Thus, double lifting has a different method of stability review, and an additional study should be carried out based on this study result. Also, as shown in Figure 1, the study reviews the ground contact stability. The range of reinforcement of soil bearing force occurs when the crane is installed on soil or ground, and this study excludes cases in which it is installed on top of a structure. The reinforcement of structure has a different procedure for reviewing bending, shearing, and sagging compared to that of soil or ground, so it requires a different model. Lastly, the study is conducted assuming that the boom, sling, and lug meet the lifting requirements, and there is no stability review for each of these materials.

Methods

To achieve the objective of this study, first, ground contact stability review items are identified and a conceptual process is developed. Second, the conceptual process for reinforcing ground contact pressure and soil bearing force is proposed. Third, a stability review method and algorithm models are created in connection with the proposed conceptual process. Finally, automatic models among the created models are developed, taking into account safety and economic feasibility. The developed processes analyzes correlations among the performance of mobile cranes, ground contact, and soil bearing force reinforcement and so on in order to assist the systematic selection of optimal mobile cranes.

Literature review

Studies related to crane selection and stability review have been conducted using a wide range of methods. Reddy and Koshy developed a lifting route plan program considering the working radius and lifting capacity of mobile cranes (Reddy and Koshy 2002). Al-Hussein et al. implemented an optimal crane selection algorithm using Visual Basic in consideration of the lifting capacity of mobile cranes (Al-Hussein et al. 2001). However, the studies mentioned above did not consider ground conditions, including the soil bearing force. Tantisevi and Burcu (2007) analyzed and controlled the work space and visual/spatial movement of mobile cranes to minimize friction with other works and delays (Tantisevi and Burcu 2007). This is useful in an effective arrangement and operation of mobile cranes, but it is far from a detailed stability review. Tamate et al. analyzed outrigger loads that cause overturn of mobile cranes. They also conducted an experiment on overturn of mobile cranes as the load is delivered to the outrigger and analyzed the result; however, they did not propose the selection of stability-secured mobile cranes

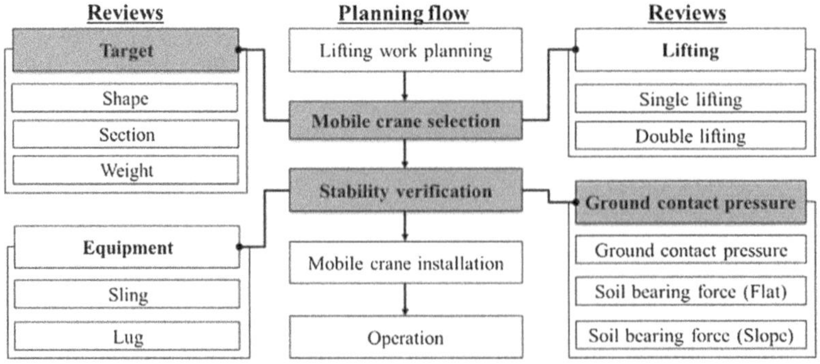

Figure 1 Study scope.

and plate design algorithms like this study (Tamate et al. 2005). Shapira et al. considered the characteristics of building projects and conducted a study in selecting whether a mobile crane or a tower crane is most suitable. Also, the items to consider in choosing the detailed model (building, site, safety, air, weather, etc.) were analyzed to identify the importance of each item (Shapira and Schexnayder 1999). However, there was no stability review or selection procedure in relation to the items considered. Alkass and Osama developed a simulation program to select cranes by using tools like database management, spreadsheet application programs, graphics, and simulation (Al-Hussein et al. 1997). Similarly, Struková and Ištvánik developed a program to select the optimal location of mobile cranes considering the lifting location, weight and building shape (Struková and Ištvánik 2011). However, there was not enough detailed review on stability of the soil bearing force and the program was aimed at selecting cranes in consideration of lifting conditions such as the weight, height, and building shape.

Furusaka and Gray used the objective function that minimizes rental, installation, and disconnection costs to select an optimal crane, proposing a model that combines various tower cranes (Furusaka and Gray 1984). However, their proposal was not a stability review of mobile cranes. Gray and Little suggested a process of selecting mobile cranes and tower cranes that suit designs at the initial design phase (Gray and Little 1985). However, there was no method of reviewing the stability of the selected cranes or of using the available crane information in the market. Ali et al. studied collision avoidance upon the double lifting of a mobile crane, and Perez et al. developed an algorithm for planning collision-free paths among polyhedral obstacles (Lozano-Pérez and Wesley 1979). Other studies include those conducted by Maczynski and Wojciech, who developed a 3D optimization algorithm of a rotating part to secure rigidity of mobile cranes (Maczynski and Wojciech 2003), and Lin and Haas (1996), who developed a double lifting simulation program (Lin and Haas 1996). These studies did not consider the ground contact stability of mobile cranes. Lastly, Ho et al. studied a method of reviewing the ground contact stability of mobile cranes, but only main items were explained and a formula was proposed (lacking detailed review items and procedure), and no automatic model was proposed (Ho, Seo and Kim 2007). Thus, this study proposes detailed review items and procedure regarding ground contact stability to secure the stability of mobile cranes, and develops an automatic model.

Review process of ground contact stability

Table 1 shows review items of automatic design algorithms for securing the ground stability of mobile cranes. The review items are divided into review of objects to be lifted, equipment information, review of ground contact pressure and reinforcement on soil bearing force of flat and slopping site. When examining the ground contact pressure, the maximum reaction of the outrigger is calculated to check stability and the available mobile cranes are identified. When the mobile cranes are selected, the reinforcement of soil bearing force is reviewed based on the ground shape. At the reinforcement planning of soil bearing force of flat site as illustrated in Plate type (steel, timber), rigidity of plate, soil bearing force, and bending moment and shearing force acting on plate should be reviewed. At the reinforcement planning of soil bearing force of slopping site, slope data such as slopping angle, slopping length, slopping height, and distributed load on soil should be examined additionally to design a plate. Here the most optimal plate is designed depending on the crane type and the ground state.

Figure 2 shows the ground contact stability review process. The crane operation plan is classified into reviewing lifting conditions, mobile crane selection, and reviewing ground contact stability. When reviewing lifting conditions, the shape, section, and weight of lifting objects should be checked.

Automatic review and design algorithms

Review of basic information in selecting cranes

The lifting plans and crane selection have complex, dynamic correlations, so the influencing factors include not only the duration and cost in terms of construction management, but also the working radius, work efficiency, interference with nearby buildings and other works, quantity, location selection, site installation, and disconnection, all of which should be taken into consideration (Lin and Haas 1996). The optimal mobile crane is selected considering these factors, and a respective plan should be set by step in order.

For the stability review of mobile cranes, the characteristics, and heights of lifting objects and the distance from a crane should be reviewed as shown in Figure 3. The characteristics of objects to be lifted refer to the shape, weight, and materials, and the crane characteristics indicate the lifting capacity, available lifting height, and turning radius of the boom. These basic factors should be considered when selecting a crane or cranes, and cranes that do not satisfy the given factors are excluded from analyses. In addition, the ground contact stability should be examined based on these factors to prevent overturn of mobile cranes.

Review of ground contact pressure

As shown in Figure 4, the ground contact model of this study calculates the maximum ground contact load according to the boom location (A, B, C). The maximum reaction is at the outrigger when the boom is at point B. However, this cannot be seen as the point where the maximum reaction is realized, because B is not the main

Table 1 Review items for ground contact stability

Step		Items reviewed
Basic Items	Review of the objects to be lifted	Type, material and weight
	Equipment Information	Working radius, the maximum height of objects, crane weight and outrigger radius
Ground contact stability	Review of ground contact pressure	Work direction, ratio of impact load and the maximum reaction of outrigger
	Reinforcement of soil bearing force (flat site)	Plate type (steel, timber), rigidity of plate, soil bearing force, bending moment, and shearing force acting on plate
	Reinforcement of soil bearing force (slopping site)	Additional data including slope data such as slopping angle, slopping length, slopping height, and distributed load on soil

location of the work, and instead is where temporary rotation and passing are demonstrated. Thus, when the boom is located at A (the main location of work), it is regarded as the maximum reaction. Also, considering that the boom rotates at B, 50% of the maximum reaction force should be added for safe design (Ho, Seo and Kim 2007). The formula is as shown in Equations 1 and 2.

When the impact load is not considered:

$$P_{max} = 0.85 \times (W + W_1) \times 0.5 \times 1.5 \quad (1)$$

where P_{max} is the maximum reaction, w is the crane weight, and W_1 is the lift weight.

When the impact load is considered:

$$P_{max} = 0.85 \times \{(W + W_1) \times 1.3\} \times 0.5 \times 1.5 \quad (2)$$

Figure 5 shows the ground contact stability review process. The crane operation plan is classified into reviewing lifting conditions, mobile crane selection, and reviewing ground contact stability. When reviewing lifting conditions, the shape, section, and weight of lifting objects should be checked. The ground contact pressure of the outrigger is then reviewed along with the lifting conditions when selecting mobile cranes. This is Simulation 1 proposed in the study. When reviewing ground contact stability, soil bearing force is reviewed based on the results of Simulation 1. Soil bearing force is reviewed separately by flat and slopping site, and each process is developed as in Simulations 2 and 3, respectively.

The generation model for stability review of ground contact stability uses the specifications provided by the crane manufacturer and the project information. Engineers and technicians should set the project information as shown in Figure 6, and the crane information is automatically imported from data that has already been implemented. The ground contact pressure can be examined based on the outrigger span, the turning radius of boom, the body weight, and the lift weight.

The simulation model for automatic stability review of ground contact stability is designed to review all MCs. As illustrated in Figure 7, the simulation model sets the installation location for all MCs and examines the ground contact pressure automatically by the simulation algorithms illustrated in Figure 7.

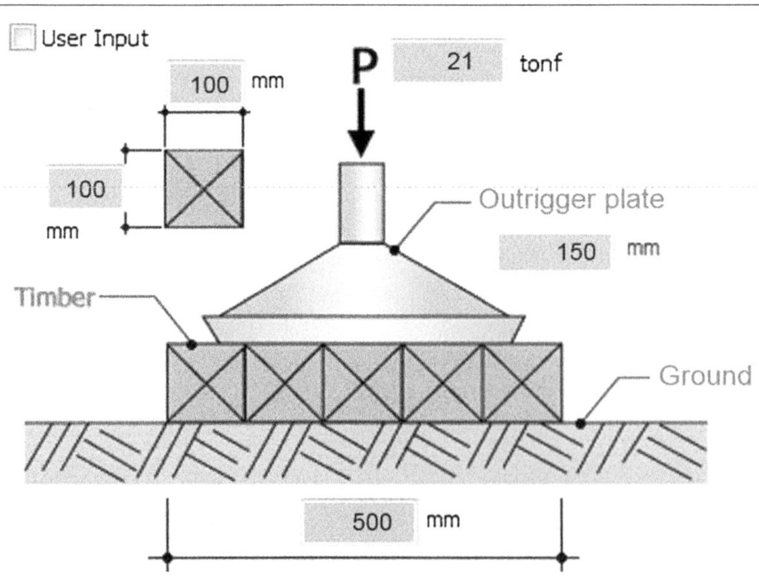

Figure 2 Ground contact stability review process.

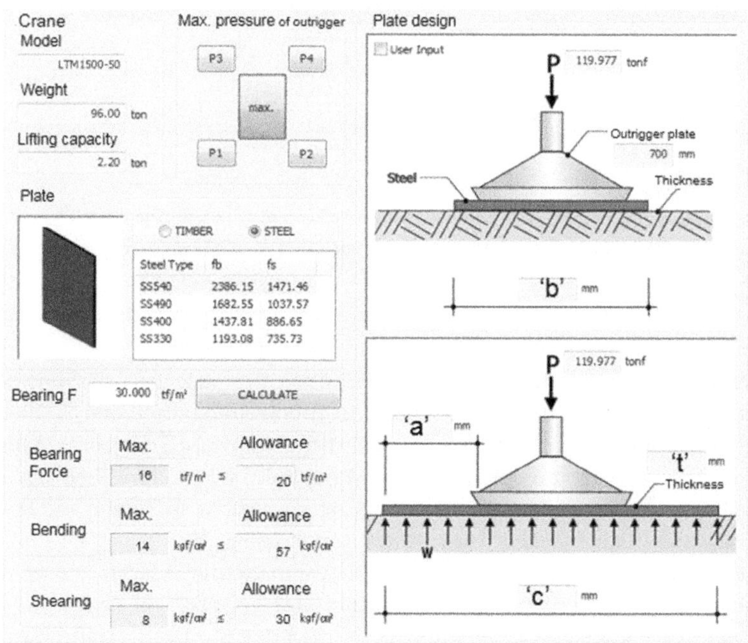

Figure 3 The working radius and height of a mobile crane.

The ground contact pressure is reviewed in the order of the maximum reaction of the outrigger, the working direction and height, and the impact load. The impact load value that is reviewed shows stability through a comparison with the ground conditions. Cranes that are regarded as stable are classified and saved automatically as the candidate cranes. When all cranes have been reviewed, the simulation examining the ground contact pressure ends.

Reinforcement of Soil Bearing Force (flat site)

After the review of ground contract pressure, the stability of soil bearing force reinforcement should be reviewed to

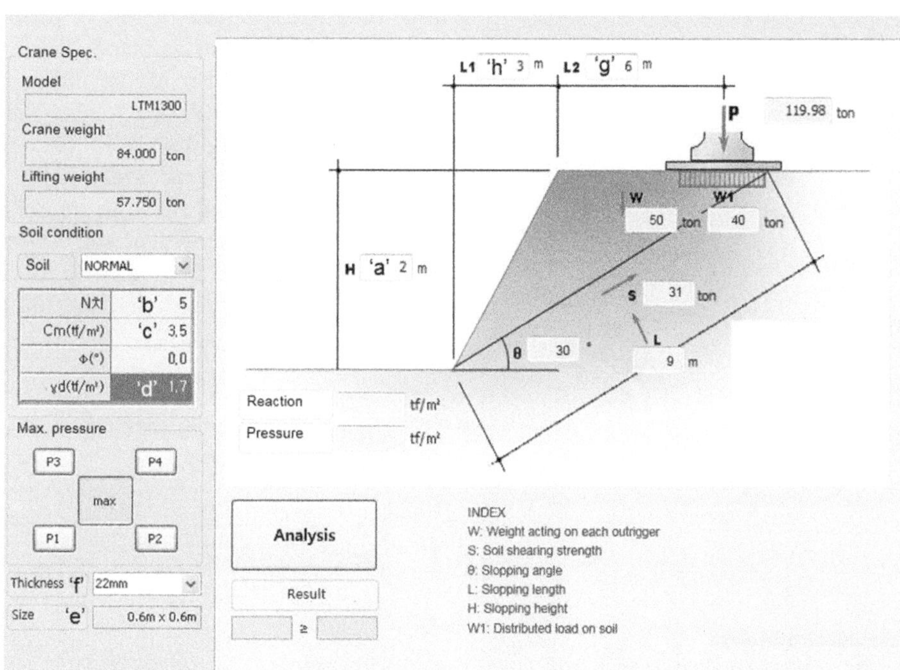

Figure 4 Identification of maximum reaction of a hydraulic crane.

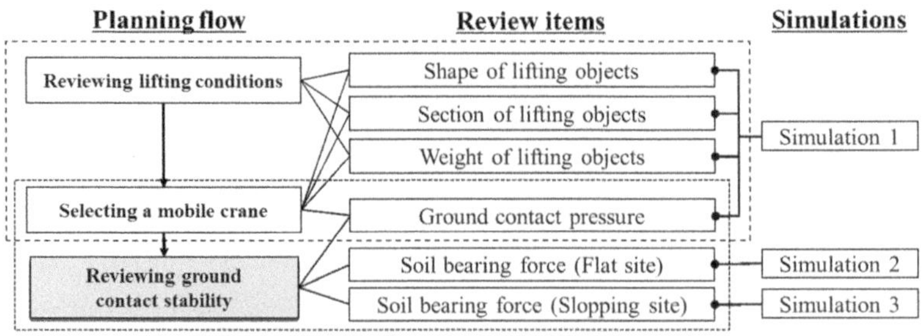

Figure 5 Generation model for ground contact stability review.

prevent overturn. The reinforcement of soil bearing force is reviewed under 2 different ground conditions: a flat and a slopping site. For the flat site, the plate material, thickness, and area should be reviewed manually by the user as shown in Figure 8 or automatically by the simulation algorithm as shown in Figure 9. Since the plate transfers the ground contact load of the outrigger to the ground, it should be designed after checking the adhesive power and the rigidity of the plate. The plate is usually made of timber and steel, and data including bending and shearing according to the thickness and area should be gathered to be automatically applied to the simulation.

As shown in Figure 10 the generation model of soil bearing force reinforcement as plate reinforcement imports the additional data on ground contact pressure of the outrigger stored in the system implemented by the proposed algorithms besides the basic information specified in Table 1. Based on such data, the plate type, rigidity, bearing soil force, and bending and shearing factors are automatically reviewed for reinforcement of soil bearing force. When the plate fails to transfer the ground contact load to the ground, the crane is likely to overturn. For a slope, the soil type, soil property value and slope information should also be examined. When stability reviews of all items are completed, the crane and the suitable plate design are saved in pairs.

The automatic plate design for reinforcement of the soil bearing force on a flat site is applied with information like the outrigger of the candidate cranes drawn from simulation 1 as shown in Figure 7 and the body weight. For the plate design in simulation 2, the outrigger of a crane as shown in Figure 9 is selected, and the plate materials and allowable soil bearing force are checked for the design of the materials. The designed plate and the crane are saved in pairs within the scope that stability is secured. Simulation 2 is completed when all the candidate cranes deduced from simulation 1 are reviewed.

As demonstrated in Figure 11, the program automatically reviews the plate design for reinforcement of soil bearing force on a flat site. The rigidity of plates by type and size is imported from the gathered data, and the maximum ground contact load of the outrigger is inserted based on the result of simulation 1. Also, the crane weight and lifting capacity imported from the crane data are inserted. The values are saved after the soil bearing force and the plate's bending and shearing are reviewed.

The plate for reinforcement of soil bearing force on a flat site is divided into timber and steel for the design depending on the material used, and the soil bearing force,

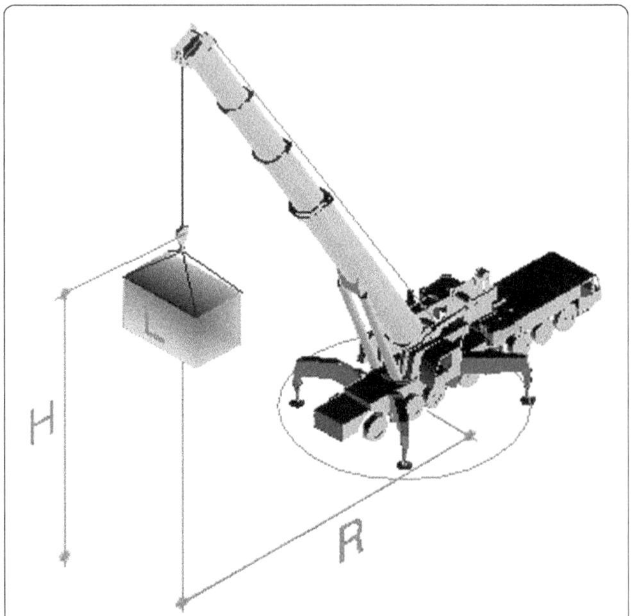

Figure 6 Input page of crane information.

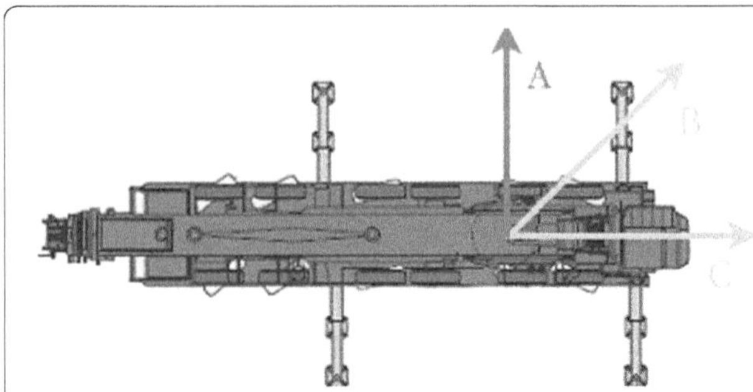

Figure 7 Automatic ground contact pressure review (simulation 1).

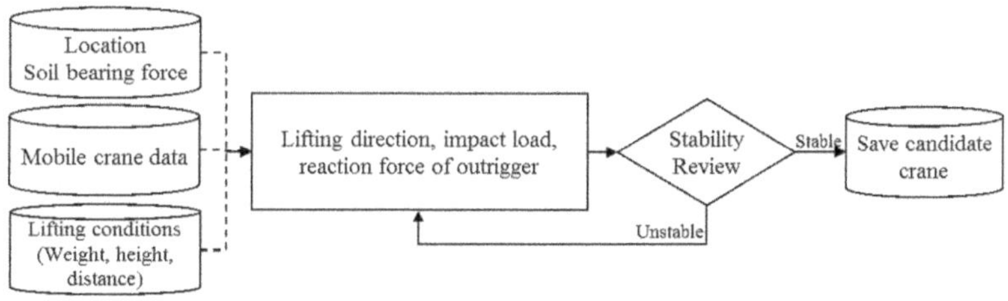

Figure 8 **Design of a reinforcement plate (flat site).**

bending, and shearing should be automatically reviewed by the equations (3) to (7). Steel plate review items can be described as follows (Shapiro et al. 2000). Figure 11 shows the variables related to plate design (a, b, c, and t), which should meet the three requirements regarding soil bearing force, bending, and shearing in the automatic design algorithm.

Soil bearing force review

$$q = \frac{P}{bc} \leq qa\text{-------O.K} \tag{3}$$

P: Maximum Reaction,
qa: Allowable Soil Bearing Force

Bending review

$$M = \frac{qba^2}{2}$$

$$f1 = \frac{3qa1^2}{t^2} \leq fb\text{-------O.K} \tag{4}$$

$$f2 = \frac{3qa2^2}{t^2} \leq fb\text{------O.K} \tag{5}$$

M: bending Moment,
f1: Transversal Bending Stress,
f2: Longitudinal Bending Stress,
fb: Allowable Bending Stress,
a1: Transversal Direction a,
a2: Longitudinal Direction a,
t: Thickness of member

Shearing review

$$v1 = \frac{1.5qa1}{t} \leq fs\text{-------O.K} \tag{6}$$

$$v2 = \frac{1.5qa2}{t} \leq fs\text{-------O.K} \tag{7}$$

v1: Transversal, horizontal shear stress
v2: Longitudinal, horizontal shear stress,
fs: Allowable shear stress of member

Reinforcement of Soil Bearing Force (slopping site)

For the reinforcement of the soil bearing force of a slope, an additional review of the slope shape and the soil property value is required. As shown in Figure 12, the reinforcement plate is designed after reviewing the vertical weight, shearing strength, sliding angle, slope length, and height. The reinforcement plate installed on a slope should consider not only the transfer of load, but also the landslide at the slope.

The plate for reinforcement of the soil bearing force on a slope requires reviews of soil bearing force, bending, and shearing. Steel plate review items are described as follows (Shapiro et al. 2000). Figure 11 shows the variables related

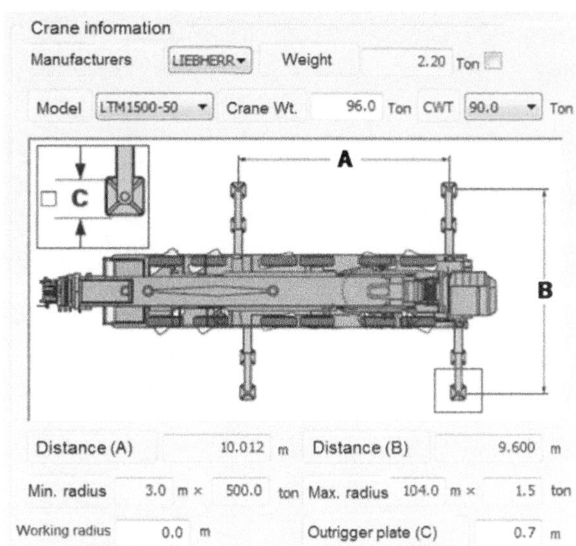

Figure 9 **Automatic plate design for flat site (simulation 2).**

to plate design and slope shape (a, b, c, d, e, f, g, h, and t), and the variables related to plate design (e, f) should meet three requirements regarding soil bearing force, bending, and shearing in the automatic design algorithm.

Subgrade reaction (q)

$$q = P_{max} \times n \ \left(t/m^2\right) \quad (8)$$

Reaction area (A)

$$A = \frac{P}{q} \ \left(m^2\right) \quad (9)$$

Therefore, the longitudinal reaction force L3 is assumed to have a square distribution.

$$L3 = \sqrt{A} \ (m)$$

Vertical sliding force

$$L2 = g + L3/2 \ (m) \quad (10)$$

$$Wa = 1/2 \times d \times a \times L2 \quad (11)$$

$$W = Wa + (q \times L3) \ (t) \quad (12)$$

Shearing strength S

$$L = \sqrt{\left(a^2 + (h + L2)^2\right)} \quad (13)$$

$$\theta = tan^{-1}\left[\frac{a}{h + L2}\right] \quad (14)$$

$$S = c \times L + W\cos\theta \times \tan\Phi m \quad (15)$$

Φm: Friction angle

Safety angle

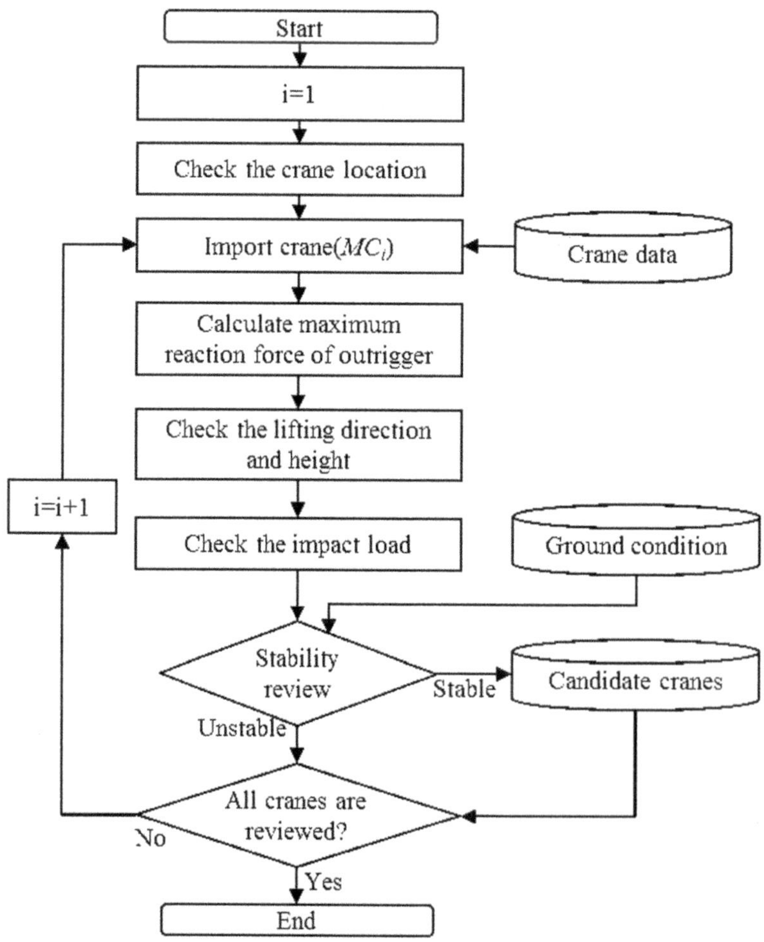

Figure 10 Generation model for plate stability review.

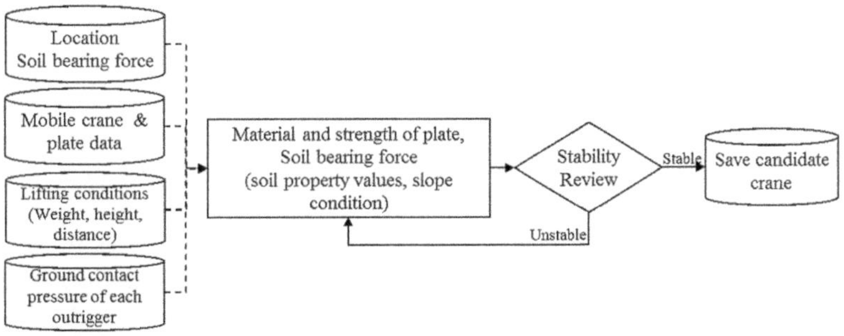

Figure 11 Plate design information (flat site).

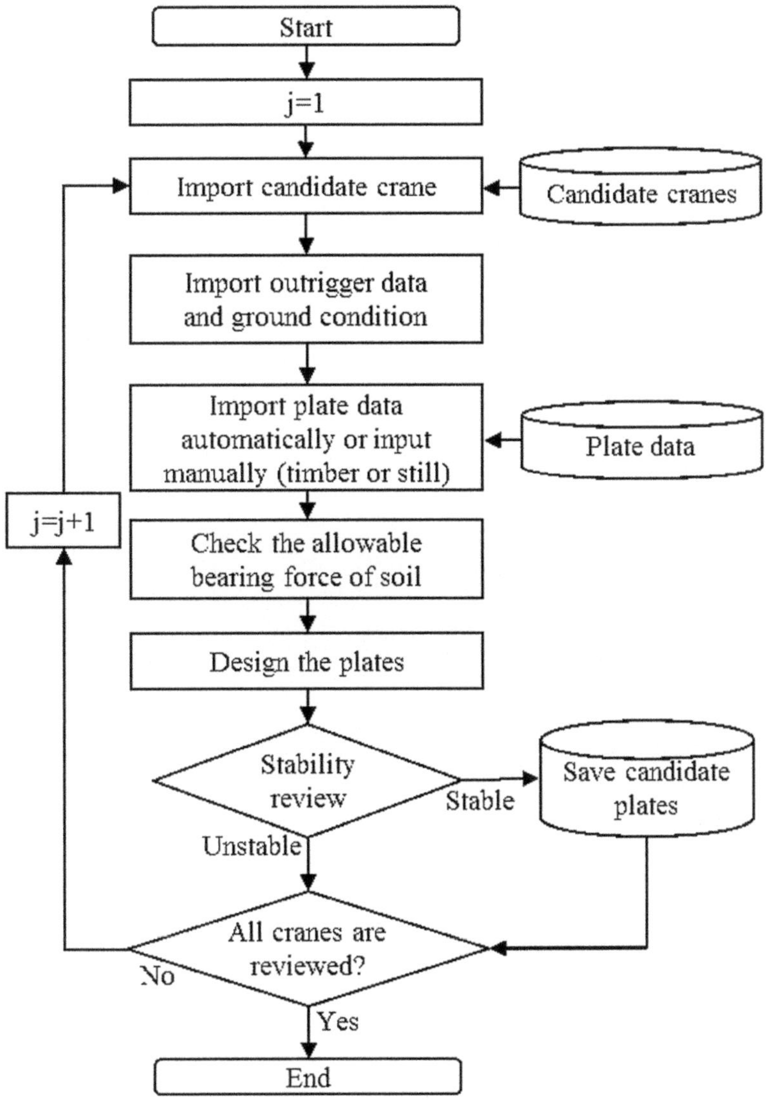

Figure 12 Plate design information (slopping site).

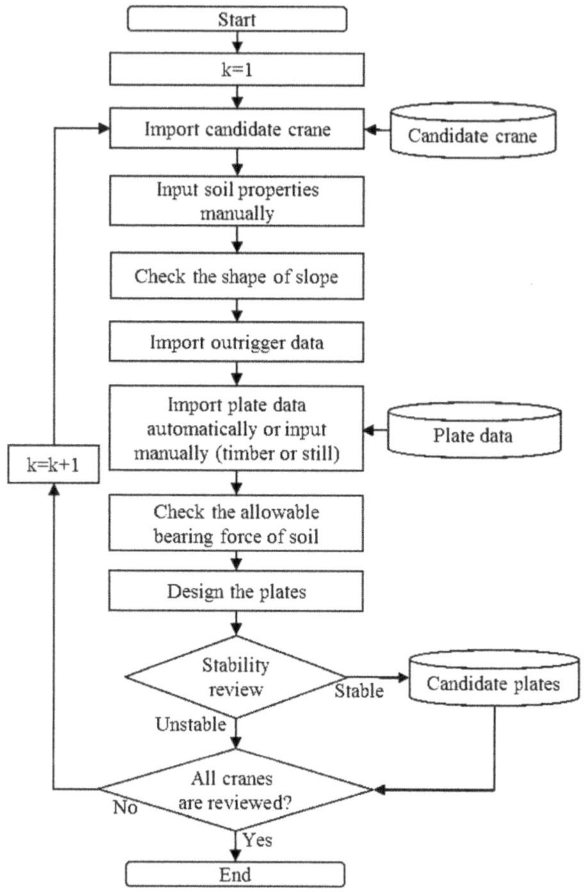

Figure 13 Automatic plate design for slopping site (simulation 3).

$$\mathrm{Fa} = \frac{S}{W \times sin\theta} > 1.2\text{-------O.K} \tag{16}$$

As illustrated in Figure 13, the plate design of a slope analyzes the candidate cranes deduced from simulation 1 as illustrated in Figure 7. In simulation 3, the plate is designed by checking the soil properties such as the N value, adhesive power, friction angle, and weight of unit volume that are input manually by the user as shown in Figure 12. Also, the slope shape is automatically analyzed by the equations (8) to (16) to review both the soil bearing force of the soil/ground and the landslide. For the plate design in simulation 3, the outrigger of a crane as shown in Figure 13 is selected, and the plate materials and the allowable soil bearing force are checked for the design of the materials. The designed plate and the crane are saved in pairs within the scope that stability is secured. Simulation 3 is automatically completed when all the candidate cranes deduced from simulation 1 are reviewed.

The top level process for mobile crane plans is comprised of 3 simulations, as illustrated in Figure 14. When selecting mobile cranes (Figure 7), the project data and crane data are used to deduce the candidate cranes within the required ground contact pressure. It should be checked whether the mobile crane is available in the market, and the selected cranes become the basis for the plate design in simulation 2 or 3.

As previously described, only 1 type is simulated according to the ground shape for both simulations (2 & 3), and the plate suitable for the candidate crane is designed. The plate design is paired with the candidate crane for a review of economic feasibility, and is then finally selected.

This study proposes a model for selecting mobile cranes with the lowest cost possible, and designing plates by identifying multiple cases and calculating cost based on securing stability, which is the prerequisite. The objective variable of mobile crane and plate design is cost (C_{mc}). The objective function of this model is to minimize the sum of mobile crane rental cost (C_{rental}) and plate cost (C_{plate}) as shown in Formula 17. The cost calculation of the model shown in Figure 14 is automatically carried out in 2 steps. First, a rental cost is calculated from the data of n candidate cranes. Second, respective cost is calculated for 'm' plates designed in simulations 2 and 3. The model aims to estimate the minimum cost, and the objective and constraint functions are as shown below.

$$Minimize\{\, C_{mc}\} \tag{17}$$

$$C_{mc} = C_{rental} + C_{plate}$$

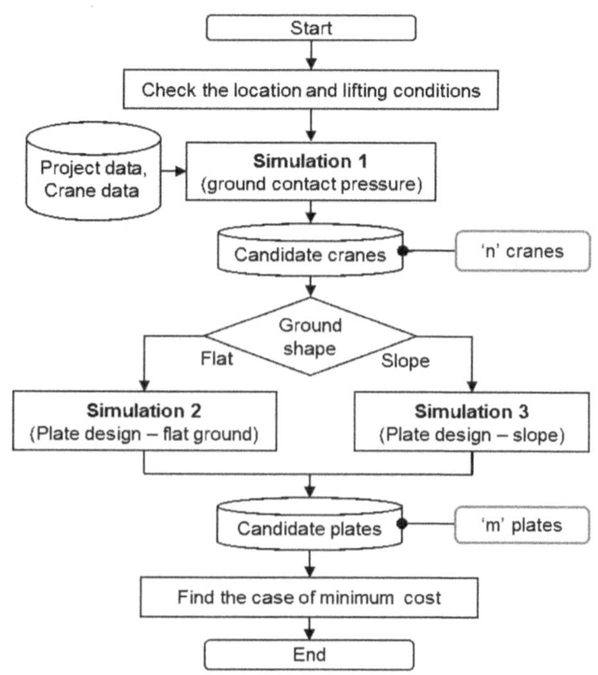

Figure 14 Automatic design model.

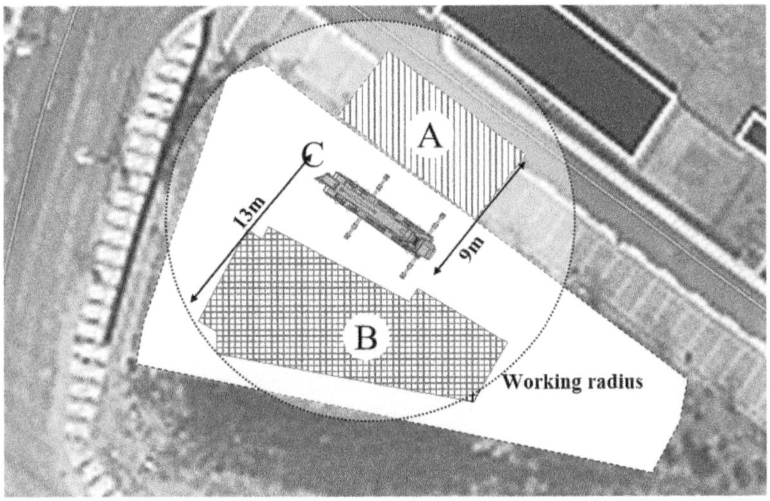

Figure 15 Site of case study project.

Subject to: $WR_{mc} \geq$ required working radius
$LC_{mc} \geq$ required lifting load
$MI_{outrigger} \leq$ allowable bearing force of soil
$SH_{plate} \geq$ shear force of plate
$BE_{plate} \geq$ bending force of plate
Where C_{mc}: cost of mobile crane work
C_{rental}: rental cost of mobile crane
C_{plate}: cost of designed plate
WR_{mc}: working radius of selected mobile crane
LC_{mc}: lifting capacity of selected mobile crane
$MI_{outrigger}$: maximum impact load of outrigger
SH_{plate}: allowable shear stress of designed plate
BE_{plate}: allowable bending stress of designed plate

Result and discussion

To verify the effectiveness of the automatic design model, a case project is selected and the algorithm is applied assuming several conditions. Above all, the case project is a site where buildings are not completed, and it is assumed that PC members are to be installed on the first floor. C in Figure 15 is the location of a mobile crane, and A shows a material storage yard. B is the building to be constructed, and the distance between a crane and a building is around 13 m. The heaviest lifting object is the PC beam that is 8 m long, which weighs 4.6 tons. The total lifting weight considering sling and lug is 5.06 tons.

In consideration of lifting conditions such as the working radius and weight of lifting objects, 18 candidate cranes are selected in total. Ground contact pressure upon operation of each candidate crane is calculated, and allowable soil bearing force is set as 20 tf/m^2 to design plates for reinforcement of soil bearing force.

As a result no information regarding self-weight, so they are not automatically designed. a, b, c, and t in Table 2 refer to the width, breadth and thickness of plates shown in Figure 3. Ground contact pressure increases in the case of large cranes due to their self-weight, and they are designed with width and breadth of 2500 mm. This restricts the use of cranes. The smallest plate is designed with width of 1000 mm and thickness of 24 mm.

When the allowable soil bearing force is set as 30 tf/m^2 under the same lifting conditions, plate design is as described in Table 3. Applicable crane types are the same, yet plate design has changed. Due to the difference in allowable soil bearing force, the plate for reinforcement of soil bearing force is designed with width and breadth of 800 mm and thickness of 15 mm. Thus, the specifications in Table 3 should be applied for reinforcement plates when using cranes to limit crane overturn risks.

Conclusion

This study develops automatic design algorithms for securing the ground contact stability of mobile cranes. In addition, the algorithms proposed in the study focus on the ground stability review process of mobile cranes. The study subjects are hydraulic cranes (or tire cranes), and single lifting is assumed for the study. Ground contact stability is reviewed based on the installation of cranes on soil or ground.

The study results are as follows.

Firstly, we found that the database built using the available tower cranes in the market for automatic design of mobile cranes can easily and quickly select candidate cranes according to lifting conditions and design plates through simulations.

Secondly, design components required for ground contact plate design are collected to build the database, which allows easier and quicker stability review. Moreover, automatic design algorithms are capable of obtaining optical

Table 2 Result of automatic design model (1)

Mobile crane	Capacity (ton)	Ground contact pressure (ton)	Steel plate			
			a (mm)	b (mm)	c (mm)	t (mm)
TM1500	80	58.786	570	1750	1750	34
TG1500E	150	-	-	-	-	-
TG1000E	100	-	-	-	-	-
TG800E	80	33.316	420	1300	1300	26
TG8500EX	50	27.451	400	1200	1200	24
TG300E	30	-	-	-	-	-
SC50H	50.5	33.316	400	1200	1200	24
SC25H	25	27.451	300	1000	1000	18
LTM1500-84	500	121.801	900	2500	2500	54
LTM1500-50	500	121.801	900	2500	2500	54
LTM1400	400	124.988	900	2500	2500	54
LTM1300	300	114.151	900	2400	2400	54
LTM1200	200	87.376	750	2100	2100	45
LTM1160	160	73.351	670	1950	1950	40
LTM1120	120	43.516	450	1500	1500	27
LTM1090	100	46.576	520	1550	1550	31
LTM1070	70	40.647	470	1450	1450	28
LTM1060	70	39.245	470	1450	1450	28

Table 3 Result of automatic design model (2)

Mobile crane	Capacity (ton)	Ground contact pressure (ton)	Steel plate			
			a (mm)	b (mm)	c (mm)	t (mm)
TM1500	80	58.786	390	1400	1400	29
TG1500E	150	-	-	-	-	-
TG1000E	100	-	-	-	-	-
TG800E	80	33.316	320	1100	1100	23
TG8500EX	50	27.451	300	1000	1000	21
TG300E	30	-	-	-	-	-
SC50H	50.5	33.316	300	1000	1000	22
SC25H	25	27.451	200	800	800	15
LTM1500-84	500	121.801	670	2050	2050	49
LTM1500-50	500	121.801	670	2050	2050	49
LTM1400	400	124.988	720	2050	2050	53
LTM1300	300	114.151	700	2000	2000	50
LTM1200	200	87.376	570	1750	1750	42
LTM1160	160	73.351	500	1600	1600	36
LTM1120	120	43.516	320	1250	1250	23
LTM1090	100	46.576	370	1250	1250	28
LTM1070	70	40.647	350	1200	1200	25
LTM1060	70	39.245	320	1150	1150	24

design plans from among those that satisfy stability review requirements after relevant costs are calculated.

Lastly, a plate design model connected to the selected mobile crane should first review ground contact pressure. It was found that system runtime is shortened when ground contact pressure to prevent overturn is reviewed before design.

The algorithms proposed in this study conduct simulations for all possible candidate cases, save the results in a database, review practical alternatives, and select the solution that minimizes cost. This is because stability is more important than economic feasibility when operating mobile cranes. Developed processes are useful for analyzing correlations of mobile crane performance, ground contact and soil bearing force reinforcement designs and so forth, and for systematically selecting optimal mobile cranes. Other studies should be conducted on stability review models of double lifting (where 2 mobile cranes are used). Furthermore, stability review models of components such as the sling and lug required for lifting must be developed to prevent the fall of lifting objects.

Competing interests

The algorithms proposed in this paper demonstrate the review for securing the ground contact stability and the design of reinforcement plates automatically after a mobile crane selection. The case study was carried out to verify the effectiveness of the proposed algorithms.

Many researchers have conducted a wide range of studies about mobile cranes such as the implementation of a lifting rout plan program (Reddy and Koshy 2002; Lozano-Pérez and Wesley 1979) and a crane selection algorithm (Al-Hussein et al. 2001; Shapira and Schexnayder 1999; Al-Hussein et al. 1997; Furusaka and Gray 1984; Gray and Little 1985), the overturn analysis of outriggers (Tamate et al. 2005), the friction with other works (Tantisevi and Burcu 2007) and even the development of a dual lifting simulation program. However, they didn't deal with a series of mobile crane management for securing both stability and economy, but dealt with the partial issue of it, whereas this paper covered the principal issues from selecting a proper crane responding to the lifting conditions to reviewing the stability of cranes automatically, followed by the economic design of reinforcement plates.

The case study of this paper illustrates the automatic reinforcement plate design for two cases of allowable soil bearing force. Site engineers can figure out the rental and plate costs of candidate cranes and, then, select the minimum cost for crane operation. A series of algorithms for mobile crane management was implemented to the software for practical use except the cost minimization algorithm that should be additionally equipped to the software for better use in practice.

Authors' contributions

DL and SK together carried out the filed study. Subsequently, the gathered data was analyzed. All authors contributed to develop the algorithms. All authors read and approved the final manuscript.

Acknowledgement

This work was supported by the National Research Foundation of Korea (NRF) grant funded by the Korea government (MSIP) (No. 2013R1A2A2A01068297).

Author details

[1]Department of Architectural Engineering, Kyung Hee University, 1732 Deogyeong-dearo, Giheung-gu, Yongin-si, Gyeonggi-do, Korea. [2]Department of Q-HSE Management Division, Technical Advisory Team, Samsung C & T Corporation, Seocho2-Dong, Seocho-Gu Seoul 1321-20, Korea.

References

Al-Hussein, M, Alkass, S, & Osama, M. (1997). Computerized crane selection for construction projects. *Journal of Construction Engineering Management, Association of Researchers in Construction Management, 2*, 427–436.

Al-Hussein, M, Alkass, S, & Osama, M. (2001). An algorithm for mobile crane selection and location on construction sites. *Construction Innovation: Information, Process, Management, 1*(2), 91–105.

Furusaka, S, & Gray, C. (1984). A model for the selection of the optimum crane for construction sites. *Construction Management and Economics, 2*(2), 157–176.

Gray, C, & Little, J. (1985). A systematic approach to the selection of an appropriate crane for a construction site. *Construction Management and Economics, 3*(2), 121–144.

Ho, JK, Han, KK, & Kim, SK. (2007a). Tower crane foundation design and stability review model. *Korea Institute of Ecological Architecture and Environment, 7*(6), 99–106.

Ho, JK, Seo, JM, & Kim, SK. (2007b). Mobile crane ground fixing system. *Korea Institute of Ecological Architecture and Environment, 7*(6), 83–90.

Kim, HH, & Lee, K. (2007). An analysis of the accident types and causes of construction cranes. *Korea Institute of Construction Engineering, 7*(1), 109–112.

Lin, KL, & Haas, CT. (1996). Multiple heavy lifts optimization. *Journal of Construction Engineering and Management, 122*(4), 354–362.

Lozano-Pérez, T, & Wesley, MA. (1979). An algorithm for planning collision-free paths among polyhedral obstacles. *Communications of the ACM, 22*(10), 560–570.

Maczynski, A, & Wojciech, S. (2003). Dynamics of a mobile crane and optimisation of the slewing motion of its upper structur. *Nonlinear Dynamics, 32*(3), 259–290.

Reddy, HR, & Koshy, V. (2002). Automated path planning for mobile crane lifts. *Computer-Aided Civil and Infrastructure Engineering, 17*(6), 439–448.

Shapira, A, & Schexnayder, CJ. (1999). Selection of mobile cranes for building construction projects. *Construction Management & Economics, 17*(4), 519–527.

Shapiro, HI, Shapiro, JP, Shapiro, LK, & Shapiro, HI. (2000). *Cranes and Derricks*. New York: McGraw-Hill.

Struková, Z, & Ištvánik, M. (2011). Tools for mobile crane selecting and locating. *International Review of Applied Sciences and Engineering, 2*(1), 69–74.

Tamate, S, Naoaki, S, & Toshiyuki, K. (2005). Analyses of instability in mobile cranes due to ground penetration by outrigger. *Journal of Construction Engineering and Management, 131*(6), 689–704.

Tantisevi, K, & Burcu, A. (2007). Automated generation of workspace requirements of mobile crane operations to support conflict detection. *Automation in Construction, 16*(3), 262–276.

The Vertikal Press. (2014). *Loader crane overturn in UK, vertikal.net, News article*. http://www.vertikal.net/en/news/story/19261/.

14

Successful education for AEC professionals: case study of applying immersive game-like virtual reality interfaces

Farzad Pour Rahimian[1*], Tomasz Arciszewski[2] and Jack Steven Goulding[1]

Abstract

Background: Global competition and the transdisciplinary nature of evolving Architecture-Engineering-Construction (AEC) activities makes it progressively important to educate new AEC professionals with appropriate skill sets. These skills include the ability and capability of not only developing routine projects, but also delivering novel design solutions and construction processes (some of which may be unknown), to feasible, surprising, or potentially patentable solutions. For example, despite recent innovations in immersive visualisation technologies and tele-presence decision-support toolkits, the AEC sector as a whole has not yet fully understood these technologies, nor embraced them as an enabler.

Methods: Given this, this paper proposes a new approach for delivering education and training to address this shortcoming. This approach focuses on doing traditional (routine) work with creative thinking in order to address these challenges. This rationale is based on the principles of *Successful Education* as a new paradigm for engineering education, which is inspired by the Theory of *Successful Intelligence*, by the *Medici Effect* and Leonardo da Vinci's *Seven Principles*. The paper presents the educating AEC professionals is presented the AEC sector. The Theory of *Successful Intelligence* and its three forms of intelligence (Practical, Analytical, and Creative), are supported by lessons learned from the Renaissance, including the *Medici Effect* and da Vinci's *Seven Principles*.

Results: Based on these theoretical pillars, a new approach to educating AEC professionals is presented with a proof-of-concept prototype that uses a game-like virtual reality (VR) visualisation interface supported by Mind Mapping is introduced as an exemplar.

Conclusion: The developed interface in this study applies Game Theory to non-collocated design teams in accordance with Social Sciences Theory (social rules) and Behavioural Science Theory (decision making). It contributes by supporting new insights into AEC actor involvement, pedagogy, organisational behaviour, and the social constructs that support decision making.

Keywords: Training; Successful education; Pedagogy; Medici effect; da Vinci Principles; Visualisation

Background

The Architecture-Engineering-Construction (AEC) sector is one of the largest industrial employers in many countries. In the European Union (EU) for example, it encompasses more than 2 million enterprises and approximately 12 million employees, representing 9.8% of the EU's Gross Domestic Product and employing over 7.1% of the workforce (NGRF 2010). This contribution and global competition makes the novelty of the AEC projects increasingly important. Therefore, AEC professionals need to be educated how to develop not only traditional, or routine projects, but also projects incorporating novel designs and construction processes. They need to be creative, and be able to develop unknown (or unproven) solutions which are feasible, surprising, and potentially patentable. Currently, AEC professionals are no longer being seen as leaders or innovators, more followers - using deductive problem solving rather than seeking opportunities, using

* Correspondence: fpour-rahimian@uclan.ac.uk
[1]Centre for Sustainable Development, The Grenfell-Baines School of Architecture, Construction and Environment, University of Central Lancashire, Preston PR1 2HE, UK
Full list of author information is available at the end of the article

their creativity and developing inventions. This resonates with thinking derived from innovation literature (Akintoye et al. 2012). As a result, designers and engineers in particular have seemingly lost their ability to innovate. This is partly attributable to 'inappropriate' education that has historically focused on production, rather than creativity. This is just the opposite of what happened in the 19th and early 20th Centuries, when designers and engineers were seen as the true 'drivers' of change. During this time, high-level education was aligned to incentives (e.g. the highest salary rates) which helped design and engineering schools attract the most talented students; and these graduates were capable of meeting all technological and socio-cultural challenges of the quickly expanding societies (Arciszewski 2006; Arciszewski and Harrison 2010a, 2010b; Arciszewski and Rebolj 2008). For instance, the construction of some monumental buildings during this period in history (e.g. Eiffel Tower, Villa Savoye, and The Bauhaus Building) created not only technological solutions, but also cultural revolutions- leading to a fundamental change in the way design and engineering was perceived.

This research posits that creativity has increasingly been underrepresented; and as such, needs to be revisited, especially in a rapidly evolving technological-driven world. For example, such challenges now include environmental and sustainability demands, increased levels of safety compliance, enhanced security issues, and whole life demands (energy, maintenance etc.). Whilst it could be argued that some of these challenges extend beyond the AEC domain *per se*, it is important to identify the key promoters and inhibitors of engineering creativity. In doing so, the profession as a whole will benefit from a new cogent way of embedding creativity into solutions; the result of which will not only benefit society, but also help inspire future AEC successors to follow this approach. Any changes, particularly those related to the ways that AEC students are educated, are extremely difficult, mostly because of the Vector of Psychological Inertia (G. Altshuller 1984) in action. This phenomenon refers to a natural tendency of individuals and communities to resist any changes, thereby delaying progress as much as possible. This is also influenced by the way in which the instructors were originally educated (mostly as highly sophisticated analysts) as this has a significant impact on the way they want to teach students. Cognisant of this, it is important to recognise the need to apply a complex systems approach to analyse the impact of this in order address the current situation.

This paper presents design and engineering leadership as three interrelated abilities: 1) to develop a vision, 2) to transform it into a strategy, and 3) to implement it. The key to leadership is the ability to develop feasible ideas or concepts (e.g., a new type of engineering system or construction process) using a set of abilities (traits) required to implement them [as opposed to using existing concepts to perform typical/routine work]. In particular, the development of a vision similar to conceptual design, to inventive design. In both cases a new idea, or a concept of an engineering system, needs to be developed. This is the area of activities in which creativity, or abductive generation of new ideas, takes place. This position is proffered, as historically, 'followers' have been seen to create stagnation, producing what has been called "Vector of Psychological Inertia" (H. Altshuller 1984), or fixation (Youmans and Arciszewski 2014). This psychological phenomenon therefore tends to makes change and progress more difficult, and in some cases often even prevents it. The emphasis therefore is to consider the development of leaders (not followers), in order to minimise the negative impact of the Vector of Psychological Inertia.

Building upon the principles of the *Theory of Successful Intelligence (Sternberg* 1985, 1996, 1997*)*, this paper describes "Success" as a relative concept, which is defined by a given person in relation to the socio-cultural context and personal desires. This study therefore posits that there is a need to develop a new paradigm that recognises the importance of both analytical and creative works. Given this, this research defines analyst learners as the people who use rote learning and deduction, eventually induction, as opposed to creative people who use also abduction for reasoning. This approach extends learning capability beyond the learners' cognitive capability. Relying on the principles of Theory of Successful Intelligence (Sternberg 1985, 1996, 1997), Positive Psychology (Schueller 2012), and Appreciative Intelligence (Barrett and Fry 2008), this paper asserts that by using the 'right' methodologies and media, general principles of creative work could be translated into an explicit knowledge form and become part of a body of knowledge; hence, enabling the "Successful Departments" (Arciszewski 2009) to teach learners the *"Creative Intelligence"* and "Appreciative Intelligence". In this context, the potential of utilising advanced visualisation tools such as immersive game-like virtual reality interfaces is deemed vital - especially for augmenting analytical and parametric thinking capacity to intuitive idea generation (which could both be supported by these interfaces).

Theory of successful intelligence

The *Theory of Successful Intelligence* (Sternberg 1985, 1996, 1997) is a major step toward understanding how individuals' abilities are interrelated with their life success. In the context of design and engineering education, this theory presents a new understanding of how education can be conceptualised, designed, and delivered. Through this theory, successfully intelligent people are defined as those being able to achieve their goals by leveraging their strengths, by compensating for their weaknesses, and those able to adapting to, shape, and select environments that will facilitate their success. This theory is underpinned by three fundamental pillars:

1. Successful intelligence can be learned;
2. Successful intelligence is a combination of three independently acquirable abilities, namely: practical intelligence, analytical intelligence, creative intelligence;
3. Successful intelligence is dynamic; both the criteria of success and the abilities the individual employs (i.e. the relative combination of the three intelligences) to achieve success may change during one's life-time.

In accordance with this theory, practical intelligence is an ability to solve simple everyday problems, and this is achieved using readily available knowledge and heuristics. Abilities to open a door or to ride a bus are good examples of practical intelligence. Analytical intelligence is an ability to solve analytical problems, and that requires using deductive skills and utilising existing knowledge (for example, analysis of traffic flow, numerical optimization, or planning a typical construction process, etc.). Analytical intelligence is acquired through the combination of rote learning and learning deductive skills. Analytical intelligence alone is what traditional IQ tests measure. In addition, traditional engineering education emphasizes analytical intelligence almost entirely. However, the Theory of *Successful Intelligence* stipulates that a balance of the three intelligences is absolutely necessary for life success, including professional success.

In the AEC context, creative intelligence is the ability to solve inventive problems, which require abductive skills and obviously the use of existing knowledge. Solving such problems requires development of unknown solutions or ideas, e.g. development of a new type of a wind bracing system in a tall building or a new type of a tunnel. Creative intelligence is acquired through the combination of rote learning with learning of both deductive and abductive skills.

The Medici effect

Johansson (2004) proposed two interrelated concepts of the "Medici Effect" and of the "Intersection". These concepts identify mechanisms driving an environment facilitating and stimulating emergence of transdisciplinary knowledge, which is the foundation of creativity in engineering. Transdisciplinary knowledge is a body of integrated knowledge with roots in two or more domains, but knowledge which is domain-independent. For this reason, both concepts are important for engineering educators who should recreate them when educating creative engineers.

The Medici Effect (Johansson 2004) was a mechanism driving the emergence of the Renaissance intellectual foundation. It was named after the Medici family, which lived in Florence, Italy, in the 15th Century. The Medicis were one of the richest families in Europe, and sponsored many artists and scientists who were members of their court. Ultimately, members of the community began developing understanding of knowledge from outside of their domains. That led to new understanding of individual disciplines and to gradual emergence of the transdisciplinary knowledge. This knowledge became the intellectual foundation of the Renaissance.

The Intersection (Johansson 2004) is a product of the Medici effect. Johansson (2004) argued that Intersection is a time and place specific integration of knowledge with elements coming from various disciplines, cultures, and personalities. When a new concept is developed within a given discipline, it usually follows the existing line of evolution (Zlotin and Zusman 2006) and is considered directional. However, when an intersection occurs, a new idea represents a radical change, or beginning of a new line of evolution. Such an idea can be called "intersectional idea". Intersection can be described as knowledge integration with knowledge coming from two or more domains and resulting in transdisciplinary knowledge, valid in all contributing domains (Sage 2000).

The Medici Effect should be used in AEC education to create an educational environment, called by Arciszewski (2009) "Successful Department". Such an environment should be not only supporting but also stimulate teaching and learning engineering creativity. This is a way to reconstruct an environment critical for the emergence of the Renaissance; and more importantly, for the creation of an environment necessary to educate creative designers and engineers.

Da Vinci's Principles

Gelb (1998, 1999, 2004) introduced the term "Da Vinci Seven Principles" and proposed seven principles describing the core characteristics of Da Vinci's approach to science, design and engineering. These seven principles which are shown in an artist's vision in (Figure 1) are as follows:

Principle No. 1, "Curiosita, means in Italian "curiosity." According to Gelb (2004), this da Vincian represents a curious and open attitude and a life-long learning accomplished by constantly asking questions about everything. Unfortunately, mostly analytical current educational materials gradually destroy *Curiosita* (Arciszewski 2014). In order to educate creative designers and engineers, it is necessary to not only maintain their *Curiosita,* but also expand it. The challenge here is to teach students the

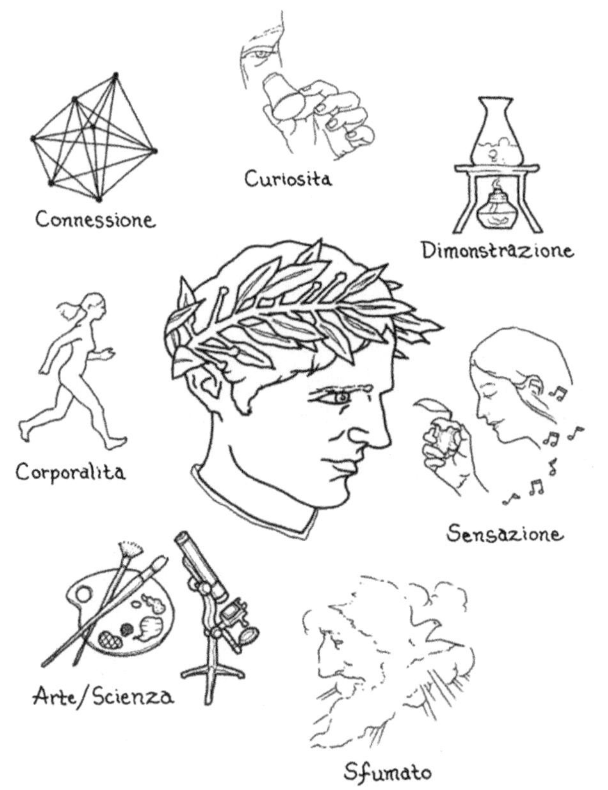

Figure 1 Da Vinci Seven Principles, Source: (Arciszewski 2009).

practical and analytical intelligence and at the same time expand their *Curiosita,* which is the key to the creative intelligence.

Principle No. 2, "Dimostrazione", means in Italian "demonstration". Gelb (2004) explained it as a unique attitude of experimentally verifying acquired knowledge. The Renaissance concept of apprenticeship is a good example of Demonstration in education. It was a combination of individual studies with extensive hands-on experience resulting in experiential learning. In this case, a master/teacher provides only guidelines and helps students how to learn on their own.

Principle No. 3, "Sensazione" means in Italian, "sensitivity to feelings". Gelb (1998) used this term to identify a complex attitude. It is development of all senses, practicing both the rational/intellectual and emotional approaches to life, and problems, integration of all abstract and physical inputs to create synaesthesia. It is a complex emotional state when an artist or a scholar is fully engaged in solving a problem, both intellectually and emotionally - using all his/her senses as using synaesthesia to acquire transdisciplinary knowledge or to create new ideas. Sensazione can be also interpreted as a practicing "whole-brain thinking" in which focus is on the emotional dimension of our cognition ultimately leading to a much more complete understanding of the world, of our environment, and of ourselves, including our consciousness and ability to transform it.

Principle No. 4, "*Sfumato,*" means in Italian "turn to mist"; or in the case of colours, "soft," or "mellow." Gelb (1998) interpreted *Sfumato* as a willingness to accept and to understand the world in its infinite complexity. It also means keeping an open mind in the face of unknown and of uncertainty, a willingness to embrace contradictions, and paradoxes, and acquiring, accepting and using ambiguous knowledge. Sfumato is a surprisingly modern notion. In the case of knowledge discovery and inventive problem solving such processes are known today as lengthy and having subsequent periods of conscious and subconscious activities. To produce inventions, all kinds of input are obviously desired in order to activate and use the entire power of the human brain, both the analytical left hemisphere and the creative right side.

Principle No. 5, "*Arte/Scienza*" means that a Renaissance person should be a "Whole-Brain Thinker". People should develop an understanding of the world using two entirely different but complementary perspectives with roots in art and science, respectively. These two perspective should be balanced since both are necessary but neither sufficient. If AEC educators are interested in creating an education producing inventors, then the Principle "Arte/Scienza" is very important. It represents a significant departure from the traditional engineering education nearly entirely focused on the rational, "scientific" approaches and knowledge.

The Principle No. 6, "*Corporalita*" means "the state of being in physical or bodily form rather than spiritual form" in accordance to MSN Encarta. It is incomplete, if not simply wrong description of the Da Vinci's Principle No. 6. The attitude of *Corporalita* is much more complex. It is described by Gelb (1998) as *"Means sana in corpora sano"* - sound mind in a sound body. Da Vinci argued that a human being should carefully maintain a balance between the intellectual and physical development in order to realize his/her full potential. It was also a reflection of the Renaissance belief that a genius must be physically superior with respect to ordinary people. *Corporalita* is particularly important for designers and engineers. They need to maintain a balance between body, mind, and spirit, but also to attain a relatively high level of physical fitness to survive long hour of climbing stairs and ladders on a construction site.

Principle No. 7, "*Connessione*" means in Italian "connection", however this principle in fact means recognition of interconnectedness of all things and phenomena in nature and life, the world is a single system with its all elements connected by direct and indirect feedbacks, the world is a complex and chaotic system, knowledge is a non-linear system. Only recently, in the second half of the 20th Century, the science of holistic understanding of the world, called "Cybernetics" emerged. It gradually led to the development of the "Systems Analysis" based on the principle of wholeness in its approach to analysis of all systems, built and natural, real and abstract, small and large. Therefore, "*Connessione*" may be interpreted as a systems view of the world.

Methods

Successful education

Successful Education (Arciszewski 2009) is a new paradigm in design and engineering education. This paradigm was inspired by the latest developments in the modern cognitive psychology, especially by the Theory of *Successful Intelligence* (Sternberg 1985, 1996, 1997). This paradigm has also been strongly influenced by a new understanding of historical and social mechanisms behind the emergence of the Renaissance, including the Medici Effect (Johansson 2004) and the Da Vinci Principles (Gelb 1998, 1999, 2004). (Arciszewski (2009)) argued that Principles are particularly important because they provide a synthesis of all attitudes practiced by Da Vinci and by the other great Renaissance engineers.

In this paradigm, the key concept is "Successful Designers and Engineers" and it describes the designers and engineers who have acquired as students not only the necessary and sufficient body of knowledge to practice engineering, but also learned *Successful Intelligence* including its all three components, i.e. practical, analytical, and creative intelligence. Such graduates are prepared to not only undertake any kind of routine work, but, if necessary, also prepared to become inventors and leaders, since in both cases the key to success is an ability to develop new ideas.

In Table 1, Successful Education is compared with a past design and engineering education paradigm, called "Master-Apprentice Paradigm", and the present one, called by us "Scientific Paradigm". The comparison is done from the perspective of the Theory of *Successful Intelligence* and of its three main components. In this context, only Successful Education is complete since only it addresses all three components of *Successful Intelligence* and consequently creates an opportunity to educate successful engineers.

Successful Education requires not only a new understanding of design and engineering education priorities and several new or modified courses, it also requires a complex environment, called "Successful Department", which will enable and stimulate the creation of successful engineers. A modern Medici Effect and the resulting intersection of ideas are crucial for the learning process. Therefore, they require a revolutionised environment (in terms of intellectual and technological structures) which is completely different that the current look of so many design and engineering departments. In essence, there are four major components of a Successful Department, namely courses, instructors, physical environment, and ambience (Arciszewski 2009). This is aligned with Salama's (2008) *"Integrating Knowledge in Design Education"* theory which argues that a responsive architectural design pedagogy giving credit to socio-cultural, and environmental needs can enable future architects to create livable environments.

Table 1 Comparison of teaching paradigms

Teaching paradigm	Practical intelligence	Analytical intelligence	Creative intelligence
Master-apprentice	Yes		Yes
Scientific	Yes	Yes	
Successful education	Yes	Yes	Yes

Traditional, analytical courses are absolutely necessary for the future successful engineers, although they are grossly insufficient for them. They require additional courses on Inventive Design and Engineering, i.e. focused on the emerging science of inventive problem solving. For the best results, such courses could/should be offered to students through their entire period of studies. A single course for seniors (the present practice at George Mason University) is a step in the right direction, but it comes too late to impact learning in other courses and to transform students into successful engineers. A much better solution is a sequence of several courses, even if the total number of credit hours is the same.

Instructors are the key component of a Successful Department. A faculty in academic units are surprisingly similar in many aspects (ergo birds of feather flock together) despite all efforts to create diversity, which is often imposed only for political reasons. A successful Department requires, however, a true diversity, which may be described as "balanced intersection". This term is understood as a selection of instructors resulting in a department in which cultural backgrounds of instructors are strongly differentiated, they represent both applied and fundamental research, have experience in analytical and exploratory research and they represent various thinking styles.

Physical environment creates a framework for learning and also send a message about the nature of a given academic unit (Hou and Ji 2010). An ideal urban design for a Successful Department should be based on the concept of the agora, as an ideal form stimulating human interactions through complex socio-psychological mechanisms. Such an urban complex should have several buildings, arranged around the central square/agora. A building should be dedicated to teaching practical intelligence and designed with all kinds of testing laboratories and workshops. Another building should be dedicated to teaching analytical intelligence and it should have various computer laboratories. A third building, "Inventors Heaven", a must, should be dedicated to teaching creative intelligence with appropriately selected laboratories and workshops specifically designed for teams working on their inventive challenges. Finally, there should be an administrative building for faculty and classrooms.

A Successful Department would never be fully effective without a proper ambience. In this case, ambience is understood as a multi-sensory experience that positively affects students, faculty, and staff helping them to learn or teach in the best way to create successful engineers. Ambience obviously has an emotional dimension, which distinguishes it from a traditional department. Ambience is a reflection of people's perception of an environment surrounding them and can be carefully created in such a way as to contribute successful designers and engineers. Arciszewski (2009) discussed various components of ambience in a Successful Department, e.g. guiding principles and stories, colours, music, art, various activities, and even the proper lighting in the Successful Department.

Building upon the theoretical bases discussed in the theory of Successful Education (Arciszewski 2009), this paper highlights the potentials of the advanced IT interfaces for leveraging all four components of such a Successful Department. The paper particularly suggests use of advanced game-like virtual workspaces in order to leverage education of successful designers and engineers for the AEC professions.

Games and virtual reality in construction engineering education

The nature and complexity of communication mechanisms within the Architecture, Engineering, and Construction (AEC) projects has changed significantly over the last ten years, especially the modus operandi and integration with core business operations. This has been reflected through the increased prevalence, use, and deployment of web-based project collaboration technologies and project extranets. Within the AEC sector, Information and Communications Technology (ICT) has revolutionised production and design (Cera et al. 2002), which has led to dramatic changes in terms of labour and skills (Fruchter 1998). However, it is also important to acknowledge that the capabilities of such applications (and implementation thereof) in predicting the cost and performance of optimal design proposals (Petric et al. 2002) should enable design engineers to compare the quality of any one tentative solution against the quality of previous solutions. This was further reinforced by Goulding et al. (2007), regarding the ability to experiment and experience decisions in a 'cyber-safe' environment in order to mitigate or reduce risks prior to construction. It is therefore crucial for the AEC industry to employ cutting-edge ICT technologies to issues related to organisational management and decision making (Friedman 2005). Furthermore, whilst advocates note that these have helped to resolve some of the aforementioned challenges, Pour Rahimian et al. (2011) noted that project teams are still facing real and signification problems and challenges regarding heterogeneous systems faced by project teams using project extranets. In this essence, the problem here is that the industry is experiencing confusion as to how to manage project information in order to support decision-making processes. This is the point where Fruchter (2004) suggested the digital integration of the whole data creation, retrieval, and management system within building industry in order to prevent tacit knowledge loss and miscommunication among various parties from different disciplines. In this respect, recent innovation in Virtual Reality (VR) technologies and AEC decision- support toolkits have now matured, enabling tele-presence engagement to occur through integrated collaborative environments. Several opportunities are now available, including significantly improved immersive interactivity with haptic support that can enhance users' engagement and interaction.

Employing cutting edge ICT tools is also expected to leverage training systems within the AEC sector (Fruchter 1998) as the implementation of effective training could make impact on the whole industry by addressing and fulfilling the needs of the different stakeholders in the industry. In this respect, advanced ICT systems are expected to address the shortcomings of 'typical' learning models that often provide the trainees with only general instructions (Laird 2003) and issues associated with unaffordable costs of the 'traditional' on -the-job trainings (Clarke and Wall 1998). Therefore, new ICT advancements that incorporate innovative proactive experiential learning approaches which link theory with practical experience, using Virtual Reality interactive learning environments can be especially effective (Alshawi et al. 2007). This research builds upon the findings of previous studies in this area and links them to the principles of Successful Education (Arciszewski 2009), with specific emphasis on supporting the decision-making process at the construction stages. The study provides a novel approach of applying Game Theory to non-collocated design teams using Game-Like VR environments blended to Social Sciences Theory (social rules) and Behavioural Science Theory (Decision Science/Communication Science). In essence, the aim of this study is to advocate the advantages of applying flexible, interactive, safe learning environment for practicing new working conditions with respect to offsite production (OSP) in general, and Open Building Manufacturing (OBM) in particular; without the 'do-or-die' consequences often faced on real construction projects (Goulding and Rahimian 2012).

As the underpinning technology, VR has been defined as a 3D computer-generated alternative environment to be immersed in, for navigating around and interacting with (Briggs 1996), or as a component of communication taking place in a 'synthetic' space, which embeds human as its integral part (Regenbrecht and Donath 1997). The definitions of VR systems usually includes a computer capable of real-time animation, controlled by a set of wired gloves and a position tracker, and using a head-mounted stereoscopic display as visual output. For instance, Regenbrecht and Donath (1997) defined the tangible components of VR as a congruent set of hardware and software, with actors within a three-dimensional or multi-dimensional input/output space, where actors can interact with other autonomous objects, in real time. VR has also been defined as a simulated world, which comprises of some computer-generated images conceived via head mounted eye goggles and wired clothing – thereby enabling the end users to interact in a realistic three-dimensional situation (Yoh 2001).

Over the last 30 years, ICT systems have matured and enabled construction organisations to fundamentally restructure and enhance their core business functions. A. Z. Sampaio and Henriques (2008) asserted that the main objective of using ICT in construction field is supporting management of digital data, namely to convert, store, protect, process, transmit, and securely retrieve datasets. They acknowledge the commencement of VR techniques as an important stepping stone for data integration in construction design and management as they are capable of holding and presenting the whole information about buildings (e.g. size, material, spatial relationships, mechanical and electrical utilities, and etc.) through a single output. Similarly, Zheng et al. (2006) proposed the use of VR to reduce time and costs in product development and to enhance quality and flexibility for providing continuous computer support during development lifecycle.

Early studies that incorporated VR into the design profession used it as an advanced visualisation medium. Since as early as 1990, VR has been widely used in the AEC industry as it forms a natural medium for building design by providing 3D models, which can be manipulated in real-time and used collaboratively to explore different stages of the construction process (Whyte et al. 1998). It has also been used as a design application to provide collaborative visualisation for improving construction processes (Bouchlaghem et al. 2005). However, expectations of VR have changed during the current decade. According to A.Z. Sampaio and Henriques (2008), it is increasingly important to incorporate VR 3D visualisation and decision support systems with interactive interfaces in order to perform real-time interactive visual exploration tasks. This thinking supports the position that a collaborative virtual environment is a 3D immersive space in which 3D models are linked to databases, which carry characteristics. This premise has also been followed through other lines of thought, especially in construction planning and management by relating 3D models to time parameters in order to design 4D models (Fischer and Kunz 2004), which are controlled through an interactive and multi-access database. In similar studies, 4D VR models have been used to improve many aspects and phases of construction projects by: 1) developing and implementing applications for providing better communication among partners (Leinonen et al. 2003), 2) supporting design creativity (Rahimian and Ibrahim 2011), 3) introducing the construction plan to stakeholders (Khanzade et al. 2007), and, 4) following the construction progress (Fischer 2000).

With regards to education, Wellings and Levine (2010) posited that there was a need to redesign the current text-based lessons into collaborative and multidisciplinary problem-based materials, expressly to take on board real world problems and solutions. They argued that this was not possible unless immersive and interactive games were employed for improving trainees' engagement. Similarly, Thai et al. (2009) asserted that pedagogical digital games offered an intact opportunity to enhance engagement of trainees and revolutionise teaching and learning. ACS (2009) summarised the benefits of the emerging educational interactive immersive game environments: 1) annotated objects could provide deeper level of knowledge on demand, 2) incorporating additional dimensions of subjects (nD), 3) supporting distance team collaboration, 4) leveraging equal opportunities by providing distance learning opportunities and, 5) simulated learning by modelling a process or interaction that closely imitates the real world in terms of outcomes.

VR applications and game engines are now increasingly being used in the teaching and learning AEC. According to Zudilova-Seinstra et al. (2009), VR as a teaching tool can contribute to the trainees' professional future by developing some learning activities beyond what is available in the conventional training systems. With respect to educational issues in the AEC industry, A. Z. Sampaio et al. (2010) argued that the interaction with 3D geometric models can lead to active learner thoughts which seldom appear in conventional pedagogical conditions. Moreover, Juárez-Ramírez et al. (2009) asserted that when augmented to 3D modelling, VR could lead to better communication in the process of AEC training. However, VR training environments have arguably not yet fully reached the potential of reducing training time, providing a greater transfer of expert knowledge; or supporting decision making. This was primarily down to the ways in which this technology was augmented. It is therefore argued that educational training tools need to 'engage' learners by putting them in the role of decision makers and 'pushing' them through challenges; hence, enabling different ways of learning and thinking through frequent interaction and feedback, and connections to the real world context (Goulding et al. 2007). Furthermore, it is postulated that paring instructional content with game features, could engage users more fully, hence, help to achieve the desired instructional goals. In this respect, this study applied an input-process-output model (Garris et al. 2002) of instructional games and learning to design an instructional program which incorporated certain features or characteristics from gaming technology; which trigger a cycle that includes user judgment or reactions, such as enjoyment or interest, user behaviour such as greater persistence or time on task, and full learner feedback (Figure 2).

Results and discussion

This section presents the developed Game-Like Virtual Reality Construction-Site Simulator (GVRSS) in this study. The aim of the developed GVRSS was to embrace 'real life' issues facing offsite construction projects in order to appeal to professionals by engaging and challenging them to find 'real life' solutions to problems often encountered on site. Given this, a real construction project was used to govern the authenticity of the learning environment. In this context, the prototype learning simulator was designed specifically to allow 'things to go wrong', and hence, allow 'learning through experimentation' or 'learning by doing'. In this respect, although the 'scenes' within the simulator take place on a construction site, the target audience was focussed primarily on construction professionals e.g. project managers, construction managers, architects, designers, commercials, suppliers, manufacturers etc. Thus, the construction site was used as the main domain through which all the unforeseen issues and problems (caused through upstream decisions, faulty work etc.) could be enacted. The key learning impact areas were to acknowledge the importance, significance and real implications of time, cost, resources etc. Learning was planned and reinforced through a debriefing session, where learners were able to demonstrate additional understanding, particularly with respect to mitigating such issues in future construction projects. In this context, learning occurred through the following:

- Learner autonomy - to make all decisions;
- Interactivity - environment provides feedback on the decisions taken, and their implications on the overall project (cost, time, resources, health and safety, etc.);
- Reflection - users are able to defend decisions on the feedback provided, and have the ability to identify means to avoid/mitigate potential problems in the future.

In essence, the main concept of the simulator was based on its ability to run scenarios through a VR environment to address predefined training objectives. In this respect, learning was designed to be driven by problems encountered in this environment, supported by a report critique on learners' choices, rationale, and defence thereof. In accordance to these objectives, the GVRSS was designed and developed as an educational web-based simulation tool comprising of both non-immersive and immersive pages for providing construction managers (and other disciplines) the opportunity to experience challenges of real-life AEC projects through simulated scenarios. In order to minimise interruption on the learners' reasoning process, the Graphical User Interface (GUI) was designed to be as simple and straightforward as possible with respect to data input. Thereby, the interface was designed as to be accessible through any standard web browser to provide users with login account details and other criteria, e.g. selection of available construction sites, projects, contractors, equipment, scenarios etc. All choices made by 'players' as well as their registration data was automatically recorded in a MySQL database, which was also accessible through the immersive application for project simulation. After completing the initial decision-making process through the interactive ASP. Net Web Forms, learners are able to commence the training session, starting with a

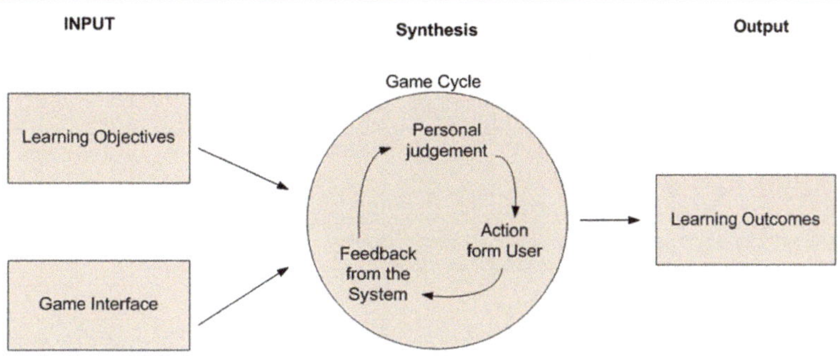

Figure 2 Educational game model input-synthesis-outcome (Garris et al. 2002).

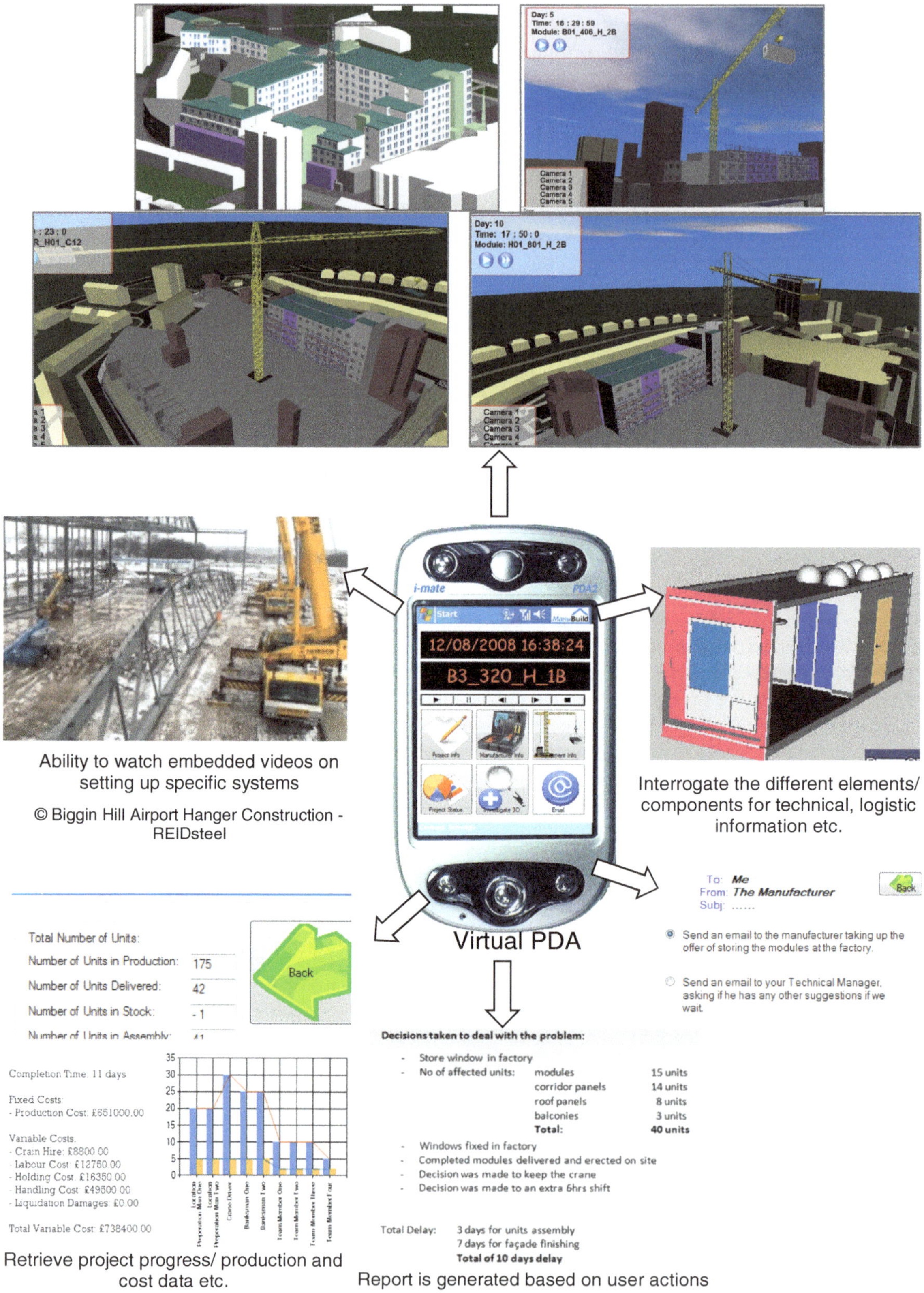

Figure 3 The VR simulation sessions.

'walkthrough' to experience and appreciate the complexity of the project. At this stage, the application provides users with a summary of the project and contract, and runs the simulation of the project within an immersive and interactive environment developed in Quest3D™ VR programming Application Programming Interface (API).

Within the simulated Quest3D environment, the users are able to experience the outcomes of all decisions made. They are also challenged by unexpected events designed according to the selected scenario, and are required to make decisions for dealing with these issues. The monitoring and communication tools are embedded in different parts of the main interface as well as the facilitated standard embedded virtual PDA or smart phone-type interface, which appears when required. The simulator ultimately records and tracks the users in the database and navigates to the conclusion page to reveal all scores of the user (together with the logic behind the marking procedure). Figure 3 illustrates a selection of the various functions available to the user of the simulator to fully interact with and retrieve information from the simulator during the VR simulation session. Further inclusion of the whole tree is considered for the exploitation phase.

Conclusions

Construction projects are increasingly becoming more complex, often engaging new business processes and technological solutions to meet ever-increasing demands. These business demands are complex and multifarious; often requiring the conjoining of high level skill sets to deliver the solutions needed. These skill sets are currently underrepresented, and seldom engage the collective ethos needed to envelop creative thinking, through such approaches as *Successful Intelligence* in order to create new innovative solutions. It is therefore paramount that the industry as a whole engages the right type (and level) of skill sets and competence needed to meet these project requirements and business imperatives. Acknowledging this, it is also important the causal drivers and influences associated with creativity and successful decision-making in global AEC teams are fully understood and supported. This however, requires a radical review in the way educational programmes and systems are designed and delivered. For example, with respect to leveraging creativity and delivering innovation, this study reflected on the Renaissance period and the creativity-oriented learning/teaching paradigm called *"Master-Apprentice Paradigm"*, as opposed to the current analysis focused *"Science Paradigm"*. The Medici Effect and the related phenomenon of intersection and seven *Da Vinci Principles* have been acknowledged as being able to revolutionise modern design and engineering education. This study then introduced the theory of *Successful Intelligence* and its three components as an underpinning platform for educating a new generation of designers and engineers.

The *"Successful Education"* paradigm (Arciszewski 2009) was presented as a new approach for educating AEC professionals was presented, including the concept of a new educational environment; the need for a new combination of courses that focus on teaching the three kinds of *Successful Intelligence* (in the context of AEC sector); including guidelines of how to properly select instructors that are capable of implementing such approach. A proof-of-concept prototype that uses a game-like virtual reality (VR) visualisation interface supported by Mind Mapping was presented as an exemplar, to demonstrate how the proposed approach could be implemented. The developed simulator offers a risk free environment where learners can evaluate how decisions they make affect their business. This includes (but is not limited to) analysing issues occurring on the construction site, such as: design concerns, process conflicts, logistics challenges, and supply chain issues etc.

This paper proffers that enhanced engagement through an immersive project environment could lead to a better understanding of the real-life AEC problems. This can be achieved by placing learners in a cyber-safe environment; specifically to leverage learners' cognitive processes to real-world issues. This study supports a novel approach of applying Game Theory to non-collocated design teams using Game-Like VR environments blended to Social Sciences Theory (social rules) and Behavioural Science Theory (decision science/communication science). This can address the need to evaluate actor involvement in order to reveal new insight into AEC organisational behaviour and the social constructs that often affect decision making. In this paper, advanced VR training and simulation tools were proffered through an exemplar in order to highlight the possibilities available, especially as this forms a conduit for aligning pivotal drivers to achieve specific learning outcomes. Future research in this area is likely to embrace the importance of pedagogy (learner styles/traits), as this has been openly acknowledged as being particularly efficacious and instrumental for delivering training material to specific learner-types.

Competing interest

All authors declare that they have no competing interests.

Authors' contributions

TA developed the theory of Successful Education. JS Goulding developed the case study of Game-Like Virtual Reality Construction-Site Simulator. FPR carried out further programming and coding for extending the Game-Like Virtual Reality Construction-Site Simulator in order to enable non-collocated AEC collaboration through this interface. All three authors worked on linking the principles of theory of Successful Education to the potentials of emerging VR interfaces. All authors read and approved the final manuscript.

Author details

[1]Centre for Sustainable Development, The Grenfell-Baines School of Architecture, Construction and Environment, University of Central Lancashire,

Preston PR1 2HE, UK. [2]Civil, Environmental and Infrastructure Engineering Department, Volgenau School of Engineering, George Mason University, Fairfax, Virginia, USA.

References

ACS. (2009). *3D Learning and Virtual Worlds. An ACS: Expertise in ActionTM White Paper.* http://www.trainingindustry.com/media/2043910/acs%203d%20worlds%20and%20virtual%20learning_whitepaper%20april%202009.pdf, Access Date, 22/02/2014.

Akintoye, A, Goulding, JS, & Zawdie, G (Eds.). (2012). *Construction Innovation and Process Improvement.* London: Wiley-Blackwell.

Alshawi, M, Goulding, JS, & Nadim, W. (2007). Training and Education for Open Building Manufacturing: Closing the Skills Gap. In AS Kazi, M Hannus, S Boudjabeur, & A Malon (Eds.), *Open Building Manufacturing: Core Concepts and Industrial Requirements.* Helsinki, Finland: ManuBuild in collaboration with VTT - Technical Research Centre of Finland.

Altshuller, G. (1984a). *Creativity as an Exact Science.* New York: Gordon and Breach, Science Publishers, Inc.

Altshuller, H. (1984b). *Creativity as an Exact Science.* New York: Gordon and Breach, Science Publishers, Inc.

Arciszewski, T. (2006). Civil Engineering Crisis. *ASCE Journal of Leadership and Management in Engineering, 6*(1), 26–30.

Arciszewski, T. (2009). *Successful Education. How to Educate Creative Engineers.* Fairfax, VA: Successful Education LLC.

Arciszewski, T. (2014). Future of engineering education. *Proceedings of the ICE - Management, Procurement and Law, 167*(1), 46–59.

Arciszewski, T, & Harrison, C. (2010a). Successful Civil Engineering Education. *ASCE Journal of Professional Issues in Engineering Education and Practice, 136*(1), 1–8.

Arciszewski, T, & Harrison, C. (2010b). *Successful Education: The Key to Engineering Creativity (Paper presented at The International Conference on Computing in Civil and Building Engineering).* Nottingham, UK: Nottingham University Press.

Arciszewski, T, & Rebolj, D. (2008). Civil Engineering Education: Coming Challenges. *International Journal of Design Science and Technology, 14*(1), 53–61.

Barrett, FJ, & Fry, RE. (2008). *Appreciative Inquiry: A Positive Approach to Building Cooperative Capacity.* Chagrin Falls, OH: Taos Institute.

Bouchlaghem, D, Shang, H, Whyte, J, & Ganah, A. (2005). Visualisation in architecture, engineering and construction (AEC). *Automation in Construction, 14*(3), 287–295.

Briggs, JC. (1996). The Promise of Virtual Reality. *The Futurist, 30,* 30–31.

Cera, CD, Reagali, WC, Braude, I, Shapirstein, Y, & Foster, C. (2002). a Collaborative 3D Environment for Authoring Design Semantics. *Graphics in Advanced Computer-Aided Design, 22*(3), 43–55.

Clarke, L, & Wall, C. (1998). UK construction skills in the context of European developments. *Construction Management and Economics, 16*(5), 553–567.

Fischer, M. (2000). *4D CAD-3D models incorporated with time schedule, CIFE Centre for Integrated Facility Engineering in Finland, VTT-TEKES, CIFE Technical Report, Helsinki.* Finland: University of Helsinki Press.

Fischer, M, & Kunz, J. (2004). *The Scope and Role of Information Technology in Construction. CIFE Technical Report* (p. 19). San Francisco: Center for Integrated Facility Engineering, Stanford University.

Friedman, TL. (2005). *The World is Flat: A Brief History of the 21st Century.* New York: Farrar, Straus and Giroux.

Fruchter, R. (1998). nternet-based Web Mediated Collaborative Design and Learning Environment, in Artificial Intelligence in Structural Engineering. In *Lecture Notes in Artificial Intelligence* (pp. 133–145). Berlin: Heidelberg: Springer-Verlag.

Fruchter, R. (2004). Degrees of Engagement in Interactive Workspaces. *International Journal of AI & Society, 19*(1), 8–21. doi:10.1007/s00146-004-0298-x.

Garris, R, Ahlers, R, & Driskell, JE. (2002). Games, Motivation, and Learning: A research and Practice Model. *Simulation Gaming, 33*(4), 441–467.

Gelb, MJ. (1998). *How to Think like Leonardo da Vinci.* New York: Random House.

Gelb, MJ. (1999). *How to Think like Leonardo da Vinci, Workbook.* New York: Random House.

Gelb, MJ. (2004). *Da Vinci Decoded: Discovering the Spiritual Secrets of Leonardo's Seven Principles.* New York: Bantam Dell.

Goulding, JS, & Rahimian, FP. (2012). Industry Preparedness: Advanced Learning Paradigms for Exploitation. In A Akintoye, JS Goulding, & G Zawdie (Eds.), *Construction Innovation and Process Improvement* (pp. 409–433). Oxford, UK: Wiley-Blackwell.

Goulding, JS, Sexton, M, Zhang, X, Kagioglou, M, Aouad, GF, & Barrett, P. (2007). Technology adoption: breaking down barriers using a virtual reality design support tool for hybrid concrete. *Construction Management and Economics, 25*(12), 1239–1250.

Hou, Y, & Ji, L. (2010). Stimulating Design Creativity by Public Places in Academic Buildings. *Journal Structure and Envirinment, 3*(2), 5–13.

Johansson, F. (2004). *The Medici Effect.* Boston, MA: Harvard Business School Press.

Juárez-Ramírez, R, Sandoval, G, Cabrera Gonzállez, C, & Inzunza-Soberanes, S. (2009). *Educational strategy based on IT and the collaboration between academy and industry for software engineering education and training* (Paper presented at the m-ICTE 2009, V International Conference on Multimedia and ICT's in Education). Lisbon, Portugal, Badajoz, Spain: FORMATEX.

Khanzade, A, Fisher, M, & Reed, D. (2007). *Challenges and benefits of implementing virtual design and construction technologies for coordination of mechanical, electrical, and plumbing systems on large healthcare project* (Paper presented at the CIB 24th W78 Conference). Maribor, Slovenia: University of Maribor Press.

Laird, D. (2003). *New Perspectives in Organisational Learning, Performance, and Change: approaches to training and development* (3rd ed.). USA: Preseus Books Group.

Leinonen, J, Kähkönen, K, Retik, AR, Flood, RA, William, I, & O'Brien, J. (2003). *New construction management practice based on the virtual reality technology. In R. R. A. Issa, I. Flood, & W. O'Brien (Eds.), 4D CAD and Visualization in Construction: Developments and Applications (pp. 75-100).* Tokyo: AA Balkema Publishers.

NGRF. (2010). *N. G. R. F.* http://www.guidance-research.org/future-trends/construction/info. Accessed 14th Aug 2010.

Petric, J, Maver, T, Conti, G, & Ucelli, G. (2002). *Virtual reality in the service of user participation in architecture* (Paper presented at the CIB W78 Conference). Aarhus Denmark: Aarhus School of Architecture.

Pour Rahimian, F, Ibrahim, R, Wirza, R, Abdullah, MTB, & Jaafar, MSBH. (2011). Mediating Cognitive Transformation with VR 3d Sketching During Conceptual Architectural Design Process. *Archnet-IJAR, International Journal of Architectural Research, 5*(1), 99–113.

Rahimian, FP, & Ibrahim, R. (2011). Impacts of VR 3D sketching on novice designers' spatial cognition in collaborative conceptual architectural design. *Design Studies, 32*(3), 255–291.

Regenbrecht, H, & Donath, D. (1997). Architectural Education and Virtual Reality Aided Design (VRAD). In D Bertol (Ed.), *Designing Digital Space - An Architect´s Guide to Virtual Reality* (pp. 155–176). New York: John Wiley & Sons.

Sage, AP. (2000). Transdisciplinarity Perspectives in Systems Engineering and Management. In MA Somerville & D Rapport (Eds.), *Transdisciplinarity: Recreating Integrated Knowledge* (pp. 158–169). Oxford: EOLSS Publishers Ltd.

Salama, AM. (2008). A Theory for Integrating Knowledge in Architectural Design Education. *Archnet-IJAR, International Journal of Architectural Research, 2*(1), 100–128.

Sampaio, AZ, & Henriques, PG. (2008). *Visual simulation of previous termcivil engineeringnext term activities: didactic virtual previous termmodels* (Paper presented at the WSCG 2008, 16th International Conference in Central Europe on Computer Graphics). Plzen, Czech Republic: Visualization and Computer Vision.

Sampaio, AZ, Ferreira, MM, Rosário, DP, & Martins, OP. (2010). 3D and VR models in Civil Engineering education: Construction, rehabilitation and maintenance. *Automation in Construction, 19*(7), 819–828.

Schueller, SM. (2012). Positive Psychology. In VS Ramachandran (Ed.), *Encyclopedia of Human Behavior* (2nd ed., pp. 140–147). San Diego: Academic Press.

Sternberg, RJ. (1985). *Beyond IQ: A Triarchic Theory of Intelligence.* Cambridge: Cambridge University Press.

Sternberg, RJ. (1996). *Successful Intelligence.* New York: Simon & Shuster.

Sternberg, RJ. (1997). A Triarchic View of Giftedness: Theory and Practice. In NN Coleangelo & GA Davis (Eds.), *Handbook of Gifted Education* (pp. 43–53). Boston, MA: Allyn and Bacon.

Thai, AM, Lowenstein, D, Ching, D, & Rejeski, D. (2009). *Game Changer: Investing in digital play to advance children's learning and health. The Joan Ganz Cooney Center.* http://www.joanganzcooneycenter.org/wp-content/uploads/2010/03/game_changer_final_1_.pdf, Access Date: 06/06/2014.

Wellings, J, & Levine, MH. (2010). *The Digital Promise: Transforming Learning with Innovative Uses of Technology. A white paper on literacy and learning in a new media age, Joan Ganz Cooney Center at Sesame Workshop.* http://dmlcentral.net/sites/dmlcentral/files/resource_files/Apple.pdf, Access Date: 06/06/2014.

Whyte, J, Bouchlaghem, N, & Thorpe, A. (1998). The promise and problems of implementing virtual reality in construction practice. In *The Life-cycle of Construction IT Innovations: Technology Transfer From Research To practice (CIB W78), Stockholm, 3–5 June, 1998.*

Yoh, M. (2001). The Reality of Virtual Reality. In *Seventh International Conference on Virtual Systems and Multimedia (VSMM'01), Organized by Center for Design Visualization*. Berkley , USA: University of California Berkley. IEEE. doi:0-7695-1402-2/01.

Youmans, R, & Arciszewski, T. (2014). *Design Fixation: Classifications and Modern Methods of Prevention. Artificial Intelligence for Engineering Design, Analysis and Manufacturing*. in print.

Zheng, X, Sun, G, & Wang, SW. (2006). *An Approach of Virtual Prototyping Modeling in Collaborative Product Design* (Paper presented at the CSCW 2005, LNCS 3865).

Zlotin, B, & Zusman, A. (2006). *Directed Evolution: Philosophy, Theory, and Practice*. Southfield, MI: Ideation International.

Zudilova-Seinstra, E, Adriaansen, T, & van Liere, R. (2009). *Trends in Interactive Visualization: State-of-the-Art Survey (Advanced Information and Knowledge Processing)*. London: Springer-Verlag.

Effectiveness of communication of spatial engineering information through 3D CAD and 3D printed models

Gabriel B Dadi[1*], Paul M Goodrum[2], Timothy RB Taylor[3] and William F Maloney[4]

Abstract

Background: Poor engineering information provided to construction crews results in inefficient communication of design, leading to construction rework, disputes, and lower worker morale. The resulting errors, omissions, and misinterpretations indicates that a significant opportunity exists to improve the traditional documentation of engineering information that craft professionals use to complete their work. Historically, physical three dimensional (3D) models built by hand provided 3D physical representations of the project to assist in sequencing, visualization, and planning of critical construction activities. This practice has greatly diminished since the adoption of three dimensional CAD (computer-aided design) and BIM (building information modeling) technologies. Recently, additive manufacturing technologies have allowed for three dimensional printing of 3D CAD models.

Methods: The effectiveness of 2D drawings, a 3D computer model, and a 3D printed model in delivering engineering information to an end-user are scientifically measured.

Results: The 3D printed model outperformed the 2D drawings and 3D computer interface in productivity and workload measures.

Conclusions: A physical model has the ability to improve communication of spatial design for certain tasks. This could lead to improved productivity, reduction of errors, and better quality for construction tasks. This paper's primary contribution to the body of knowledge is that it identifies how different mediums of engineering information impact the communication of spatial engineering information.

Keywords: Communication; Visualization; 3D printing; Labor productivity

Background

While the U.S. construction industry comprises 5.4% of the overall U.S. Gross Domestic Product (GDP) (Bureau of Economic Analysis 2013; and Huesman et al. 2013), it lags the rest of the economy in terms of productivity measurement and improvement (Goodrum et al. 2002). Due to changes in real output and differences in accounting procedures, there is no isolated industry level measure of labor productivity (Bureau of Labor Statistics 2012), although recently the U.S. Bureau of Labor Statistics has developed an aggregate multifactor productivity measure for the combined agricultural, mining, and construction industries (Bureau of Labor Statistics 2013). Regardless of this recent advancement, an isolated productivity index for the U.S. construction industry makes it difficult to track progress, benchmark, and measure the effects of governmental and industrial policies across the construction industry.

However, at a project level, productivity figures are more diligently kept, although still inconsistently company to company. With profit margins near 3%, firms must do what they can to track their performance and make necessary changes (Cooper and Lee 2009). A major contributor to the overall productivity is execution of workface practices. If construction practitioners are not equipped with the necessary tools, information, materials, and equipment to effectively perform their tasks, the productivity of the project will be negatively affected. This research presents

* Correspondence: gabe.dadi@uky.edu
[1]Department of Civil Engineering, University of Kentucky, 151C Raymond Building, Lexington, KY 40506-0281, USA
Full list of author information is available at the end of the article

insights into how different mediums, specifically two dimensional drawings, three dimensional computer models, and three dimensional physical models, of engineering information influence the performance of task execution.

Information delivery and its effect on construction productivity

Traditionally, two dimensional drawings (commonly referred to as blueprints) are the spatial and technical communication media for all project participants (Gould and Joyce 2013). Drawings are presented in a variety of formats including plan views, elevations, detailed sections, and isometrics. Individual drawings are often scaled, list dimensions, and frequently reference other sheets to help provide the user a representation of the final design intent from all viewpoints.

One of the main challenges in processing engineering graphical information is that the design may be well intended, but for a variety of reasons, the message received differs from the original intent. This process of creating a message, disseminating, and then processing essentially describes the well published theory of the linear standard communication process. Many models have been created to describe the process with varying stages, however, the essential elements are outlined in Figure 1 (Shannon and Weaver 1949; Schramm 1954; and Berlo 1960).

In this standard, linear model of communication, the sender must encode their interpretation of the desired end user information. This interpretation is then translated in the message medium, whether it is verbal or non-verbal. Receivers must decode the message into their individual interpretation, where the final message is processed and understood. However, the intermediary steps of encoding the message, the creation of the message, and decoding the message increases the potential for noise that can affect the ultimate outcome of the communication. The message channel is the actual flow of the message, verbal or nonverbal, from the sender to the receiver. In the linear model, there is very little, if any, feedback from the receiver to the sender, where ultimately only a downstream process occurs (Shannon and Weaver 1949; Schramm 1954; and Berlo 1960). In the communication of construction documents, poor design and poor design communication are two distinct opportunities for errors (Eckert and Boujut 2003). Poor design is the result of errors made by designers in the encoding process, while poor design communication are errors made in the interpretation or decoding of the message.

Errors made through poor design have been studied and identified as having an impact on project performance. Several studies discuss errors in the design outputs concluding that many practitioners feel information delivery is a deterrent to efficiently performing their job (Goodrum and Dai 2006; Dai et al. 2009a; Dai et al. 2009b; Mourgues and Fischer 2008; and Rojas 2008). The main inefficiencies from drawing management according to a survey of construction craft workers are due to errors in the drawing, availability of the drawings, slow management response to questions, legibility, and omission of necessary information on the documents (Goodrum and Dai 2006; Dai et al. 2009a; and Dai et al. 2009b). In a focus group study, Dai et al. (2005) found that craft professionals believe 32% of their negative productivity is due to inaccuracy and poor quality of information, with individuals claiming that "some drawings could not be physically built". Rojas (2008) discusses inefficiencies from design drawings ultimately leading to increased rework on the project. Borcherding and Samelson (1980) found that rework was one of the three most significant drivers to poor productivity and decreased morale, oftentimes as a result of poor engineering information design. Haas et al. (2011) found that design, engineering, instruction, and monitoring accounted for 29.08% of the total amount of rework on an analysis of records from the industrial sector. Supervisors and foremen then become focused on correcting engineering errors and rework instead of planning future work and focusing on crew performance. In the highway construction sector, an analysis

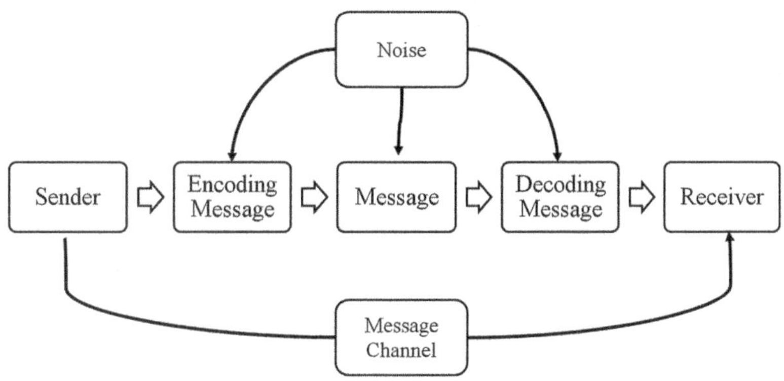

Figure 1 Standard model of communication.

of change orders showed that omissions of information led to a 4.53% increase in original contract amount among a sample of roadway projects (Taylor et al. 2012). With 40% of the total construction cost being in direct and indirect craft labor, there is a need to maximize efficiency and reduce non-value adding activities of the workers (Haas et al. 2011).

Similarly, poor design communication can lead to errors and confusion in task execution. Some of the issues result from the difference in the message intended versus the message received. The format and intent of drawings is easier to comprehend by the designer that creates the drawing than it is for the contractor and his/her workforce that has to interpret the message (Emmitt and Gorse 2003). This problem is magnified when the contractor must reference several different drawings to understand the design intent for a particular building element. Further, different symbols and terminology used by various designers can also lead to confusion and complications (Emmitt and Gorse 2003). Mourgues and Fischer (2008) argue that communication of project information to the workface is ineffective and can negatively impact quality, safety, and productivity. Studying the understanding and communication of design throughout all project participants can greatly improve project outcomes by increasing project morale, improving productivity, and reducing rework.

Physical models and additive manufacturing

Physical scale models have had several uses in the construction industry, for tasks such as sequencing of concrete and steel or for planning rigging, reinforcing steel, or excavation operations. Oglesby et al. (1989) surveyed managers who concluded that physical models were a useful tool for planning and communications, and that modeling pays for itself easily. However, no direct benefit was quantified. Physical models, as a supportive piece of information delivery, can be a useful information tool. A study estimates that labor costs can be reduced by 25% minimizing productivity losses on indirect work through a detailed execution planning strategy (O'Brien et al. 2011). The use of physical models in construction was prevalent historically, but their current use is diminished (Emmitt and Gorse 2003). The development of 3D CAD technologies is cited as one of the main causes behind the fading use of physical scale models (Emmitt and Gorse 2003).

3D printing technologies have developed and advanced to the point where 3D CAD models can be effectively printed (Breen et al. 2003; and Simondetti 2002). These modeling techniques are defined to be "the process of joining materials to make objects from 3D model data, usually layer upon layer" (Buswell et al. 2007; and Lim et al. 2012). Having the ability to create physical models through a printing process presents an opportunity to support traditional information delivery with an accurate, scaled representation of a three dimensional space. There are many models of 3D printers with similar technologies. The printers work by essentially building up an object with individual thin layers of a material, such as plastics or starch. ABS (acrylonitrile butadiene styrene) plastics or PLA (polylactic acid) plastics are two of the more common plastic outputs. A 3D printed model was selected to represent a physical model to test its applicability as a format for information display. Ultimately, the means to achieve the end result of a physical model are not relevant to the outcomes of the study, as the primary objective was to test a physical model as opposed to blueprints or a CAD model.

Mental workload

A critical cognitive component to the design of information delivery is in the mental workload requirements. Assuming that everyone has a fixed maximum cognitive capacity, mental workload is the amount of mental resource required compared to the total resources available to that person (Carswell et al. 2005). An effective method of information delivery should reduce the mental workload requirements while allowing the necessary information to be conveyed. Typically, this involves reducing the amount of time the user must retain the information in their working memory and reduce the irrelevant, distracting mental operations that may occur. Ultimately, a desirable level of mental workload leads to improvements in design communication as information is able to be retained.

Research in cognitive design has identified three classes of mental workload metrics used to evaluate the outcome of the study. The classes are physiologic, secondary task, and subjective measures (Carswell et al. 2005). Physiologic measures use indirect measures of mental workload by studying ocular and cardiac responses. These physiologic responses have a relationship with cognitive activity in the brain. Increased cognitive activity has been found to result in small increases in pupil dilation, slowing blinking patterns, more consistent heart rates, and heightened heart rates (Carswell et al. 2005). Secondary task measures look to identify the remainder of the mental workload, which is not occupied by performing the desired task. These secondary tasks are developed for certain applications such as aviation and high-demand environments. For this study, objective and subjective measures are used as they are readily available, universally accepted, nonintrusive, and the simplest to administer.

One of the most widely used standardized subjective measures of mental workload is the National Aeronautics and Space Administration Task Load Index (NASA-TLX) (Hart and Staveland 1988). This tool has been used to measure mental workload in over 1,200 studies since its inception (Hart 2006). Although its use is widespread across fields (e.g. aviation, transportation, and other display oriented fields), its use within construction research has been limited. Mitropoulos and Memarian (2013) investigated task demands in masonry work using the NASA-TLX to identify factors affecting activity performance and propose strategies to improve performance.

However, there are drawbacks to current subjective measures. The subjects must self-evaluate their performance and their cognitive capacity. Response bias could factor into the results if the subjects are stakeholders in the study (Carswell et al. 2005). For instance, if conducting this study with a veteran journeyman electrician, he or she may be inclined to prefer the traditional 2D drawing set because of their familiarity with the medium.

Methods

This research focuses on spatial information format as an initial mechanism for improving understanding and communication of design and evaluates three different formats using several measures. The ability to evaluate cognitive abilities of practitioners in using various information delivery formats requires defined performance metrics. In a discussion of construction communication deliverables, Emmitt and Gorse (2003) suggest that information formats must yield quick, simple, and easily interpretable results. Using those guidelines along with the cognitive principles and measures previously discussed, a series of evaluations have been created for assessment.

Cognitive task

The main portion of the experiment is a building task using scale model elements to recreate a structure based on given information. The basis of design for the experiment's model must be simple enough to solely capture the cognitive aspects of spatial information processing, yet complex enough to where there is difficulty and mistakes can be made. If the model is too complex, the design would bias the results away from understanding the performance of medium. If the model is too simple, there will not be a differentiation in the measures. The structure design was created through a standard set of two dimensional construction drawings (blueprints), a three dimensional computer aided design (CAD) model, and a physical model created by a three dimensional printer from the CAD model. These forms of media were selected as they are typical display formats and are accepted as the forms used to communication spatial design (Emmitt and Gorse 2003). Subjects were presented with information in one of the three formats, from a randomized sequence, and then instructed to build the structure from simple building elements. The same structure was used for each experiment but the participants were not told that the three structures were identical. To control and identify a potential learning curve, the sequence that the mediums were presented was rotated. The subjects were free to manipulate the models as needed, and the order in which the subjects received the models was randomized and were available to consult throughout the experiment. Understanding that the subjects worked with the same model, the authors wished to identify if there was a learning curve present. Due to the scope of this paper and length restrictions, this phenomena was studied and reported in another publication (Dadi et al. 2014), where the authors examined the cognitive outcomes of the experiment in significant depth. The objective of the previous manuscript was to study the cognitive outcomes of various information formats, where this current manuscript identifies the performance of the formats in many key areas such as speed to completion, productivity, and errors made. Samples of the 2D drawings are illustrated in Figures 2 and 3. The 3D CAD model is pictured in Figures 4 and 5. Finally, the physical model can be seen in Figure 6. Figures 7 and 8 show the building elements, the Tekton Tower Girder and Panel Building Set by Bridge Street Toys, before and after model creation respectively. The amount of building elements in the set outnumbered the pieces necessary to complete the model.

Sample group

Practitioners were recruited to provide a sample of individuals that regularly use design and construction drawings for the purposes of field installation of the final design intent. The subjects from this sample group were recruited from engineering and construction firms through an e-mail invitation with a range of experience from one year to over thirty years. The scope of the study was presented to management and executive level employees at several companies throughout the state of Kentucky. In this notification of study, the employees were asked to make engineers and craft workers available to complete the study, and the companies provided 26 individuals to complete the experiment. Job titles included carpenter, laborer, electrical, and mechanical foreman, as well as project engineer and design engineer. The study attracted the participation of 26 subjects and are outlined in Table 1.

Five-minute rating analyses have been performed on construction field projects to measure the effectiveness of a construction crew and identify sources of delay, travel, and other non-effective work activities (Oglesby et al. 1989). For this experiment, a five-minute rating yielded the percent of the task time that was spent on direct or effective work and on non-effective work or rework. The percentage can be applied to the overall time to completion to give the amount of time spent on each activity category. The data yields effective work percentages of each

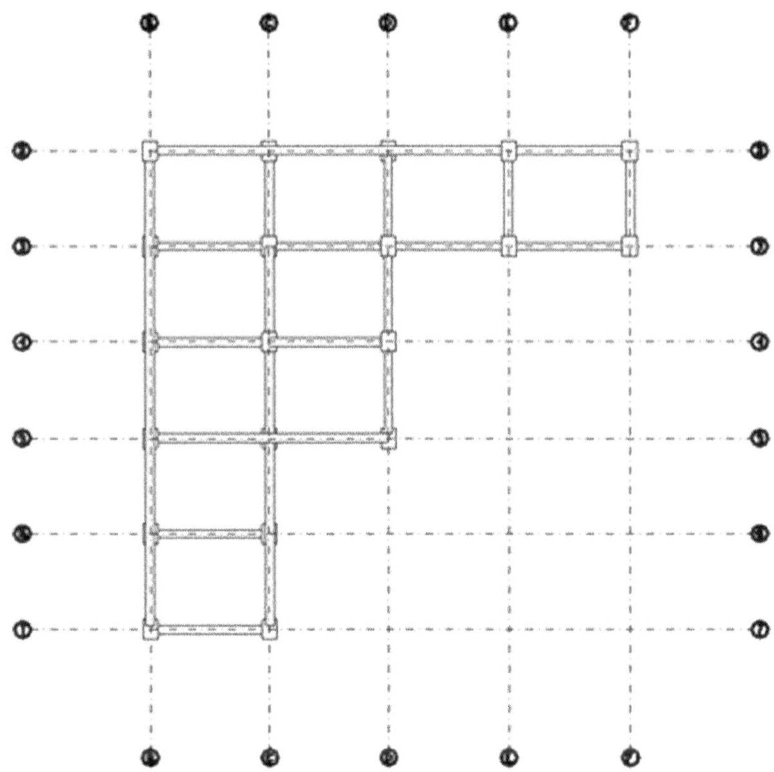

Figure 2 2D plan view of experiment model.

information delivery format. To conduct a five-minute rating, a time sheet was prepared and divided into subsets of time and then columns for notation of the activity classification. The classification categories are direct work, indirect work, rework, and delay due to rework. Direct work is defined as any physical building of the model towards the final product. Indirect work is defined as any activities performed towards the end result that is not physically building the model. This includes time getting familiar with the building elements, and manipulating and processing the information delivery format. Rework includes any disassembling or reassembling of a previously built portion of the model. Finally, delay due to rework includes time spent reprocessing the information delivery medium after rework occurs.

To measure the cognitive performance of the subjects, the National Aeronautics and Space Administration Raw Task Load Index (NASA-rTLX) was utilized to measure mental workload. Mental workload is a measure of the amount of mental resources required to complete a task compared to the total amount of mental resources available to that individual (Carswell et al. 2005). Often, the desired workload imposed by a task has a balance. Too much mental workload and the user may not have the

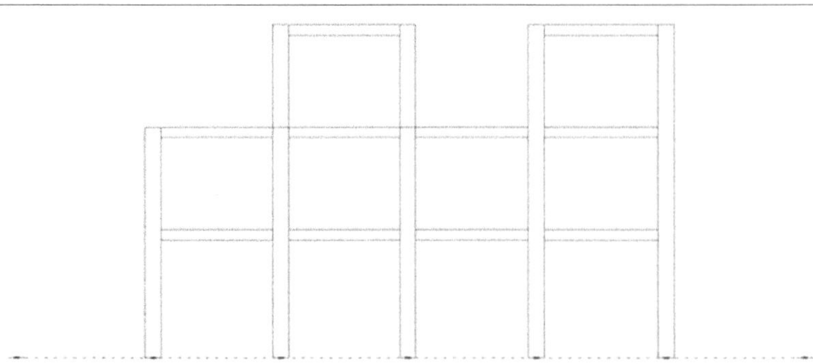

Figure 3 2D elevation view of experiment model.

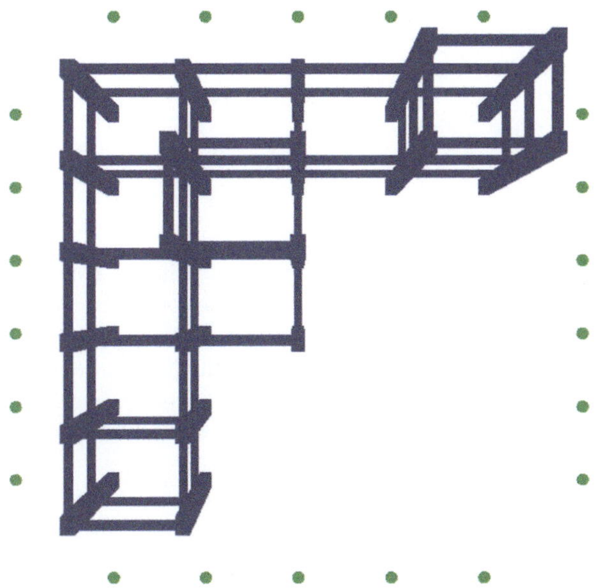

Figure 4 3D computer model top view.

capacity to maintain a proper level of performance. Too little mental workload and the user may not have the focus and diligence required to complete the task appropriately (Hart and Staveland 1988; and Mitropoulos and Memarian 2013).

The NASA-TLX workload measure is a multidimensional tool that rates responses in accordance to six factors: 1) mental demand, 2) physical demand, 3) temporal demand, 4) effort, 5) performance, and 6) frustration (Table 2). The traditional NASA-TLX obtains pairwise weights for each of the six factors prior to identifying a scaled response for the factors. These sub-factors then combine based on the weight from the pairwise scores to form a composite workload score. A derivative of the NASA-TLX is the NASA Raw Task Load Index (NASA-rTLX) that does not weight the subscales by their pairwise comparisons, resulting in a composite workload score from an average of the subscales. Several studies

Figure 5 3D computer model isometric view.

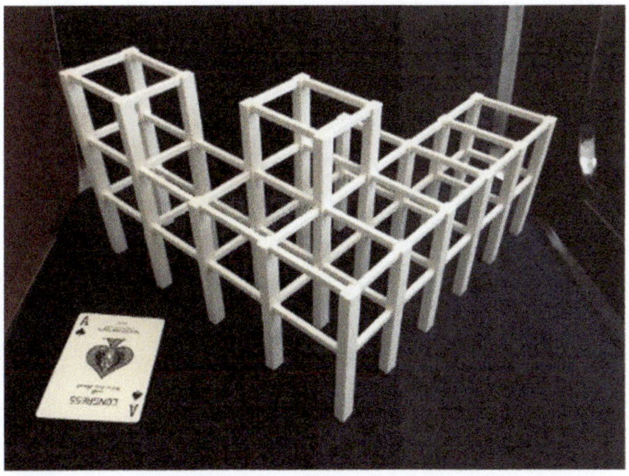

Figure 6 3D printed model isometric view (shown next to standard size playing card for scale).

have found a strong correlation between the NASA-TLX and NASA-rTLX, which lends towards the adoption of the simpler NASA-rTLX tool (Moroney et al. 1995; Moroney et al. 1992; and Byers et al. 1989). This study utilized the NASA-rTLX tool to measure workload for the participants. Table 2 provides a description of the NASA-rTLX factors and the measurement scale.

Statistical analysis procedure

A key statistical measure used in the study's experiments is the analysis of variance (ANOVA) procedure. ANOVA models seek to test whether there is a difference between means of several populations (Dielman 2005; Fellows and Liu 2008). The often performed procedure estimates statistically significant differences between the means through an F value, while also measuring the amount of variation in the dependent variable that is explained by

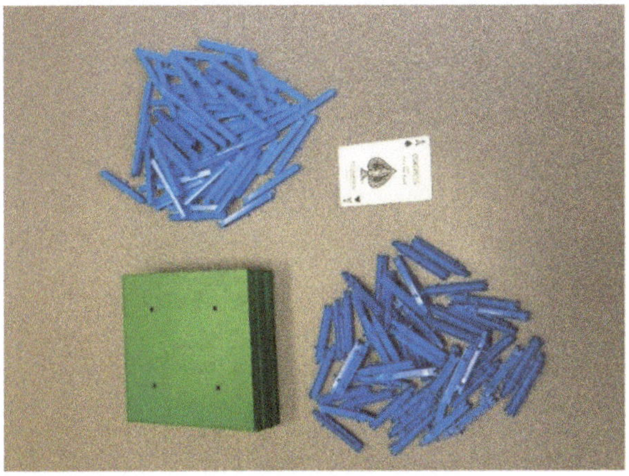

Figure 7 Scale model building elements disassembled (shown next to standard size playing card for scale).

Figure 8 Scale model building elements appropriately assembled (shown next to standard size playing card for scale).

the independent variables (η^2) (Dielman 2005; and Fellows and Liu 2008).

For this study, the three populations tested are individuals completing the experiment using the two dimensional drawing set, individuals completing the experiment using the three dimensional computer model, and individuals completing the experiment using the three dimensional printed, physical model.

To quantify these statistically significant differences, there are several post hoc tests available to compare multiple means. The original post hoc test was Fisher's Least Significant Difference (LSD) test. This test compared multiple means through a series of *t*-tests. However, no adjustment was made to the error rate for the comparisons. In the assumptions of a *t*-test, the sampling distribution was intended for only one test. When multiple comparisons are made, the true alpha value for significance is lower than 0.05, which is the value assumed in the LSD test (Dielman 2005). Another, more reasonable post hoc test is the Bonferroni method.

Table 1 Sample demographics

Demographics	Practitioners
Number	26
Age range (Mean/Median)	20-62 (40.7/40.5)
Number of engineering course work hours in post-secondary education	9-190
Years of experience (Mean/Median)	1-33 (16.5/15)
Position titles (Number,%)	Carpenter foreman (11, 42%)
	Laborer foreman (5, 19%)
	Electrical foreman (2, 8%)
	Mechanical foreman (2, 8%)
	Project fngineer (4, 16%)
	Design fngineer (2, 8%)

Bonferroni uses *t* tests to perform pairwise comparisons but sets the critical alpha value as the experiment-wise error rate divided by the total number of tests. This corrects for the effect that multiple tests has on the tested *t* value (Dielman 2005).

Results and discussion

For the model with all subjects, there were 26 participating subjects, resulting in 77 complete building experiments. One individual could not complete a third model, which prevented the sample size from reaching 78. Each subject completed the task with all three model types in a randomized order. Table 3 reports the average results across all participants for the four key obtained variables.

The physical model had preferred outcomes in time to complete the experiment, composite mental workload, and direct work rate. The 2D drawing set resulted in the lowest rework rate percentage. Of significance, individuals spent more time interpreting the information from the 3D computer interface, whereas the physical model required little processing time. As with all information processing, the work task will demand the optimal display format.

Preferences for engineering and construction tasks

In a post-test questionnaire, the subjects were presented with real work scenarios that require the use of spatial information and asked which format should be used to complete the task. The scenarios were selected based on previous research on the advantages of display format (Boyer and Wickens 1994; Cockburn and McKenzie 2002; Hickox and Wickens 1999; and St John et al. 2001). Table 4 summarizes the desired information traits. Relative positioning presents better in two dimensions, as the specific planar dimensions can be focused on, and the third, and unnecessary dimension, is eliminated. When projective ambiguity is a concern, a two dimensional format is superior. Projective ambiguity exists when three dimensions are recreated in a two dimensional format, resulting in a distorted third dimension. 3D displays better represent shape understanding as a full 360° viewing angle can be achieved. In a similar fashion, a 3D display allows the user to focus on a plane while still having quick reference to a third dimension. When understanding a layout or terrain, a profile view can be accessed while also having the depth (or width, depending on the chosen profile) dimension readily available. Finally, a 3D display allows for depth cues to be referenced. This means that a 2D sheet can be studied while also having the third (depth) dimension represented to give a point of reference for depth and location.

Table 2 NASA-rTLX factors and descriptions (Hart and Staveland, 1988)

NASA-rTLX factors	Rating scale	Description
Mental demand	1-100 (Low-High)	How much mental and perceptual activity was required (e.g., thinking, deciding, calculating, remembering, looking, searching, etc.)? Was the mission easy or demanding, simple or complex, exacting or forgiving?
Physical demand	1-100 (Low-High)	How much physical activity was required (e.g., pushing, pulling, turning, controlling, activating, etc.)? Was the mission easy or demanding, slow or brisk, slack or strenuous, restful or laborious?
Temporal demand	1-100 (Low-High)	How much time pressure did you feel due to the rate or pace at which the mission occurred? Was the pace slow and leisurely or rapid and frantic?
Performance	1-100 (Good-Poor)	How successful do you think you were in accomplishing the goals of the mission? How satisfied were you with your performance in accomplishing these goals?
Effort	1-100 (Low-High)	How hard did you have to work (mentally and physically) to accomplish your level of performance?
Frustration	1-100 (Low-High)	How discouraged, stressed, irritated, and annoyed versus gratified, relaxed, content, and complacent did you feel during your mission?

There were four tasks presented in the post-test questionnaire to identify preferences of the practitioners. The tasks were:

- You are a structural steel subcontractor and need to plan and present an erection sequence, which information delivery format (s) would you use to complete the task (2D, 3D Interface, Physical Model)?
- If you are calculating the necessary cubic yards of concrete for an upcoming slab pour, which information delivery format (s) would you use to complete the task (2D, 3D Interface, Physical Model)?
- If you are a mechanical, electrical, or plumbing (MEP) engineer and need to design piping runs with sufficient access space, which information delivery format (s) would you use to complete the task (2D, 3D Interface, Physical Model)?
- If you are estimating the quantity of earthwork that will have to be cut and/or filled on a project, which information delivery format (s) would you use to complete the task (2D, 3D Interface, Physical Model)?

In a construction setting, a structural steel erection plan requires an understanding of relative positioning, as it involves coordination of the construction of steel shapes in two directions or dimensions. Therefore an ideal information format choice would be the 2D drawings. Calculating the required yardage of concrete for a future placement event requires an understanding of the shape and the ability to measure distances. Shape understanding presents well in three dimensions, which would point towards the 3D computer model or the physical model. Being that distances are represented and automatically calculated in the computer software, the 3D computer model provides the best representation. MEP runs are typically associated with having sufficient access and coordination between the trades to fit the pipes in the allowable space provided. This requires depth cues and shape understanding without projective ambiguity. The depth cues and shape understand lends itself towards a 3D model, while projective ambiguity concerns

Table 3 ANOVA results: model type by dependent variables, all subjects

Dependent variables		Mean	Overall mean	N	df	F	p	η^2
Time to completion (secs)	2D	10.02	10.70	77	74	1.73	0.19	0.045
	3D	11.75						
	Physical	10.33						
NASA composite workload (1-100 scale)	2D	33.72	34.26	77	74	0.47	0.63	0.012
	3D	36.63						
	Physical	32.34						
Direct work rate (%)	2D	75.55	76.53	77	74	16.77*	<0.00	0.312
	3D	66.69						
	Physical	87.78						
Rework rate (%)	2D	3.41	4.38	77	74	0.68	0.51	0.018
	3D	5.82						
	Physical	3.90						

*Significant above 95%.

Table 4 2D versus 3D display comparisons

Tasks where 2D displays are advantageous	Tasks where 3D displays are advantageous
Relative positioning	Shape understanding
Projective ambiguity concern	Layout understanding
	Depth cues

lead the user towards a 2D representation. However, a physical model provides the necessary depth cues and shape understanding in a proper and efficient 3D representation. Finally, estimating the quantity of earthwork for cut and fill requires project information and layout understanding of the terrain. Similar to the concrete placement scenario, a 3D computer has the necessary display, information, and calculating tools to complete the task.

Having reviewed the scenarios and proper information format displays, Figures 9, 10, 11, and 12 display the preferences for practitioners to complete the steel erection plan, concrete placement, MEP coordination, and earthwork quantity calculation tasks respectively.

For the steel erection sequence plan, subjects preferred the 3D computer model 58% of the time, 2D drawings 23%, and a physical model 19%. Literature suggests the 2D drawings would be preferred as it gives a proper viewing of relative positioning of the steel members. A 3D computer model would distort distances due to projective ambiguity and does not provide addition information that would be desirable. In addition, the practitioners did not perform a simple steel erection sequence during the task completion. It would be a reasonable assumption that a more complex project with more moving parts would prove even more difficult.

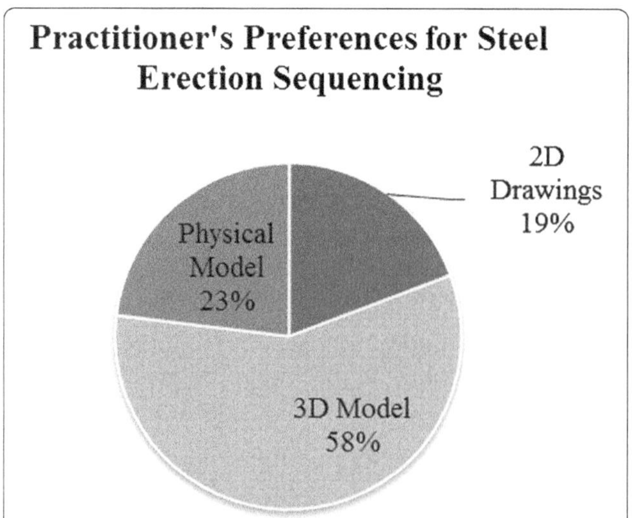

Figure 9 Practitioner's preferences for planning steel erection sequence.

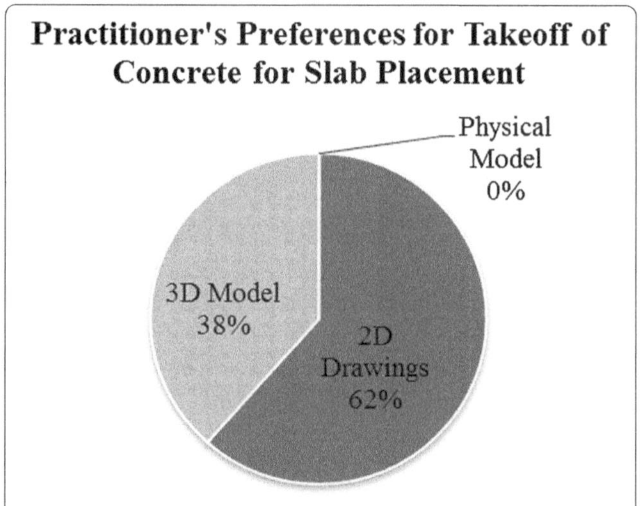

Figure 10 Practitioner's preferences for quantity takeoff of concrete for slab placement.

When calculating concrete quantities for a slab placement, 62% of practitioners preferred using 2D drawings compared 38% preferring a 3D computer model and 0% for a physical model. This task requires shape understanding and understanding of necessary dimensional properties, which makes a 3D computer model a superior choice. Given this information, subjects likely prefer the 2D drawings due to their limited experiences with CAD technologies. In the current CAD software packages, a concrete slab element can be clicked on and exact quantities will immediately be presented. Without this knowledge and experience, practitioners revert to their familiarity with quantity takeoffs from two dimensional drawings.

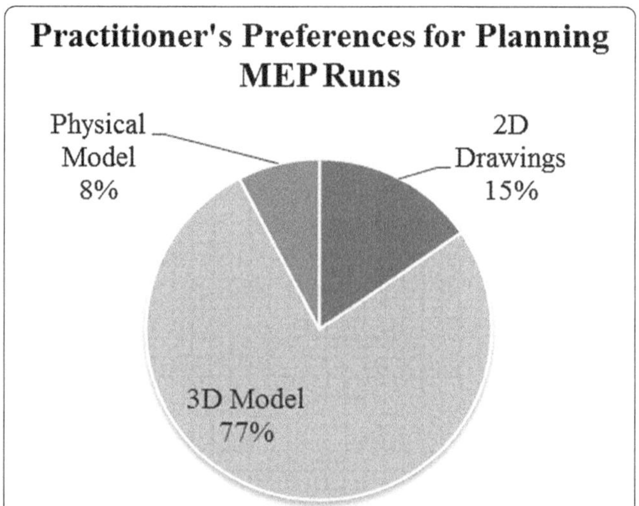

Figure 11 Practitioner's preferences for planning MEP piping runs.

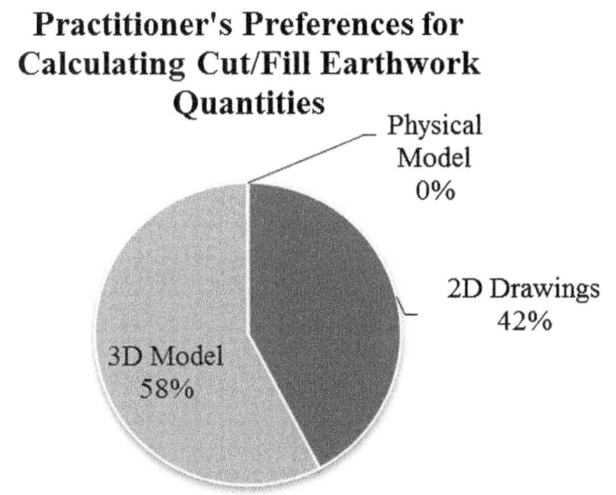

Figure 12 Practitioner's preferences for calculating cut and fill earthwork quantities.

With the need for depth cues, shape understanding, and avoidance of projective ambiguity, coordinating the locations of mechanical, electrical, and plumbing pipes is a demanding task. On one hand, depth cues and shape understanding require a 3D display, while a standard 3D display presents issues of projective ambiguity. The issue is averted in a physical model where subjects benefit from depth cues and shape understanding of a 3D display and avoiding projective ambiguity from a true three dimensional, haptic output. However, 77% of practitioners preferred a 3D computer model, while 15% and 8% chose 2D drawings and a physical model respectively.

Calculating cut and fill earthwork quantities requires a knowledge of the terrain and layout, and ideally, the ability to quickly calculate volumes. 3D CAD software packages are readily equipped with this capability and provide a 3D display that is optimal to complete the task. 58% of practitioners appropriately identify the 3D computer model as the information format of choice for this operation, while 42% would use the standard 2D drawing set and 0% would reference a physical model.

When these responses are aggregated (see Figure 13), 58% of practitioners would use a 3D model for the construction tasks. 34% and 8% would use 2D drawings and a physical model respectively. These numbers are interesting, as the results of the experimental task showed that practitioners perform better with a physical model. Practitioners had difficulty manipulating the computer model to a proper and efficient orientation. In fact, several practitioners could not turn the computer model towards a desired display and ended up turning their work platform to match the orientation on the screen. Furthermore, the participants indicated a significantly higher NASA Composite Workload Demand with the 3D computer model than either of the 2D or physical model (Table 3). With this much difficulty with a simple structural model, a more complex and layered computer model, as are the ones currently populating the industry, would prove to be too burdensome and laborious for efficient field interpretation. Despite the participants' difficulty in manipulating the 3D model, the participants still indicated a significantly stronger preference for using 3D digital information to perform these common construction tasks. Moving forward, these suggests provide compelling data that improving the delivery of 3D digital information to construction crews is both desired and would be beneficial in improving craft worker performance.

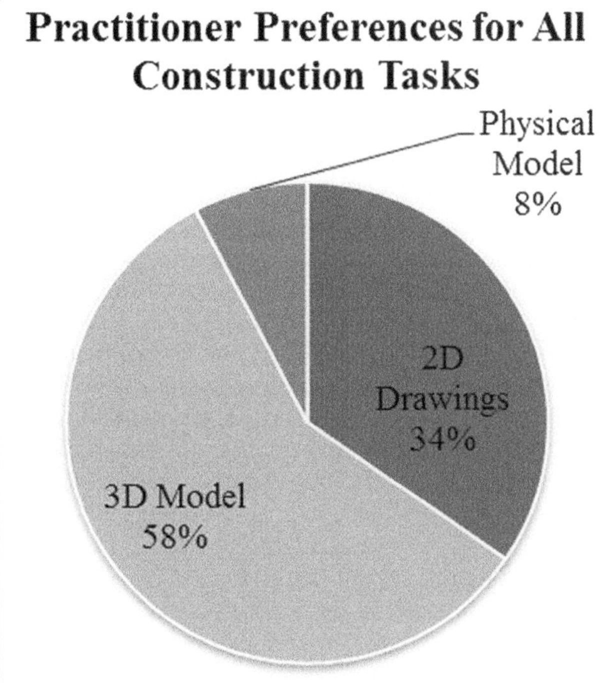

Figure 13 Practitioners' preferences for all engineering and construction tasks.

Conclusion

The current work tested subjects using various formats of information delivery (2D, 3D interface, and a physical model) to complete a building task and evaluated performance through cognitive and productivity measures. This paper's primary contribution to the body of knowledge is identification of how different mediums of engineering information influences the performance of construction task execution. By conducting a cognitive task experiment, the performance and preferences of subjects with a 3D computer interface and a physical, haptic model was evaluated.

The cognitive workload mean scores were lower with the physical model than the two dimensional drawings and three dimensional computer model average composite workload. These findings suggest that 3D printed models present a promising alternative as at lease a supplemental form of spatial information to the traditional forms of information delivery. There are forms of engineering information, such as dimensions and material specifications, that are impractical to be imposed on a 3D printed model with current printing capabilities. While a 3D computer model may be a more convenient method to representing 3D design due to its current prevalence in industry, it does not appear to be an effective method for spatial designs. This can negate the benefits of the third dimension, since space is the critical addition to the information presentation.

For construction practitioners, the current work implies that physical models present a potential alternative for spatial representation than a 3D computer model and even 2D drawings for spatial tasks. The physical models have higher direct work rates and lower composite workload scores than a 3D computer model and 2D drawings. The task dependent nature of information displays, along with the performance of subjects in the experiment suggest specific situations control the advantages and disadvantages of various mediums. For example, 2D drawings can be useful in tasks involving layouts and relative positioning of objects. A 3D computer model can be useful for dimensioning 3D elements, repetitive calculations, and shape properties due to the intelligence contained within the model. Physical models provide an advantage for tasks that require visualization of spatial elements, coordination of space, and a scaled understanding of depth such as critical lift planning, structure steel sequencing, and planning staging areas for material. Limitations of the current work include a limited sample size and the limited applicability of the model building task to an actual construction task. With a larger sample size and replication of actual construction tasks, the implications for construction practitioners become more impactful. Future work in this area includes further analysis of mental workload in construction professionals to begin the process of understanding the cognitive work demands associated with construction tasks. The outcomes of such studies can identify components of tasks that are mentally demanding and inhibit understanding and execution of the task efficiently. Furthermore, there is a need to examine the use of 3D printed models for more complex designs that can be physically assembled and dissembled by craft workers and other end users and the influence this would have on construction performance. This in essence would replicate the function that traditional modeler built physical models provided on jobsites decades ago. In addition, the current cost to 3D print a project model is substantial, and the print sizes can be relatively small. A cost-benefit analysis on the use of a 3D printed model as a means of information presentation is needed.

Competing interests
The authors declare that they have no competing interests.

Authors' contributions
GBD conducted the literature review, administered the experiments, collected the data, and wrote the initial draft of the manuscript. PMG revised the manuscript for intellectual content, was instrumental in the concept of the experiment, and assisted in the design of the experiment. TRBT revised the manuscript for intellectual content, assisting in drafting the scope of the manuscript, and assisted in the design of the experiment. WFM revised the manuscript for intellectual content and assisted in the design of the experiment. All authors read and approved the final manuscript.

Author details
[1]Department of Civil Engineering, University of Kentucky, 151C Raymond Building, Lexington, KY 40506-0281, USA. [2]Department of Civil, Environmental, and Architectural Engineering, University of Colorado Boulder, 428UCB, 1111 Engineering Drive, Boulder, CO, USA. [3]Department of Civil Engineering, University of Kentucky, 151A Raymond Building, Lexington, KY, USA. [4]Department of Civil Engineering, University of Kentucky, 151B Raymond Building, Lexington, KY, USA.

References

Berlo, DK. (1960). *The Process of Communication: An Introduction to Theory and Practice*. New York: Holt, Rinehart and Winston: The University of Michigan.

Borcherding, S, & Samelson. (1980). Improving motivation and productivity on large projects. *Journal of the Construction Division, 106*, No. CO1, 77.

Boyer, BS, & Wickens, CD. (1994). *3D weather displays for aircraft cockpits* (Technical Report ARL-94-11/NASA-94-4). Savoy, IL: Aviation Research Laboratory.

Breen, J, Nottrot, R, & Stellingwerff, M. (2003). Tangible virtuality – perceptions of computer-aided and physical modelling. *Journal of Automation in Construction, Elsevier, 12*, 649–653.

Bureau of Economic Analysis. (2013). *National Income and Product Accounts Tables*: U.S. Department of Commerce. http://www.bea.gov/iTable/iTable.cfm?ReqID=9&step=1#reqid=9&step=3&isuri=1&903=317 (July 16th, 2013).

Bureau of Labor Statistics. (2012). *Overview of BLS Productivity Statistics*: U.S. Department of Labor. http://www.bls.gov/bls/productivity.htm (April 21st, 2013).

Bureau of Labor Statistics. (2013). *News Release: Multifactor Productivity Trends for Detailed Industries, 2011* (USDL-13-1941): U.S. Department of Labor. http://www.bls.gov/news.release/pdf/prin3.pdf.

Buswell, RA, Soar, RC, Gibb, AGF, & Thorpe, A. (2007). Freeform construction: mega-scale rapid manufacturing for construction. *Journal of Automation in Construction, Elsevier, 16*, 224–231.

Byers, JC, Bittner, AC, & Hill, SG. (1989). "Traditional and raw task load index (TLX) correlations: Are paired comparisons necessary?". In A Mital (Ed.), *Advances in Industrial Ergonomics and Safety* (pp. 481–485). London: Taylor and Francis.

Carswell, CM, Clarke, D, & Seales, WB. (2005). Assessing mental workload during laparoscopic surgery. *Surgical Innovation, 12*(1), 80–90.

Cockburn, A, & McKenzie, B. (2002). Evaluating the Effectiveness of Spatial Memory in 2D and 3D Physical and Virtual Environments. In *CHI 2002*. Minneapolis, MN: ACM, New York.

Cooper, K, & Lee, G. (2009). "Managing the Dynamics of Projects and Changes at Fluor." Conference of the System Dynamics Society. In *Proceedings from the 27th International Conference of the System Dynamics Society* (pp. 1–27). Albuquerque, New Mexico.

Dadi, GB, Goodrum, PM, Taylor, TRB, & Carswell, CM. (2014). Cognitive workload demands using 2D and 3D spatial engineering information formats. *Journal of Construction Engineering and Management, 140*(5), 04014001.

Dai, J, Goodrum, PM, Maloney, WF, & Sayers, C. (2005). Analysis of Focus Group Data Regarding Construction Craft Workers' Perspective of the Factors Affecting Their Productivity. In *Proceedings from the 2005 Construction Research Congress* (pp. 1–10): Published by the American Society of Civil Engineers.

Dai, J, Goodrum, PM, & Maloney, WF. (2009a). Construction craft workers' perceptions of the factors affecting their productivity. *Journal of Construction Engineering and Management, 135*(3), 217–226.

Dai, J, Goodrum, PM, Maloney, WF, & Srinivasan, C. (2009b). Latent structures of the factors affecting construction labor productivity. *Journal of Construction Engineering and Management, 135*(5), 397–406.

Dielman, TE. (2005). *Applied Regression Analysis: a Second Course in Business and Economic Statistics 4th ed.* Mason, OH: South-Western Cengage Learning.

Eckert, C, & Boujut, JF. (2003). The role of objects in design co-operation: communication through physical or virtual objects. *Computer Supported Cooperative Work, 12*, 145–151.

Emmitt, S, & Gorse, C. (2003). *Construction Communication*. Oxford: Blackwell Publishing.

Fellows, R, & Liu, A. (2008). *Research Methods for Construction* (3rd ed.). Malden, MA: Wiley-Blackwell Publishing.

Goodrum, PM, & Dai (CII), J. (2006). *Work Force View of Construction Productivity* (Research Summary 215-1). University of Texas at Austin: The Construction Industry Institute.

Goodrum, PM, Haas, CT, & Glover, R. (2002). The divergence in aggregate and activity estimates of U.S. construction productivity. *Journal of Construction Management and Economics E & F N Spon, 20*(5), 415–423.

Gould, F, & Joyce, N. (2013). *Construction Project Management* (4th ed.). New Jersey: Prentice Hall.

Haas, CT, Goodrum, PM, Caldas, CH, Granger, R, & Zhang, D. (2011). *A Guide to Construction Rework Reduction* (Implementation Resource 252-2b). University of Texas at Austin: The Construction Industry Institute.

Hart, SG. (2006). "NASA-task load index (NASA-TLX); 20 years later". In *Proceedings of the Human Factors and Ergonomics 50th Annual Meeting* (pp. 904–908). Santa Monica, CA: Human Factors and Ergonomics Society.

Hart, SG, & Staveland, LE. (1988). *Development of the NASA-TLX (Task Load Index): Results of empirical and theoretical research.*

Hickox, JG, & Wickens, CD. (1999). Effects of elevation angle disparity, complexity and feature type on relating out-of-cockpit field of view to an electronic cartographic map. *Journal of Experimental Psychology: Applied., 5*, 284–301.

Huesman, J, Holland, L, & Langley, T. (2013). *December 2012 Construction at $885.0 Billion Annual Rate*: U.S. Census Bureau News. U.S. Department of Commerce. http://www.census.gov/construction/c30/pdf/pr201212.pdf. (July 16th, 2013).

Lim, S, Buswell, RA, Le, TT, Austin, SA, Gibb, AGF, & Thorpe, T. (2012). Developments in construction-scale additive manufacturing processes. *Journal of Automation in Construction, Elsevier, 21*, 262–268.

Mitropoulos, P, & Memarian, B. (2013). Task demands in masonry work: sources, performance implications and management strategies. *Journal of Construction Engineering and Management., 139*(5), 581–590.

Moroney, WF, Biers, DW, Eggemeier, FT, & Mitchell, JA. (1992). *A comparison of two scoring procedures with the NASA Task Load Index in a simulated flight tasks* (pp. 734–740). Dayton, OH: Proceedings from NAECON.

Moroney, WF, Biers, DW, & Eggemeier, FT. (1995). Some measurement and methodological considerations in the application of subjective workload measurement techniques. *International Journal of Aviation Psychology., 5*(1), 87–106.

Mourgues, C, & Fischer, M. (2008). *A Work Instruction Template for Cast-in-Place Concrete Construction Laborers*: Center for Integrated Facility Engineering. Stanford University. Working Paper #109.

O'Brien, B, Leite, F, & Meeks, S. (2011). *Enhanced Work Packaging: Design through Workface Execution* (Implementation Resource 272-2). University of Texas at Austin: The Construction Industry Institute.

Oglesby, CH, Parker, HW, & Howell, GA. (1989). *Productivity improvement in construction.* New York: McGraw-Hill.

Rojas, EM. (2008). *Construction productivity: a practical guide for building and electrical contractors*. Fort Lauderdale: J. Ross Publications.

Schramm, W. (1954). *The Process and Effects of Communication*. Champaign, IL: University of Illinois Press.

Shannon, CE, & Weaver, W. (1949). *The Mathematical Theory of Communication* (Vol. 1). Champaign, IL: University of Illinois Press.

Simondetti, A. (2002). Computer-generated physical modelling in the early stages of the design process. *Journal of Automation in Construction, Elsevier, 11*, 303–311.

St John, M, Cowen, MB, Smallman, HS, & Oonk, HM. (2001). The use of 2D and 3D displays for shape-understanding versus relative-position tasks. *Human Factors, 43*(1), 79–98. Human Factors and Ergonomics Society.

Taylor, TRB, Uddin, M, Goodrum, PM, McCoy, A, & Shan, Y. (2012). Change orders and lessons learned: knowledge from statistical analyses of engineering change orders on Kentucky highway projects. *Journal of Construction Engineering and Management, ASCE, 138*(12), 1360–1369.

Uncertainty-aware visualization and proximity monitoring in urban excavation: a geospatial augmented reality approach

Xing Su[1], Sanat Talmaki[2], Hubo Cai[3*] and Vineet R Kamat[4]

Abstract

Background: This research aims to improve the urban excavation safety by creating an uncertainty-aware, geospatial augmented reality (AR) to visualize and monitor the proximity between invisible utilities and digging implements. Excavation is the single largest cause of utility strikes. Utility strikes could be prevented if the excavator operator were able to "see" buried utilities and excavator movement, and judge the proximity between them in real time. Geospatial augmented reality (AR) is an enabling technology for such knowledge-based excavation. It synergizes the geospatial utility locations and the excavator movement into a real-time, three-dimensional (3D) spatial context accessible to excavator operators. The key to its success is the quality of the utility location data.

Methods: This paper describes a dynamic approach to incorporate positional uncertainties of buried utilities into an uncertainty-aware, geospatial-AR system for real time visualization and proximity analysis. Uncertainties are modeled as probability bands (e.g. spatial bands with certain probabilities of enclosing the "true" location of utilities). Positional uncertainties are derived in real time by referring to its determinant, data lineage, the genesis and processes used to collect and interpret data.

Results: A computational framework, and a generic data model and its XML-format implementation are developed and tested. A method is developed to analyze the proximity in the context of positional uncertainties of both the utilities and the excavator movement.

Conclusions: This newly created approach is expected to contribute to the safety in urban excavation via the integration of Geoinformatics and construction informatics into an uncertainty-aware, geospatial-AR, with real time visualization and analytical capabilities.

Keywords: Buried utilities, Error modeling, GIS, Safety, Uncertainty modeling, Virtual reality

Background

Underground utilities are critical components of the massive utility networks that provide basic services to the society. It is estimated that the total length of underground utilities including water, sewer, gas, electrical, and telecom in the US is in excess of 35 million miles. The largest single threat to the safety of underground utilities is excavation (National Transportation Safety Board (NTSB) (2000) 1998; National Transportation Safety Board (NTSB) 1997). In the US, underground utilities are hit or damaged by excavation every 60 seconds (Spurgin et al. 2009; Common Ground Alliance (CGA) 2010).

Besides the high frequency of its occurrence, a hit on utilities by an excavation operation often leads to disastrous consequences in aspects of disruption to services, property damage, deaths, and serious injuries (Felt 2007; Nelson and Daly 1998; Doctor et al. 1995). For instance, the natural gas pipeline rupture and subsequent explosion caused by excavation in St. Cloud, Minnesota on December 11, 1998 caused four fatal injuries, one serious injury, and 10 minor injuries; and destroyed six buildings (National Transportation Safety Board (NTSB) (2000) 1998; National Transportation Safety Board (NTSB)

* Correspondence: hubocai@purdue.edu
[3]School of Civil Engineering, Purdue University, Civil Engineering Building, Room 1235, 550 Stadium Mall Drive, West Lafayette, IN 47907, USA
Full list of author information is available at the end of the article

1997). In 2007, the excavation strike on a high pressure gas main in Cary, North Carolina resulted in a 100 feet high fireball that burned for nearly six hours and consequently, the evacuation of nearby residents and closing of major roads (WRAL archives 2011). The Office of Pipeline Safety's Pipeline and Hazardous Materials Safety Administration (PHMSA) reported a total of 2770 serious incidents from year 2001 to 2010, of which nearly 20% (544 incidents) were excavation related and caused a total of 37 fatalities, 152 injuries, and $200 million in property damage (PHMSA 2011).

Despite the implementation of the 811 One-Call System that requires excavation contractors to call the state One-Call center that in turn, informs utility owners to mark utility locations with spray paint or flags, excavation remains the single largest cause of pipeline accidents. For instance, a UNCC (Utility Notification Center of Colorado (UNCC) 2005) study reported that 55.7% of the 9,371 incidents in Colorado in 2005 occurred even though the excavators followed the One-Call procedure. These incidents occur due to two primary reasons: (1) reliable data regarding the true location of underground utilities is missing or incomplete, i.e., utilities are often NOT at locations where the records specify (Sterling et al. 2009); and (2) uncertainty in the utility location is not communicated to excavator operators in real-time to help them objectively perceive the digging machine's position relative to the buried utilities (Sterling et al. 2009; CGER - Commission on Geosciences, Environment and Resources 2000).

The authors have created a framework that synergizes geospatial informatics and construction informatics (Figure 1) to visualize and monitor the interaction between buried utilities and excavation implements to overcome these limitations. Recognizing the pivotal effects of the positional accuracy/uncertainty of buried utilities to downstream visualization and proximity analysis, the newly created framework is intended to be uncertainty-aware. This paper presents the technical approach in achieving this intention by modeling the uncertainty of geospatial utility data, quantifying the parameters of the uncertainty model based on data lineage, visualizing both the geospatial utility data and its associated uncertainty in a Geospatial Augmented Reality (GAR) environment, and analyzing the proximity between buried utilities and the digging implements. Based on this analysis, appropriate quantitative warnings can be issued and together with visual displays, be brought to excavation operators in real-time. The newly created framework is expected to contribute to safe urban excavation and the improvement of construction productivity.

The remainder of this paper is organized as follows. The authors first review the current practice and related studies. Following this review, the authors describe the technical details of their methodology in modeling the positional uncertainty of geospatial utility data, monitoring the excavator movement, and synergizing them into a geospatial AR environment for real-time visualization and proximity analysis. The authors then illustrate the test and validation of the newly developed framework. Finally, the authors draw their conclusions and point out future research directions.

Related studies

This section reviews related studies in modeling the geospatial underground utilities and the associated positional uncertainties, the methods of locating invisible underground utilities, and emerging Augmented Reality-based approaches in visualizing buried, invisible utilities in the context of their spatial context. The current practice of the 811 One-Call procedure and its limitation are also reviewed.

Geographic information systems (GIS) for utility data

Geographic information science (GISci) is the discipline that focuses on understanding the world by describing, analyzing, and explaining human relationships with the earth (Huxhold 1991). GIS is an information system built upon GISci to manage, analyze, and report spatial data, describing phenomena above, on, and underneath the earth's surface. A GIS is both a database system to manage spatial and non-spatial data and a set of spatial operators for working on spatial relations (Poku and Arditi 2006).

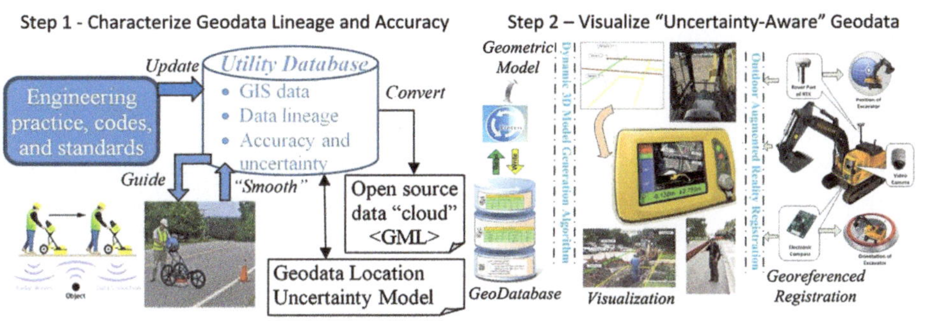

Figure 1 Computational framework for uncertainty-aware visualization and proximity monitoring in urban excavation.

Integration of spatial and non-spatial information on a database platform, registration of locations to the real world coordinates, and spatial analytical capabilities are the distinguishing merits of GIS, leading to the proliferation of its applications in civil engineering (Poku and Arditi 2006; Miles and Ho 1999).

GIS allows a utility owner to have a complete utility inventory stored in a single repository that is easy to update and extract (Corbley 2007). Since its emergence, GIS has been steadily replacing paper-based as-built drawings and digital Computer Aided Design and Drafting (CADD) drawings to inventory and manage utility data. It has become the de facto tool of choice for utility owners for creating, organizing and managing geospatial utility information (Sipes 2007). Many utility owners have gone through the transition from paper maps to CADD files or GIS databases via digitization (Cypas et al. 2006).

GIS has been historical two-dimensional (2D), modeling the geometries of objects into georeferenced points, polylines, and polygons. Buried utilities are predominantly modeled as 2D polylines in GIS, missing vertical location information. Some utilities might have their buried depths (e.g. "depth of cover") stored as attributes and their vertical location might be derived by consulting the reference surface. However, utility depths are rarely referenced to a recognized elevation datum (Federal Highway Administration (FHWA) 1999; Anspach 1995) and any changes of the reference surface make the buried depth a very unreliable source for deriving vertical utility locations.

Locating utilities in the field

When utilities are first installed, their locations are captured in as-built drawings that vary in terms of information richness, positional accuracy, and storage format (e.g. paper-based versus digital). The advancement in tracking technologies such as Global Positioning Systems (GPS) and Radio Frequency Identification (RFID) has greatly facilitated the collection of accurate utility locations for new installations in both horizontal and vertical dimensions (Dziadak et al. 2008; North 2010).

After utilities are installed and covered, the procedure of locating them in the field typically starts with the as-built drawings. Geophysical surveys based on a variety of sensing and locating techniques might be performed to locate buried utilities with good accuracies at the levels of A and B, based on the generally accepted definitions of quality levels in Subsurface Utility Engineering (SUE) (Stevens and Anspach 1993; Lew 1996; American Society of Civil Engineers (ASCE) 2002). Locating techniques include radio frequency (RF) detection techniques, electromagnetic techniques, magnetic methods, vacuum extraction, ground penetrating radar (GPR), and terrain conductivity (Anspach 1995). A number of studies have been conducted to apply the GPR technique in detecting buried utilities (Hereth et al. 2006; National Research Council (NRC) 2000; Butler 2001; Lanka et al. 2001). GPS, though not a detecting technique, has been frequently combined with locating techniques to register the location of detected utilities to real-world spatial referencing system and thus, forms the foundation of integrating with GIS to automate the inventory and update of utility locations in GIS and guide future field location of utilities (Common Ground Alliance (CGA) 2010; Ellis et al. 2009; Manacorda et al. 2007; Bakhtar 2006; Ishikawa et al. 2006; Goldstein 1997).

Uncertainty in geospatial data

Uncertainty has been a major issue in GIS for many years (Heuvelink and Burrough 2002; Goodchild 1998), and is one of the top ten research priorities in GISci (Cobb et al. 2000). Uncertainty can be generally defined as the discrepancy between what a database indicates and what actually exists in the real world (Goodchild 1998) and be described in aspects of positional inaccuracy, errors, vagueness, ambiguity, fuzziness, scale, and sampling (Goodchild 1998; Cobb et al. 2000; Fisher 1999). The literature on the topic makes use of a range of terms related to uncertainty (Crosetto and Tarantola 2001; Duckham et al. 2001), including quality, accuracy, reliability, error, ignorance, precision, clearness, distinctiveness, etc. (Foody 2003). A large amount of research articles have been included in the publications of two international conferences of Symposia on Spatial Accuracy Assessment and Symposia on Spatial Data Quality.

Current practice uses metadata (e.g. data about data) standards such as ISO 19113 (ISO 2001), ISO 19115 (ISO 2003), and Federal Geographic Data Committee (FGDC) (Federal Geographic Data Committee (FGDC) 1998) to measure and record the quality of spatial data upon a common platform (Fisher et al. 2010). Spatial data quality is generally measured in aspects of lineage (e.g. data collection methods and data sources), accuracy, consistency, and completeness as the assessment of results (Fisher et al. 2010). The concept of "fitness for use," e.g. how well a certain data set meets the needs of an application, has also been proposed as the overall measure of the quality/certainty of geospatial data (Fisher et al. 2010; Devillers et al. 2010).

While metadata provides a common platform for recording and communicating data uncertainty information, it does not provide methods to handle and mitigate the inherent uncertainty in geospatial data. The current practice of GIS is deterministic. All visualization and analyses are being performed as if the underlying GIS data were correct without any incompleteness, errors, and inaccuracy. To cope with this limitation, Burrough (Burrough 1992) discussed an "intelligent GIS" to benefit from available metadata to support the use of uncertain data. Unwin (Unwin 1995) introduced the concept of "error-sensitive GIS" for error management such as data verification and validation, visualization of errors/uncertainties, and simulation and sensitivity analysis to obtain a range of potential results as well as associating a sense of credibility to each scenario. Duchham and McCreadie (Duckhma and McCreadie 2002, 1999) proposed the concept of "error-aware GIS" as an extension to "error-sensitive GIS" by adding techniques to understand errors and integrate errors into decision-making. Devillers et al. (Devillers et al. 2005) designed a prototype multidimensional database system to assist users in assessing the fitness for use of geospatial data, considering errors and error propagation issues.

Error modeling for linear objects

For linear, geospatial utility data, uncertainty is mostly concerned with the positional discrepancy between the records-indicated object locations and their real world locations and thus, uncertainty might be interchangeable with positional error/inaccuracy. Accuracy assessment and error modeling of linear objects have been active research topics. A number of studies have proposed and tested several error models for geospatial linear objects such as roads, utilities, and streams (Mozas and Ariza 2011; Shi and Liu 2000; Goodchild and Hunter 1997; Caspary and Scheuring 1993; Perkal 1956).

Utility lines in GIS are typically modeled as straight line segments that connect two end points. In 2D, the uncertainty of a straight line can be captured as an uncertainty epsilon band that encloses the "true" location of the utility centerline (Mozas and Ariza 2011). This concept was initially proposed by Perkal (Perkal 1956) who used an epsilon band, the 2D space enclosed by two parallel lines that are also tangents to circular errors at the ending points as the probability range for a line. The epsilon band has been discussed frequently in the literature (Blakemore 1984; Aspinall and Pearson 1995) and has been implemented in 2D GIS in various algorithms in the form of a tolerance (Goodchild and Hunter 1997). Caspary and Sheuring (Caspary and Scheuring 1993) and Shi and Liu (Shi and Liu 2000) suggested that intermediate points on a line have smaller errors than the end points and thus, the error band will be "slim" in the middle, leading to a genetic band, or a G-band. Probabilities can be determined for G-bands with various sizes to model the uncertainty in lines (Heuvelink et al. 2007; Wu and Liu 2008). Goodchild and Hunter (Goodchild and Hunter 1997) pointed out that epsilon was often interpreted in a deterministic sense as the minimum buffer width that enclosed the true location of the objects under testing/assessing and was very sensitive to outliers. They proposed a simple buffering approach to evaluate the positional accuracy of linear objects by simultaneously referring to the buffer width and the percentage of lines within this buffer (Goodchild and Hunter 1997). The main limitations of current error modeling for linear objects are (1) being 2D, (2) being deterministic (e.g. the epsilon band is determined to enclose the true location and given the existence of outliers, the band width is unreasonably large), and (3) the lack of a method to estimate the most probable location of linear objects given their recorded location and associated positional uncertainty, the reverse procedure of accuracy assessment.

Augmented reality (AR) for utility visualization

A relatively new technological advancement in AR has enabled the visualization of invisible, buried utilities in the context of their real-world surroundings (Kamat 2003; Behzadan 2008; Roberts et al. 2002a). In an AR environment, buried utilities can be visualized as floating lines on the correct locations relative to background images and photos. If three-dimensional (3D) reference data is available, the buried utilities can be offset downward to compose an interactive 3D display (Talmaki and Kamat 2012). The positional uncertainties associated with geospatial utility data might be visualized as 3D buffers/"halos" to provide an uncertainty-aware visualization and proximity analysis (Talmaki et al. 2012).

Current practice of the One-call system

Recognizing excavation damage as the largest single cause of pipeline accidents and associated deaths and injuries, the National Transportation Safety Board (NTSB) initiated the development of the One-Call (811) notification system in 1970s (National Transportation Safety Board (NTSB) 1998). Before digging, it is required by Federal Law to have the location of all buried utilities in the vicinity of the excavation area to be pre-marked. Excavation contractors are required to contact state-wide one-call agencies 48 to 72 hours prior to the start of the operations. One-call agencies, in turn contact their member companies with the location of the excavation site. If the member companies determine an overlap between the job site and their utility lines' location, they mark the location of the utility lines using spray paint, flags, stakes or any combination of these.

The markings are typically made referring to data from as-built drawings. An obvious limitation of this one-call procedure is that the markings are typically the very first things being removed when excavation starts. The excavation operators then have to rely on their memory to estimate the utility location and their imagination to compose mental images of the proximity of the digging implement to utilities and how the excavation operation might interact with utilities. This is an error-prone procedure that in turn, is the main reason of the large number of utility strikes even after the one-call procedure is followed.

The overarching goal of this study is to prevent unintended collisions between a digging implement and buried utilities via uncertainty-aware visualization and proximity analysis of excavation operations in a virtual environment. The premises for the research study presented in this paper are that (1) increased spatial awareness of excavator operators (e.g. being able to "see" the buried utilities and the movement of the excavator bucket and judge the proximity between utilities and the bucket) leads to improved excavation safety, and (2) such an increased in spatial awareness can be achieved through the combination of visual perception and analytical proximity monitoring of excavation operations. Considering the pivotal role of data quality and the tremendous uncertainty in utility locations, the research hypothesis is that uncertainty, particularly spatial uncertainty, can be brought into the decision-making process in an easy-to-understand manner for both visualization and analytical proximity monitoring to prevent unintended collisions and increase excavation safety.

Methods

To achieve the aforementioned goal, the authors designed a framework (Figure 1) to synergistically incorporate inherent uncertainties associated with geospatial utility data into a geospatial AR environment for error-aware visualization and proximity monitoring. This section describes the technical details in modeling error/uncertainty in geospatial utility data and quantifying the error model by linking accuracy to data lineage. This section also presents an error-aware utility data model in the format of Universal Modeling Language (UML) and its implementation of Extensible Markup Language or XML (Cypas et al. 2006) model for data transfer and sharing with downstream visualization and analysis, given its flexibility, extensibility, and compatibility with open-source requirements. The mechanism of proximity monitoring and uncertainty-aware reasoning is also provided as the base for appropriate warning messages.

Uncertainty modeling of geospatial utility data

The authors modeled utility lines as 3D straight line segments connected at turning points (Talmaki et al. 2012). In practice, the 3D locations of turning points are first obtained and 3D straight lines can be constructed by connecting these 3D turning points. Of all positional accuracy/error models, the most related are those that apply to points and lines. Uncertainty, when expressed as probabilities associated with geospatial extents that contain the "true" location of that object, can be derived from the recorded utility location and its positional accuracy.

Figure 2 illustrates the 3D uncertainty model that was designed specifically for geospatial utilities points

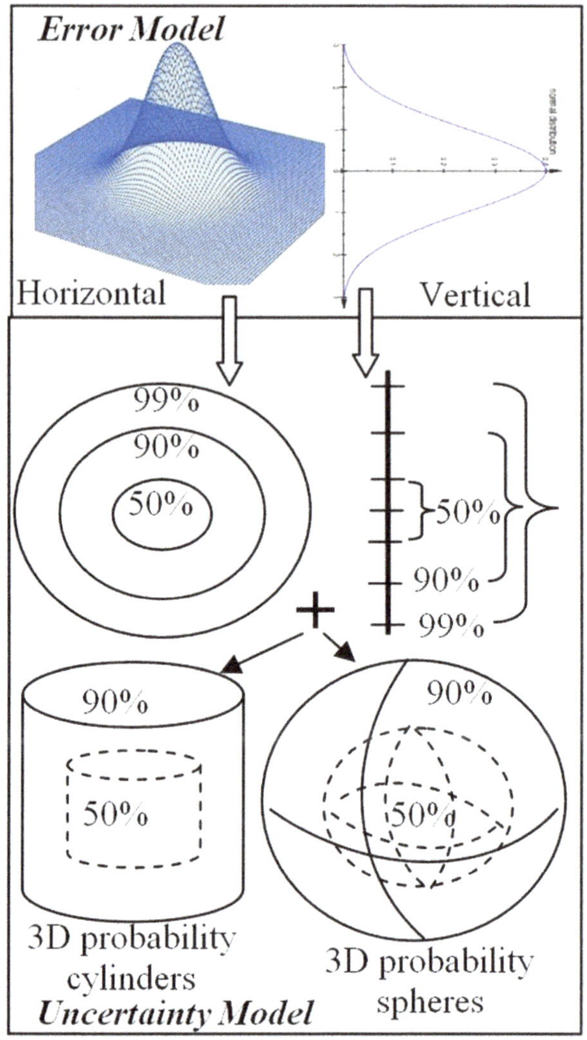

Figure 2 Error and uncertainty model for points.

Probabilities were introduced to represent uncertainties. The assumptions for this model include: (1) the positional error consists of a systematic error and a random error, (2) the random error is normally distributed in all dimensions, (3) random errors in X, Y, and Z directions are independent, and (4) horizontal errors are not different in X and Y dimensions. The systematic error can be remedied via a translation in a 3D coordinate system. The random error in the Z dimension follows a normal distribution and probability ranges can be determined as linear ranges. The random errors in X and Y dimensions are combined into a horizontal radial error calculated as square root of $(e_x^2 + e_y^2)$, where e_x and e_y are the random errors in the X and Y dimensions, respectively. The horizontal radial error can be treated as a normal distribution (Greenwalt and Shultz 1968) or a χ^2 distribution with two degrees of freedom (Caspary and Scheuring 1993). Probability circles can then be determined as pairs of a radius in the unit of standard deviation (SD) and a probability expressed as a percentage.

The horizontal probability circles and the vertical linear probability ranges can be combined into 3D probability ellipses/spheres (Greenwalt and Shultz 1968) or probability cylinders. A probability cylinder is constructed by taking a probability circle and extruding it vertically. The resulting probability is the product of the corresponding circle probability and the linear probability. For instance, the horizontal 50% circle is extruded to reach the vertical 90% linear probability, resulting in a 45% probability cylinder. Unlike probability ellipse/circle, many cylinders can have the same probability, making implementation impractical. A random simulation was conducted, of which the results confirmed that 3D ellipsoids closely follow the point clouds and thus, were chosen to model the uncertainty of utility turning points.

In this study, utility lines were modeled as straight line segments that connect two end points. In 2D, the uncertainty of a straight line is captured as an uncertainty epsilon band (Mozas and Ariza 2011). The epsilon band took a shape that is "slim" (Shi and Liu 2000; Caspary and Scheuring 1993) in the middle to indicate the positional error at turning points is larger than that at intermediate points on the line. Such a band model in 2D is commonly referred to as a genetic band, or G-band. Since this study extended the 2D G-band into 3D and introduced probabilities to represent uncertainties, the resulting uncertainty model for utility lines was named 3D Probability G-band. Figure 3 illustrates the uncertainty model for utility lines by extending the epsilon concept from 2D into 3D and incorporating probabilities to represent uncertainty. Only one probability band (the outermost surface) is shown in Figure 3. The 3D space enclosed by this boundary represents the space that encloses the true location of the utility line at a particular probability. The shape and the size of the band can be described via a mathematical function (Equation (1)):

$$r_{x,p\%} = f(x,\ p\%, \text{data lineage}), \tag{1}$$

where $r_{x,\ p\%}$ represents the radius of the p% probability band at location x along the centerline.

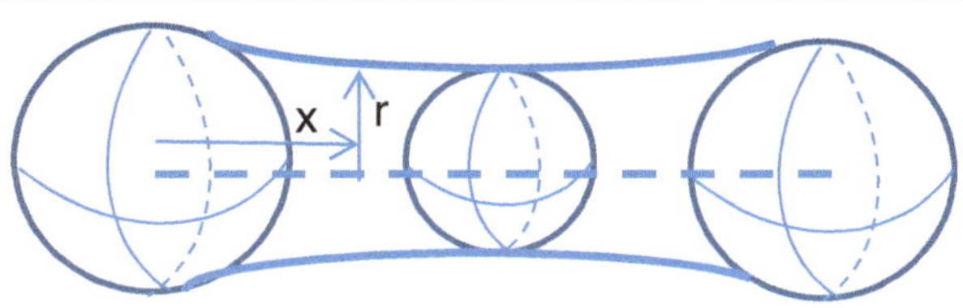

Figure 3 Uncertainty model for lines.

The function f indicates that radius is dependent on the location along the centerline, the probability of interest, and the data lineage. The determining effect of the data lineage on uncertainty models will be described detail in the section of Construction of Uncertainty Models for Geospatial Utility Data. When the locations of utility lines are collected directly such as in GPR, the 3D Probability G-band can be simplified into 3D Probability Bands that can be derived via the 3D buffering approach. Rather than being "slim" in the middle, a probability band is of uniform size along the line.

Validation of uncertainty models

Monte Carlo simulations were performed to validate both the point and line error models. Figure 4 illustrates the distributions for end points and the centerline. The Monte Carlo simulations generated random 3D locations for end points. Corresponding end points were connected to form centerlines. Corresponding 3D ranges of the 50% and 90% probabilities are highlighted in Figure 4, where the white line indicates the 3D extent that is associated with a probability of 50% to enclose the true utility location and the black line indicates the 3D extent that is associated with a probability of 90% to enclose the true utility location. Figure 4 clearly illustrates that the point distribution follows a sphere shape and the 3D Probability G-bands for the line are "slim" in the middle. Such 3D ranges (e.g. 3D Probability G-bands) compose the 3D uncertainty models for utility centerlines in this study.

Construction of uncertainty models for geospatial utility data

The authors constructed uncertainty models by quantifying the 3D Probability bands in pairs of probability

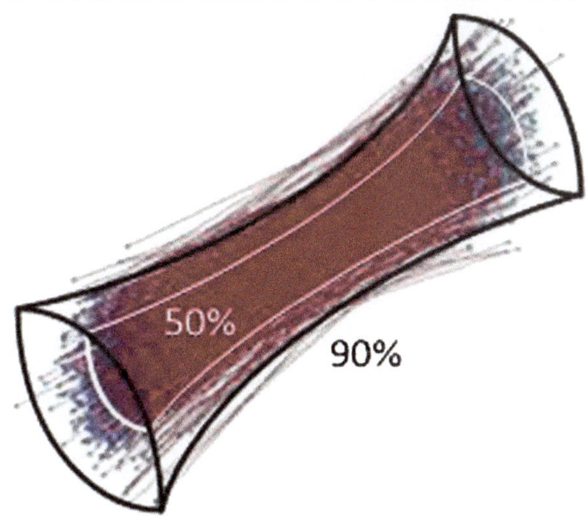

Figure 4 Uncertainty model for lines.

and band size. The probability expressed in percentage refers to the probability of a specific 3D geospatial volume determined by a particular band size enclosing the "true" location. The band size was further detailed via a mathematical function that describes the shape and extent of the 3D G-band. The function might be simplified into a cylinder function when describing a 3D Probability band that does not "slim" in the middle along the line.

In implementation, the probability bands for centerlines must be transformed into utility circumferences. This is because GIS models 3D utility pipes/lines as lines that correspond to utility centerlines. Figure 5 illustrates this transformation needs for circular shape utilities. The size for the transformed band is the sum of the pipe radius and the model band size. This transformation works with cross-sectional dimension and is straightforward. The size of a probability range that is linear (1D), radial (2D), or 3D G-band, is quantified in the unit of standard deviation (SD), a commonly used statistical index for positional accuracy. Such a mechanism establishes a direct link between positional accuracy models (typically described via the deviation of true location from observed location) and uncertainty models (typically expressed as the probability of being true, e.g. the probability of enclosing the true utility location). Creating uncertainty models via 3D probability eclipses/spheres and probability G-bands for utility points and lines facilitated the adoption of proximity (buffer) process of GIS and formed the foundation for visualizing and analyzing positional uncertainties in utility lines and points.

The premise for this dynamic approach for model construction is the linkage between data lineage and positional uncertainty, i.e. the genesis and the process to collect and interpret data determines the positional accuracy and uncertainty, a widely accepted scenario in Subsurface Utility Engineering (SUE) (Sterling et al. 2009) and GIS. In order to estimate the positional accuracy associated with the data lineage, utility data sources are first semantically categorized into four Level 1 quality groups according to (American Society of Civil Engineers (ASCE) 2002), detailed in Table 1. The level 1 groups are then detailed to Level 2 groups based on spatial measurement technologies. For instance, group A can be categorized based on subsequent measurement method and equipment. The use of real time kinematic (RTK) GPS could result in a positional accuracy of 6 mm (horizontal) and 12 to 18 mm (vertical). Similarly, Groups B, C, and D are further categorized based on geophysical methods used; density and spatial coverage, and the methods used to survey and record above-ground utility features; and original data collection methods; respectively. Since data lineage is now categorized based on spatial measurement technologies that in turn, determine the achievable levels of accuracy, the correlation between data lineage and positional accuracy is logically established. Consequently, uncertainty models, following the link between accuracy models and uncertainty models, can then be constructed to quantify uncertainty into probability 3D G-bands.

The dynamic nature of the uncertainty model construction approach is highlighted in Figure 6 that also illustrates the system architecture as well as the flow chart in managing the uncertainties of geospatial utilities data in a synergetic geospatial AR environment. The system is composed of three modules. The Uncertainty Modeling Module contains accuracy models, uncertainty models, the link between data lineage and accuracy, and the link between accuracy and uncertainty. The Open Source <UML> & <XML> Module includes data models employed to facilitate the management and sharing of geospatial utility data with downstream engineering applications. The Visualization and Proximity Analysis Module focuses on visualizing invisible utilities and the excavator movement, and analyzing the proximity between a digging implement (e.g. the excavator) and buried utilities, given inherent uncertainties in the utility location and excavator movement. The workflow (indicated by arrows in Figure 6) is such that: (1) utility data is enhanced with data lineage information (based on an expanded version of Table 1), (2) based on the links between data lineage and accuracy, and between accuracy and uncertainty, utility data is enhanced with uncertainty information, e.g. probability bands, and (3) uncertainty-aware utility data feeds into the geospatial AR for visualization and proximity analysis. Throughout the workflow,

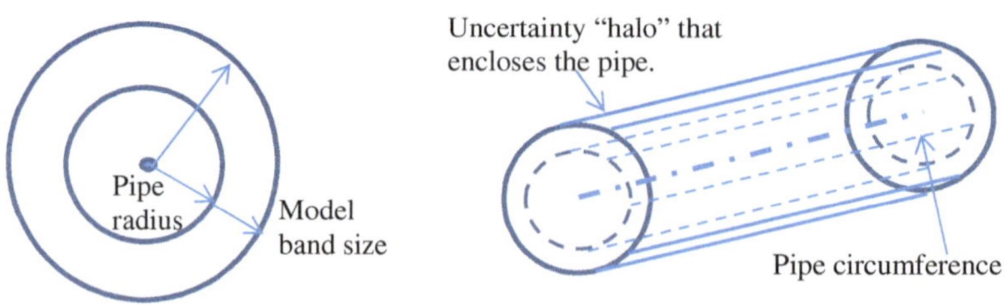

Figure 5 Transformation to circumferences.

Table 1 Quality Groups of Utility Location Data

Quality groups	*Data lineage*
A	Precise horizontal and vertical location obtained by actual exposure and measurement.
B	Location information obtained through the application of surface geophysical methods
C	Location information obtained by inferring from above-ground utility features.
D	Location information derived from existing records or oral recollections.

probability bands are derived rather than stored as properties of the corresponding utilities. This dynamic nature allowed easy update of the uncertainty model as well as extension and expansion. Any changes/updates in the uncertainty model and/or data lineage are automatically updated during future data utilization as utility data is extracted. For instance, the data lineage of a utility might change from Quality Group B to A. Simply updating the data lineage information automatically takes care of the uncertainty bands. Similarly, when the uncertainty model of a specific type of data lineage is changed, the uncertainty bands for all utilities that have that particular type of data lineage are automatically updated. Such a design is similar to the CASCADE rules in databases to ensure data integrity and consistency (Date 2000).

Uncertainty-aware geospatial data model for utilities

To facilitate the sharing of uncertainty-aware geospatial utility data in downstream applications, the authors created an uncertainty-aware utility data model illustrated in Figure 7. A solid box represents a *class*. A line represents an *association* between two classes. Special symbols such as diamonds and triangles are added to lines to indicate specific associations. For instances, the inclusion of a solid diamond represents *composition* – the relationship of "owns"; the inclusion of a hollow diamond represents *aggregation* – the relationship of "has"; and the inclusion of a hollow triangle represents *generalization/inheritance* - the relationship of "is a". In this model, the utility network is composed of utility lines, straight line segments that are spatially constructed by connecting ending points, or utility vertices. The uncertainty model is specialized into linear uncertainty model for utility lines and ellipsoid model for utility vertices. The linear uncertainty model is further specialized into 3D G-band model and 3D cylinder model to suit different application needs. All uncertainty models are quantitatively (e.g. probabilities and band sizes derived) associated with their corresponding utility objects (e.g. utility lines and vertices) by referring to the linkage model. Such a model echoes the dynamic nature highlighted in Figure 6.

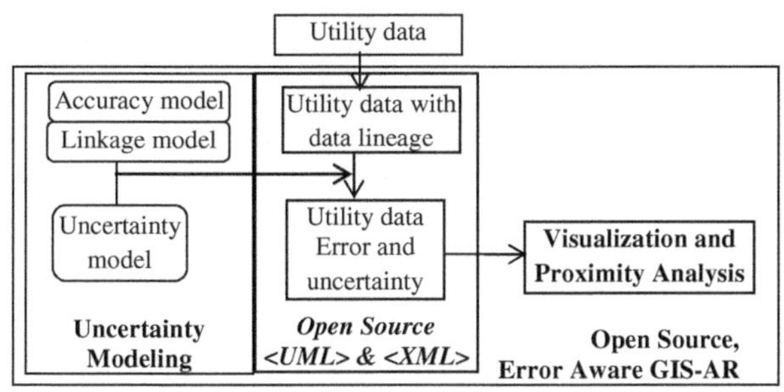

Figure 6 The dynamic procedure in the construction of uncertainty model.

The authors explored XML based data schemes such as the Geography Markup Language (GML) (Burggraf 2006) and CityGML, an extension to GML for the built environment (Kolbe et al. 2008); and created an XML-based model (Figure 8) as an implementation of the uncertainty-aware utilities data model in Figure 7. The newly created XML schema is an open source data model that is flexible and extensible to facilitate modeling, sharing, exchange, and integration of geospatial data. In the newly created model, each utility line is composed of straight line segments that are fixed in the space via their two ending points. A vertex is only needed when the utility changes direction, e.g. a bend. All spatial and non-spatial lineage information is inherent in utility vertices and lines to dynamically construct probability 3D-bands for uncertainty-aware visualization and proximity monitoring in excavation, and the sharing of uncertainty-aware data in downstream applications.

Geospatial augmented reality environment for visualization and proximity analysis

In order to visualize uncertainty-aware geospatial data in Augmented Reality, the data must first be converted to a format suitable for 3D graphical visualization. More importantly, the geodata accuracy and its uncertainty must be graphically characterized and displayed for it to be useful in excavator operation and control. This research developed methods to characterize utility geodata in terms of its lineage and accuracy. Together, the accuracy and lineage help characterize the errors of a geodata set and the reliability that can be associated with its source. This information can be usefully exploited during excavation by displaying not only the expected locations of utilities to an operator, but also the degree of uncertainty (or "buffer") associated with the expected locations in the form of a "halo". In a 2D projection, the buffer calculated by interpreting the geodata's lineage and accuracy is represented as a "band" whose width represents the uncertainty associated with the utility's location. In a 3D projection, the buffer is identified by increasing the

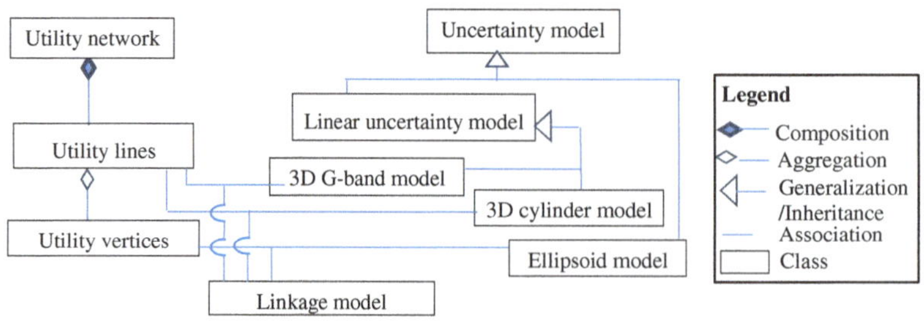

Figure 7 **The uncertainty-aware utilities data model.**

diameter of the cylindrical geometry representing the utility line.

The proximity analysis included three scenarios illustrated in Figure 9, with increased capabilities of handling positional uncertainty. Figure 9(a) illustrates a deterministic proximity measure that is the closest distance between an underground utility and the digging implement, e.g. the excavator bucket, assuming both the utility pipe and the excavator bucket are at their "true" locations with no positional uncertainty. Figure 9(b) introduces the positional uncertainty in the geospatial utility data, but the excavator bucket is still deterministic.

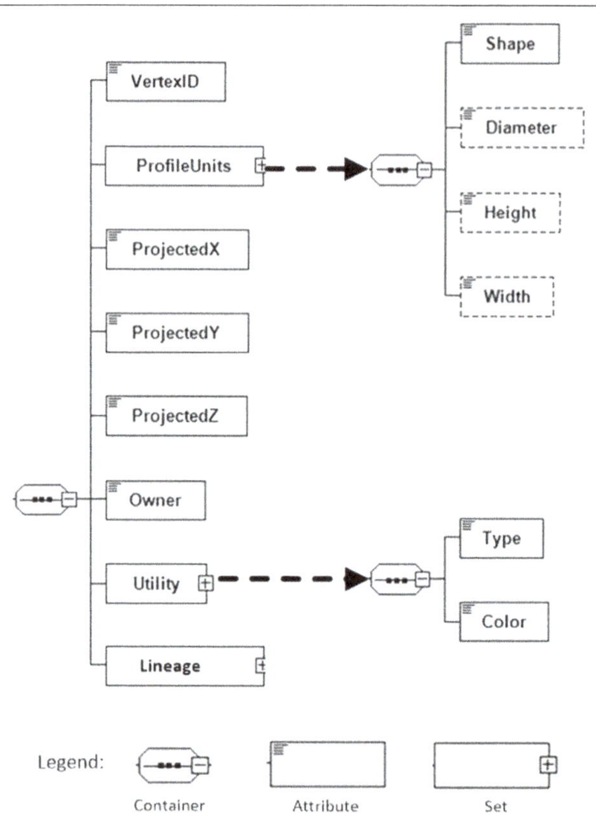

Figure 8 **XML schema for uncertainty-aware geospatial utility data.**

Figure 9(c) introduces the positional uncertainty in the excavator bucket in addition to the positional uncertainty in the utility location. For the illustration purpose, both Figure 9(b) and (c) are presented in 2D.

The original positional uncertainty "propagates" into the proximity, reflected as a probability associated with the resulting 3D distance. However, it is inappropriate to interpret the propagated uncertainty as "an x% probability of proximity of d." Rather, it shall be interpreted as "an x% probability that the proximity is at least d." Table 2 provides interpretations of Figure 9(b) scenarios. Note that 40% is exactly (100%-60%) and 15% is exactly (100%-85%). In implementation, interpretations, together with the corresponding probability bands, could be displayed to an operator as warning messages when both the probability and the proximity thresholds are reached.

When the uncertainty in the bucket location is also introduced as in Figure 9(c), similar analyses can be carried out with corresponding interpretations. The difference is that instead of a single 3D geometry representing the bucket, the analysis applies to a series of 3D geometries representing the probability bands of the bucket. Also, the probability to be used is the product of the probabilities in the corresponding utility and excavator bucket bands. For instance, assuming in Figure 10, the bucket band has a 60% probability and one interpretation is that there is at most a 6% ((100%-85%)*(100%-60%)) probability that the excavator WILL hit the utility. Compared to underground utilities, the uncertainty associated with the bucket is significantly smaller. This is because most of the time, the bucket is above ground or has a line of sight. Thought it could be outside the operators view (e.g. blocked by the terrain or cabin), sensing technologies such as GPS, video cameras, and accelerometers can accurately track its location and movement. Thus, the uncertainties associated with the bucket might be ignored in visualization and proximity analysis, as illustrated in Figure 9(a) and Figure 9(b). When the utility data is acquired via highly accurate sensing technologies such as RTK GPS with an uncertainty at the same magnitude to the uncertainty of bucket, then the uncertainty associated with the bucket

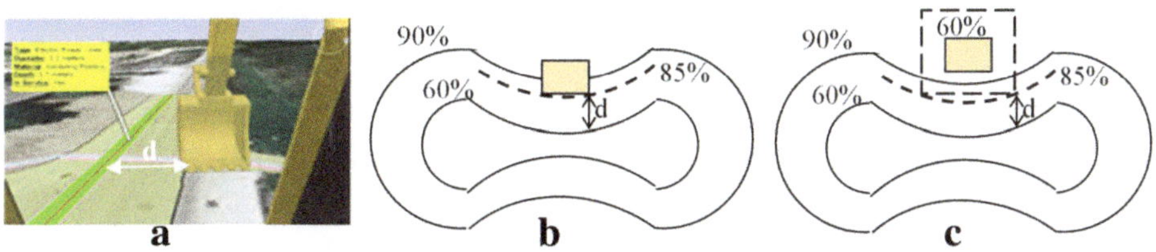

Figure 9 Interpretation of resulting uncertainty in proximity analysis; (a) Deterministic proximity measure; (b) Proximity measure incorporating uncertainty in pipe location; (c) Proximity measure incorporating uncertainty in pipe location and bucket movement.

matters and shall be incorporated in the analysis, as illustrated in Figure 9(c).

Results and discussion

The authors validated the developed geospatial uncertainty models in a controlled environment. Monte Carlo simulations for validating the uncertainty models for points and lines were described in an earlier section and therefore, will not be repeated here. The technical feasibility of the proposed geospatial AR ideas was evaluated by implementing a visualization and proximity analysis framework designed to visualize "error-aware" subsurface utilities during ongoing excavation operations for improved context awareness and accident avoidance.

The data used in the experiments was provided by DTE Energy, which is the largest provider of electricity and gas in southeast Michigan and by consequence owns significant underground distribution assets. The first step in preprocessing the data was assigning a specific lineage to chosen geospatial data sets. A data set assumed to have been recorded following a Ground-Penetrating Radar (GPR) survey was selected for subsequent analysis. The uncertainty model was chosen next. Given the data lineage, a cylindrical uncertainty model was chosen with quantitative error magnitudes in both horizontal and vertical directions. The error characterized data set was then archived in the developed XML schema.

The current practice of excavation damage prevention followed by state one-call centers, and the limitations in current practice were noted in an earlier section. Geospatial Augmented Reality can help to accurately visualize a proposed excavation area and digitize the located underground utilities, thus helping bridge the communication gaps among excavation contractors, one-call centers, utility owners, field locators and excavator operators. First, a contractor can issue a "visual" ticket to the one-call center and utility owners by superimposing a semi-transparent layer above a proposed excavation area. This in turn enables dispatched locators to come to the field, "see" the proposed excavation area (Figure 10), and precisely mark the surveyed area.

The following procedure was adopted to interpret error-aware geospatial data files and build conduit (e.g. pipe) models in the augmented space. First, the spatial and attribute information of pipelines was extracted by parsing the data model. For example, the geographical location of pipelines is recorded under the Geometry element as "LineString". A cursor was designed to iterate through the XML file, locate "LineString" elements, and extract the geographical locations. Second, consecutive vertices within one "LineString" were converted from the geographical coordinate to the local coordinate in order to raise computational efficiency during the registration routine. The first vertex on the line string was chosen as the origin of the local coordinate system, and the local coordinates of the remaining vertices were determined by calculating the relative 3D vector between the rest of the vertices and the first one, using the Vincenty algorithm (Vincenty 1975). In order to save memory, a unit cylinder is shared by all pipe segments as primitive geometry upon which the transformation matrix is built.

Table 2 Interpretation of System Level Uncertainty in Proximity

Probability		*Distance*	*Interpretations*
>=	60%	1 m	At ***least*** a 60% probability the bucket is at ***least*** d (1 m) away from the utility
<=	40%	1 m	At ***most*** a 40% probability the bucket is at ***most*** d (1 m) away from the utility
>=	85%	0 m	At least an 85% probability the bucket will **NOT** hit the utility
<=	15%	0 m	At ***most*** an 15% probability the bucket **WILL** hit the utility

Figure 10 Digital field location and visualization of error-aware geospatial utility data.

Figure 11 Visualization of excavator proximity to expected buried utility locations.

Third, the primitive cylinder geometry was scaled, rotated, and translated to the correct size, attitude, and position. For simplicity, the normalized vector between two successive vertices was named as the pipeline vector. First, the primitive cylinder was scaled along the X- and Y-axis by the radius of the true pipeline, and then scaled along the Z-axis by the distance between two successive vertices. In addition, the scaled cylinder was rotated along the axis—formed by the cross product between vector <0, 0, 1> and the pipeline vector—by the angle of the dot product between vector <0, 0, 1> and the pipeline vector. Finally, the center of the rotated pipeline was translated to the midpoint between two successive vertices. This step was applied to each pair of two successive vertices to extract the complete geospatial data set.

The extracted data is stored digitally and contains information about the utility type and geometry, as well as the computed buffer zones indicating the geospatial uncertainty predicted by the developed models. Upon extraction and conversion of the geospatial data into 3D models, excavator operators are able to persistently visualize what utilities lie buried in a digging machine's vicinity, and consciously avoid accidental utility strikes as excavation progresses by estimating the evolving distance between the digging machine and vicinal utilities (Figure 11).

Conclusions

This paper described a dynamic approach to incorporate the uncertainties associated with buried utilities data into a geospatial-AR system for real time visualization and proximity analysis. This research modeled uncertainties of buried utilities data as probability bands, described by pairs of band size and the probability of the 3D space constructed by "buffering" at the band size to enclose the "true" location of utilities. Given the research hypothesis of that positional uncertainty of utilities data is dependent on data lineage, e.g. the genesis and processes used to collect and interpret data, the positional uncertainty of utilities data was derived in real time by referring to the data lineage model. Consequently, not only the 3D shapes and locations of utility lines and vertices, but also the associated uncertainties could be visualized, as 3D probability bands in a geospatial AR environment. This newly created approach is expected to contribute to the safety in urban excavation via the integration of Geoinformatics and construction informatics in real time in an uncertainty-aware manner.

A framework, a generic data model, and a sample XML implementation of the data model were developed and tested in this study. The impacts of the uncertainties in the utilities data on proximity analysis, e.g. analyzing the closeness between a digging implement and the underground utilities, were also discussed. A method was also developed for analyzing the proximity and interpreting the results in the context of uncertainties that could come from both the utilities and the excavator movement. Visualizing the uncertainty associated with utility location data and appropriately interpreting the resulting proximity were found to be key elements in any visual and analytical guidance furnished to excavator operators or field personnel to prevent utility strikes. It was found that uncertainty-aware, geospatial AR was the enabling technology to bring the interaction between a digging implement and the buried utilities to the excavator operator both visually and analytically (with appropriate interpretations). It was also found that having a practical object-oriented, open access, and uncertainty-aware utility data model was critical to the sharing of utilities data and its inherent uncertainties with downstream applications. Such a data model could also serve as the base for management and sharing of uncertainty-aware utilities data from a life cycle perspective.

The authors' future goal in this uncertainty-aware, geospatial AR direction is to create advanced uncertainty qualifying and quantifying algorithms, and the further integration of technologies such as GPR and machine control and guidance (MAC).

Competing interests
All authors declare that they have no competing interests.

Authors' contributions
XS developed the error and uncertainty models. ST created the XML model and implemented the augmented reality visualization platform. HC and VK created the framework and oversaw the research. All authors participated in drafting the manuscript. All authors read and approved the final manuscripts.

Author details
[1]School of Civil Engineering, Purdue University, 550 Stadium Mall Drive, West Lafayette, IN 47907, USA. [2]Department of Civil and Environmental Engineering, University of Michigan, 2340 G.G. Brown, 2350 Hayward, Ann Arbor, MI 48109, USA. [3]School of Civil Engineering, Purdue University, Civil Engineering Building, Room 1235, 550 Stadium Mall Drive, West Lafayette, IN 47907, USA. [4]Department of Civil and Environmental Engineering, University of Michigan, 2340 G.G. Brown, 2350 Hayward, Ann Arbor, MI 48109, USA.

References

American Society of Civil Engineers (ASCE). (2002). *Standard Guidelines for the Collection and Depiction of Existing Subsurface Utility Data*. New York, N.Y: ASCE Codes and Standards Activity Committee (CSAC).

Anspach, JH (1995). *ASCE Continuing Education Class*. Reston VA: CI/ASCE 38–02.

Aspinall, RJ, & Pearson, DM (1995). Describing and Managing Uncertainty of Categorial Maps in GIS. In P Fisher (Ed.), *Innovations in GIS 2* (pp. 71–84). Taylor & Francis: London.

Bakhtar, K (2006). Demonstration of BakhtarRadar Buried Utility Detection and 3-Dimensional Imaging Capabilities. In *PowerPoint Presentation, 2006, 32 slides*.

Behzadan, AH (2008). ARVISCOPE: Georeferenced Visualization of Dynamic Construction Processes in Three-Dimensional Outdoor Augmented Reality. In *Ph.D. Dissertation, Department of Civil and Environmental Engineering*. Ann Arbor, MI: University of Michigan.

Blakemore, M. (1984). Generalization and Error in Spatial Databases. *Cartographica, 21*, 131–139.

Burggraf, DS. (2006). "Geography Markup Language". *Data Science Journal, Vol 5*, 178.

Burrough, PA. (1992). Development of Intelligent Geographical Information Systems. *International Journal of Geographical Information Systems, 6*, 1–11.

Butler, DK (2001). "Potential Fields Methods for Location of Unexploded Ordnance", The Leading Edge, 20(8). *Aug., 2001*, 890–895.

Caspary, W, & Scheuring, R. (1993). Positional Accuracy in Spatial Databases. *Computers, Environment, and Urban Systems, 17*, 103–110.

CGER - Commission on Geosciences, Environment and Resources. (2000). *"Seeing Into The Earth: Noninvasive Characterization of the Shallow Subsurface for Environmental and Engineering Applications"*. Washington, DC: National Academy Press.

Cobb, M, Petry, F, & Robinson, V. (2000). Uncertainty in Geographic Information Systems and Spatial Data. *Fuzzy Sets and Systems, 113*, 1.

Common Ground Alliance (CGA). (2010). *"Locate Accurately, Dig Safely"*. Available: http://www.commongroundalliance.com/Content/NavigationMenu/Publications_and_Resources/Educational_Programs/LocateAccuratelyBrochure.pdf. Accessed 15 September 2010.

Corbley, K (2007). *"Can you dig it?" The American City & County; Feb 2007; 122, 2* (p. 32). <http://americancityandcounty.com/technology/government_dig/> [Accessed: November 19, 2011].

Crosetto, M, & Tarantola, S. (2001). Uncertainty and Sensitivity Analysis: Tools for GIS-Based Model Implementation. *International Journal of Geographic Information Science, 15*, 415–437.

Cypas, K, et al. (2006). Storage of underground utilities data in three dimensional geoinformation system. *Geodetski Vestnik, 50*, 481–491.

Date, CJ (2000). *An Introduction to Database Systems*. Addison Wesley Longman: Seventh edition.

Devillers, R, Bédard, Y, & Jeansoulin, R. (2005). Multidimensional Management of Geospatial Data Quality Information for Its Dynamic Use within Geographical Information Systems. *Photogrammetric Engineering and Remote Sensing, 71*(2), 205–15.

Devillers, R, Stein, A, Bédard, Y, Chrisman, N, Fisher, P, & Shi, W. (2010). Thirty Years of Research on Spatial Data Quality: Achievements, Failures, and Opportunities. *Transactions in GIS, 14*(4), 387–400.

Doctor, RH, Dunker, NA, & Santee, NM (1995). *Third-Party Damage Prevention Systems. GRI-95/0316* (p. 199). Houston, Texas, USA: NICOR Technologies.

Duckham, M, Mason, K, Stell, J, & Worboys, M. (2001). A Formal Approach to Imperfection in Geographic Information. *Computers, Environment and Urban Systems, 25*, 89–103.

Duckhma, M, & McCreadie, JE (1999). *"An Intelligent, Distributed, Error-Aware OOGIS", Proceedings of the 1st International Symposium on Spatial Data Quality* (pp. 496–506). Hong Kong: The Hong Kong Polytechnic University.

Duckhma, M, & McCreadie, JE (2002). Error-Aware GIS Development. In W Shi, PF Fisher, & MF Goodchild (Eds.), *Spatial Data Quality* (pp. 63–75). Taylor & Francis: London.

Dziadak, K, Sommerville, J, & Kumar, B (2008). "RFID based 3D buried assets location system", *ITcon* Vol. 13. In *Special Issue: Sensors in Construction and Infrastructure Management* (pp. 155–165). http://www.itcon.org/data/works/att/2008_11.content.06106.pdf.

Ellis, R, Venner, M, Paulsen, C, Anspach, J, Adams, G, & Vandenbergh, K (2009). *SHRP 2 Report S2-R15-RW: Integrating the Priorities of Transportation Agencies and Utility Companies*. Washington, D.C.: TRB.

Federal Geographic Data Committee (FGDC). (1998). *Content Standard for Digital Geospatial Metadata*. Virginia: FGDC-STD-001-1998, Reston. http://www.fgdc.gov/standards/projects/FGDC-standards-projects/metadata/base-metadata/v2_0698.pdf [Accessed: April 21, 2012.

Federal Highway Administration (FHWA). (1999). *Cost Savings on Highway Projects Utilizing Subsurface Utility Engineering*. Washington DC: Project Report of FHWA Contract DTFH61-96-00090. http://www.fhwa.dot.gov/programadmin/pus.cfm. Accessed 10 December 2011.

Felt, T (2007). "Pipeline Safety: Momentum on Damage Prevention", In the Pipe, Association of Oil Pipe Lines. *American Petroleum Institute, 1*(2), 2007.

Fisher, PF (1999). "Models of Uncertainty in Spatial Data". In PA Longly, MF Goodchild, DJ Maguire, & DW Rhind (Eds.), *Geogrpahical Information Systems: Principles and Technical Issues* (2, Volume 1st ed., pp. 191–205). New York: John Wiley & Sons.

Fisher, PF, Comber, A, & Wadsworth, R (2010). "What's in a Name? Semantics, Standards and Data Quality". In R Devillers & H Goodchild (Eds.), *Spatial Data Quality: From Process to Decisions* (pp. 3–16). Boca Raton, FL: CRC Press.

Foody, GM. (2003). Uncertainty, Knowledge Discovery and Data Mining in GIS. *Progress in Physical Geography, 27*(1), 113–121.

Goldstein, H. (1997). "Mapping Convergence: GIS Joins the Enterprise". *Civil Engineering ASCE, 67*(6), 36–39.

Goodchild, FM (1998). "Uncertainty: The Archilles Heel of GIS?". In *Geo Info Systems, Academic Pursuits, the University Consortium for Geographic Information Science (USGIS), November* (pp. 50–52).

Goodchild, MF, & Hunter, GJ. (1997). A Simple Positional Accuracy Measure for Linear Features. *International Journal of Geographical Information Science, 11*(3), 299–306.

Greenwalt, CR, & Shultz, ME (1968). *Principles of Error Theory and Cartographic Applications, ACIC Technical Report No. 96*. St. Louis, Missouri: United States Air Force, Aeronautical Chart and Information Center.

Hereth, M, Selig, B, Leewis, K, & Zurcher, J (2006). *Compendium of Practices and Current and Emerging Technologies to Prevent Mechanical Damage to Natural Gas and Hazardous Liquids Transmission Pipelines*. GRI 8747.

Heuvelink, GBM, & Burrough, PA (2002). Development in statistical approaches to spatial uncertainty and its propagation. *International Journal of Geographical Information Science, 16*, 111–113.

Heuvelink, GBM, Brown, JD, & Van Loon, EE (2007). Representing and Simulating Uncertain Environmental Variables in GIS. *International Journal of Geographic Information Science, 21*, 497–513.

Huxhold, WE (1991). *An introduction to urban geographic information systems*. New York: Oxford University Press.

Ishikawa, J, Kiyota, M, & Furuta, K. (2006). "Test and Evaluation of Japanese GPR-based AP Mine Detection Systems Mounted on Robotic Vehicles". *Journal of Mine Action, 10.1*, 16.

ISO. (2001). *ISO 19110 Geographical Information – Methodology for Feature Cataloguing* (p. 53). Geneva: International Standards Organization.

ISO. (2003). *ISO 19115 Geographical Information – Metadata* (p. 140). Geneva: International Standards Organization.

Kamat, VR (2003). VITASCOPE: Extensible and Scalable 3D Visualization of Simulated Construction Operations. In *Ph.D. Dissertation*. Blacksburg, VA: Department of Civil and Environmental Engineering, Virginia Tech.

Kolbe, TH, Groger, G, & Pllumer, L (2008). "CityGML – 3D City Models and Their Potential for Emergency Response". In Zlatanova & Li (Eds.), *Geospatial Information Technology for Emergency Response* (pp. 257–274). London: Taylor & Francis Group.

Lanka, M, Butler, A, & Sterling, R. (2001). Use of Approximate Reasoning Techniques for Locating Underground Utilities. *A Supplement to Tunneling and Underground Space Technology, 16*(1), 13–31.

Lew, JJ (1996). Subsurface Utility Engineering - An initial step in Project Development. *The Proceedings of the Associated Schools of Construction, Texas A&M University, Texas, 1996*, 217–222.

Manacorda, Scott, GH, Rameil, M, Courseille, R, Farrimond, M, & Pinchbeck, D (2007). *"The ORFEUS Project: A Step Change in Ground Penetrating Radar Technology to Locate Buried Utilities"*. Rome: Proc., Presented at ISTT NoDig 2007.

Miles, SB, & Ho, CL. (1999). "Applications and Issues of GIS as Tool for Civil Engineering Modeling". *Journal of Computing in Civil Engineering, 13*(3), 144–152. ASCE.

Mozas, AT, & Ariza, FJ. (2011). New Method for Positional Quality Control in Cartography Based on Lines. A Comparative Study of Methodologies. *International Journal of Geographical Information Science, 25*(10), 1681–1695.

National Research Council (NRC). (2000). *Seeing into the Earth: Noninvasive Characterization of the Shallow Subsurface for Environmental and Engineering Applications* (p. 148). Washington, D.C.: National Research Council.

National Transportation Safety Board (NTSB). (1997). *"Protecting Public Safety through Excavation Damage Prevention"*. Washington D.C.: Safety Study NTSB/SS-97/01.

National Transportation Safety Board (NTSB). (1998). *We are all safer: NTSB-inspired improvements in transportation safety*. Washington, D.C: National Transportation Safety Board (NTSB).

National Transportation Safety Board (NTSB) (2000). (1998). *"Natural Gas Pipeline Rupture and Subsequent Explosion in St. Cloud, Minnesota"*. Washington, D.C: Pipeline Accident Report NTSB/PAR-00/01.

Nelson, R, & Daly, M (1998). *"Creating a Major Emphasis on Damage Prevention,"*. Atlanta, GA: Proc., Damage Prevention Convention.

North, D (2010). *"Marking the Spot" CE News Archives*. http://www.cenews.com/print-magazinearticle-marking_the_spot-8115.html, Accessed 11 January 2012.

Perkal, J. (1956). On Epsilon Length. *Bulletin de l'Académie Polonaise des Sciences, 4*, 399–403.

PHMSA (2011). *"Significant Pipeline Incidents through 2010 By Cause"*. http://primis.phmsa.dot.gov/comm/reports/safety/sigpsi.html, Accessed 31 October.

Poku, SE, & Arditi, D. (2006). "Construction Scheduling and Progress Control Using Geographical Information Systems". *Journal of Computing in Civil Engineering, 20*(5), 351–360. ASCE.

Roberts, GW, Evans, A, Dodson, AH, Denby, B, Cooper, S, & Hollands, R (2002a). *"The Use of Augmented Reality, GPS and INS for Subsurface Data Visualization"*. Washington, D.C: FIG XXII International Congress. April 19–26.

Shi, W, & Liu, W. (2000). A Stochastic Process-Based Model for the Positional Error of Line Segments in GIS. *International Journal of Geographical Information Science, 14*(1), 51–66.

Sipes, L (2007). *"GIS for the Utilities Industry"*. <http://www.cadalyst.com/gis/gis-utilities-industry-spatial-tech-column-9178> [Accessed: November 22, 2011].

Spurgin, JT, Lopez, J, & Kerr, K (2009). Utility Damage Prevention – What Can Your Agency Do? *APWA Congress Session, September, 15*, 2009.

Sterling, RL, Anspach, J, Allouche, E, Simicevic, J, Rogers, CDF, Weston, K, et al. (2009). *Encouraging Innovation in Locating and Characterizing Underground Utilities, SHRP 2 Report S2-R01-RW*. Washington, D.C.: Transportation Research Board.

Stevens, RE, & Anspach, JH (1993). *"New technology overcomes the problems of underground system interferences on power projects"* (v 55 part 1st ed., pp. 323–326). Chicago, IL, USA: Proceedings of the American Power Conference, Published by Illinois Institute of Technology.

Talmaki, SA, & Kamat, VR (2012). "Real-Time Hybrid Virtuality for Prevention of Excavation Related Utility Strikes". *Journal of Computing in Civil Engineering, American Society of Civil Engineers, Reston*. In Review.

Talmaki, SA, Kamat, VR, & Cai, H (2012). "Geometric Modeling of Geospatial Data for Visualization-Assisted Excavation". *Advanced Engineering Informatics,* 27(2), 283–298.

Unwin, DJ. (1995). Geographical Information Systems and the Problem of Error and Uncertainty. *Progress in Human Geography, 19*(4), 549–558.

Utility Notification Center of Colorado (UNCC). (2005). *Perspectives on Facility Damage—2005* (p. 76). Golden, Colorado: UNCC.

Vincenty, T (1975). "Direct and inverse solutions of geodesics on the ellipsoid with application of nested equations.". *Survey Review, 23*(176), 88. Maney Publishing, Leeds, UK.

WRAL archives (2011). *"Who's to Blame for Cary Gas Line Rupture?"*. http://www.wral.com/news/local/story/1916911/, Accessed 17 June 17.

Wu, HS, & Liu, ZL (2008). *"Simulation and Model Validation of Positional Uncertianty of Line Feature on Manual Digitizing a Map"*. Beijing China: The International Archives of the Photogrammetry, Remote Sensing and Spatial Information Sicences. XXXVII, Part B2.

Chip-based adaptive skin color detection using trajectory constraints on hue

Ying-Hao Yu[1*], Tsu-Tian Lee[2,5], Ngai-Ming Kwok[3] and Pei-Yin Chen[4]

Abstract

Background: In a construction site, the safety of workers can be better secured if vehicle or robot can be properly maneuvered by using image-based gesture guidance. This kind of semi-automatic motion control is based on the features extracted in colors, outlines, textures, and motion intentions. Among these features, detecting worker s skin color on face and hands could serve as an essential method for enabling robot to determine the region of interest (ROI). However, the performance of skin detection is usually unreliable due to interferences from reflections and shadows. Although many efficient skin color detection methods had been developed in past years, those are mostly based on complicated learning and statistical processes and are unsuitable for embedded systems. The exploration of a concise and efficient skin color detector, therefore, becomes a challenge for applications in mobile robots.

Method: In this paper, we propose a novel adaptive skin detector on face and hands as a fundamental capability of a gesture tracking system. This approach enhances the detection performance of traditional HSV color space but only requires a low computing power. For the design criteria of small size, low-power, low-cost, and minimum computing resource usages on mobile robots, the entire detecting system is built on single field-programmable-gate-array (FPGA) chip. Meanwhile, besides the contributions of adaptive algorithms and FPGA chip designs, the proposed skin detector also employs a touch screen to designate expected skin color of worker as the human-robot interaction (HRI) in real-time.

Results: The chip design of the FPGA is based on hardware circuits in register-transfer-level (RTL) for real-time image processing. A reasonable amount of hardware resource usages of FPGA have consumed 8% of logic elements (LEs) and 1.4% for embedded random access memory (RAM). Demonstrations with various pictures had indicated that sufficient ROI can be efficiently identified by using the proposed adaptive skin color detector.

Conclusions: According to our experimental results, the proposed adaptive skin detection can work well for non-ideal illumination. It has demonstrated the reliability and feasibility on supporting the embedded robotic control in the future.

Keywords: Gesture tracking; Skin color detection; Color space; Field-programmable-gate-array (FPGA); Trajectory constraints on hue (TCH)

Background

In past years, skin color detection has been regarded as an essential component in human-robot interaction systems for analyzing human intentions (Kim et al. 2006; Luo et al. 2011). This methodology had become popular because it requires a lower computing effort than other image processing approaches even though skin color analysis is a fundamental area of pattern recognition. Recently, we have also developed a chip-based gesture tracking system by cooperating with motion detection to control mobile robots on construction sites (Yu et al. 2014). Our system simulated a robot to receive worker s instructions by detecting a traffic light baton on his hand. Once our gesture tracking system confronts the situation of similar lighting sources on the same field, localizing traffic light baton involving regional information of worker s face and hands will be an efficient and feasible scheme.

In this paper, we propose an adaptive skin color detection strategy by using single FPGA chip. We replaced image collection and training processes with a touch screen as the

* Correspondence: yinhaun@gmail.com
[1]Department of Electrical Engineering, National Chung Cheng University, No. 168, University Rd, Min-Hsiung Township, Chiayi County 62102, Taiwan
Full list of author information is available at the end of the article

man machine interface after considering the characteristics of skin colors of different human races and working environments. The worker can designate his skin color as a target through the interface, then the proposed adaptive skin color detection system, based on HSV color space, can define a preliminary threshold as a skin map. With the input of images, a novel adaptive algorithm named trajectory constraints on hue (TCH) is proposed to dynamically adjust the thresholds of skin color detection.

The contributions of our work rest on the economic circuit resource usages on chip, satisfactory real-time performance, high detection rate under non-ideal illumination, and convenience in installation. Comparing with the state-of-the-art technologies either based on PC simulations and FPGA chip designs, our skin color detection system can achieve superior detecting rate under severe tests. While nowadays efficient skin color detection mostly relies on the complicated algorithms, intelligent learning, skin texture analysis, working on high-end computers or a large amount of FPGA gates count; our concise skin color detection provides the answer to improve detection rates on embedded systems and miniature robots.

In the rest of this paper, the background of skin color detection will be first reviewed. Next, the relationship between HSV and RGB color spaces and the mechanism of the proposed skin color detection by using TCH will be presented. The configuration of chips, hardware resource usages, and the performance of the proposed skin color detection will be described in fourth section. The final sections contain the discussion and the conclusion.

Background

In designing a skin color detection system, there are many color spaces that can be chosen to obtain the components of skin color. The initial idea comes from the use of thresholds to examine human skin color with the red (R), green (G), and blue (B) color space. Human skin color can be determined by the specific magnitudes of R, G, and B. This kind of RGB thresholding method has been recognized as unreliable in that the imaged skin color is highly dependent on the combination of luminance on the skin and results in poor resistance to non-ideal illumination, e.g., shadows and reflections (Yang et al. 2010). Although the design by using multi-intensity thresholds can mitigate the interferences of shadows and reflections, complicated calibrating processes for different intensity thresholds become another problem for implementation (Yu et al. 2011). Consequently, modern color detection systems prefer to use color attributes for detection, which are improved from the use of RGB color space for independent color components and discarding intensity, e.g., Normalized RGB, YCbCr, YUV, YIQ, and HSV (Chaves-Gonzłez et al. 2010; Liu et al. 2009). With these kinds of color spaces, it is expected to detect skin without the being affected by intensity. Better detecting performance can be achieved by combining different color spaces in order to support each other (Kumar 2014; Liu and Shi, 2011).

However, the hue of skin color may vary in real-life operations. The interferences from shadows and reflections usually shift skin color beyond thresholds and degrade detection performance if intensity-independent color spaces were used (Zhu and Cai, 2011). This leads to skin color detection methods developed on the basis of adaptive or learning methodologies. For example, Liu employed a variable as the learning rate of histogram to adapt to different environments and then improved skin color detection by using morphological operations (Liu and Peng 2010). Similar work can be also found on (Li et al. 2013); where authors simply adopted a ratio to evaluate the deviation distance from centroid value of skin color. This proportion played the role as a threshold to refine their skin color maps in the database. Moreover, Khan et al. (2011) designed a skin detector by using RGB color space and Bayesian-based classifiers training, which contained about 5000 samples of skin samples. Chenaoua and Bouridane (2006) employed the Markov Random Field to estimate the distribution of skin pixels in images. Fotouhi et al. (2009) combined with YC_bC_r, texture extraction, Gaussian mixture models, multi-layer perceptron, and artificial neural networks to perform a comprehensive skin detector. In summary, adaptive skin color detection usually has higher detection rate than methodologies based on traditional color spaces. Particularly for the system designed with learning capabilities, the regions of skin color can be reliably detected except small areas on chin, neck, and fingers.

Although a sufficient skin detection rate can be achieved by probability and learning, the real-time detection performance while running with limited computing power is also another critical issue in robotic applications. Intensive-computing algorithms may achieve a superior detection rate for a static picture, but it is unsuitable for a high resolution and operating at video frame rate. In the works of Toledo et al. (2006), Zheng et al. (2010), and Liu and Shi (2011), the detection of skin color was obtained from rules table, probability, normalized RGB, YCbCr and $I_1I_2I_3$ color spaces for low computing effort. However, the shadow regions on skin could not be detected efficiently or they could only work under ideal illumination by lacking of adaptive capability. Zafarifar proposed a skin detector by using YUV color space, texture-adaptive analysis, and DSP slices on chip (Zafarifar et al. 2012). Their works featured additional capability of background eliminations, which were colored as skin. However, a large amount of hardware circuits on chip were required and detection performance in non-ideal illuminating environments was difficult to verify.

In general, skin color detection based on high-level statistics and machine learning usually requires huge computing resources. In spite of the sufficient performance with static image analysis, these methodologies usually result in low frame rate and therefore are unsuitable for tracking with embedded video systems without any modification (Fotouhi et al. 2009). On the other hand, traditional skin color detection working on basic color space thresholds cannot provide enough quantity of detection for real-life applications. These design dilemmas had motivated our work both on computer- and embedded-based systems, which provides excellent detection rate to provide ROI of targets for further pattern recognition processes.

Methods

Relationship between RGB and HSV

Many intensity independent color spaces can be utilized for skin color detection. Here we have adopted HSV color space after considering the advantage of separating skin color in hue, saturation, and intensity. The distribution of hue and saturation on a circular plan is convenient to develop an algorithm for examining the trajectory variation of skin color.

The HSV color space is known as an improvement from traditional RBG color space. The term of H denotes the hue (or color-depth), and S is for saturation (or color-purity). Intensity is represented as V (or brightness). The algorithm for converting RGB into HSV color space can be expressed as the follows (Chaves-Gonzľez et al. 2010):

$$H \begin{cases} Undefined, & if \quad MAX \quad MIN \\ \frac{60\ G-B}{MAX-MIN}, & if \quad MAX \quad R \ and \ G \geq B \\ \frac{60\ G-B}{MAX-MIN} \quad 360, & if \quad MAX \quad R \ and \ G < B \\ \frac{60\ B-R}{MAX-MIN} \quad 120, & if \quad MAX \quad G \\ \frac{60\ R-G}{MAX-MIN} \quad 240, & if \quad MAX \quad B \end{cases} \qquad 1$$

and

$$S \begin{cases} 0, & if \quad MAX \quad 0 \\ \frac{MAX-MIN}{MAX} & Otherwise \end{cases} \qquad 2$$

$$V \quad MAX \qquad 3$$

where *MIN* and *MAX* in equations (1) to (3) denote the RGB-intensity bound in [0,1]. The distribution of hue is usually plotted as a circular plan in four quadrants, and the plan radius is equivalent to the saturation value S. Hue is composed of red, green, and blue colors changing by every 120˚. When the HSV color space is plotted as a 3D model, the part right under the H-S circular plan is the intensity V of color space, which is in an inverted conical shape.

As shown in Figure 1, by changing the magnitudes of RGB from the scale of 0 to 255 with constraints for $R \geq G \geq B$, the variation of blue are approximately the S, and adjusting green will obtain H. On the other hand, the trajectory of red color combines with the features of H and S at the same time. According to the studies of Vezhnevets et al. (2003) and Lin et al. (2011), human skin color can be roughly classified as $R > G > B$, and different skin colors can be derived by adjusting R, G, or B magnitude arbitrarily within the constraints. By avoiding the use of intensity value, the variation of skin color is the same as the changes of H and S.

Based on above discussion, most colors can be simulated by altering the positions of colors in constraints. For example, the relationship of $R > B > G$ results in a pink color and the B trajectory becomes H instead. If we continually increase the magnitude of B, the output of color will be closer to vivid blue in a clockwise direction for $B > R > G$ towards the plan boundary. Since the behavior of color between upper and lower constraints is similar to hue, here we define the color in the middle as the dominant color. Therefore, it can be seen that human skin color is dominated by green.

The definition of H-S plan

According to the discussion in the last sub-section, the intensity V could be neglected in skin color detection. Although skin color may look different under dim environments, one can achieve skin color discrimination by only examining the variations on hue and saturation. When detecting skin color by using HSV color space, the H-S plan can be plotted as in Figure 2(a). The coordinate of centroid *T* of skin color map can be expressed as:

$$T_x \quad S\cos H \qquad 4$$

$$T_y \quad S\sin H \qquad 5$$

thus

$$S \quad \sqrt{T_x^2 \quad T_y^2} \qquad 6$$

where the hue H in equations (4) and (5) is equivalent to the angle suspended by vector *T*, and *S* is the radius of the circular plan. Based on the coordinate of *T*, an initial skin color map can be defined by four points of as:

$$T_A\ H, S \qquad H_T \quad E_H, S_T - E_S \qquad 7$$

$$T_B\ H, S \qquad H_T - E_H, S_T - E_S \qquad 8$$

$$T_C\ H, S \qquad H_T - E_H, S_T \quad E_S \qquad 9$$

$$T_D\ H, S \qquad H_T \quad E_H, S_T \quad E_S \qquad 10$$

where H_T and S_T denote the H and S of *T*, and the variables E_H and E_S both are tolerances in determining

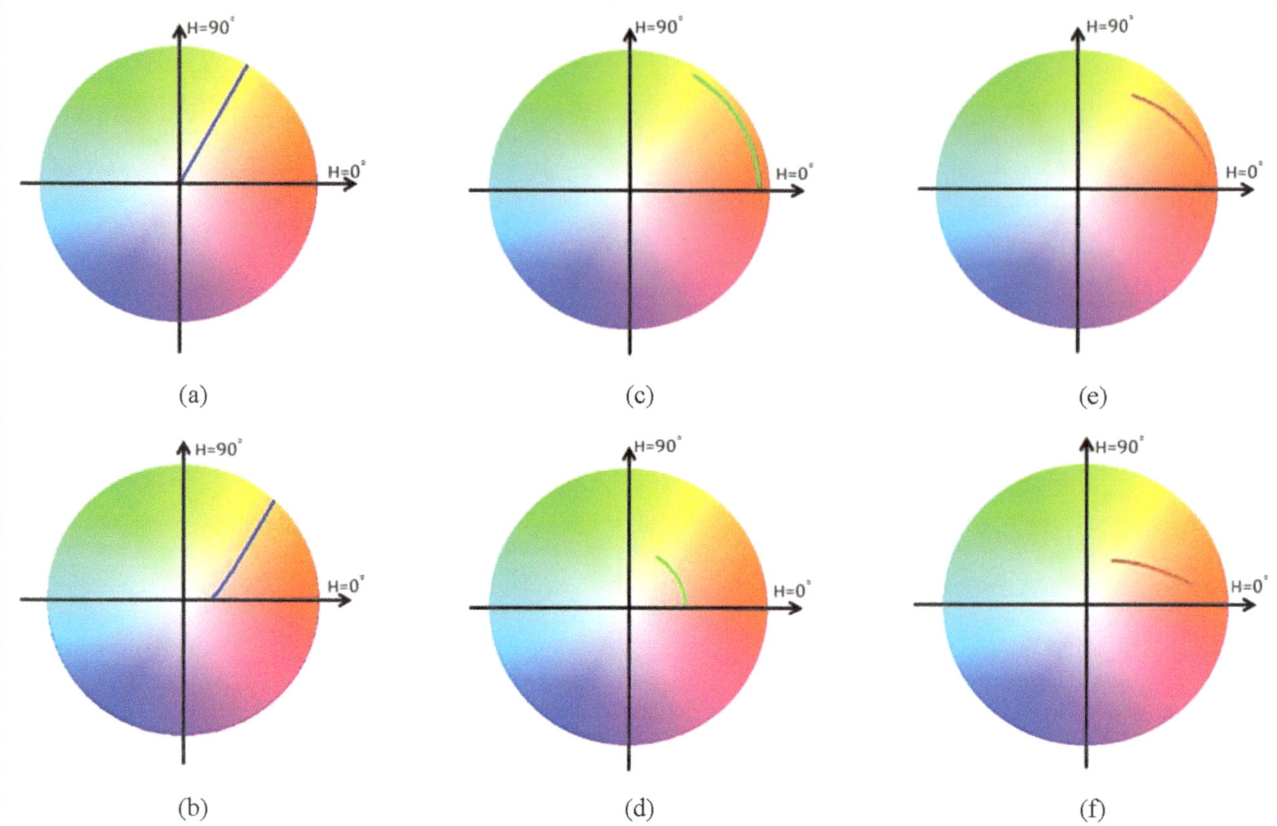

Figure 1 The trajectories in HSV color space by adjusting R-G-B color individually; (a) R = 255,G = 250,B = 0 ~ 255, (b) R = 146,G = 115, B = 0 ~ 115, (c) R = 240,G = 20 ~ 240,B = 20, (d) R = 170,G = 100 ~ 170,B = 100, (e) R = 52 ~ 255,G = 52,B = 12, (f) R = 92 ~ 255,G = 92,B = 60.

the range of initial skin color map *F*, as shown in Figure 2(b). The intensity scale of H-S plan is [0, 255] for an 8-bits digital system. Thus the coordinates of initial skin color map are:

$$T_{A\;x,y} \quad S_{T_A}\cos H_{T_A}, S_{T_A}\sin H_{T_A} \qquad 11$$

$$T_{B\;x,y} \quad S_{T_B}\cos H_{T_B}, S_{T_B}\sin H_{T_B} \qquad 12$$

$$T_{C\;x,y} \quad S_{T_C}\cos H_{T_C}, S_{T_C}\sin H_{T_C} \qquad 13$$

$$T_{D\;x,y} \quad S_{T_D}\cos H_{T_D}, S_{T_D}\sin H_{T_D} \qquad 14$$

According to equations (7) to (10), various skin color can then be obtained by adjusting the range of H and S. Chen et al. (2012) had defined humans skin colors by using constraints of $0 \le H \le 50$ and $51 \le S \le 173$. Before

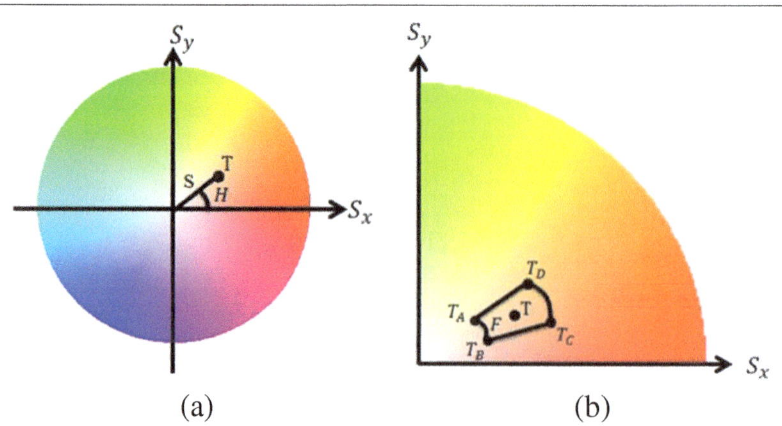

Figure 2 The definition of H-S plan; (a) the H-S plan, (b) the initial region of skin color map.

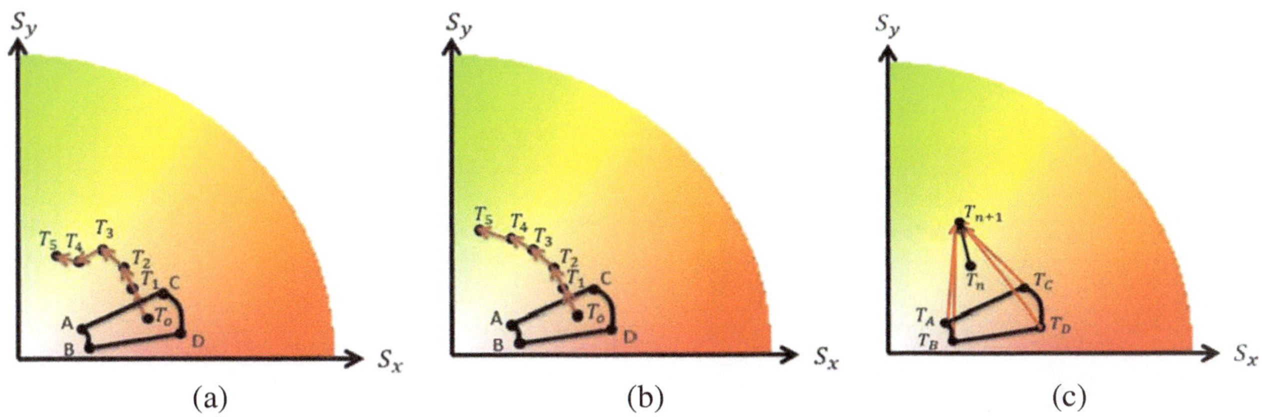

Figure 3 The operations of TCH algorithm; (a) incoming image pixels colors exceeding out of skin region *F*, (b) updating colors saturations with homogenous value of T_0, (c) examining slope using line $\overline{T_nT_{n\ 1}}$.

this, Yang et al. (2011) had also defined the skin color for the east Asian races by using the constraints of $14.4 \leq H \leq 46.8$ and $69.615 \leq S \leq 149.55$. Here we define the centroid point *T* by using $3 \leq H \leq 18$ and leave *S* as undefined in that value is varied to different human races and can be dynamically determined by touch screen. Finally, the tolerance of skin hue is set for $E_H = 25$ and the range of $E_S = 30$ is for saturation testing based on our experiments. This arrangement leads to TCHs adaptability on H which will mostly stop around yellow and only a small amount of sampled greenish color will be passed at low saturation.

Skin color detection by using TCH algorithm

After described the H-S plan characteristics, next we can determine an adaptive skin color threshold by using the TCH algorithm, which can be discussed as in the following cases.

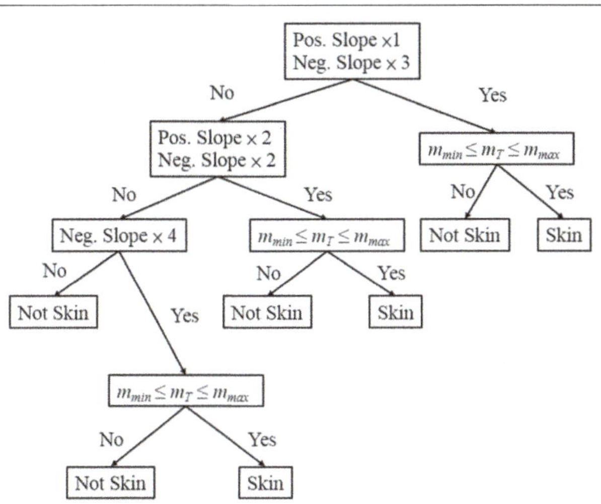

Figure 4 The decision tree to judge skin color.

1. $|H_i - H_T| \leq 25$ and $|S_i - S_T| \leq 30$:

If the input of sampled image pixel color (H_i, S_i) always falls into the region of initial skin color map *F* as shown in Figure 2(b), this pixels is defined as corresponding to human skin.

2. $|H_i - H_T| \leq 25$ and $|S_i - S_T| > 30$:

These kinds of input colors indicate that pixels in image have exceeded saturation values of the initial skin color map *F* but are still with acceptable hue. Since the saturation value is impossible for human skin color, the locations of these pixels in picture are defined as the non-skin areas.

3. $|H_i - H_T| > 25$ and $|S_i - S_T| \leq 30$:

Skin color may vary slowly in hue by reflections and illumination. It makes it difficult to determine skin color without learning capability, as shown in Figure 3(a). Once any sampled image pixel color first conforms to this case, the adaptive algorithm using TCH is triggered and then proceeds in the following steps:

a. Record the pixel saturation value S as a constant, which is from the last color T_0 before exceeding the initial skin color map *F*, as shown in Figure 3(a) and (b). This procedure ignores the effects from saturation besides *F*. In other words, S will be unvalued if H is too far away from centroid T.
b. If incoming image pixels colors are all outside the region of *F*, the proposed TCH will continually replace these image pixels saturations from T_0 otherwise the system resets its operation to the initial condition. Such arrangement leads to varying trajectories of image pixels colors beyond both sides

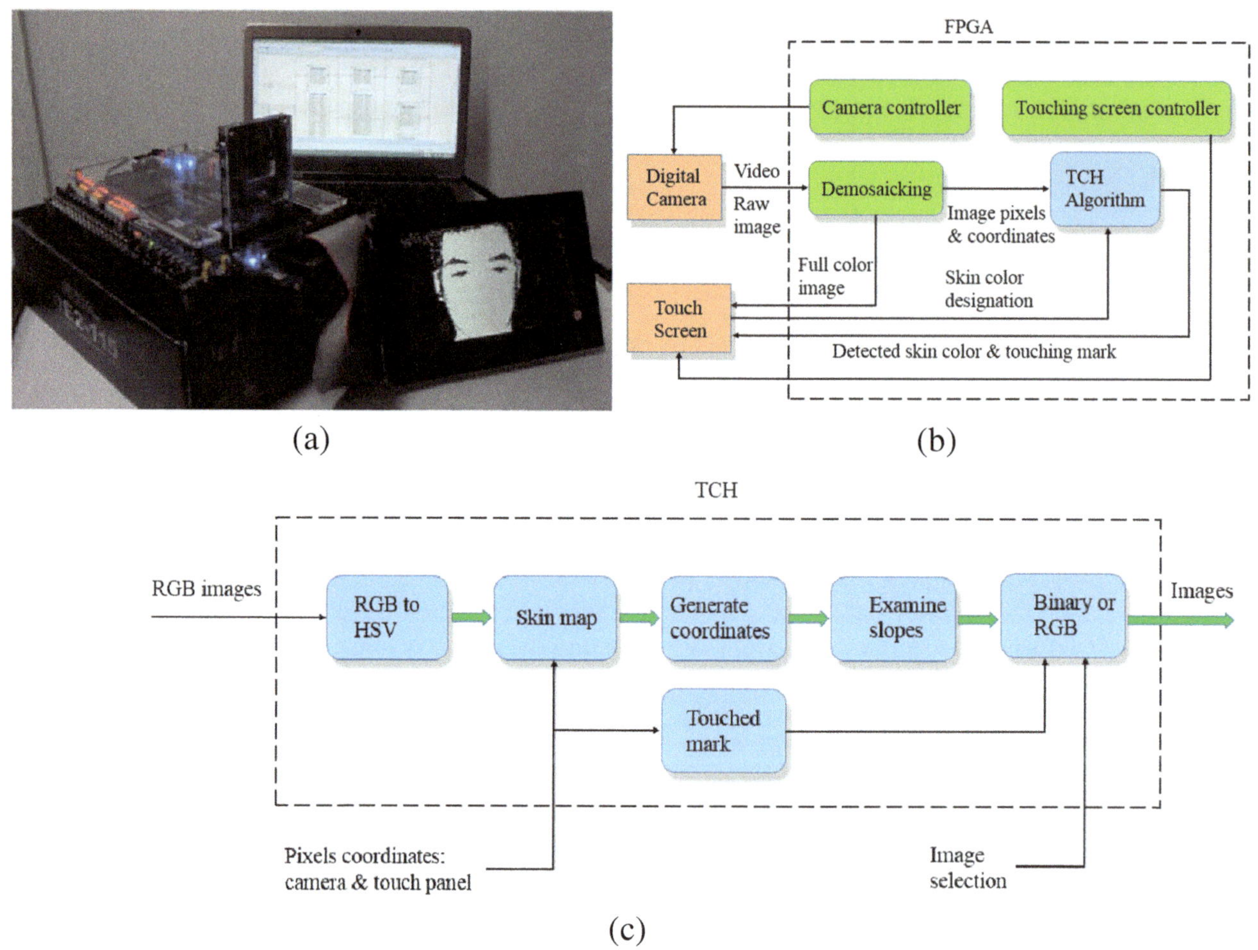

Figure 5 The proposed chip-based skin detection; (a) System setup, (b) FPGA block diagram, (c) TCH block diagram.

of F and will all follow a concentric circle, as shown in Figure 3(b). This results in a simple and adaptive skin color detection process which only needs to examine the variances on the pixels hue.

c. Examining the slope of line $\overline{T_nT_{n\ 1}}$ after obtained the coordinates of points T_A to T_D, as shown in Figure 3(c); T_n, and T_{n+1} are derived based on

Table 1 FPGA resource usages of proposed skin color detection

Chip family	Cyclone IV E
Device	EP4CE115F29C7 N
Total logic elements	9,216 LEs/114,480 (8%)
De-mosaicking and camera controller	1,275 (1.1%)
TC algorithm	7,646 (6.7%)
Touch screen control	295 (0.2%)
Embedded multiplier 9-bit elements	21/532 (3.9%)
Total embedded memory bits	57,400 bits/3,981,312 (1.4%)
Total PLLs	2/4 (50%)

equations (4) to (14), and the corrected hue trajectory of line $\overline{T_nT_{n\ 1}}$ will run on a circle as tangent. Once the slope m_T of tangent $\overline{T_nT_{n\ 1}}$ exceeds the slopes constraints of four lines $\overline{T_{n\ 1}T_A}$, $\overline{T_{n\ 1}T_B}$, $\overline{T_{n\ 1}T_C}$, and $\overline{T_{n\ 1}T_D}$, the pixel color at point T_{n+1} is defined as a non-skin color and then the system is reset to the initial condition. The examination mechanism can be represented in a decision tree shown in Figure 4. The m_{min} and m_{max} in the tree denote the upper and lower constraints of four lines slopes respectively. The slopes of $\overline{T_{n\ 1}T_A}$, $\overline{T_{n\ 1}T_B}$, $\overline{T_{n\ 1}T_C}$, and $\overline{T_{n\ 1}T_D}$ on both sides of skin color map F can be summarized for three conditions at $H_i \geq 0$:

State_1: One positive slope (1) and three negative slopes (3).
State_2: Two positive slopes (2) and two negative slopes (2).
State_3: The slopes of the four lines are all in negative (4).

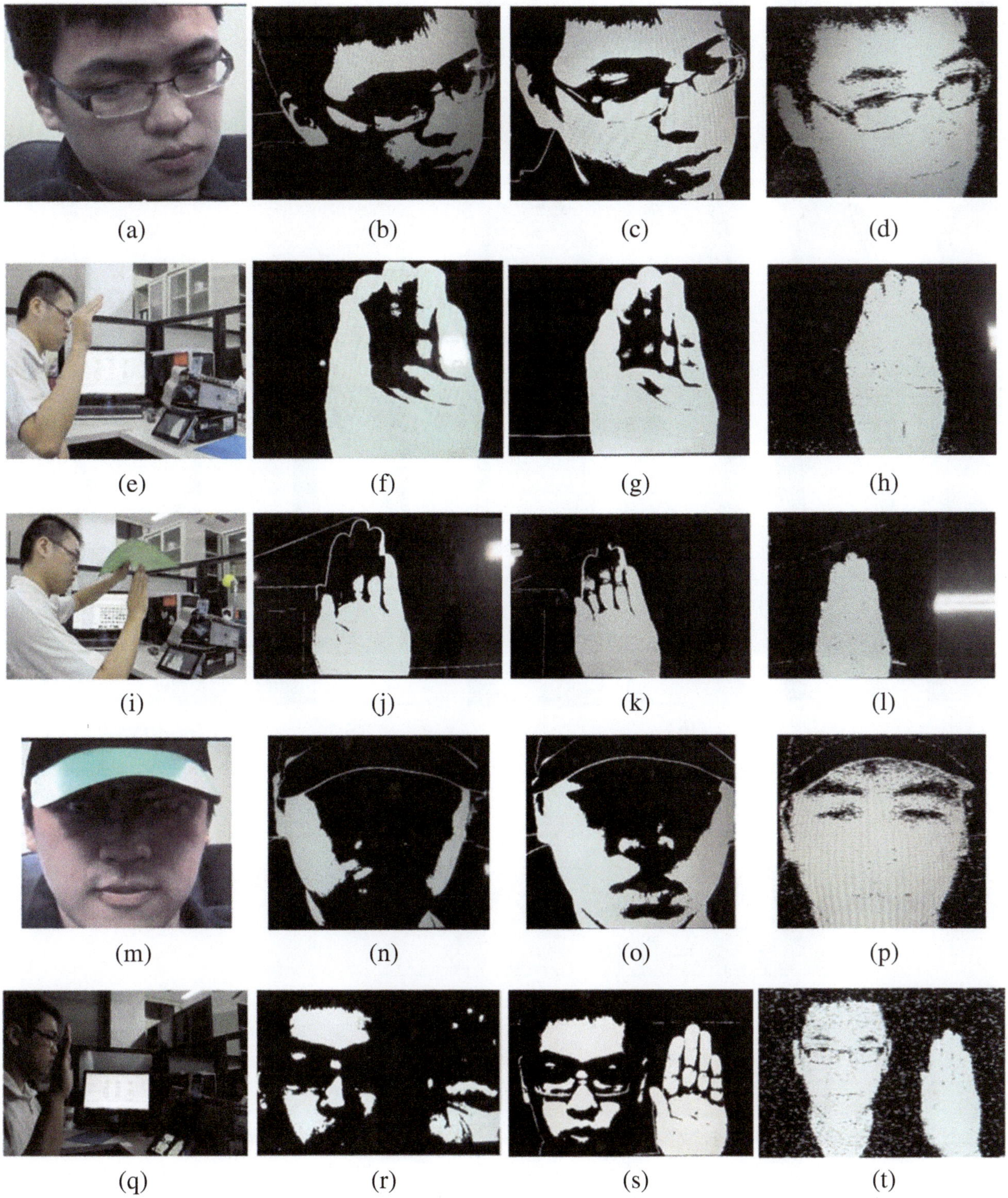

Figure 6 Comparisons among traditional RGB and HSV thresholds and proposed TCH algorithm; (a)(e)(i)(m)(q) the test scenarios, (b)(f)(j)(n)(r) tested with RGB threshold, (c)(g)(k)(o)(s) tested with HSV threshold, (d)(h)(l)(p)(t) tested with proposed TCH.

The detected skin color below line $\overline{T_B T_D}$ represent lip and blood vessel colors which can be judged as a part of the skin.

Results

Figure 5(a) shows the hardware setup of the proposed skin color detection system. A touch screen is employed

Figure 7 Proposed TCH without noise filter for working on various human races and illumination conditions with imperfect reproduction of pictures; (a)-(d) are from Vanity fair (2014), (e) is from Gamebase (2008), (f) and (g) are from Sina house (2011), (h) is from Gigacircle (2014), (i) is from 10202blog (2013), (j) is from Sohu sport (2010), (k) and (l) are from Eastday (2013), and (y) is from Nextmedia 2013.

to designate expected skin color of worker and examines detection performance. The digital camera module is vertically installed on a slot of the FPGA platform. Synthesis of FPGA circuits and summary of chip resources usages are based on the Quartus II from Altera software. The block diagram of FPGA chip is shown in Figure 5(b). Raw video images from the digital camera are converted into full color images by the real-time de-mosaicking of our previous works (Yu et al. 2011). Full color images from de-mosaicking are then sent to the touch screen for user to designate expected skin color. Once any skin region has been designated (touched) by the user s finger, the information of skin color and screen coordinate of the touched point are sent to the TCH algorithm together for skin detection. As shown in Figure 5(c), output images can be controlled by the signal denoted as Image Selection. A full color was first chosen to observe the region of skin color. After the user touching screen, a touched mark is generated and shown on the screen to indicate the location of the touching point. This action also determines the initial skin color map F in HSV color space, which is transferred from RGB. Next the user switch image output to detection mode. Skin color map coordinates T_A to T_D are then derived and examined for slopes constraints as aforementioned. In the end, the skin color detected by TCH is represented in a binary image and displayed on the touch screen.

Resource usages of the proposed FPGA chip designs are shown in Table 1. Consumption of logic elements (LEs) are 9,216 (8%), which involves 7,646 LEs for TCH. The FPGA chip is the Altera Cyclone IV EP4CE115F29C7 N. The proposed TCH was designed in RTL circuits in order to achieve real-time image processing. It can be seen that the TCH algorithm consumed the largest amount of LEs in the proposed skin color detection system that the algorithm required 21 9-bit embedded multipliers for RGB-HSV conversion and slope derivation with trigonometric functions. The embedded memory consumption was 57,400 bits (1.4%). This consumption was for the image buffer of touch screen and full color de-mosaicking.

The digital camera adopted in our design has a maximum of 5 M pixels resolution, but it was set for 800 ? 480 pixels for use with the touch screen at a frame rate of 15fps and 30 MHz pixel clock. Meanwhile, the proposed TCH required 300 MHz clock speed and 37 clocks to run TCH during image pixel scans. Such arrangement leads to about four pixels delay showing on the touch screen.

Snapshots of real-life operations for proposed skin color detection are shown in Figure 6 without a video noise filter. Five severe test scenarios were designed to demonstrate the reliability of the proposed adaptive skin color detection system. We focused on the resistances to shadows and reflections at forehead, chin, hands, and the part of the face under the brim of a cap. Figure 6 demonstrates excellent skin color detection of TCH, which provides satisfactory ROI area for pattern recognition, object tracking, and image processing purposes. In these test scenarios, the skin color detections by using traditional RGB and HSV thresholds were also conducted to compare with our TCH algorithm using constant skin color map. We could see that the skin color detection using RGB threshold in these tests exhibited the worst detection rate while the proposed TCH algorithm had the best detection performance in tests.

Table 2 Comparisons of resource usages among different skin detector designs

Items	(1)	(2)	(3)	(4)	(5)	(6)
Resources	Chenaoua and Bouridane (2006)	Fotouhi et al. (2009)	Khan et al. (2011)	Liu and Shi (2011)	Zafarifar et al. (2012)	Proposed TC
Platforms	PC	PC	PC	FPGA	FPGA	FPGA
				EP2C20	Xilinx	Cyclone IV E
						EP4CE115F29C
						7 N
Methods	Markov random field	Texture analysis & ANN classifier	Cascaded skin training & classifier	YC_bC_r & $I_1I_2I_3$ Classifiers	YUV & Adaptive texture analysis	Adaptive HSV constraints
Tested images	Static picture	Static picture	Static picture	Video	Video	Video
Adaptability	Yes	Yes	Yes	No	Yes	Yes
Resource usages (LEs/LUTs)	-	-	-	4,091	30,720	7,646
DSPs	-	-	-	No	192	No
Embedded memory (bits)	-	-	-	36,512	3,456,000	57,400

Figure 8 The comparison with skin detection methodologies from references; (a)-(e) are test scenarios from Table 2, (1)-(5) with their results in (f)-(j), and (k)-(o) are for pixel-based TCH tested without noise filter.

In the first test scenario of Figure 6, see Figures 6(a) to (d), the interferences of shadows and reflections existed at the same time on tester s face. It can be seen that the proposed TCH skin color detection system demonstrated excellent adaptive capability to detect skin areas with shadows and reflections. For the test result using HSV threshold, although it has been a popular method on embedded systems, the detection rates at testers chin and neck were still deteriorated by shadows. The detection performances between RGB and HSV thresholds had not shown much difference in this test. In the second test scenario, we tested TCH algorithm with extreme interference

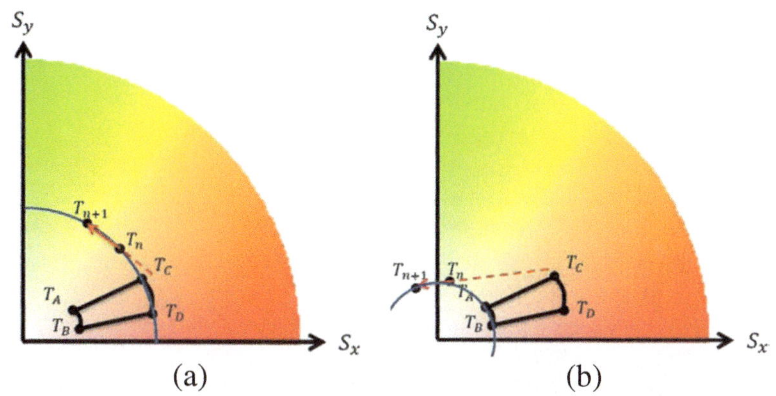

Figure 9 Operations of TCH at upper and lower boundaries of adaptability; (a) upper saturation boundary, (b) lower saturation boundary.

of shadow by huddling up a palm. The RGB and HSV thresholds both still had low detection rate in test. Comparatively, proposed TCH algorithm in this test could detect the skin color reaching the inner part of palm perfectly. In the third test scenario, we tested our skin detector by using green color to shift skin color moderately. The tester covered his palm with a green paper to obtain some green reflection on palm, as shown in Figure 6(i). It can be seen that the proposed TCH algorithm operating in such environment could still resist to green shadow and reflection. Again, RGB and HSV constant thresholds in this test still had low detection rate at fingers, where was covered by green shadow. After illustrated the success in resistance to shadows and reflections, the fourth test scenario simulated the condition of wearing a cap, as shown in Figure 6(m). In this test, the tester s face was covered by shadow and some yellow reflections from inner brim of cap. The RGB threshold with this test scenario nearly failed to detect the tester s face, and even the HSV threshold could only improve little. Nevertheless, our TCH algorithm could still work quite well in this test except a small part right under the brim of cap. In the last test scenario of Figure 6, the capability of skin detection under a dark environment was also tested, as shown in Figure 6(q). The lights in laboratory were nearly turned off in this test scenario. It can be seen the RGB and HSV thresholds could surprisingly detect skin region of the subject, and the silhouette of HSV threshold was sufficient for recognition on the face and palm. The proposed TCH skin color detection system under such severe environment was apparently superior to traditional HSV threshold for full scope detection rate of the subject s face and palm.

For quantitative analysis, we tested TCH with different human races and illumination conditions. Tested images were all downloaded from the internet (Eastday 2013; Gamebase 2008; Gigacircle 2014; Nextmedia 2013; Sina house 2011; Sohu sport 2010; Vanity fair 2014; 10202blog 2013) and references and then printed by a color printer so that digital camera could capture these pictures into FPGA. However, it should be noted that this test method also incurred chromatic distortions due to imperfect reproduction in paper and from the printer. Figure 7 indicates that TCH could detect these candidate faces efficiently even in outdoor environments. With the same test method, we also compared with state-of-the-art methodologies given in Table 2, and experimental results are shown in Figure 8. The images for comparison were copied from articles and enlarged so that the TCH system was fed with images of larger chromatic distortions than that given in Figure 7. It can be seen that TCH with such severe test condition can also provide sufficient detection accuracy for ROI.

Discussions

The salient features of the proposed skin color detection system involve simple calibration processes, moderate hardware resource usage on chips, and efficient adaptability without complicated training algorithms. Comparing with the state-of-the-art technologies listed in Table 2, the proposed TCH adaptive algorithm cooperating with a touch screen is comparable to the methodologies using statistical and training classifier for high detection rate and low implementation effort. Besides, the combination of different color thresholds may achieve sufficient detection rate on embedded system by using a small amount of hardware resources (Liu and Shi, 2011), but such system can only work under adequate illumination and tedious calibration is still unavoidable because of the lack of adaptive mechanisms. After inspecting the skin detector results in Figure 8, the best skin color detection design for embedded system is the texture-based skin color analysis; see the performance in Figure 8(j). This kind of system is not only applicable for skin color detection but also provides additional functionality on background elimination, where it is colored with skin color. However, a large amount of hardware resources usages is required for complicated texture analysis. Consequently, it dramatically increases design difficulty on real-time processing speed and the implementation in a small chip.

In addition, the adaptability of TCH s skin color detection is determined by the pixel saturation before exceeding the initial skin color map *F*. As shown in Figure 9, the pixel color before exceeding the initial skin color map at a high saturation value will result in a low adaptability to variation in skin color, and a null adaptability will be set if the exceeded color is at the upper saturation boundary of the initial skin color map, as shown in the Figure 9(a). In contrast, TCH s adaptability will be increased if the exceeded color is at a low saturation value, and it results in maximum adaptability at the lower saturation boundary of initial skin color map, as shown in the Figure 9(b). The proposed non-linear adaptive mechanism is similar to human s vision in that people can distinguish different colors easily between high saturation colors but not the low saturation ones.

Finally, the minor drawback of TCH arises from the delay of image processing and incurs coarse binary picture quality, as shown in Figures 6 to 8. With the pixel-based skin color detection, the image quality is sensitive to random noise in the camera, then the noise is enlarged horizontally due to the late update of pixels information during image acquisition. Although the coarse binary picture quality seems not affecting detection rate, the authors will determine how to improve processing speed with refined hardware circuits in the future, and a real-time and simplified pixel-based noise filter from our previous work may be also implemented (Yu et al. 2011).

Conclusion

In this paper, we have proposed an efficient adaptive skin color detection system. The proposed algorithm can derive an initial skin color map using traditional HSV color space after captured images from touch screen. Once incoming pixel colors exceed the region of default skin color map, the non-linear adaptive algorithm of trajectory constraints on hue (TCH) can dynamically adjust the range of thresholds for detection. Various severe test scenarios had been arranged to verify the capability of proposed skin detection. Our experimental results indicate that the proposed adaptive skin color detection is efficient and applicable to various effects from illumination conditions. A reasonable amount of hardware resource usages in the FPGA chip can allow the TCH skin color detection to be implemented on single FPGA chip and suitable for miniature electronic devices. The future works involve the refinement of circuits in economical gates count and processing delay improvement.

Competing interests

The authors declare that they have no competing interests.

Authors contributions

The first and second author contributed to 75% of the work, while the other authors share equally the rest of the work. All authors read and approved the final manuscript.

Acknowledgments

Supports from Ministry of Science and Technology, funded by the government of Taiwan under Grant NSC 99-2221-E-027-057-MY3, Ministry of Education Taiwan for the Top University Project to the National Cheng Kung University (NCKU), and Mr. En-Shi Lin are gratefully acknowledged.

Author details

[1]Department of Electrical Engineering, National Chung Cheng University, No. 168, University Rd, Min-Hsiung Township, Chiayi County 62102, Taiwan. [2]Department of Electrical Engineering, Tamkang University, No. 151, Yingzhuan Rd, Tamsui Dist, New Taipei City 25137, Taiwan. [3]School of Mechanical and Manufacturing Engineering, The University of New South Wales, Sydney, NSW 2052, Australia. [4]Department of Computer Science and Information Engineering, National Cheng Kung University, No. 1, University Rd, Tainan City 701, Taiwan. [5]Department of Electrical Engineering, Chung Yuan Christian University, 200 Chung Pei Rd, Chung Li City 32023, Taiwan.

References

10202blog. (2013). *Photos collection*. http://swjhgrade10202.blogspot.tw/2013_10_01_archive.html. Accessed 10. July. 2014.

Chaves-Gonz?ez, JM, Vega-Rodr?guez, MA, G?mez-Pulido, JA, & S?nchez-P?ez, JM. (2010). Detecting skin in face recognition systems: a colour spaces study. *Digital Image Processing, 20*, 806 823.

Chen, YH, Hu, KT, & Ruan, SJ. (2012). Statistical skin color detection method without color transformation for real-time surveillance systems. *Engineering Application of Artificial Intelligence, 25*(7), 1331 1337.

Chenaoua, K, & Bouridane, A. (2006). *Skin Detection Using a Markov Random Field and a new Color Space (IEEE international conference on image processing (ICIP 2006))*. Atlanta, USA: Georgia Institute of Technology.

Eastday. (2013). *Photos collection*. http://big5.eastday.com:82/gate/big5/msnphoto.eastday.com/hdqxb2013/20130717_9/index.html. Accessed 10. July. 2014.

Fotouhi, M, Rohban, MH, & Kasaei, S. (2009). *Skin Detection Using Contourlet-Based Texture Analysis (4th international conference on digital telecommunication (ICDT 2009))*. Colmar, France: University of Haute Alsace.

Gamebase. (2008). *Photos collection*. http://www.gamebase.com.tw/forum/5351/topic/83232878/1. Accessed 10. July. 2014.

Gigacircle. (2014). *Photo collection*. http://tw.gigacircle.com/1039866-1. Accessed 10. July. 2014.

Khan, S, Gupta, GS, & Demidenko, S. (2011). *Adaptive Classifier for Robust Detection of Signing Articulators Based on Skin Colour (6th IEEE international symposium on electronic design, test and application (DELTA 2011))*. Queenstown, New Zealand: Institute of Electrical and Electronics Engineers Incorporated.

Kim, KK, Kwak, KC, & Chi, SU. (2006). *Gesture Analysis for Human-Robot Interaction (8th international conference on advanced communication technology (ICACT 2006))*. Gangwon-Do, Korea: Institute of Electrical and Electronics Engineers Incorporated.

Kumar, A. (2014). *An Empirical Study of Selection of the Appropriate Color Space for Skin Detection: a Case of Face Detection in Color Images (2nd International conference on issues and challenge in intelligent computing techniques (ICICT 2014))*. Ghaziabad, India: Krishna Institute of Engineering and Technology.

Li, C, Liu, YH, & Shih, HC. (2013). *Adaptive Skin Color Tone Detection With Morphology-Based Model Refinement (9th international conference on information, communication, and signal processing (ICICS 2013))*. Tainan, Taiwan: National Cheng Kung University.

Lin, C, Su, CH, & Huang, HS. (2011). Colour image segmentation using relative values of RGB in various illumination circumstances. *Int J Comput, 5*(2), 252 261.

Liu, Q, & Peng, GZ. (2010). *A Robust Skin Color Based Face Detection Algorithm (2nd international conference on information in control, automation, and robotics (CAR 2010))*. Wuhan, China: Hubei University of Education.

Liu, G, & Shi, Z. (2011). *Embedded Implementation of Real-Time Skin Detection System (International conference on transportation, mechanicl, and electrical engineering (TMEE 2011))*. Changchun, China: Jilin University.

Liu, L, Sang, N, Yang, S, & Hunag, R. (2009). Real-time skin color detection under rapidly changing illumination conditions. *IEEE Trans Consum Electron, 57*(3), 1295 1302.

Luo, RC, Chang, SR, & Yang, YP. (2011). *Tracking With Pointing Gesture Recognition for Human-Robot Interaction (IEEE/SICE international symposium on system integration (SII-2011))*. Kyoto, Japan: Kyoto University.

Nextmedia. (2013). *Photos collection*. http://hk.apple.nextmedia.com/realtime/news/20130826/51675805. Accessed 10. July. 2014.

Sagheer, A, & Aly, S. (2012). *An Effective Face Detection Algorithm based on Skin Color Information (8th international conference on signal image technology and internet based systems (SITIS 2012))*. Naples, Italy:

Sina house. (2011). *Photos collection*. http://sx.bbs.house.sina.com.cn/thread-851846640456301-2.html. Accessed 10. July. 2014.

Sohu sport. (2010). *Photos collection*. http://sports.sohu.com/20100320/n270968863.shtml. Accessed 10. July. 2014.

Toledo, FJ, Mart?nez, JJ, Garrig?s, J, Ferr?ndez, J, & Rodellar, V. (2006). *Sink Color Detection for Real Time Mobile Applications (16th international conference on field programmable logic and applications (FPL 2006))*. Madrid, Spain: Universidad Aut?noma de Madrid.

Vanity fair. (2014). *Photo: The 2014 Vanity Fair Hollywood Portfolio*. http://www.vanityfair.com/hollywood/photos/2014/03/chuck-close-hollywood-portfolio#6. Accessed 10. July. 2014.

Vezhnevets, V, Sazonov, V, & Andreeva, A. (2003). *A Survey on Pixel-Based Skin Color Detection Techniques (13th international conference on computer graphics (GraphiCon 2003))*. Moscow, Russian Federation: Moscow State University.

Yang, G, Ren, J, Huang, C, & Li, H. (2010). *Research on a Skin Color Detection Algorithm Based on Hybrid Color Model (3rd international congress on image signal processing (CISP 2010))*. Yantai, China: Yantai University.

Yang, L, Li, H, Wu, XY, & Zhao, D. (2011). *An Algorithm of Skin Detection Based on Texture (IEEE international congress on image and signal processing (CISP 2011))*. Shanghai, China: Donghua University.

Yu, YH, Kwok, NM, & Ha, QP. (2011). Color tracking for multiple robot control using a system-on-programmable-chip. *Autom Constr, 20*, 669 676.

Yu, YH, Yeh, CH, Lee, TT, Chen, PY, & Shiau, YH. (2014). *Chip-based Real-time Gesture Tracking for Construction Robots Guidance (31st international symposium on automation and robotics in construction and mining (ISARC 2014))*. Sydney, Australia: The University of Technology Sydney.

Zafarifar, B, van den Kerkhof, T, & de With, PHN. (2012). Texture-adaptive skin detection for TV and its real-time implementation on DSP and FPGA. *IEEE Trans Consum Electron, 58*(1), 161 169.

Zheng, Y, Yang, G, & Wu, L. (2010). *Fast Face Detection in Field Programmable Gate Array (International conference on digital manufacturing and automation (ICDMA 2010))*. ChangSha, China: Institute of Electrical and Electronics Engineers Incorporated.

Zhu, JQ, & Cai, CH. (2011). *Region Growing Based High Brightness Skin Detection (10th international symposium on signals, circuits, and systems (ISSCS 2011))*. Lasi, Romania: Gheorghe Asachi Technical University.

18

Demand response visualization tool for electric power systems

Michael Negnevitsky* and Koon Wong

Abstract

Background: Demand response (DR) is referred to programs designed to manage and control electric loads. DR represents one of the vital tools utilized in power distribution networks to improve network efficiency. Effective implementation of DR programs delivers operational benefits such as reduced peak demands and relieved overloads, which are essential in a power system with growing penetration of fundamentally intermittent renewable energy sources.

Methods: This paper presents a visualization tool for optimising DR programs for domestic hot water systems in distribution power networks. The tool accurately models and predicts potential peak demand reductions through direct load control of domestic hot water systems. It employs a multi-layer thermally stratified hot water cylinder model and Monte Carlo simulations to generate hot water load profiles of domestic customers. To meet peak reduction targets set by the tool user, switching programs found via iterative optimizations are applied to hot water systems.

Results: The structure and individual components of the tool are described, and case studies are presented. Impacts of different switching programs on customer's comfort are evaluated and discussed.

Conclusion: The visualization tool is designed to recommend optimum DR switching programs for domestic water heating systems. The tool can assess the performance of a DR switching program by estimating potential peak load reductions and customer comfort characterized by the probability of cold showers. A power system operator can use this tool to determine the available domestic water heating load in a controlled area, and predict the potential reduction in peak load.

Keywords: Visualization tool; Demand response; Smart grid; Electric power system

Background

Electric power systems are undergoing a profound change. This change is driven by several factors that include technical, economic and environmental factors. We need to deal with an aging infrastructure of power systems and maintain the required level of grid reliability. We need to integrate renewable energy sources, particularly wind and solar, and provide secure power supply to our customers, and at the same time improve operational efficiency. The emerging changes and challenges are particularly significant for distribution grids, where the level of automation or "smartness" is relatively low. Manual and "blind" operations along with old electromechanical relays are to be transformed into a "smart grid". This transformation is necessary to meet environmental targets, accommodate distributed generation, and support plug-in electric vehicles. In fact, these needs present the power industry with the biggest challenge it has ever faced. On one hand, the transition to the "grid of the future" has to be evolutionary – we still need to supply electricity to our customers to keep the lights on. On the other hand, the challenges associated with the smart grid are significant enough to expect revolutionary changes in power system design and operation.

In the past, power utilities aimed to accurately forecast consumer demand for electricity, and then planed the growth of the power supply accordingly. But the energy crisis in the 1970s interfered in this process. In the 1970s the economies of the major industrial countries, particularly the United States, Canada, Western Europe,

* Correspondence: Michael.Negnevitsky@utas.edu.au
School of Engineering and ICT, University of Tasmania, Tasmania, Australia

Japan, Australia, and New Zealand were heavily affected and faced substantial petroleum shortages. As a result, predictable demand and low-cost supply, the foundations of traditional planning, became harder and harder to achieve (Wayne and Gellings, 1984). It was the time for power companies to turn their attention to the demand side.

Demand response (DR) was introduced more than 30 years ago for shifting or reducing peak electricity demand (Strbac, 2008). Power systems have been built to meet peak demand – the maximum demand dictates the size of generators and ratings of transmission lines and transformers even if peaks last just a few hours per year. Billions have been spent on extra capacity and infrastructure to cater for these few hours. On the other hand, power utilities could avoid costly investments by shifting or reducing peak demand. Thus, it made perfect sense to offer lower "interruptible" or "curtailable" rates to large industrial and commercial customers for the right to temporarily reduce their electricity consumption.

With the push for energy conservation, DR is becoming a vital tool under the broad smart grid paradigm. This paper outlines some experience obtained at University of Tasmania, Australia in the development and implementation of a visualisation tool for estimating impacts of demand response programs on loading of power transformers and transmission lines in electric grid. Section The concept of demand response in electric power systems provides a definition of DR and discusses its benefits for power utilities and electricity consumers. Section Methods describes the development of the RD visualisation tool for domestic water heating systems, and Section Results and discussion presents results of the tool implementation.

The concept of demand response in electric power systems

Demand response implies many activities such as direct load control, peak shaving, peak shifting, and various load management strategies. According to the US Federal Energy Regulatory Commission, DR is defined as:

> *"Changes in electric usage by end-use customers from their normal consumption patterns in response to changes in the price of electricity over time, or to incentive payments designed to induce lower electricity use at times of high wholesale market prices or when system reliability is jeopardized".*

Effective implementation of DR programs delivers operational benefits such as reduced peak demands and relieved overloads, which are essential in a power system with growing penetration of fundamentally intermittent renewable energy sources (Strbac, 2008). Successful DR programs also provide economic gains such as deferrals of costly network upgrades as well as network security enhancements. Moreover, in a deregulated electricity market, DR programs offer opportunities for aggregation of demand reduction to support market and network operations of a power system (Nguyen et al. 2011). In addition, consumers receive financial incentives through participation in DR programs.

The fundamental benefit offered by demand response is flexibility. In power systems, flexibility is necessary to address critical situations and contingencies when they occur. Today, flexibility is mostly provided by conventional generators because renewable energy sources have very limited capacity to deliver flexibility to the power system. Demand response can provide additional flexibility from the demand side. And it is not always about disconnecting loads. Flexibility also means increasing demand when there is surplus energy and the electricity price is very low or even negative. In general, demand response promotes efficient pricing and efficient utilisation of resources.

However, before we consider all potential benefits of demand response, a word of caution. Flexibility on the demand side has to compete with flexibility offered by

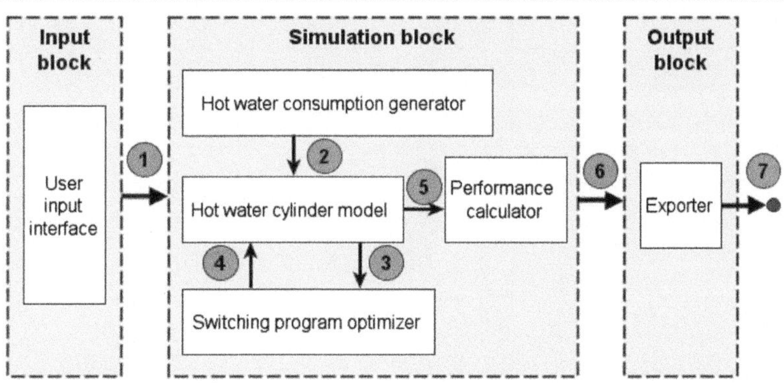

Figure 1 The structure of the DR visualization tool.

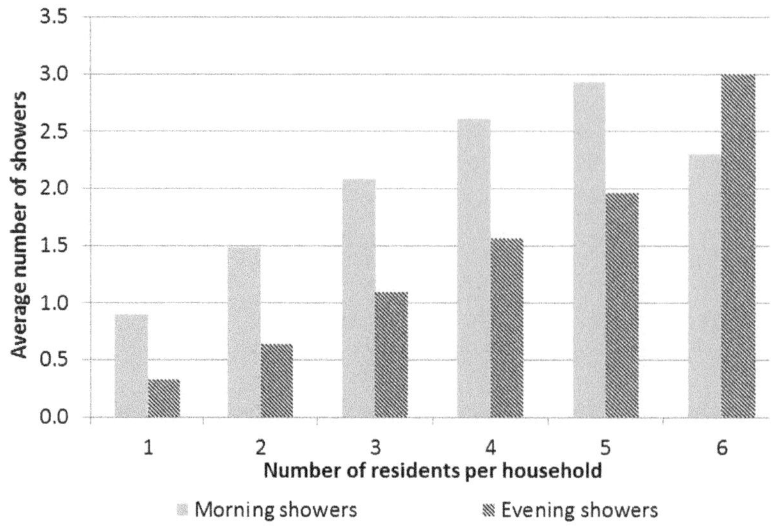

Figure 2 The average number of showers versus family size.

new generation technologies, grid investments and energy storage solutions. Demand response still needs to overcome market rules, regulatory issues, consumer behaviour, and technology and infrastructure barriers before it can reach its real potentials. In the long run, however, investments in generation, consumption and grid may all be just alternative ways to address the need for flexibility in a sustainable energy system of the future.

So, what are benefits provided by demand response? They can be found in the improved economic efficiency of electricity markets, increased reliability of power supply, financial benefits received by electricity consumers and enhanced sustainability of power systems. All these benefits are related to the flexibility offered on the demand side.

Economic benefits are associated with the ability of demand response to defer and even avoid new capital investments in infrastructure (both in generation and grid) by flattening the demand curve. Demand response can also displace the most expensive peak generators. All this leads to lowering wholesale market prices and improving efficiency of the electricity market. Consumers participating in demand response programs also receive financial benefits in the form of the bill savings and incentive payments.

Reliability benefits are associated with changes in electricity usage, which enable power utilities to meet peak loads, and avoid blackouts when there is not enough generation to satisfy demand. In addition, demand response can provide ancillary services such as contingency reserves, regulation and load following. Demand

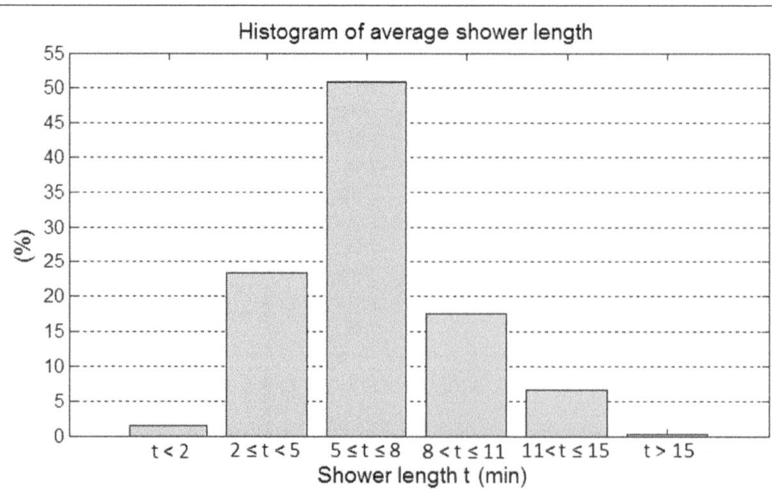

Figure 3 The histogram of average lengths of showers.

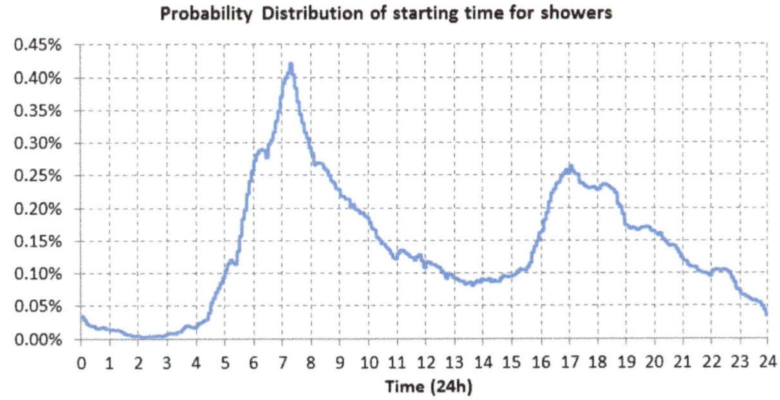

Figure 4 The probability distribution of the shower starting time.

response can react in under a second or within minutes, depending on the type of ancillary service required.

Environmental benefits are provided by enabling power utilities to avoid the use of peaking power plants or "peakers". Peakers typically have higher greenhouse gas emissions. However, most important environmental benefits can be achieved via a large-scale integration of renewable resources such as wind and solar. Because demand response can act as a form of energy storage, it can facilitate in managing periods of excessive generation through the provision of load-following and regulation services that can both increase or decrease demand.

There are three different methods to implement DR in a power distribution network. In indirect load control, consumers manually adjust their consumption in response to incentive programs such as time-of-use (TOU) tariffs (Heussen et al. 2012). In autonomous load control, devices autonomously adjust their consumption in response to detected changes in the power system or to commands sent from the control centre. In direct load control (DLC), devices are centrally controlled by the utility operator (Kondoh, 2011).

Water heating load forms a significant share of the total domestic demand. For example, Elphick (2009) estimated that water heating load accounts for up to 40% of domestic energy consumption in Australia, and around one third in Tasmania, Australia. Moreover, domestic water heating systems represent insulated thermal energy storages that continually supply hot water even during periods of power interruption. Hence, they are commonly targeted for DR programs to reduce peak loads and improve the load factor. Well-designed DR programs minimize customer discomfort due to cold showers.

Many schemes for DLC of domestic hot water systems have been proposed. A practical approach introduced in McKelvie et al. (1992) uses a ripple injection system to issue switching signals to households grouped under different modulation codes. Studies in Nehrir et al. (2007) focused on voltage control to reduce domestic hot water loads. In Gomes et al. (1999), hot water load profiles were simulated using physical models of domestic loads. Households were grouped by the family size to study the effect of DR switching programs on peak load reduction and customer comfort level. In van Tonder and Lane

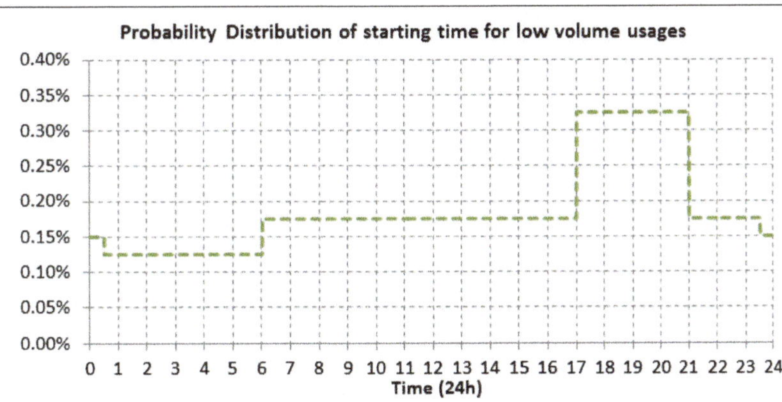

Figure 5 Probability distribution of the starting time for low volume usages.

Table 1 Family types and their distributions

Family type	1	2	3	4
Family size	Very small	Small	Average	Large
Number of residents	1	2 to 3	4 to 5	6 and above
Distribution in a population	25%	50%	22.5%	2.5%

(1996), peak load reduction was studied by considering the number of switching groups, target value and a single time-triggered control. In Gomes et al. (2004), evolutionary algorithms formed the basis for optimizing DR switching programs while maintaining customer satisfaction. A smart grid based control algorithm performing DR on a modified domestic hot water system was proposed by Kondoh et al. (2011) to regulate the aggregated power consumption. Linear programming was used by Lee and Wilkins (1983) to find optimal DR strategies.

The next section presents the development of a visualisation tool that assists in designing a DR program to deliver desired peak load reductions while maintaining satisfactory level of comfort for all customers. The tool estimates the available domestic water heating loads in a controlled area, determines optimal switching programs and presents results to the user in a graphical form. A switching program refers to a direct load control schedule applied to domestic water heating systems (to strategically switch them on and off) in order to achieve a desired load reduction during peak periods.

Methods

General structure of the DR visualization tool

Main modules of the tool are shown in Figure 1, as demonstrated in Negnevitsky and Wong (2015). The modules are grouped in three main functional blocks. The numbered grey circles represent inputs and outputs (I/O).

The Input block represents the user interface, which allows the tool user to enter parameters required for simulation (the number of households in the controlled area, the number of Monte Carlo simulations, the desired peak reduction, etc.) as well as to view default parameters and change them if necessary. The Simulation block is the main block of the tool; it contains four modules: the hot water consumption generator, hot water cylinder model, switching program optimizer, and performance calculator. The Output block contains the exporter, which exports the data to an external (Excel) file.

Default parameters and parameters entered by the user via the user input interface are represented by I/O 1. The hot water consumption generator receives I/O 1 and determines hot water consumption profiles for individual households; these profiles are represented by I/O 2. The hot water cylinder model receives I/O 2 and calculates uncontrolled water heating loads and shower temperatures for the households; the results are represented by I/O 3. The user can observe the aggregate uncontrolled load curve of the households in the controlled area, and proceed with the optimization of switching programs. The switching program optimizer receives I/O 3 and produces switching programs based on the user-defined parameters (the desired peak reduction target, control periods etc.). The best switching programs are presented to the user, so that he/she can select the most suitable switching program. The hot water cylinder model then calculates controlled water heating loads (I/O 5) by applying the user-selected switching program (I/O 4) to the hot water consumption profiles (I/O 2). The performance calculator receives I/O 5 and determines key performance indicators such as peak reductions and customer's comfort. Results in the form of 24-hour load curves are presented to the user (I/O 6), and exported to an external file (I/O 7) via the exporter.

Table 2 Shower probabilities for different families

	Number of shower					
Family type	0	1	2	3	4	5
Type 1	5%	95%	0%	0%	0%	0%
Type 2	0%	41%	53%	6%	0%	0%
Type 3	0%	20%	60%	19%	1%	0%
Type 4	0%	7%	40%	47%	5%	1%

Table 3 Shower lengths and gaps between showers

Parameter	Min (min)	Max (min)	Mean (min)	Standard deviation (min)
Shower length	5	15	8	4
Shower gap	5	7	6	1

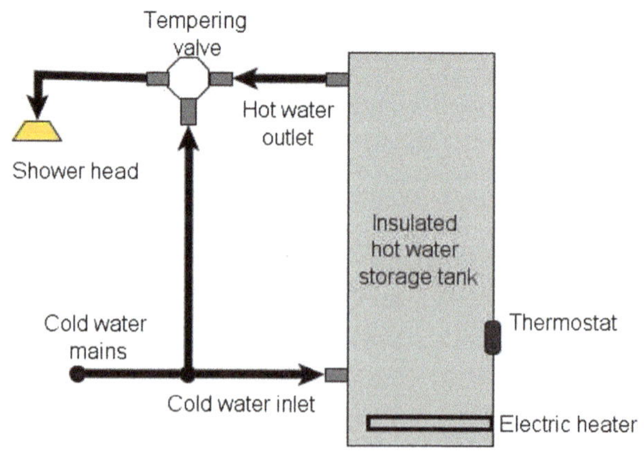

Figure 6 Block diagram of a domestic electric water heating system.

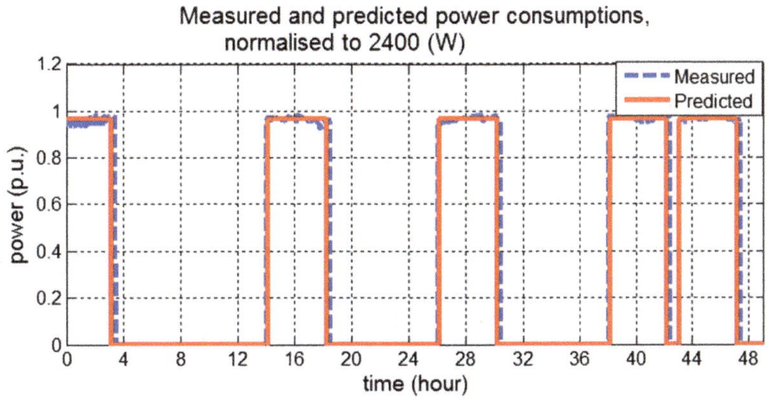

Figure 7 Measured and predicted power consumptions, normalized to 2400 (W).

Hot water consumption generator

The first step in the development of the hot water consumption generator was to acquire knowledge of hot water consumption patterns of households in the controlled area. To achieve this objective, a telephone survey was conducted on 1000 randomly selected households across Tasmania. It recorded demographic data (e.g. number of usual residents, combined income etc.) and details of hot water usages (e.g. average number of showers per day, average shower length etc.) of the surveyed households. This survey focused on two peak periods in the Tasmanian power distribution network, i.e. morning and evening peaks from 6 am to 10 am and from 5 pm to 8 pm, respectively. Figures 2 and 3 show major results of the survey. Figure 2 suggests a positive correlation between the average number of showers and the family size, in the morning and evening peaks. An unexpected drop in the average number of morning showers in households with six or more residents can be explained by a relatively small sample size of this household type (just 2.3% of the total surveyed households).

As can be seen in Figure 3, the length of a shower can vary from 2 min to 15 min, however, a great majority of showers (about 51%) last from 5 min to 8 min.

To estimate domestic hot water consumption profiles, we also acquired energy metering data of 279 households across Tasmania. These data were obtained from meters dedicated for metering electric water heating alone, and represented water heating energy consumption of individual households recorded in 5-minute intervals. We considered two types of hot water usages: high volume usage that lasts for more than 5 min and low volume usage that lasts for 5 min or less. Based on the modelling, 1 min of hot water usage requires approximately 10 min of heating to restore the temperature set by the thermostat. Thus, a continuous energy consumption (a switched-on condition of the electric water heater) for a period of more than 50 min is regarded as a high volume usage (represented by showers), and a consumption of less than or equal to 50 min is regarded as a low volume usage. Using weekday data only, we derived probability distributions of

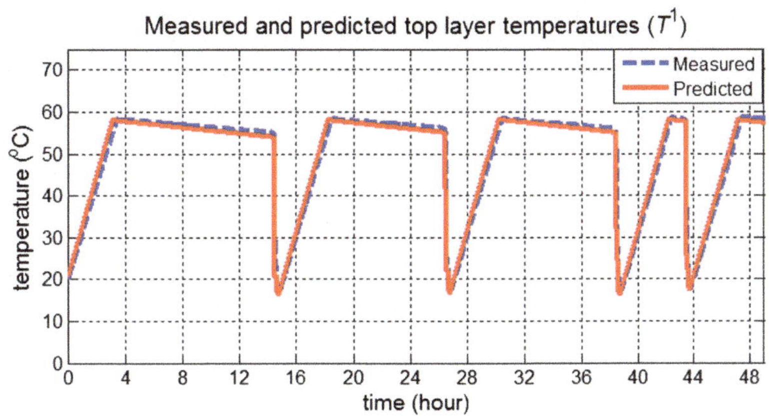

Figure 8 Measured and predicted top layer temperatures of the storage tank.

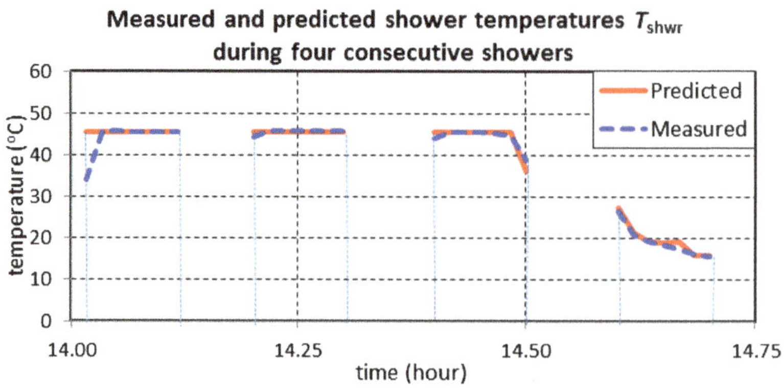

Figure 9 Measured and predicted shower temperatures in a shower schedule.

the starting time for showers (Figure 4) and low volume usages (Figure 5).

Both surveys and energy metering data revealed that domestic hot water consumption depends mostly on the family size. Therefore, all households in a controlled area are divided into four groups according to the family type based on the number of residents in a household. Table 1 shows a typical distribution of families in a controlled area.

We need also specify probabilities of household occupants taking morning showers only, evening showers only, or both. Demographic data (Australian Bureau of Statistics, 2012), and household energy consumption records are used to estimate probabilities in Table 2, which determine the number of showers each family type take in the morning, evening, or morning and evening. Similar to showers, the probability of a low volume usage depends on the family size of a household. The tool uses multipliers to scale this probability up based on the family type. Default values of the multipliers are 1.0, 1.2, 1.6 and 2.0, respectively. The tool user can redefine these values, if required. Figure 5 gives the probability of a low volume usage occurring in a household at a given time.

Shower lengths and gaps between consecutive showers are specified by their mean, maximum and minimum values. We define minimum and maximum to discard unrealistic values (e.g., a one-minute shower) in probabilistic simulations. Normal distributions are assumed. Default values used by the tool are shown in Table 3. A low volume usage is denoted as a single 5-min draw. If required, the user can redefine these values.

The starting time of each hot water usage is specified based on probability distributions derived from actual energy metering data.

The tool employs a Monte Carlo approach to generate hot water consumption profiles for each household. First, the tool generates random values to determine specific parameters for a single household: family type, when showers are taken (morning, or evening, or morning and evening), number of showers, number of low volume usages, length of each shower and each gap between consecutive showers, starting time for each shower and each low volume usage. Next, using these parameters, the tool generates a 24-hour hot water consumption profile for a single household. The tool then repeats the profile generation process for a specified number of households using a new set of random values each time. Finally, the whole process is repeated for the required number of Monte Carlo iterations. Based on the generated hot water profiles, we can now proceed with calculating loads associated with household hot water usages. However, we need to develop a hot water cylinder model first.

Hot water cylinder model

The block diagram of the system with a single heating element is shown in Figure 6. For predicting the shower

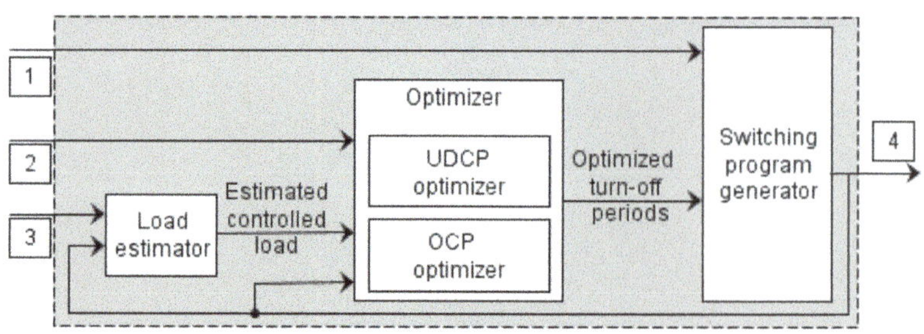

Figure 10 The block diagram of the switching program optimizer.

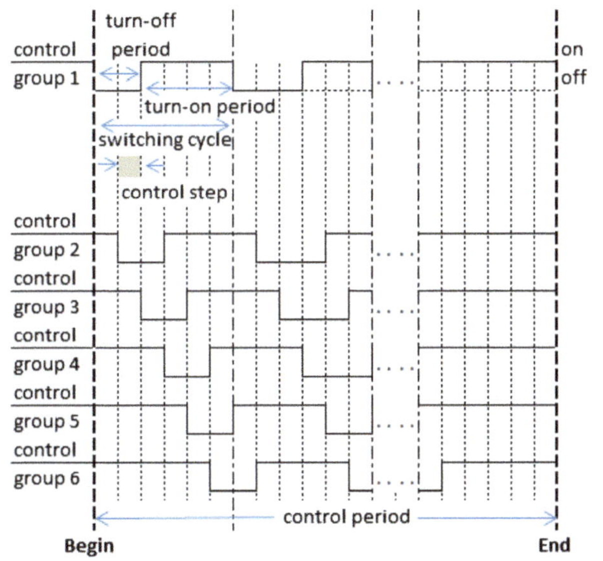

Figure 11 A switching program and its control management system parameters.

temperature and power consumption for domestic water heating systems accurately, we develop a hot water cylinder model based on the most common domestic water heating system in Tasmania, which has a 165 L cylindrical storage tank and a single 2.4 kW heating element. We validated the model with experimental data and found that predicted and measured values were closely matched. The measured and predicted values of normalized power consumption and top layer temperature over 48 hours are shown in Figures 7 and 8. Figure 9 shows the measured and predicted shower temperatures during four successive showers.

We found that the mean prediction error in the total energy consumption was less than 6%, while the mean absolute error in predicted shower temperature was less than 3°C. It was considered acceptable for the model to be used in the tool.

Performance calculator

The performance calculator has two main functions: calculating peak reductions in the water heating load and estimating the customers comfort level.

First, it determines an average uncontrolled load profile for each household. The average uncontrolled load profile for a household represents an average profile of the household obtained over a specified number of Monte Carlo iterations.

Then, it determines an aggregate uncontrolled load curve L_U by aggregating uncontrolled load profiles for all households. An aggregate controlled load curve L_C is obtained in a similar manner after a switching program is applied to the uncontrolled loads produced by the hot water cylinder model for individual households. The peak load reduction R_τ of the control period τ is defined as

$$R_\tau = 1 - \frac{max[L_C(\tau)]}{max[L_U(\tau)]} \tag{1}$$

where $\max[L_C(\tau)]$ and $\max[L_U(\tau)]$ are the peaks of L_C and L_U the control period τ, respectively.

The customer's comfort level depends on the frequency (or probability) of getting a "cold shower"— the event when the shower temperature drops below the comfort temperature (e.g. 43°C) specified by the tool user. Preferred shower temperatures range from 40°C to 44°C, Ohnaka (1994). Because of a large number of households in the controlled area, we can assume the same comfort temperature for all customers. The tool allows the user to change the comfort temperature if required.

Switching program optimization

Figure 10 shows a block diagram of the switching program optimizer. Here I/Os are depicted as numbered blocks. I/O 1 represents parameters of the control management system, I/O 2 optimization parameters, I/O 3 uncontrolled loads generated by the hot water cylinder model, and I/O 4 represents optimized switching programs.

The switching program generator uses user-specified control management system parameters and optimized turn-off periods from the optimizer to create switching programs, as shown in Figure 11. Here a control step is the smallest switching time interval, and a turn-off period is the time interval where the water heating system is turned off for a number of consecutive control steps. A switching cycle consists of the turn-off period followed by the turn-on period. A control period consists of multiple switching

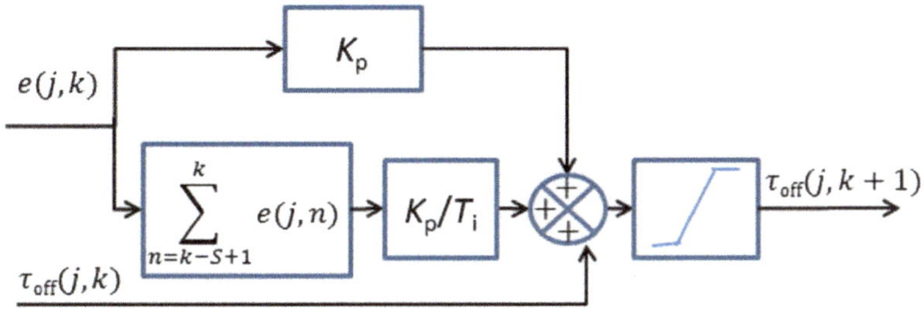

Figure 12 Block diagram of the UDCP optimizer.

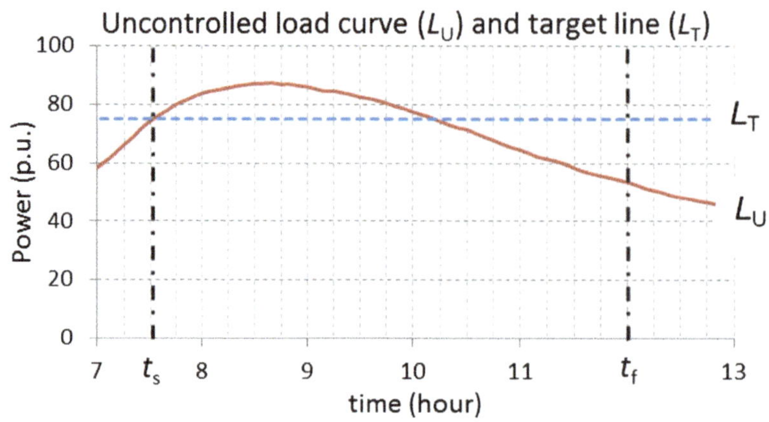

Figure 13 Initial control period in relation to LT and LU.

cycles (there are two control periods – for the morning peak and for the evening peak). Control groups are formed by shifting the switching cycles by one or more control steps. To ensure the time-shifted switching cycles are contained within a control period, each control group has one switching cycle less than the control period. In Wong and Negnevitsky (2013), it was demonstrated that division of households based on the family type does not significantly affect the comfort level of household residents. Therefore, the entire set of households can be divided into control groups of approximately the same size regardless of the family type of a household.

The load estimator determines the total controlled water heating load by applying a switching program to uncontrolled loads of individual households. The load estimator sets the load to zero during the turn-off periods of the applied switching program and restores the load during the turn-on periods. Water temperature is not considered in the load estimation.

The main function of the optimizer is to optimize turn-off periods of a switching program. It consists of the user-defined control period (UDCP) optimizer and the optimized control period (OCP) optimizer.

The UDCP optimizer determines turn-off periods based on the user-defined control periods and the peak load reduction targets. The control periods remain unchanged throughout the optimization process. The UDCP optimizer implements an iterative process to minimize the mean error between the user-defined target L_T and the estimated aggregate controlled load L_C in each switching cycle of a switching program. To calculate required changes in the turn-off period for each switching cycle, it applies proportional and integral (PI) functions to the errors. In Figure 12, $e(j,k)$ and $\tau_{off}(j,k)$ are the mean error and the turn-off period of switching cycle j in iteration k, respectively; K_p is the proportional gain and T_i the integral time of the PI functions.

The proportional function multiplies the error by K_p. The integral function sums the errors of switching cycle

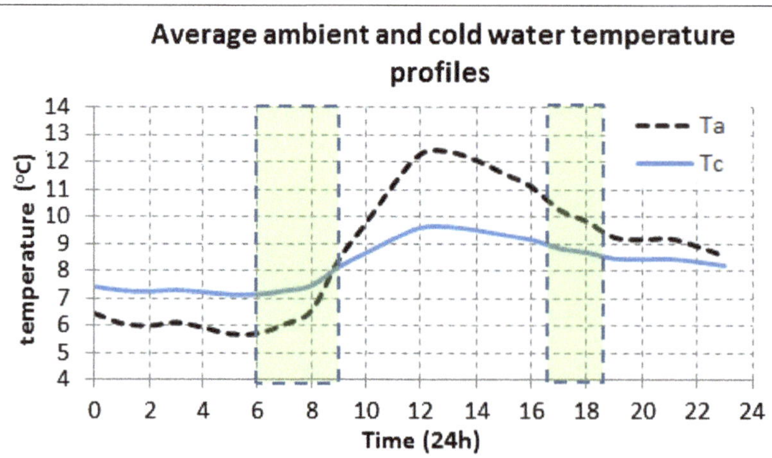

Figure 14 Average ambient and cold water temperatures in winter time.

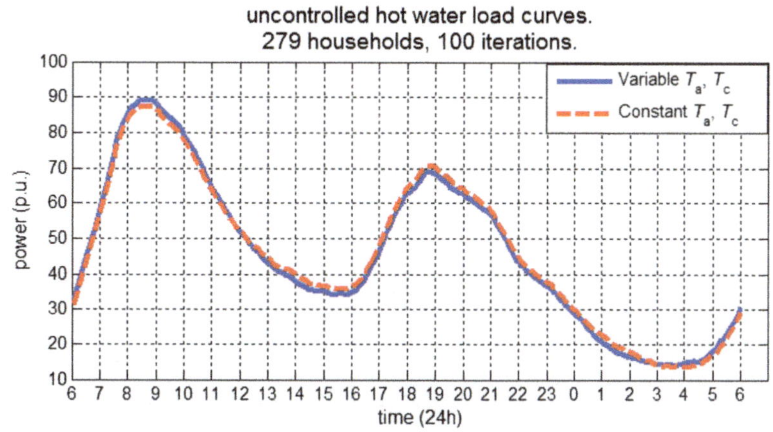

Figure 15 Uncontrolled load curves for constant and variable values of ambient and cold water temperatures.

j from the previous (S-1) iterations to the current one, and multiplies the result by K_p/T_i. The sum of the current turn-off period and outputs from PI functions is converted by the limiter function into an integer between the minimum and maximum values. The final result is the turn-off period for the next iteration.

The OCP optimizer determines turn-off periods and control periods of a switching program based on the user-defined peak load reduction target L_T. First, it finds the starting time t_s and finishing time t_f of the initial control period. The time t_s is found as the first intersection of the aggregate uncontrolled load L_U and the target L_T, as shown in Figure 13. To avoid a high payback peak after the control period, the finishing time t_f is found by solving the following equation:

$$\int_{t_s}^{t_f} L_U(t)\cdot dt = L_T\cdot(t_f - t_s) \tag{2}$$

where the left hand term represents the total uncontrolled energy consumption between t_s and t_f.

To further minimize the error between L_C and L_T, the OCP optimizer iteratively tunes the switching program optimized by the UDCP optimizer. The OCP optimizer increases or decreases the turn-off period τ_{off} of each switching cycle to minimize the error between L_T and L_C. We define three tolerance levels: L_1 and L_2 are, respectively, 1% and 2% above L_T, and $L_3(j)$ is the difference between L_T and the estimated maximum restored load in switching cycle j, if $\tau_{off}(j)$ is decreased by one control step:

$$L_3(j) = L_T - max[L_U(j-2), L_U(j-1), L_U(j)]\cdot\frac{\tau_{step}}{\tau_{sc}} \tag{3}$$

where τ_{step} is the control step; τ_{sc} is the switching cycle; $max[L_U(j-2), L_U(j-1), L_U(j)]$ is the maximum value of the aggregate uncontrolled load L_U over three switching cycles (j-2), (j-1) and j.

The OCP optimizer tunes the τ_{off} of all but the last switching cycle within a control period based on the

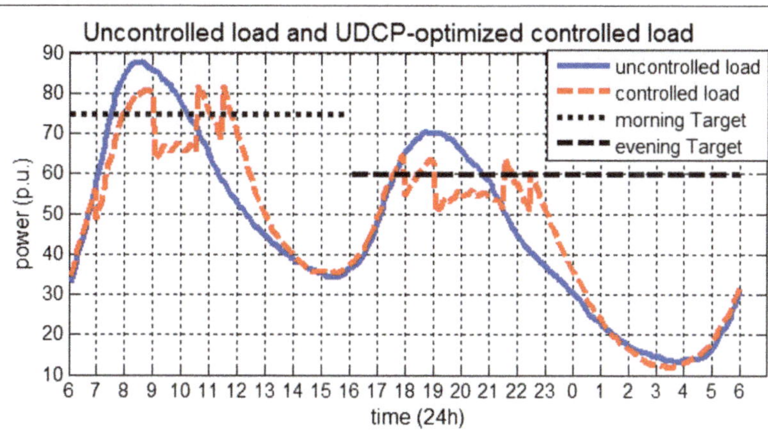

Figure 16 Result of the UDCP optimization.

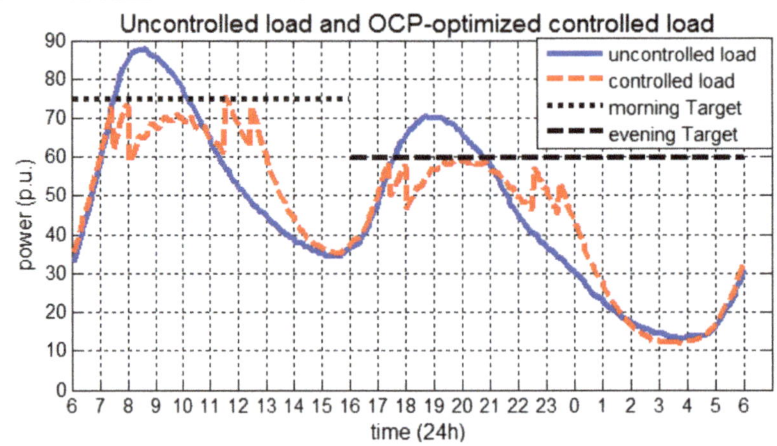

Figure 17 Result of the OCP optimization.

three scenarios shown below, where $L_C(j)$ denotes values of L_C within switching cycle j.

- Scenario 1. The peak of $L_C(j)$ is above L_2.
- Scenario 2. $L_C(j)$ stays between L_1 and L_2 for more than 15 min.
- Scenario 3. The peak of $L_C(j)$ is below $L_3(j)$.

Scenarios 1 and 2 represent overshooting, whereas Scenario 3 indicates over-control that can potentially create higher payback peaks. The OCP optimizer reduces $L_C(j)$ by increasing $\tau_{off}(j)$ by one τ_{step}, if either Scenario 1 or Scenario 2 is met. If Scenario 3 is met, $\tau_{off}(j)$ is decreased by one τ_{step}. No change is made on $\tau_{off}(j)$ if none of the above conditions are met.

Before changing $\tau_{off}(j)$, the OCP optimizer considers the current value of τ_{off} (expressed as the number of control steps) and the location of the peak of $L_C(j)$ within switching cycle j. For a peak located within control step n of the switching cycle, increasing τ_{off} of this switching cycle will reduce the peak only if the current value of τ_{off} is below or equal to (n-1); decreasing τ_{off} of this switching cycle will increase the peak only if the current value of τ_{off} is below or equal to n.

If j is the last switching cycle of a control period, and either Scenario 1 or Scenario 2 is met, the control period is extended by one switching cycle; $\tau_{off}(j)$ is then set to a value equal to a multiple of τ_{step} and proportional to the error between the peak of $L_C(j)$ and L_T. Through iterations, the OCP optimizer tunes the switching program so that the aggregate controlled load stays below or as close as possible to the user-defined target.

Table 4 Control periods and achieved peak reductions in case study 2

	Morning		Evening	
	Control period	*Peak reduction*	*Control period*	*Peak reduction*
UDCP optimizer	07:00–12:00	7.1%	18:00–23:00	9.3%
OCP optimizer	07:30–13:00	14.3%	17:30–00:00	15.0%

Results and discussion

We conducted several case studies to evaluate the performance of the DR visualisation tool under various scenarios for 279 households. This set of households provided us the opportunity to use actual energy metering data in the developed tool. We used the tool to randomly generate hot water consumption profiles for 279 households and obtained an aggregate uncontrolled water heating load curve, which matched the actual data. In case studies 1 and 2, we investigated potential impacts of using constant values of ambient temperature, cold water temperature and thermostat settings on the simulation results. In subsequent studies, we evaluated the performance of switching programs produced by the optimizer in terms of the peak load reduction and customer comfort level. We used 43°C as the preferred shower temperature for all households. The default switching program configuration had 30 min switching cycles and 5 min control steps. The turn-off period in a switching cycle varied from 5 min to 25 min in the 5-minute step. The households were divided into six control groups of almost equal size.

Table 5 Probabilities of cold showers in case study 2

	Uncontrolled	UDCP optimizer	OCP optimizer
Family type 1	0.02%	0.02%	0.03%
Family type 2	4.37%	4.52%	4.63%
Family type 3	7.96%	8.27%	8.44%
Family type 4	13.85%	14.07%	14.36%
Overall	5.06%	5.23%	5.34%

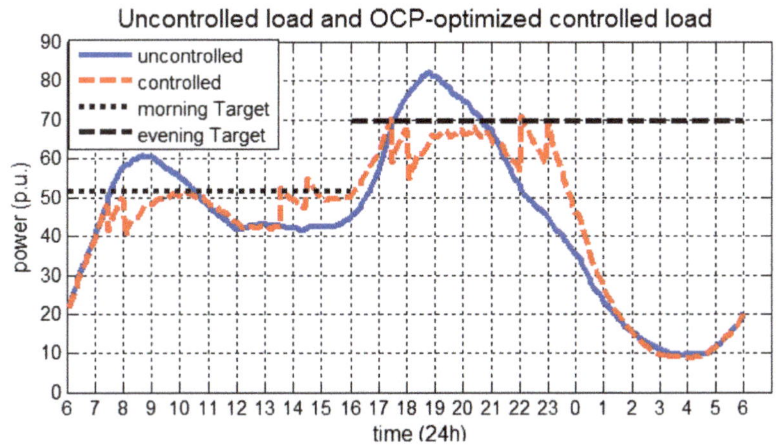

Figure 18 The OCP optimization of a water heating load profile with the dominant evening peak.

Case study 1

This case study compares results of two simulations. In the first simulation, we use actual values of ambient temperature T_a and cold water temperatures T_c, shown in Figure 14. Shaded areas indicate peak periods of hot water usage (06:00 – 09:00 and 16:30 – 18:30). The profile of T_a is obtained from historical climate data for Tasmania (October 2, Bureau of Meteorology and Australia 2012); T_c usually has a positive correlation with T_a , van Harmelen and Delport (1999), but has a smaller range of variation. As can be seen in Figure 14, values of T_a and T_c vary considerably over the 24-hour period (particularly, values of T_a), but their variations during peak periods are rather small. Therefore, in the second simulation, T_a and T_c are set to constant value of 8°C.

Figure 15 shows two aggregate uncontrolled water heating load curves obtained using variable and constant values for ambient and cold water temperatures. We find insignificant difference between the two curves. The difference in the total energy consumption is about 1%, and the mean absolute error (MAE) is about 1.3p.u. The results can be explained by the fact that a great majority of hot water usages occur during peak periods when variations of actual cold water temperature are rather small (within ± 1°C, in shaded areas of Figure 14). On the other hand, although T_a varies significantly during the day, its variation has negligible overall effect on the rate of hot water tank heat losses. An insulated hot water tank idles for a long period (usually from 13 to 15 hours) between two consecutive recharges due to heat loss. During this period, the effect of T_a variation is smoothed, and the average value of T_a produces similar results as its variable values. Thus, variations of T_a and T_c can be represented with their respective average values in further studies.

Case study 2

This case study compares the performance of the UDCP optimizer and the OCP optimizer. Both use the default switching program configuration to produce optimized switching programs that are applied to the same set of water heating loads. The peak reduction target is 15% in both cases. Figures 16 and 17 show the aggregate controlled load curves produced by the UDCP and OCP optimizers, respectively. Table 4 shows the control periods and peak reductions achieved. The UDCP optimizer keeps user-specified control periods constant in its optimization process. Probabilities of cold showers for each family type are shown in Table 5 – for the uncontrolled scenario, and scenarios controlled by the UDCP-optimized and OCP-optimized switching programs.

Comparing the aggregate controlled load curves produced by both optimizers, we find that the OCP optimizer performs much better in terms of peak load reduction.

The starting and finishing times of control periods in a switching program are vital for peak load reduction. A

Table 6 Probabilities of cold showers for case study 3

	Uncontrolled	Controlled
Family type 1	0.03%	0.05%
Family type 2	4.11%	4.48%
Family type 3	7.50%	8.31%
Family type 4	14.32%	15.81%
Overall	4.82%	5.30%

Table 7 Switching program configurations in case study 4

	Configuration 1 (default)	Configuration 2
Control groups	6	3
Switching cycle	30 (min)	30 (min)
Control step	5 (min)	10 (min)
Turn-off periods	5, 10, 15, 20, 25 (min)	10, 20 (min)

delayed control period produces an initial peak above the target line, as in the evening period of Figure 16. Starting a control period too early defers loads needlessly and creates slightly higher peaks in subsequent switching cycles of the same control period, as in the morning control period of Figure 16. Control periods with sufficient length allow a gradual restoration of loads below the target line. Ending a control period prematurely creates an unwanted high payback peak at the end of the control period, as seen at around 11:30 of Figure 16. Similar results were reported in Kondoh (2011), and Lee and Wilkins (1983). Due to shorter than required control periods used for the UDCP optimization, reducing the peaks at 10:30 and 21:30 will produce higher payback peaks at the end of the respective control periods.

While both controlled scenarios produce higher probabilities of cold shower than in the uncontrolled scenario, the OCP optimizer degrades the comfort level more than the UDCP optimizer due to its longer control periods (Table 5).

Case study 3

In this case study, we evaluate the tool's ability to optimize switching programs for two different water heating load profiles. The first one has a dominant morning peak (this load profile was used in the case study 3) and the second – a dominant evening peak. The default switching program configuration (30 min switching cycle with 5-minute control steps and six control groups) is used. The peak reduction target is 15%. Figure 18 shows the aggregate uncontrolled load curve of the second water heating load profile, and the aggregate controlled load curve after the OCP-optimized switching program is applied.

Optimized morning and evening control periods are from 07:30 to 15:00 and from 17:30 to 23:30, respectively.

Table 8 Probabilities of cold showers for case study 4

	Uncontrolled	Controlled
Family type 1	0.02%	0.08%
Family type 2	4.37%	4.80%
Family type 3	7.96%	8.67%
Family type 4	13.85%	14.55%
Overall	5.06%	5.51%

A 9.1% peak reduction is achieved for the morning control period, and 13.4% for the evening. Table 6 shows probabilities of cold showers estimated for each family type under uncontrolled and controlled scenarios. As can be seen from Figure 18, the tool cannot further reduce the payback peak detected at 14:30 as the morning control period has reached the maximum limit of 7.5 hours.

Comparison of the results produced by the OCP optimizer in the case studies 2 and 3 (Tables 5 and 6) reveals that customers experience similar comfort under different load profiles.

Case study 4

In this case study, we use the water heating load profile of case study 2 and compare two different switching programs represented in Table 7. Results produced by the OCP optimizer for case study 3 represent the implementation of the default configuration. Results shown in Figure 19 represent the implementation of the second switching program (configuration 2), and Table 8 shows probabilities of cold showers estimated for each family type. The optimized control period is from 07:30 to 13:30 in the morning and from 17:30 to 00:00 in the evening. Peak reductions for morning and evening control periods are 14.8% and 13.2%, respectively.

The default switching program configuration performs slightly better in peak reduction as it has smaller control

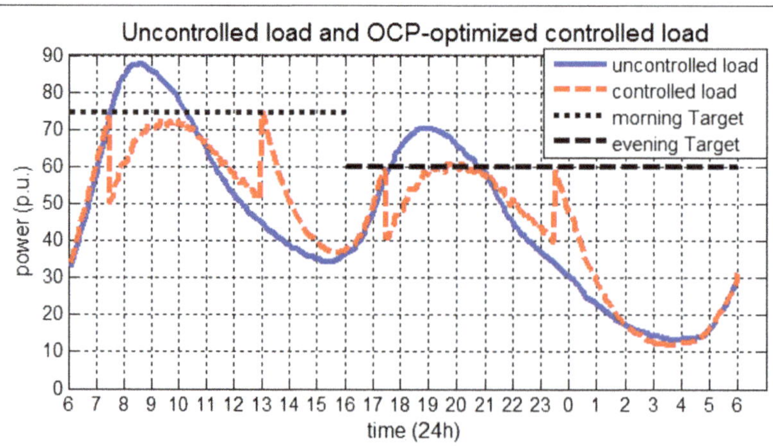

Figure 19 The OCP optimization with the switching program configuration 2.

steps and higher number of control groups. Switching program configuration 2 degrades the customer comfort level further as water heating systems are switched off for longer periods of time.

Conclusions

With a strong drive for energy conservation, demand response is becoming vital for the implementation of the smart grid concept. This paper outlines some experience obtained at University of Tasmania, Australia in developing a DR visualization tool.

The visualization tool is designed to recommend optimum DR switching programs for domestic water heating systems. The tool assesses the performance of a DR switching program by estimating potential peak load reductions and customer comfort characterized by the probability of cold showers. The starting time and the length of control periods are crucial in peak reduction. However, the length of control periods must be limited to minimize negative impact on customer comfort. The developed tool aims to assist distribution system operators in designing their DR programs. The results are presented in a graphical form. An operator uses this tool to determine the available domestic water heating load in a controlled area, and predict the potential reduction in peak load. The tool described in this paper has been implemented in the Tasmanian power system since June 2013.

The developed tool has a modular structure, which enables it to be extended to cover air conditioners, pool pumps and electric vehicle charging load.

Competing interests

The authors declare that they have no competing interests.

Authors' contribution

Both authors share equally the work. MN led the research and drafted the manuscript. KW developed the tool and carried out the case studies. Both authors read and approved the final manuscript.

Acknowledgements

This work was supported in part by TasNetworks, Tasmania Australia. The authors gratefully acknowledge contributions of Peter Milbourne, James O'Flaherty, Daniel Capece, Cherry Wynn and Dr. Thanh Nguyen of TasNetworks, Australia for productive discussions of the results presented in this paper.

References

Australian Bureau of Statistics. (2012). Available: www.ausstats.abs.gov.au.

Bureau of Meteorology, Australia. (2012). Available: www.bom.gov.au.

Elphick, S, Ciufo, P, & Perera, S. (2009). Supply current characteristics of modern domestic loads. In *Proceedings of the Australasian Universities Power Engineering Conference* (pp. 1–6).

Gomes, A, Martins, AG, & Figueiredo, R. (1999). Simulation-based assessment of electric load management programs. *International Journal of Energy Research, 23*, 169–181.

Gomes, A, Antunes, CH, & Martins, AG. (2004). A multiple objective evolutionary approach for the design and selection of load control strategies. *IEEE Transactions on Power Systems, 19*, 1173–1180.

Heussen, K, You, S, Biegel, B, Hansen, L, & Andersen, K. (2012). Indirect control for demand side management-A conceptual introduction. In *Proceedings of the 3rd IEEE PES ISGT Europe* (pp. 1–8).

Kondoh, J. (2011). Direct load control for wind power integration. In *Proceedings of the IEEE PES General Meeting* (pp. 1–8).

Kondoh, J, Lu, N, & Hammerstrom, DJ. (2011). An evaluation of the water heater load potential for providing regulation service. In *Proceedings of the IEEE PES General Meeting* (pp. 1–8).

Lee, S, & Wilkins, C. (1983). A practical approach to appliance load control analysis: a water heater case study. In *IEEE trans. power apparatus and systems* (pp. 1007–1013).

McKelvie, P., Hoddinott, Boland, G. and Kahl, G. (1992). "Load Management Tools for Optimal Control of Hot Water Loads," *Proceedings of the Electric Energy Conference*, 1992, pp. 153–159.

Negnevitsky, M, & Wong, K. (2015). Demand-side Management Evaluation Tool". *IEEE Transactions on Power Systems, 30*(1), 212–222.

Nehrir, MH, Jia, R, Pierre, DA, & Hammerstrom, DJ. (2007). Power management of aggregate electric water heater loads by voltage control. In *Proceedings of the IEEE PES General Meeting* (pp. 1–6).

Nguyen, DT, Negnevitsky, M, & de Groot, M. (2011). Pool-based demand response exchange—concept and modeling. *IEEE Trans Power Systems, 26*, 1677–1685.

Ohnaka, T, Tochihara, Y, & Watanabe, Y. (1994). The effects of variation in body temperature on the preferred water temperature and flow rate during showering. *Ergonomics, 37*, 541–546.

Strbac, G. (2008). Demand side management: benefits and challenges. *Energy Policy, 36*, 4419–4426.

van Harmelen, G, & Delport, GJ. (1999). Multi-Level Expert-Modelling for the Evaluation of Hot Water Load Management opportunities in South Africa". *IEEE Trans Power Systems, 14*(4), 1306–1311.

van Tonder, JC, & Lane, IE. (1996). A load model to support demand management decisions on domestic storage water heater control strategy. *IEEE Transactions on Power Systems, 11*, 1844–1849.

Wayne, M., and Gellings, C.W. (1984). "Demand Planning in the '80s", *EPRI Journal*, Electric Power Research Institute.

Wong, K, & Negnevitsky, M. (2013). Development of an evaluation tool for demand side management of domestic hot water load. In *Proceedings of the IEEE PES GM*. Canada: Vancouver.

19

A step-by-step construction site photography procedure to enhance the efficiency of as-built data visualization: a case study

Hossein Jadidi[1*], Mehdi Ravanshadnia[1], Mujtaba Hosseinalipour[2] and Fatemeh Rahmani[3]

Abstract

Introduction: Visualization of as-built data may change the future of construction project management to a more efficient area of knowledge if appropriate and easy photography and reconstruction tools would be developed and used by practitioners. Most of the current reconstructed 3D point cloud models use unordered photograph collections to generate 4D as-built models. Some of these photographs are not used in the final model mostly because of either a possible overlap with other photos or some faults in photography procedure. Computation time increases exponentially as the number of photos in a photo collection increases. Therefore, the unstructured processes may reduce the performance of a point-cloud model representation. This work shows how the available application of unordered photograph collections are regularly inefficient by measuring the performance of some important criteria, such as the registration success score and the computation time.

Case description: The case study is the construction of a gas compressor station. Such industrial projects involve several building and work areas (e.g., substation, control building, and piping area). The construction site covers approximately 20 hectares. The case study was conducted in two stages. In the first stage, preexisting images have been used for image based modelling (IBM). In the second stage, images captured based on a step-by-step photography procedure (SPP) have been used for IBM.

Discussion and evaluation: IBM performance in the first stage of the case study has been compared with the performance of the second stage by comparing the registration success scores. The IBM in the first stage of the case study results in sparse models, which hardly show the geometry of construction scenes. By contrast, capturing images based on the SPP in the second stage of the case study significantly changed the performance of IBM and increased the registration success score.

Conclusion: This study provides an easy applicable on-site photography procedure. By adopting the proposed approach and by training the photographers, the model would be more desirable for application in more construction projects. The application of the SPP in the case study shows a significant improvement in the final reconstructed 3D point cloud model and as-built data visualization criteria.

Keywords: Image-based modelling; As-built data visualization; 3D point cloud model; Photography procedure; Construction site

*Correspondence: hossein.jadidi@gmail.com
[1] Department of Construction Engineering and Management, Science and Research Branch, Islamic Azad University, Tehran, Iran
Full list of author information is available at the end of the article

Introduction

The effective decision making of a project manager in selecting corrective actions during the construction phase significantly depends on the immediate detection of schedule delay; such corrective actions can prevent delays and budget deficiencies (Golparvar-Fard et al. 2009a). Figure 1 shows that the project control process mainly occurs during the construction phase. This process comprises three steps, namely, monitoring, comparing, and corrective action selection. In the first and second steps, members of the project management team prepare information products that visualize important data to facilitate decision making by the project manager (Jadidi et al. 2014). Figure 1 shows that decision making during the construction phase significantly depends on accessible as-built and as-planned data. Figure 1 also shows the importance of how to visualize these data to make quick decisions during the construction phase. Golparvar-Fard et al. 2011a reported that the identification, processing, and communication of progress discrepancies are enhanced by the integration of the visualization of as-built and as-planned performance and can thus serve as a powerful remote project management tool for remote decision making in architecture, engineering, construction, and facility management industrie (Golparvar-Fard et al. 2011a).

Several researchers highlighted the limitations of current manual as-built data collection approaches in terms of speed and accuracy. According to Akinci et al. 2006, the field staff in construction sites spend 30 to 50 percent of their time recording and analyzing field data. Digital photo is one of the most usable sources of field data, which can be easily captured without additional cost and time for construction projects. According to Section 4.21-b of the International Federation of Consulting Engineer (FIDIC) book series (Red Book), Conditions of Contract for Construction (FIDIC 2006), photographs are among the requirements for the progress report that a contractor should regularly (e.g., daily, weekly, and monthly) send to the owner (Jadidi et al. 2014). Aside from progress documents, photographs that show as-built scenes have more applications. One of these applications is in image-based modelling (IBM) by computer vision algorithm, wherein these photographs are overlaid on 4D as-planned models and BIMs in an augmented reality environment to visualize the discrepancies between as-built and as-planned information and for automation of construction progress monitoring (Golparvar-Fard et al. 2009a; Kamat et al. 2010; Kim et al. 2013).

This study focuses on the data capturing approach for improving the quality of image-based 3D reconstruction models. By this way, decision makers could understand the progress of a construction project better and, thus, make quicker and more accurate decisions. Moreover, to facilitate the automation of progress monitoring (Golparvar-Fard et al. 2011b) when these models are overlaid on 4D models or BIMs, the probability of observing expected as-built element increases and related errors of inaccuracy decrease because these models are denser. Thus, probability functions can detect discrepancies with high probability that is closer to reality. To enhance the quality of captured photos, this study does not mention any image processing technique, such as histogram adjustment or filtration. In addition, this study does not mention any post-processing technique for the densification of reconstructed point cloud model, such as meshing and texturing. This study focuses on how to take photos in construction sites with several buildings and work areas to enhance the quality of IBM for the visualization of as-built data. This study answers the following question: Is it possible to visualize as-built scenes by 3D point cloud models using usual photography in construction sites or should the photographs be captured in special manner to result in such models?

To respond to this question, we hypothesize that the approach for capturing photos by common photography in construction sites does not lead to appropriate 3D point cloud models for the visualization of as-built data and progress monitoring of construction projects, which include several buildings and work areas. A case study is developed in two stages to test this hypothesis. An unordered photo collection, which was previously captured for purposes other than IBM, is used in the first stage. In this stage, photographers do not know that the photographs they captured will be used for the visualization of as-built data or progress monitoring. The photographers are field staff who capture photos for regular progress report to the owner (FIDIC 2006) and the project management office of the contractor. After conducting IBM from the collection, a step-by-step photography procedure (SPP) is assumed according to the results and based on some previous studies in the field of IBM. In the second stage of the case study, photographs are captured based on an assumed SPP. Figure 2 shows the hypothesis of this research, as well as two approaches in capturing data to test the hypothesis.

Background

IBM in construction

Field engineers in construction sites inspect work areas and check the progress, quality, and safety of construction work every day. They also take photographs to monitor and document progress, quality, safety, site layout management, and productivity (Golparvar-Fard et al. 2011a) without considering the use of these photographs for IBM. Several studies have been conducted recently on the use of daily construction site photographs taken by common

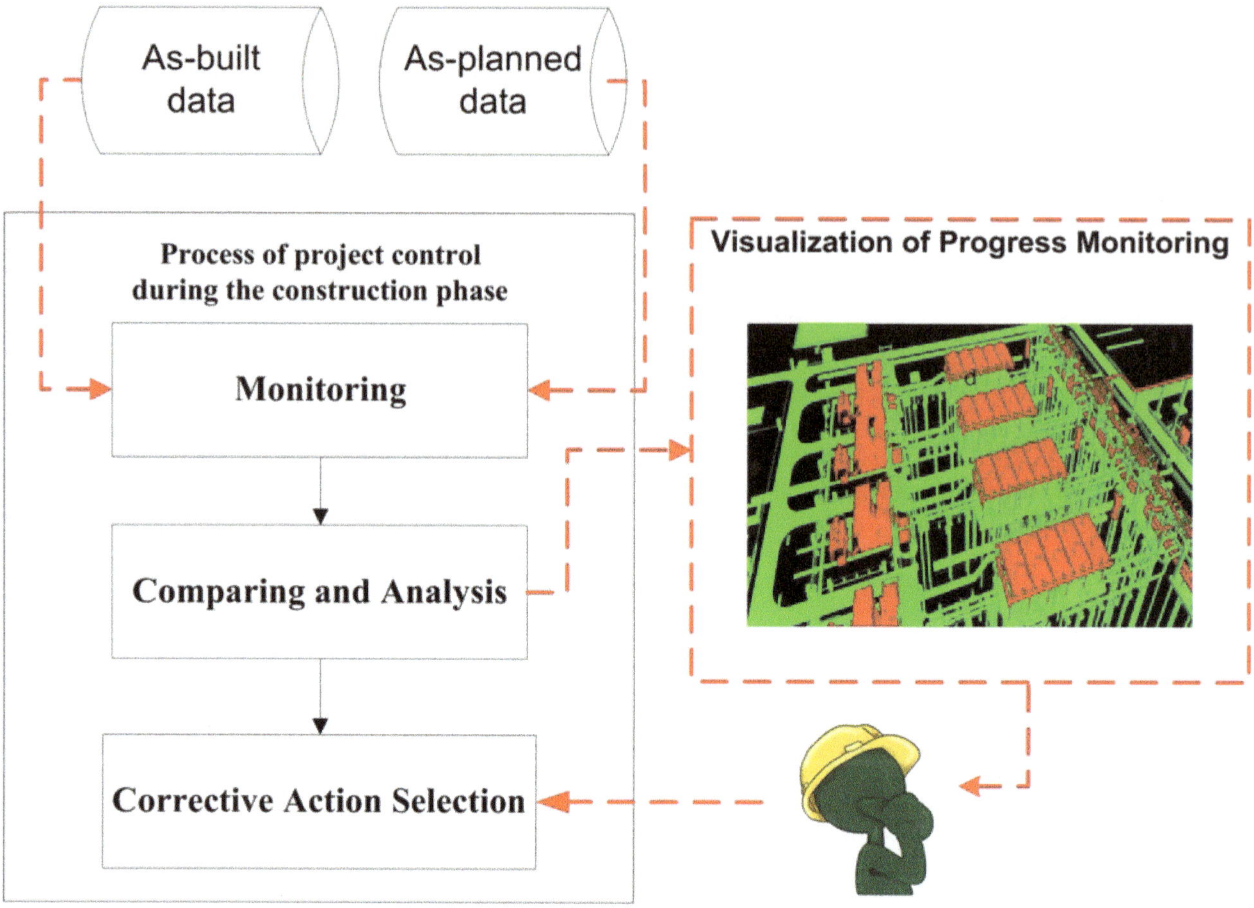

Figure 1 Decision-making process during the construction phase. Visualization of discrepancies between as-built and as-planned data could be useful for decision maker.

digital cameras for image-based modeling (IBM) such as Golparvar- Fard et al. 2009a, 2011a, Bae et al. 2013, and Jadidi et al. 2014. These models are then superimposed on 4D as-planned models or Building Information Models (BIMs) to visualize progress deviations and/or to automate construction progress monitoring (Bae et al. 2013; Golparvar-Fard et al. 2009a; 2011a; Kamat et al. 2010; Yang et al. 2013).

McCoy et al. 2014 investigated the use and access of IBM techniques to reduce uncertainty in construction planning for remote construction sites. According to their research, the use of image-based 3D reconstruction techniques reduced the risks and uncertainties in remote sites as well as the risks and uncertainties related to high-technology usage in construction planning (McCoy et al. 2014). Golparvar- Fard et al. 2009a proposed a visualization system called D4AR for the automatic visualization and determination of construction progress monitoring using daily site photographs. These photographs were captured using a common digital camera by field engineers in construction projects (Golparvar-Fard et al. 2009a). By using computer vision techniques such as structure from motion (SfM), the location of photographers are estimated using the photographs captured in various illumination, resolution, zoom, and quality conditions (Golparvar-Fard et al. 2009b). The scene of the construction site is then reconstructed and shown using 3D point cloud models and finally overlaid on a 4D as-planned model to automatically visualize and detect construction progress monitoring (Golparvar-Fard et al. 2009a, 2011a).

This paper does not report a detailed discussion of the computer vision techniques and the related algorithms used in previous studies and in the current study. We encourage readers to check Snavely et al. 2008; 2010, Golparvar-Fard et al. 2009b, and Yang et al. 2013. The computer vision techniques used in this study are applied using the VisualSfM software (Wu 2011; Wu et al. 2011). In this software, we do not use the available feature detection function, but instead import Scale-Invariant Feature Transform (SIFT) key point detector of Lowe 2004; 2014 in the VisualSFM software to conduct feature detection

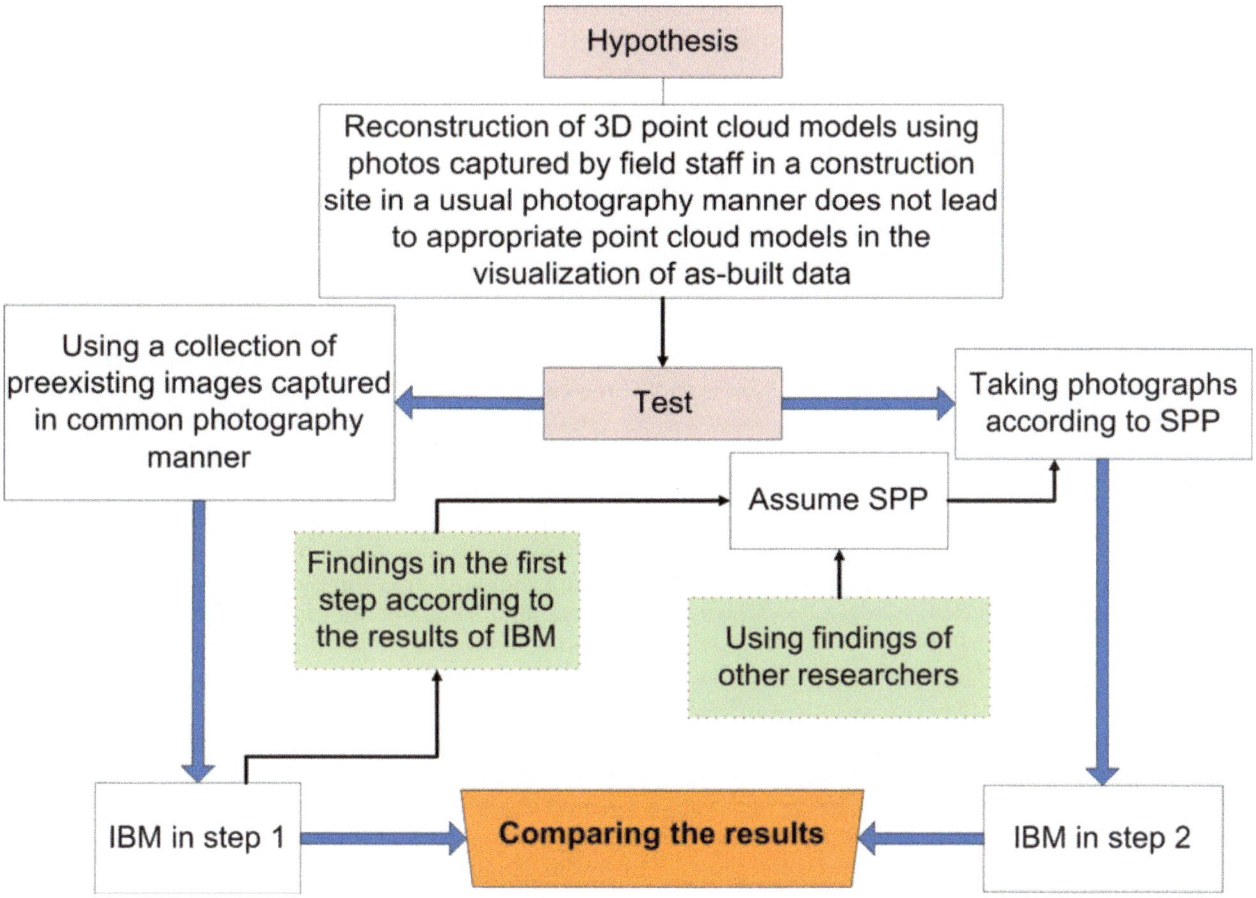

Figure 2 Research hypothesis and two approaches for testing it.

and matching. We then use available functions in the software such as SfM and bundle adjustment to reconstruct 3D scenes. We changed the sizes of the original images using Xnview 2014 to maintain the EXIF tags, which include the focal length of each photographs. These tags is useful for the estimation of intrinsic parameters of a camera by SfM procedure.

Factors that affected the quality of IBM

McCoy et al. 2014 considered the quality of image-based 3D reconstruction models by investigating the impact of the number of images on the density of reconstructed point cloud model and its completeness in contrast to real buildings. They found that a fault in data collection can significantly increase the risk on the usefulness of the outcome of image-based 3D reconstruction models. McCoy et al. 2014 addressed the risk of low-quality IBM outcome by suggesting that field staff or engineers should capture as many photographs as possible to optimize the usefulness of IBM applications for remote project planning (McCoy et al. 2014). They also suggest that low-quality photographs can decrease the quality of IBM (McCoy et al. 2014). However, the most important aspect of unsuccessful IBM is the lack of sufficient overlap between images and not on the number of photographs (Ducke et al. 2011). Snavely et al. 2006 indicated that some photos in a collection are not used in the final IBM because of too little overlap with other photos in the collection. McCoy et al. 2014 attempted to guarantee the quality of outcome of IBM by capturing images, wherein each structural component of the scene being photographed was visible from three different viewpoints McCoy et al. 2014. Ducke et al. 2011 suggested capturing overlapping images with no more than approximately 25 to 30 degrees of angular difference between them. To enhance the quality of IBM, they also emphasized the importance of walking around the object and capturing photos from different angles, viewpoints, and zoom settings instead of taking many images from the same location and using the same viewing angles and/or zoom settings (Ducke et al. 2011). In addition, cropping images results in incorrect estimations of the location of the photographer by SfM algorithm (Ducke et al. 2011); hence, the quality of IBM is decreased significantly. Xiao et al. 2008 stated that the density and accuracy of the reconstructed points in IBM vary depending on the distance between the camera and the objects. McCoy et al. 2014 chose a distance of nine meters from the buildings they studied and reported satisfactory results.

Considering the gap of previous studies

In this study, we considered industrial projects with several building and work areas. However, the authors in the aforementioned studies mostly consider a building. The data collection approach in industrial projects with several buildings and work areas is the key factor for the optimal application of IBM. The studied by McCoy et al. 2014 was a building. By intentionally deleting some images from the collection, they found that the outcome of image-based 3D reconstruction model changed significantly. The negative effect attributed to the lack of sufficient overlapping images in industrial projects such as gas compressor stations should be considered. We conducted our case study in a gas compressor station, where several buildings, piping areas, and equipment are located.

Capturing as many images as possible to guarantee the success of IBM is rational for small cases. However, this approach results in a huge collection of photos for these industrial projects; computation time grows exponentially as the number of photos increase because of pairwise matching step in the SfM algorithm (Golparvar-Fard et al. 2011a). Therefore, taking sufficient overlapped photographs from the entire site becomes more difficult as construction projects grow and become more complex. Thus, the problem of insufficient overlap will continue to affect the quality of the final point cloud model and will increase computation time.

This study mostly focuses on the data capturing stage (1) for selecting an approach to take images from industrial projects, and (2) for suggesting SPP for an image-based 3D reconstruction of construction cases to achieve better IBM. The first stage of the case study, which is presented in the following section, attempts to respond to the following question: Is it possible to reach high-quality IBM by capturing photos randomly in a usual manner of photography?

In summary, this study considers the following gaps: (1) Is common usual photography for preparing regular reports to the owner of construction projects sufficient to achieve good image-based 3D reconstruction models? (2) What is the appropriate photographic strategy to achieve the appropriate IBM that clearly visualizes as-built data for remote project planning or visualization and automation of construction progress monitoring?

Case description

A case study was conducted on a construction project to test the hypothesis mentioned in the introduction. The project is the construction of a gas compressor station in a remote location from the city. Such industrial projects involve several building and work areas (e.g., substation, control, and laboratory buildings, piping area, air coolers, and gas compressors). The construction site covers approximately 20 hectares, which is larger than common construction projects. The case study was conducted in two stages as described in the following sections.

First stage

An unordered photo collection that includes 3,000 photos, which were captured for purposes other than image-based 3D reconstruction modeling, was used in the first stage. The photographers did not know that the photographs they captured would be used for the reconstruction of 3D point clouds. These photographers are field staff or engineers who captured photos for daily, weekly, and monthly report to the project management office of the contractor. Some of these photographs are unrelated to progress monitoring and are captured from scenes other than structures or buildings that exist in the construction site. For example, one of the photographers captured several images of trucks that transport equipment to the site. Another photographer captured some scenes that show collected materials for painting, plumbing, and others. Another photographer captured several images from interior scenes such as the electrical boards in the control room. Some of these images are irrelevant to the aim of point cloud modeling for the visualization of as-built data shown in Figure 3. All of these images were manually deleted from the photo collection because it extend computation time without any result in the final point cloud model. Other remaining images were taken from a series of days with no significant change in construction. The total number of images collected was 399, which were captured with different resolutions and under different illumination conditions. Different field staff members captured these images using different digital cameras.

This study used a publicly available software called VisualSfM (Wu 2011; Wu et al. 2011) to solve the correspondence problem and to bundle adjustment for the reconstruction of the 3D point cloud model of the as-built scenes. Figure 4 shows the number of SIFT features for each image in the collection in the first stage, which were determined by the SIFT detector of Lowe 2004; 2014. Figure 5 shows the reconstructed point cloud model with 99 estimated camera locations. The registration success score (portion of the number of images used to number of all images) of the photo collection in the first stage of the case study is 0.25, a value that is lower than that obtained by Golparvar-Fard et al. 2011b and McCoy et al. 2014.

Sixteen models are reconstructed from the collection when VisualSfM was set to generate multiple models in

Figure 3 Examples of irrelevant images to 3D point cloud modeling for the visualization of as-built data.

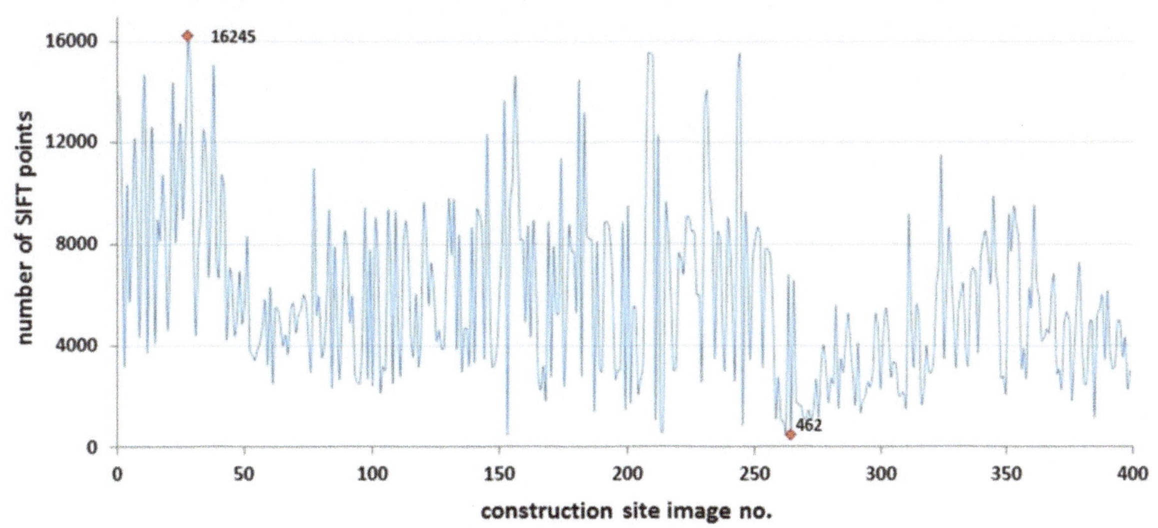

Figure 4 Number of SIFT features in 399 images of various sizes.

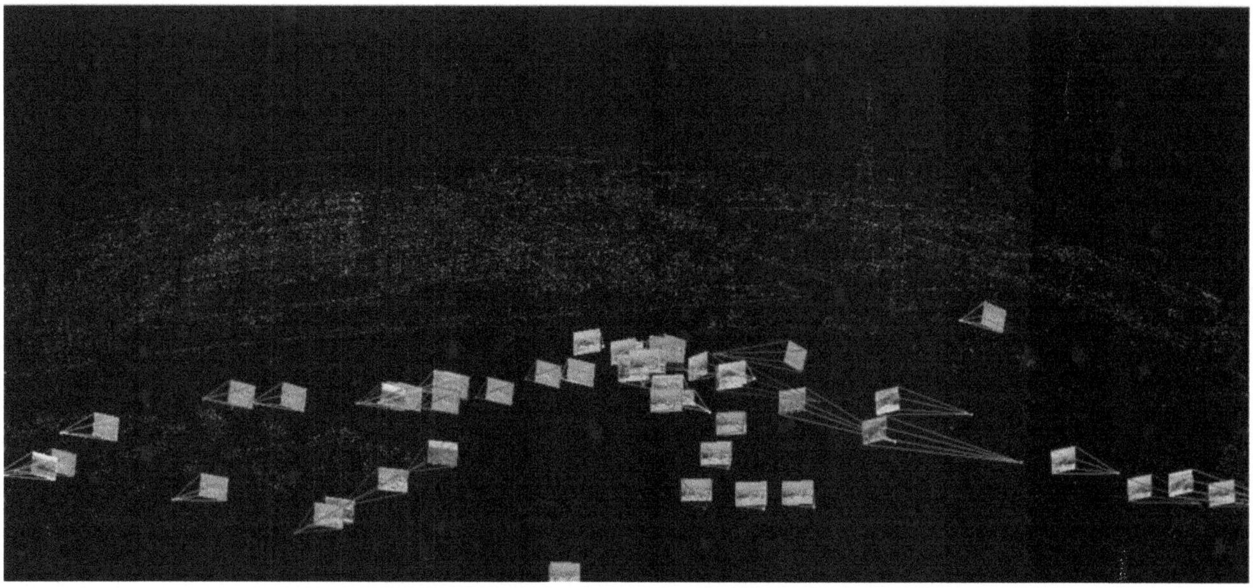

Figure 5 Sparse 3D model reconstructed from 99 images (from a total of 399 images) with camera frusta.

the first stage of the case study. The numerous numbers of reconstructed model show that the photographs in the collection are captured from several areas in the construction site. However, these reconstructed models are vague and sparse, except for the one shown in Figure 5, which is not as good as the other results conducted by other researchers (Golparvar-Fard et al. 2011a; McCoy et al. 2014). Figure 6 shows some of these sparse models. According to Snavely et al. 2010, the SfM techniques used in their study are not likely to register photos that do not have sufficient overlap, unless they are captured more frequently than in usual photography (Snavely et al. 2010). Therefore, in the first stage of the case study, the baseline between the images captured by field staff or engineers seems to be too wide. In other words, the images do not have enough overlap with other images. The lack

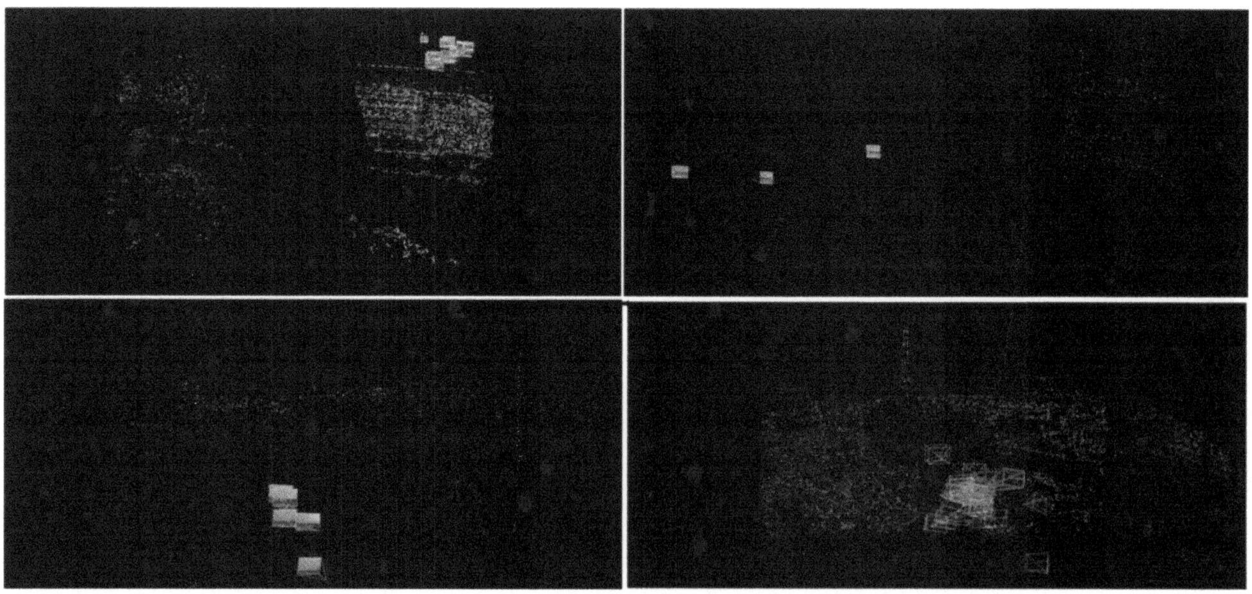

Figure 6 Sixteen reconstructed sparse models obtained from the first stage of the case study (four of them are shown).

of overlap between images results in either a significant decrease in correspondence matches or the removal of the images during the final stage of the SfM procedure (bundle adjustment) by the software.

Thus, failure to overlap the images properly hinders the establishment of a good 3D point cloud model with a high score of registration success. Therefore, the low quality of these 3D sparse models can be attributed to the small number of images relative to the large area of the construction site (20 hectares) for the gas compressor station. To show the lack of captured photos related to the size of the construction site, the absolute reconstructed 3D point cloud model is overlaid on the gas compressor station plan, as shown in Figure 7; these points were measured using SfMGeoref v.2.3 (James et al. 2012) by defining coordination points (X, Y, Z) for at least three points. Figure 7 shows that the reconstructed scene only covers a small area of the construction site where the majority of photos have sufficient overlap in the collection related to that part of the construction site. However, the photo collection includes a scene that represents the entire construction site.

In addition to the lack of photos with sufficient overlap in the collection and the lack of photos in relation to the large construction site, the fault in photography in some photos is also obvious. For example, Figure 8 shows that some images in the collection of photos used in the first stage have been captured in poor conditions. The image was taken from afar and with the glare of the sun, which resulted in the smallest number of SIFT features (462). Thus, few correspondence matches were obtained, which led to a low-quality 3D point cloud model.

SPP for IBM

The most important requisite for a successful image-based 3D reconstruction is the stage of capturing data. Better quality input data result in better reconstructed 3D model outcome (Ducke et al. 2011). Some results about errors in photography, such as those mentioned in the first stage, are used to prevent error in capturing data. The following step-by-step procedure is presented and shown in Figure 9 according to the results in the first stage and the findings of other researchers:

1- Sensing device selection: Choose an appropriate device. An expensive camera is not necessary. However, by experience, SLR cameras have better results than other devices (McCoy et al. 2014). The resolution of the camera does not need to be high, but it should be more than 3 megapixels. Even the camera of smartphones could be used.

2- Time of photography: Take pictures during daytime and whenever the sun is not in front of the camera. Do not take pictures around sunset or sunrise because the glow of the sun results in a significant decrease in SIFT features. Figure 8 shows this photography error that occurred in the first stage of the case study. In this image, the photographer captured the image with the glare of the sun, which resulted in a significant decrease in the number of SIFT feature that consequently led to a small number of matches between images and a weak reconstruction of the 3D point cloud model.

3- Object selection: Choose an object according to the approved work breakdown structure in construction projects because the image-based 3D reconstruction model in this study is used for the visualization of as-built

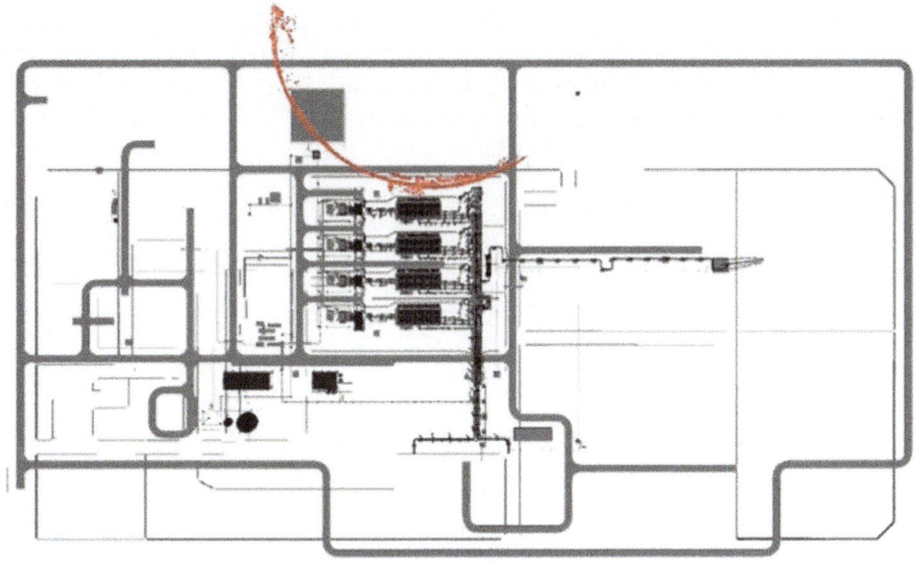

Figure 7 Plan of the gas compressor station overlaid by 36,082 reconstructed points.

Figure 8 Image of the collection with minimum number of SIFT features visualized by red points.

data for construction progress monitoring.

4- Location: Maintain a distance of approximately 9 meters from the building [according to the suggestion of McCoy et al. 2014].

5- Selection of appropriate frame: Choose the appropriate frame to capture structural elements, which should not too high nor too low. Capture the whole elements in vertical overlapping images to avoid dividing elements horizontally.

6- Motion: Walk slowly around a building and capture each structural element from three different views [according to the suggestion of McCoy et al. 2014].

Second stage

In this stage, the author chose three buildings in the construction site: the control building (B1) and the laboratory building (B2) and the prefabricated building (B3). A total of 118 photographs were captured in 6 minutes from B1 as Experiment 1 and 80 photographs are captured in 4 minutes from B2 as Experiment 2 and 26 photographs captured in a minute from B3 as Experiment 3. Table 1 shows the conditions under which these images were captured. Figure 10 shows the process of conducting image-based 3D reconstruction from captured images to generate 3D point cloud models after running the SfM procedure.

Figures 11 shows one of the images in the collection of each experiment, the reconstructed point cloud model with a reconstructed camera frusta for every three experiments, and the 3D point cloud model of B1,B2, and B3 in dense form, as determined by the CMVS/PMVS module in VisualSfM (Wu 2011; Wu et al. 2011). The registration success scores for the collection of photos are 0.85 in B1 and 1 in B2 and 1 in B3.

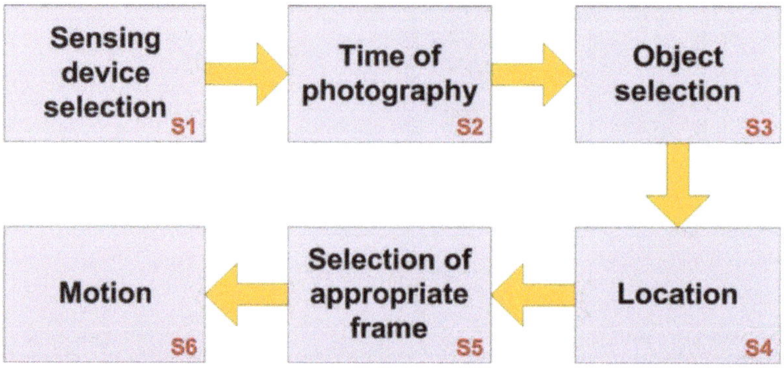

Figure 9 Step-by-step photography procedure.

Table 1 Characteristics of the captured images

	Experiment 1 (B1)	Experiment 2 (B2)	Experiment 3 (B3)
Camera model	Nikon Coolpix P510	Sony-DSC-W310	Sony-DSC-W310
Number of captured photos	118	80	26
Image resolution(pixels)	4608x3456	2592x1944	4000x3000
Photography duration	6 minutes	4 minutes	1 minute
Lighting condition	Sunny	Sunny	Sunny

Discussion and evaluation

According to the results of the first stage of the case study, capturing as many photos as possible from construction sites led to satisfactorily reconstructed point cloud models. However, the time of processing, especially for the pairwise matching stage for all of the images in a collection, is considerable. For instance, the computation time for reconstructing the point cloud model from 399 photos in the first stage is about 375 minutes, which led to a total of 99 images. This duration does not include the time wasted in capturing more photos in a construction site. This period does not include the time and effort exerted to delete several images that are irrelevant to 3D point cloud modeling for as-built visualization and construction progress monitoring.

Table 2 summarizes the results of the case study. The results are benchmarked on a computer with 2.4 GHz Intel® Core i7 CPU, 6.00 GB of RAM, and a Windows 64-bit platform.

According to the results shown in the Table 2, the IBM performance in the first stage of the case study, wherein preexisting images are captured to prepare regular reports to the owner, could be compared with the performance of the second stage by comparing the registration success scores (recall). Recall in the first stage was 0.25. By contrast, capturing images based on SPP in the second stage of the case study significantly changed the performance of IBM and increased the registration success score for Experiment 1 to 0.85 and a complete registration score for Experiments 2 and 3.

Conclusions and future work

This work focuses on the data capturing stage of IBM and presents step-by-step photography procedure (SPP)

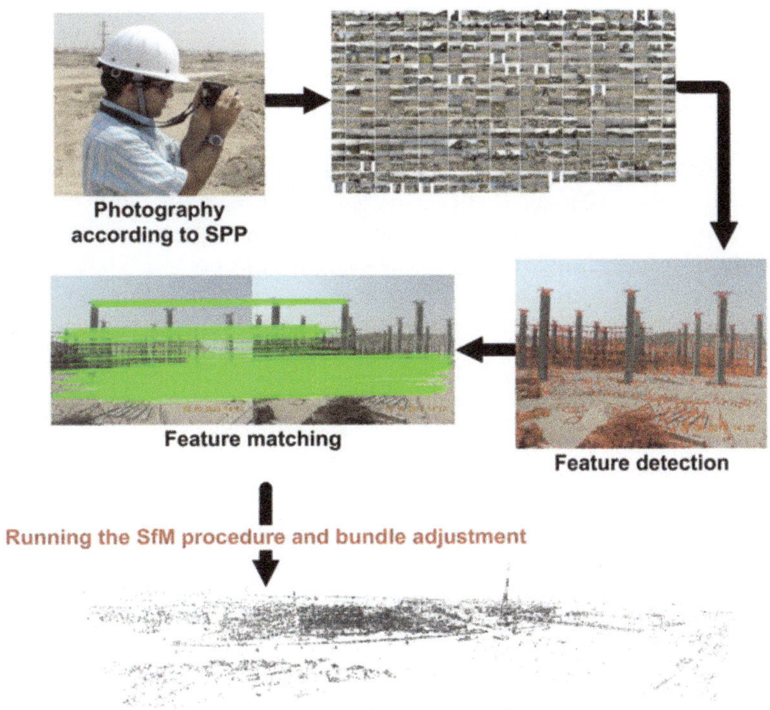

Figure 10 **The process of collecting as-built data resulting in the reconstructed 3D as-built point cloud model.**

Figure 11 Experimental buildings in captured images according to SPP, sparse 3D point cloud model, and dense 3D point cloud mode.

for image-based 3D reconstruction modeling for the visualization of as-built data in construction projects. SPP is proposed based on the findings of the first stage of the case study and on the findings of previous studies in the field of IBM. To validate SPP, the second stage of the case study is conducted and the images are captured based on SPP.

The IBM in the first stage of the case study results in sparse models, which hardly show the geometry of construction scenes. Only a particular view of the site is reconstructed when IBM was overlaid on the plan of the construction site despite the existence of images taken from another view. The results indicate that error in photography and failure to capture photos with sufficient overlap with other photos led to the unsuccessful IBM in the first stage of the case study. After capturing images based on SPP and running the process of IBM, the registration success score in the second stage increase an average of 0.95 between three experiments.

The authors do not refute the capability of the computer vision algorithm and the publicly available software used in this study to establish 3D point cloud models for visualizing as-built data from unordered preexisting construction site photographs. However, numerous images should be taken compared with the usual photography in the construction sites when using this method. This condition is impossible in industrial projects such as gas compressor stations or oil refinery plants, which have several buildings and work areas.

In summary, the research hypothesis that the approach of capturing photos in usual photography does not lead to appropriate 3D point cloud models is supported by the low registration success score (recall) in the first stage of the case study in contrast to the high score in the second stage of the case study. In addition, the results show that the proposed SPP significantly increases the efficiency of as-built data visualization.

Table 2 Summary of experimental data extract from the case study

	First stage	Second stage		
		Experiment 1	Experiment 2	Experiment 3
Total no. of images	399	118	80	26
Image resolution (pixels)	various	2304x1728	2592x1944	2400x1800
No. of used images	99	100	80	26
Recall	0.25	0.85	1.00	1.00
No. of points recovered	36,072	36,221	17,345	3,646
Computation time	375 minutes	330 minutes	135 minutes	13 minutes

The SPP proposed in this study, which was validated by the case study, seems applicable for industrial projects, such as the case in which this research was conducted. This approach has good potential for use in other construction projects because some steps of the SPP are general and not relevant to particular construction projects. However, the case study has some limitations. For instance, the buildings in our case study are not more than two storeys. In addition, in our case, it is easy to walk around a building and capture photos without occlusions. However, in some cases it is not possible to walk around a building to capture photos easily. Hence, more case studies should be conducted in the future for high rise building and buildings or objects with no possibility of walking around them. Finally, we suggest that prospective researchers in the field of IBM in the construction should train field staff and field engineers using either SPP or other guidelines in capturing construction site photos for IBM to enhance the efficiency of as-built data visualization.

Competing interests
The authors declare that they have no competing interests.

Authors' contributions
HJ conducted the literature review, developed the case study, captured images from the construction site, and wrote the initial draft of the manuscript. MR supervised the entire processes of this study. MH and FR reviewed and revised the manuscript. All authors read and approved the final manuscript.

Acknowledgements
The writers would like to thank Oil Turbo Compressor Construction Company (OTCC) and Department of Construction Engineering and Management in Science and Research Branch of Islamic Azad University for their contributions to this research project. A special thanks is extended to Dr. Hosseini, OTCC's managing director, and Mr. Azimi, OTCC's project manager, for providing the first author with the opportunity of capturing images from the construction site. Any findings and conclusions expressed in this paper are those of the authors and do not reflect the views of the companies and individuals mentioned. Moreover, we would like to thank the reviewers who suggested some points to improve the quality of this paper.

Author details
[1]Department of Construction Engineering and Management, Science and Research Branch, Islamic Azad University, Tehran, Iran. [2]Department of Construction Management, Shahid Beheshti University, Tehran, Iran. [3]Department of Information Technology Engineering, College of e-learning, Shiraz University, Shiraz, Iran.

References
Akinci, B, Kiziltas, S, Ergen, E, Karaesmen, IZ, Keceli, F (2006). Modeling and analyzing the impact of technology on data capture and transfer processes at construction sites: a case study. *Journal of construction engineering and management*, *132*(11), 1148–1157.

Bae, H, Golparvar-Fard, M, White, J (2013). High-precision vision-based mobile augmented reality system for context-aware architectural, engineering, construction and facility management (aec/fm) applications. *Visualization in Engineering*, *1*(1), 1–13.

FIDIC (2006). *Conditions of contract for construction,* for building and engineering works designed by the employer, multilateral development bank harmonized edition: Fédération Internationale des Ingénieurs-Conseils (FIDIC).

Ducke, B, Score, D, Reeves, J (2011). Multiview 3d reconstruction of the archaeological site at weymouth from image series. *Computers & Graphics*, *2*, 375–382.

Golparvar-Fard, M, Peña-Mora, F, Savarese, S (2009a). D4ar–a 4-dimensional augmented reality model for automating construction progress monitoring data collection, processing and communication. *Journal of information technology in construction*, *14*, 129–153.

Golparvar-Fard, M, Peña-Mora, F, Savarese, S (2009b). Sparse reconstruction and geo-registration of site photographs for as-built construction representation and automatic progress data collection, In *26th International Symposium on Automation and Robotics in Construction (ISARC 2009), Austin TX, U.S.A.* (pp. 535–543).

Golparvar-Fard, M, Peña-Mora, F, Savarese, S (2011a). Integrated sequential as-built and as-planned representation with d 4 ar tools in support of decision-making tasks in the aec/fm industry. *Journal of Construction Engineering and Management*, *137*(12), 1099–1116.

Golparvar-Fard, M, Pena-Mora, F, Savarese, S (2011b). Monitoring changes of 3d building elements from unordered photo collections, In *Computer Vision Workshops (ICCV Workshops), 2011 IEEE International Conference On, Barcelona, Spain* (pp. 249–256): IEEE.

Jadidi, H, Ravanshadnia, M, Alipour, MH (2014). Visualization of as-built progress data using construction site photographs: two case studies, In *31st International Symposium on Automation and Robotics in Construction and Mining (ISARC 2014), Sydney, Australia* (pp. 706–713).

James, M, & Robson, S (2012). Straightforward reconstruction of 3d surfaces and topography with a camera: accuracy and geoscience application. *Journal of Geophysical Research: Earth Surface*, *117*(F3), 1–17.

Kamat, VR, Martinez, JC, Fischer, M, Golparvar-Fard, M, Peña-Mora, F, Savarese, S (2010). Research in visualization techniques for field construction. *Journal of construction engineering and management*, *137*(10), 853–862.

Kim, C, Son, H, Kim, C (2013). Automated construction progress measurement using a 4d building information model and 3d data. *Automation in Construction*, *31*, 75–82.

Lowe, DG (2004). Distinctive image features from scale-invariant keypoints. *International Journal of Computer Vision*, *60*(2), 91–110.

McCoy, AP, Golparvar-Fard, M, Rigby, ET (2014). Reducing barriers to remote project planning: Comparison of low-tech site capture approaches and image-based 3d reconstruction. *Journal of Architectural Engineering*, *20*(1), 1–13.

SIFT Demo Program (2014). http://www.cs.ubc.ca/~lowe/keypoints/.

Snavely, N, Seitz, SM, Szeliski, R (2006). Photo tourism: exploring photo collections in 3d. *ACM Transactions on Graphics (TOG)*, *25*(3), 835–846.

Snavely, N, Seitz, SM, Szeliski, R (2008). Modeling the world from internet photo collections. *International Journal of Computer Vision*, *80*(2), 189–210.

Snavely, N, Simon, I, Goesele, M, Szeliski, R, Seitz, SM (2010). Scene reconstruction and visualization from community photo collections. *Proceedings of the IEEE*, *98*(8), 1370–1390.

Wu, C (2011). Visualsfm: A visual structure from motion system. http://ccwu.me/vsfm/.

Wu, C, Agarwal, S, Curless, B, Seitz, SM (2011). Multicore bundle adjustment (pp. 3057–3064): IEEE.

Xiao, J, Fang, T, Tan, P, Zhao, P, Ofek, E, Quan, L (2008). Image-based façade modeling, In *ACM Transactions on Graphics (TOG), New York, NY, U.S.A.*, 27 (p. 161): ACM.

Xnview (2014). http://www.xnview.com.

Yang, M-D, Chao, C-F, Huang, K-S, Lu, L-Y, Chen, Y-P (2013). Image-based 3d scene reconstruction and exploration in augmented reality. *Automation in Construction*, *33*, 48–60.

Sensing the room: an integrated implementation process to visualize indoor temperature data on floor plans

Minkyu Shin, Sang-ik Lee, Hyunsoo Lee and Jin-Kook Lee*

Abstract

Background: Recently multi-use environmental sensors became affordable and easy to control. This improvement enabled us to measure the broad range of indoor environments by using low-priced and controllable sensor devices. As the data set acquired by widely installed environmental sensors increased significantly, the need for making effective use of the data has become of importance. We developed an integrated visualization mechanism to express larger amounts of sensor data.

Method: The mechanism for visualization on floor plan described in this paper consists of the following modules: 1) the sensor module; 2) the data collection module; 3) the data processing module; 4) the spatial data module; 5) the sensor location module; and 6) the data visualization module. A demonstration following the mechanism is introduced in this paper for evaluating the integrated visualization approach. We set up a test-purposed and versatile toolkit that is cheaper, smaller and more controllable than conventional tools. We collect indoor environmental data set composed of sequential numeric data so as to use them as parameters for visualization. We inspected three major issues in the process: 1) indoor temperature data of a specific room collected at a second interval; 2) such a data set can be varied by subdivided spots of interest using multiple toolkits; 3) as a result, the collection of data is regarded as one type of parameter for visualization on top of the room's floor plan, e.g. a sudden change of sequential numbers.

Results: Indoor environment such as temperature changes are shown as colors overlapped on the floor plan. In this way it may be easier to understand the state of indoor environment and factors which influence the environment. The floor plan based visualization of an indoor environmental element seems intuitive compared to just listing the numeric information.

Conclusion: This paper introduces and demonstrates an integrated approach for visualizing the indoor environment on a floor plan with an actual test case. Combined with the building model, the visualized data can be used for recognition of the factors affecting to the target environmental element. We expect floor plan-based visualization to be used for decision-making.

Keywords: Sensor data; Data visualization; Floor plan-based visualization; Indoor temperature changes

Background

People spend most of their daily lives indoors, and this simply explains why indoor temperature is such an important environmental factor (Hyo Joo et al. 2011; Yoon and Spengler 1995). Therefore, controlling comfortable indoor environment is essential for the people to keep their emotional and physical states healthy. An additional benefit is better work productivity (Hee-geon et al. 2013; Moon et al. 2009). Conventionally, to measure and collect the indoor environmental data such as temperature data set used to be inconvenient and expensive due to several reasons (Sik and Bae 2012). The multi-use environmental sensors, however, became affordable and easy to control recently, and this enabled us to measure the broad range of indoor environments by using such low-priced and controllable sensor devices (Yick et al. 2008). However, as the data set acquired by widely installed environmental sensors increased significantly, the need for making

* Correspondence: designit@hanyang.ac.kr
Department of Interior Architecture Design, Hanyang University, 222Wangsimni-ro, Seongdong-gu, Seoul 133-791, Republic of Korea

effective use of the data has become of importance (Jisun 2013; Yongdai and Kwang 2013; Yang Koo and Keun Ho 2008). The more sensor data increases, the more potentiality for managing indoor environment and necessity to manage and represent amount of them grow at the same time. Conventionally, charts are frequently used for representing sensor data (Bo-ram et al. 2012; Chauk and Kyung-Ae 2008; Kwon et al. 2008; Sam-gil et al. 2010). Chart is useful to visualize and compare few sensor data but for expressing a number of sensor data, it is ineffective (Seung-heon et al. 2007) and hard to understand the condition of indoor environment. We, therefore, suggest an integrated mechanism to express such a number of sensor data in an effective way. This paper aims to figure out an integrated approach to visualize such conditional factors using indoor temperature changes. The baseline of this approach is to measure the temperature and analyze given conditions (Seung-Chul et al. 2008). This research aims to develop an integrated visualization mechanism using accumulated data from sensors, rather than the issue for data integrity or the analysis of indoor environment. To verify the feasibility of the proposed mechanism in this paper, consequently, we propose an integrated approach to the visualization of indoor temperature changes using several pragmatically implemented software and hardware modules. This is one of the graphical representations on the existing building geometry using sensor devices and sequential processing mechanism. The entire process from sensing indoor environment to processing and visualizing acquired data can be subdivided into three parts. Each part is composed of submodules. After dealing with the functions of each module inside three parts, this paper shows a demonstration on an actual space. The target environmental element to be visual ized is hourly variation of temperature changes.

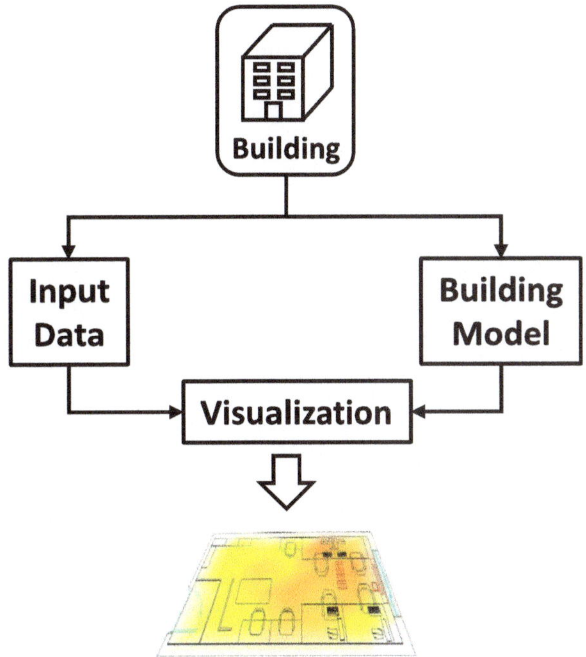

Figure 1 Overview of the system architecture for the visualization of indoor temperature changes.

Methods

System architecture

The system architecture for visualizing the indoor environment on floor plans is composed of three parts, as shown in Figure 1: 1) Input data part, 2) Building model part, and 3) Visualization part. In the building model part, the 2D floor plan or 3D building model is loaded which becomes the background for visualization. Also, all the influence factors of the target environment and the reference point of the installed sensors in the building model are included for more accurate analysis. The input data part deals with the environmental data. In this stage, the environmental data are sensed by installed sensors and collected, retrieved, and processed in order to generate the data to be visualized. The outcome data generated in the Input data part are transferred to an infographic and combined with a loaded building model in the visualization part.

Implementation

The data flow during the entire visualization process is summarized in Figure 2. Input Data Part is composed of three modules: 1) Sensor Module, 2) Data Collection Module, and 3) Data Processing Module. These three modules measure the target environmental element and process environment raw data. The input data to be visualized generates the outcome data to be visualized. The Building Model Part consists of two modules: 1) Building Model Module, and 2) Sensor Network Module. The Building Model Part loads the building model such as 2D floor plans or 3D building model and determines the reference point of installed sensors. The input data and loaded building model are combined by the Visualization Module in the Visualization Part.

Sensor module

The Sensor Module measures the target indoor environmental element by multiple sensors and generates sets of raw data. Each sensor senses the numeric value of the target indoor environment at an interval of time, and the data are collected into a set of raw data in the Data Collection Module. The set of raw data is used by visualizing not only condition of indoor environment at realtime but also specific purpose, such as the average value that the user or manager wants to know. Sensing and collecting sufficient number of datasets by installing the appropriate number of sensors is important because the number of input data is closely related with research

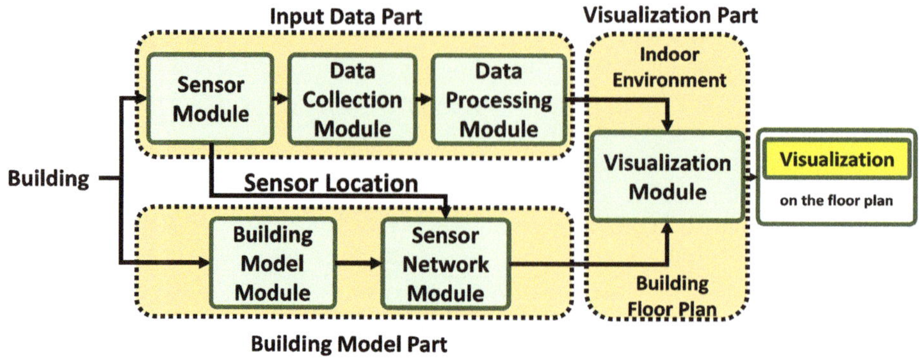

Figure 2 Data flow diagram of the six major modules of implementation.

background and effect of visualization. The greater the input data the more precise and abundant the visualization outcome. As mentioned earlier in scope, the matter of missing or erroneous data will be covered through further work. There are various types of target indoor environments that can be measured such as temperature, humidity, brightness, air quality, and movement. In this paper, the target indoor environment visualized is limited to temperature.

Data collection module

The Data Collection Module collects the target environmental data measured by the Sensor Module. In this module, the data sensed by the installed sensors each time is accumulated in real-time. The accumulated data is so significant because it enables users to freely modify data according to their intention. This module makes a material difference from conventional monitoring system which can only express environmental data in real-time or a specific moment. Depending on the compatibility between the installed sensors and the visualization software, there are two possible methods to collect data. The two methods are depicted on Figure 3. If the sensors are compatible with Visualization software, the data are able to be collected directly by the Visualization software. Otherwise, if the sensor is incompatible with Visualization software, the collected data should be transferred to a suitable format which is compatible with Visualization software such as csv, DB, or Excel file. If the visualization software is going to be used without function to process accumulated data, user should follow the latter method in order to collect data.

Data processing module

In order to generate input data, the Data Processing Module retrieves a particular data among the collected raw dataset and calculates the retrieved data according to the purpose of visualization. Defining the input data should be done before selecting a particular part of the data. The target and algorithm for the data processing are decided by the definition of the input data. As in selecting a particular part of the data from a dataset of

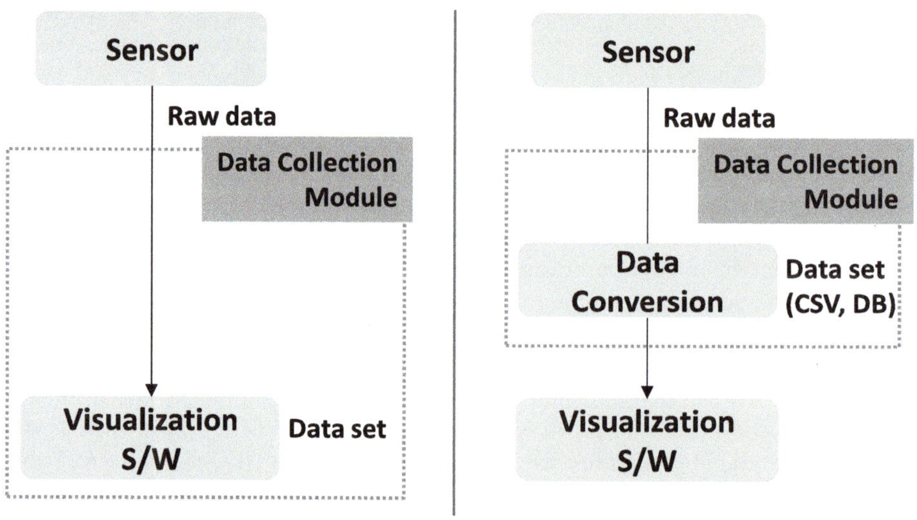

Figure 3 The role of data collection module: without data conversion (left) and with data conversion (right).

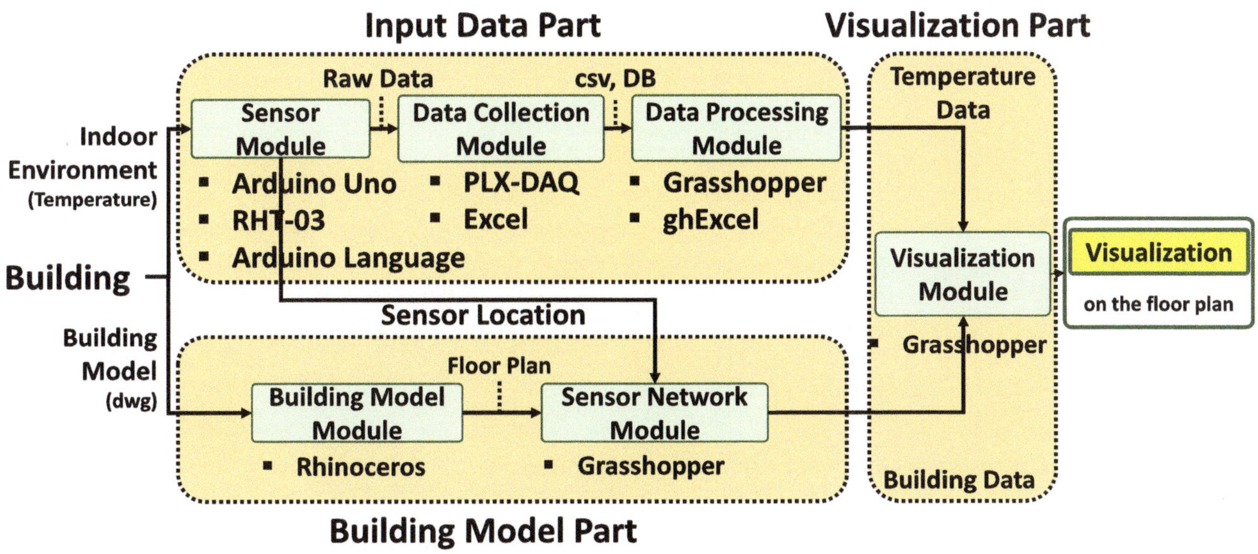

Figure 4 The data flow and software/hardware used in each module.

measured environmental data, it is also possible to decide what environmental element is visualized when there are plural indoor environmental elements collected by the installed sensors.

Building model module

The Building Model Module loads the building geometry, which becomes the back-layer of the visualization. The loaded building model is used as two ways. One is the reference for designating the boundary of the sensor network layer and another is the data for analyzing the visualization outcome. The building model should meet three requirements: 1) drawing a floor plan with precise scale and shape of the indoor space where sensors are installed; 2) including all the factors that influence the target indoor environment within the space; and 3) designating the sensor location on the floor plan accurately. The use of loaded building model regarding designation of the boundary of the sensor network layer is described on the Sensor Network Module. When the building model is combined with the visualization outcome, the building model becomes the basis of analysis of the visualized indoor environmental data.

Sensor network module

The Sensor Network Module generates the sensor network layer which becomes the basis for visualization on building model. Based on the loaded building model, the shape and the size of the boundaries of the sensor network layer are defined. The generated sensor network layer is visualized in Visualization Module and mapped with the input data processed by Data Processing Module. Each sensor location, which is mapped with the input data, can be acquired by Indoor Positioning System (IPS). There are several possible IPS systems such as using Wi-Fi, Global Positioning System antenna, or Geo-Magnetism. Without IPS, the sensor location can be designated manually comparing given building model with the sensor location in actual space. The process of making sensor network layer consists of two steps: 1) defining the boundary of sensor network layer by designating edge of floor plan; and 2) designating the sensor points on the boundary.

Visualization module

Visualization Module transforms the numeric input data into visual information using sensor network layer. Each input data of an installed sensor is connected to each sensor point on the sensor network layer. Connected with the input data, the sensor points move along the z-axis. As the sensor points move depending on input data, sensor network is transformed accordingly. The transformed sensor network enables input data to be expressed in any type

Figure 5 Arduino Uno Board (left) and RHT03 sensor (right).

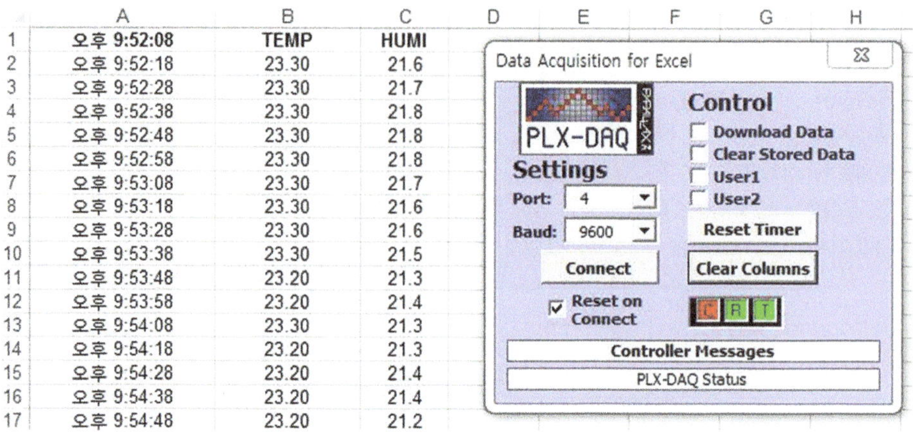

Figure 6 Exporting the measured data into Excel spreadsheet file using PLX-DAQ plug-in.

of graphical image including color or geometry. In this step, the visualization method is decided by the building model and the definition of the input data, considering how to visualize the interested environmental data in the most intuitive way. When the visualization method is decided and the input data is connected to each sensor point, the sensor network layer will be transformed into a certain form of visual information, according to the decided visualization method and the processed input data.

Demonstration

To examine feasibility of proposed mechanism We demonstrated an integrated approach for visualizing the changes of indoor temperature in the laboratory room in which the researchers are working. For this demonstration, we installed 12 sensors on the ceiling by considering how people use this room. This room is located in building of College of Human Ecology, Hanyang University. Figure 4 represents the overall flow of the demonstration including the software/hardware used in each stage.

Measurement of temperature

Arduino Uno Board and RHT03 sensors shown in Figure 5 were used as instruments measuring the temperature, which are cheap and easy to control. RHT03 is a penny-sized temperature/humidity sensor that is able to measure from −40 to 80 degrees C temperature range and also has +/− 2% error range in humidity (Maxdetect). Arduino

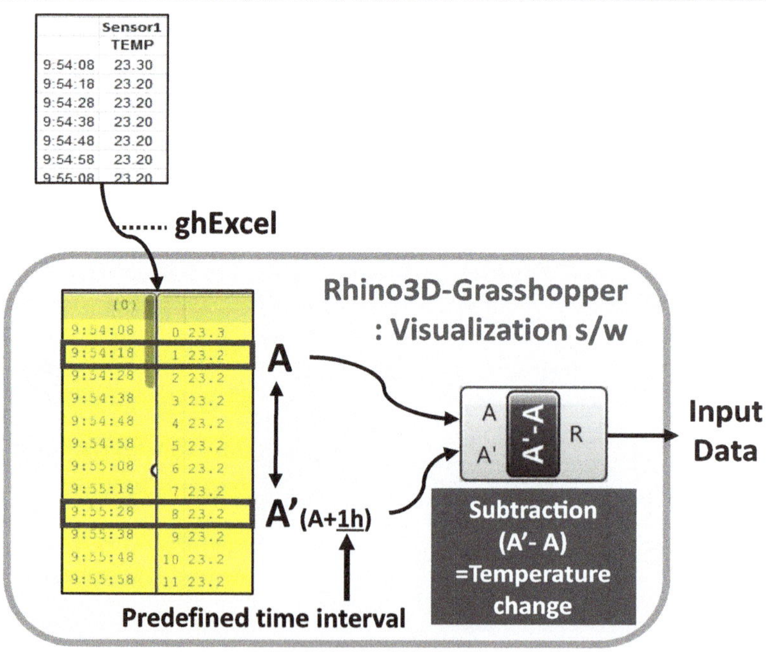

Figure 7 Generating temperature change data by the data processing module.

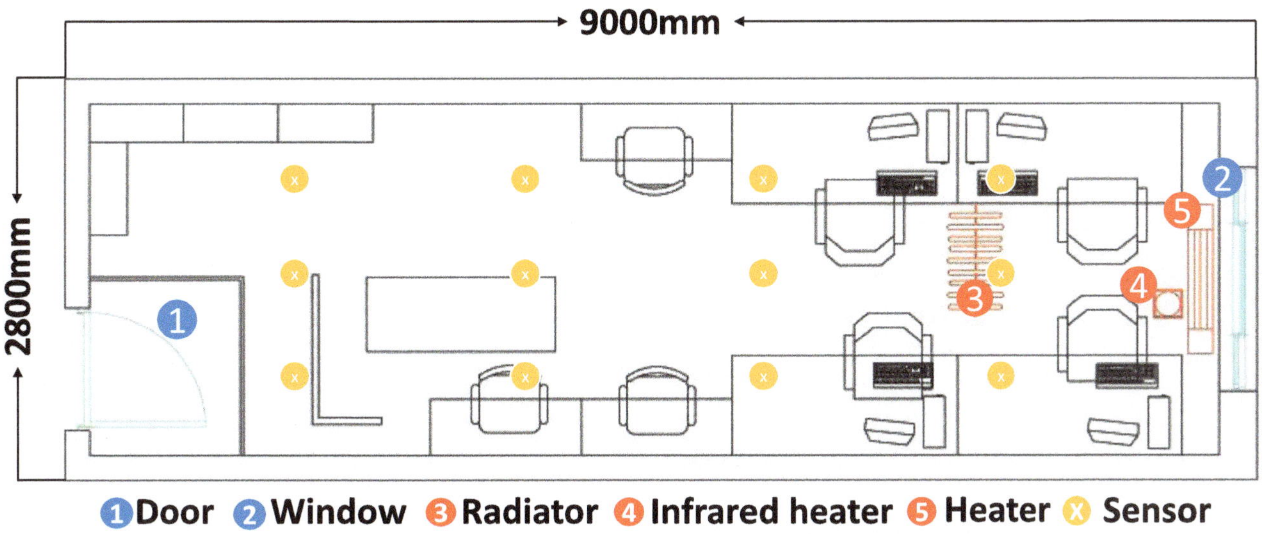

Figure 8 Floor plan for the demonstration.

Uno Board is a microcontroller board controlled by Arduino Language (Arduino). Using the language, it is possible to control how the sensors operate and in what form the data will be collected. We easily revised source code for operating RHT03 by using Arduino Integrated Development Environment which is provided from the website for free because almost all the source codes are provided by manufacturer. The revised source code was compiled and uploaded from computer to Uno board through USB. In this paper, the sensors operated with 60 second intervals and temperature was collected as degrees Celsius.

Converting and collecting the measured data

The measured environmental data was collected in an Excel spreadsheet due to the issue of compatibility with Rhino3D-Grasshopper (Rhinoceros; Grasshopper) on which the environmental data will be visualized. As illustrated in Figure 6, an Excel plug-in software named PLX-DAQ (Parallax Data Acquisition tool) was used for exporting the measured data from the Sensor Module into an Excel file (Parallax). We use PLX-DAQ because this software and manuals are offered free on the websites. Users just download and follow the manuals for operating this software. Each dataset measured by the sensors is collected in each spreadsheet and a dataset includes temperature, humidity and the time the sensor operated. The collected dataset was used for multipurpose like calculating the average or subtracting between specific times in Processing Module.

Temperature change data processing

The process from data processing to visualization were implemented using Rhino3D-Grasshopper in this demo, as described in Figure 7. The Grasshopper is compatible with Excel spreadsheet file using a plug-in program named ghExcel (Xiaoming). The datasets consist of time, humidity and temperature values in 1 minute intervals. The target environmental element is temperature. In order to generate the final input data from the collected datasets, the input data needs to be defined before selecting a particular part of data. In this demo, we defined the input data as the temperature changes at hourly intervals. That means the subtraction of two temperature values between two moments at an hour interval. As one interested moment is selected, the temperature values of the selected moment and 1 hour later are retrieved. After retrieving two temperature values to be compared, the temperature difference is generated for the final input

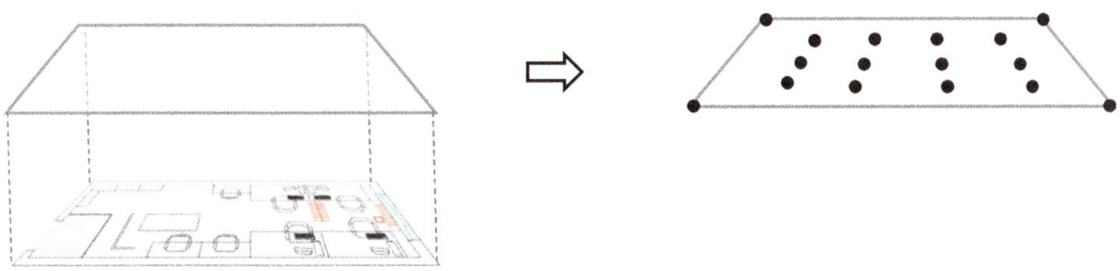

Figure 9 Generation of sensor network layer.

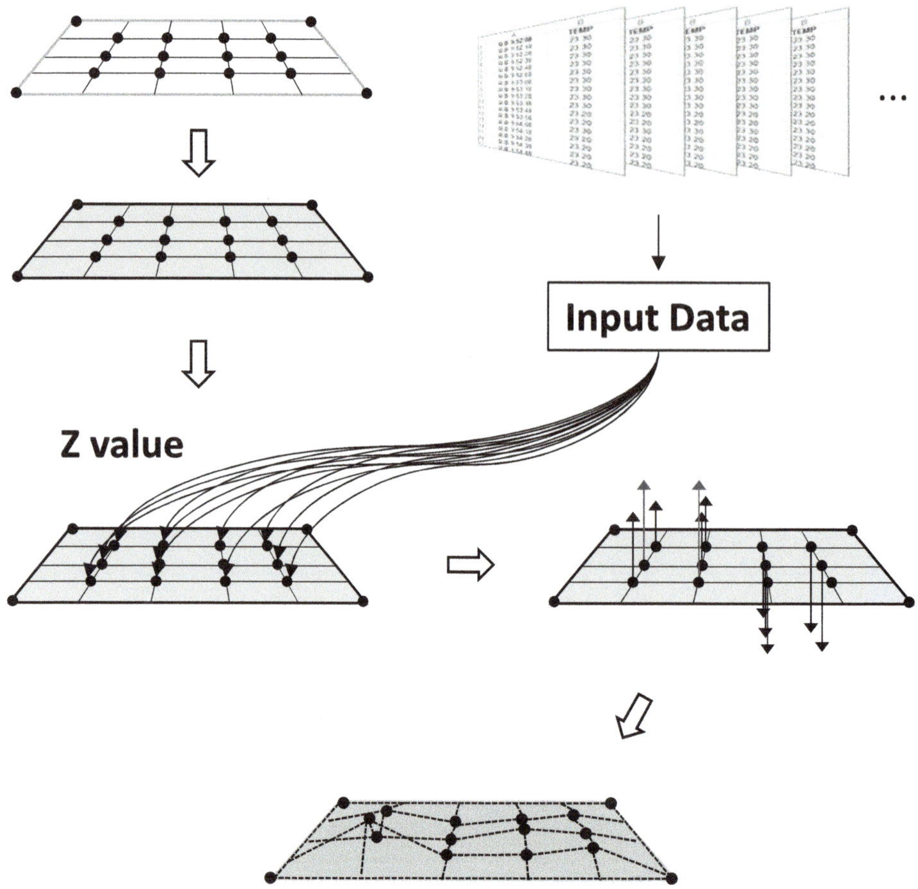

Figure 10 Process of generating geometry using grid segments.

data. This demo used Rhino3D-Grasshopper "slider" function for selecting interface for retrieval.

Importing floor plan

As the building model for this demo, the floor plan was imported to Rhino3D. The size and layout of the space are shown in Figure 8. In addition, sensors used in this demonstration and all the building elements affecting temperature such as heaters or openings are expressed too. Those elements are crucially used in analysis for visualization outcome.

Generating sensor network layer

Before the visualization stage, sensor network layer was made above the floor plan. The sensor network layer becomes the basis when transferring the input data into

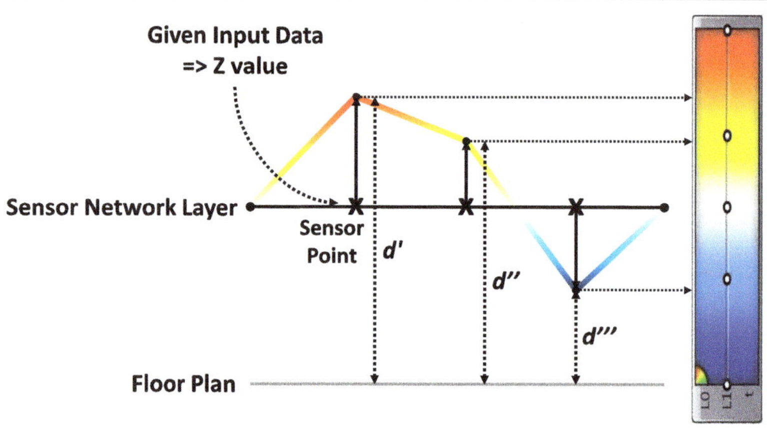

Figure 11 Visualization mechanism of converting z value to color gradation.

visual information. The series of process to generate sensor network are represented in Figure 9. Pointing at the corners of the floor plan's outline, the boundary of sensor network layer was defined. On the boundary of layer, each sensor point was created on each sensor location. The sensor points were designated on the layer manually based on the comparison of floor plan with the actual space, because IPS, by which the sensor location is acquired, is not used in this demo.

Visualizing input data

For the visualization of the given input data above the floor plan, the sensor network layer is divided into grid segments with each sensor point as the center. The

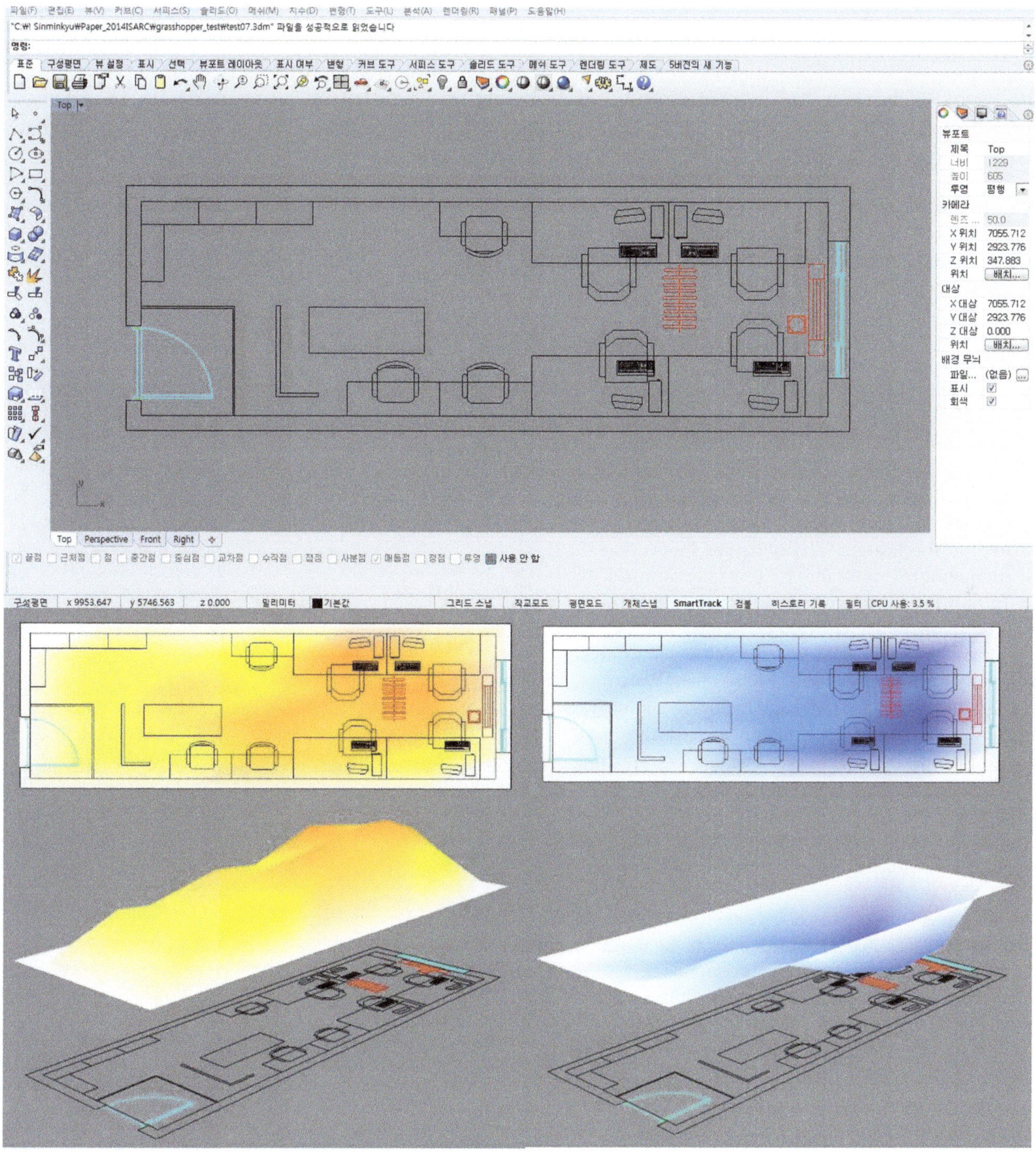

Figure 12 Result of temperature changes visualization.

divided layer forms a flat surface which is able to be connected with the input data on each sensor point. The given input data are assigned to each sensor point as z value of Rhino3D. As the sensor points move to the z axis direction, the surface is transformed in accordance with the sensor points moving into an uneven surface. The entire process of transforming sensor network layer is illustrated in Figure 10. At the same time, it is possible to express the temperature change as color as well as the geometry.

As shown in Figure 11, according to the distance to the floor plan, the surface is colored with red-to-blue color gradation. As a result, input data assigned as z value are transformed into visual information such as geometry and color.

Results and discussion

Figure 12 shows the final visualization outcome on Rhino3D viewport. In top view, the temperature changes are shown as colors overlapped on the floor plan. In this way it is easier to realize the state of indoor environment and factors which influence the temperature changes. In isometric view, the degree of temperature changes appears in the form of geometry which looks like a colored uneven plane. The three-dimensional visualization of an indoor environmental element is intuitive compared to two-dimensional or numeric information. The input data of the left visualization result is the temperature changes between 9 am – 10 am when researchers went to lab and turned on the heaters. The outcome shows that the overall temperature was raised centrally around the heater. In contrast, the input data of right visualization result is the temperature changes between 9 pm ~10 pm when researchers got off and opened window intentionally in order to make drastically temperature changes. As a result, the temperature decreased rapidly during 1 hour with the heater and window as the center because the heater was turned off and the window was opened.

Conclusions

This paper introduces and demonstrates an integrated approach for visualizing the indoor environment on a floor plan with an actual test case. The entire system consists of three major parts, which are able to be subdivided into 6 modules. Several prerequisites exist in each module such as precise floor plan. As shown in Section 3 Demonstration, which is visualizing temperature changes in a position for an hour, visualization mechanism proposed in this paper is able to process the numeric data and to convert processed data into visualized data. As a result, we find out that the visualized environmental data are more intuitive and easier to figure out the state of an indoor environment compared to numeric data. Combined with the building model, the visualized data can be used for recognition of the factors affecting the target environmental element. This information enables floor plan-based visualization to be used for improved decision-making. When changing the layout of an existing building, we can use not virtual data acquired by simulation but also the visualized data acquired by real sensor as evidence for design decision-making. In terms of monitoring, intuitive visualized data can help people evacuate under emergency, showing where an accident happened. Applicability can be expanded as the number of type of sensor. To visualize not only temperature changes but also other indoor environment, we need to develop the visualization mechanism, considering various kinds of available sensors and their network and visualization methods. We will continue to enhance the capability of the data visualization mechanism depicted in this paper on a given building model, based on various types of available sensor modules, their network, or alternated data collection.

Competing interests

The authors declare that they have no competing interests.

Authors' contributions

JL defined the topic and scope of the work, directed the entire processes of this study. MS and SL co-developed the visualization process, conducted the experiment and drafted the manuscript. HL supported the experiment. All authors read and approved the final manuscript.

Acknowledgement

This research was supported by a grant (13AUDP-B068892-01) from Architecture & Urban Development Research Program funded by Ministry of Land, Infrastructure and Transport of Korean government.

References

Arduino. Arduino Uno board. In http://arduino.cc/en/Main/ArduinoBoardUno. Accessed 10 June 2014.

Bo-ram, K, Soo-Hyun, P, & Bong-Hee, H. (2012). Indoor environment monitoring system. *Proceedings of the KIISE Korea Computer Congress, 39*(1B), 57–59.

Chauk, K, & Kyung-Ae, C. (2008). Implementation of a monitoring software using environment sensor data. *Proceedings of the KIISE Korea Computer Congress, 35*(2B), 309–313.

Grasshopper. Grasshopper. In http://www.grasshopper3d.com. Accessed 10 June 2014.

Hee-geon, P, Myeong-jin, K, & Yong-shik, K. (2013). Measurement and assessment of indoor air quality in newly built university buildings. *Journal of the Korean Society of Living Environment System, 20*(2), 216–224.

Hyo Joo, K, Geun Young, Y, & Jeong, K. (2011). A field survey of thermal comfort in office building with thermal environment standard. *Journal of the Korea Institute of Ecological Architecture and Environment, 11*(3), 37–42.

Jang Sik, B. (2012). *Efficient energy management utilizing prediction model in the sensor network based environment monitoring system*. Yeungnam University: Ph.D Dissertation.

Jisun, L. (2013). A Study on visualizing method and expression of information design for big data. *Korean Society of Basic Design and Art, 14*(3), 261–269.

Kwon, S, Park, D, Cho, Y, Park, E, Kim, S, & Jeong, M. (2008). Temperature and humidity monitoring using Ubiquitous Sensor Network (USN). In *Proceeding of the 46th Meeting of KOSAE Korean Society for Atmospheric Environment* (pp. 457–458).

Maxdetect. RHT03 Digital Humidity & Temperature Sensor. In http://www.humiditycn.com/cp22.html. Accessed 10 June 2014.

Moon, S-W, Kim, T-W, Jang, D-S, & Hong, W-H. (2009). A study on occupants' subjective evaluation and a measurement of indoor environments of an office building. *Journal of the Architectural Institute of Korea, 25*(5), 279–286.

Parallax. PLX-DAQ Data Acquisition for Excel. In https://classic.parallax.com/tabid/393/Default.aspx. Accessed 10 June 2014.

Rhinoceros. Rhino3D. In http://www.rhino3d.com. Accessed 10 June 2014.

Sam-gil, C, Ki-tael, K, & Dong-il, K. (2010). A study of environment monitoring system based on USN. *Journal of Information and Convergence Communication Engineering, 14*(6), 1488–1492.

Seung-Chul, L, Sang-Joong, J, Young-Dong, L, & Wan-Young, C. (2008). Design and implementation of flooding-based query model in wireless sensor networks for indoor environmental monitoring system. *Journal of the Korean Sensors Society, 17*(3), 168–177.

Seung-heon, S, Kyeong-jin, B, & Eung-kon, K. (2007). A study on visualization technology for environmental sensor data. *Proceedings of Korea Institute of Information and Communication Engineering, 1*, 192–197.

Xiaoming. GhExcel - Interface with Excel. In http://www.food4rhino.com/project/ghexcel. Accessed 10 June 2014.

Yang Koo, L, & Keun Ho, R. (2008). Historical sensor data management using temporal information. *Journal of Korea Spatial Information Society, 10*(4), 97–102.

Yick, J, Mukherjee, B, & Ghosal, D. (2008). Wireless sensor network survey. *Elsevier Computer Networks, 52*, 2292–2330.

Yongdai, K, & Kwang, H, C. (2013). Big data and statistics. *Journal of the Korean Data Information Science Society, 24*(5), 959–974.

Yoon, D-w, & Spengler, JD. (1995). Standards for indoor air pollutant levels and ventilation rates. *Journal of the Architectural Institute of Korea Special Issue, 39*(6), 12–17.

Permissions

All chapters in this book were first published in VE, by Springer; hereby published with permission under the Creative Commons Attribution License or equivalent. Every chapter published in this book has been scrutinized by our experts. Their significance has been extensively debated. The topics covered herein carry significant findings which will fuel the growth of the discipline. They may even be implemented as practical applications or may be referred to as a beginning point for another development.

The contributors of this book come from diverse backgrounds, making this book a truly international effort. This book will bring forth new frontiers with its revolutionizing research information and detailed analysis of the nascent developments around the world.

We would like to thank all the contributing authors for lending their expertise to make the book truly unique. They have played a crucial role in the development of this book. Without their invaluable contributions this book wouldn't have been possible. They have made vital efforts to compile up to date information on the varied aspects of this subject to make this book a valuable addition to the collection of many professionals and students.

This book was conceptualized with the vision of imparting up-to-date information and advanced data in this field. To ensure the same, a matchless editorial board was set up. Every individual on the board went through rigorous rounds of assessment to prove their worth. After which they invested a large part of their time researching and compiling the most relevant data for our readers.

The editorial board has been involved in producing this book since its inception. They have spent rigorous hours researching and exploring the diverse topics which have resulted in the successful publishing of this book. They have passed on their knowledge of decades through this book. To expedite this challenging task, the publisher supported the team at every step. A small team of assistant editors was also appointed to further simplify the editing procedure and attain best results for the readers.

Apart from the editorial board, the designing team has also invested a significant amount of their time in understanding the subject and creating the most relevant covers. They scrutinized every image to scout for the most suitable representation of the subject and create an appropriate cover for the book.

The publishing team has been an ardent support to the editorial, designing and production team. Their endless efforts to recruit the best for this project, has resulted in the accomplishment of this book. They are a veteran in the field of academics and their pool of knowledge is as vast as their experience in printing. Their expertise and guidance has proved useful at every step. Their uncompromising quality standards have made this book an exceptional effort. Their encouragement from time to time has been an inspiration for everyone.

The publisher and the editorial board hope that this book will prove to be a valuable piece of knowledge for researchers, students, practitioners and scholars across the globe.

List of Contributors

Suyang Dong
Department of Civil and Environmental Engineering, University of Michigan, 2350 Hayward Street, Suite 1380, Ann Arbor, MI 48105, USA

Vineet R Kamat
Department of Civil and Environmental Engineering, University of Michigan, 2350 Hayward Street, Suite 2340, Ann Arbor, MI 48105, USA

Ming-Fung Francis Siu
Department of Civil & Environmental Engineering, The University of Alberta Markin/CNRL Natural Resources Engineering Facility, 9105 116th St, Edmonton, Alberta T6G 2W2, Canada

Ming Lu
Department of Civil & Environmental Engineering, The University of Alberta Markin/CNRL Natural Resources Engineering Facility, 9105 116th St, Edmonton, Alberta T6G 2W2, Canada

Simaan AbouRizk
Department of Civil & Environmental Engineering, The University of Alberta Markin/CNRL Natural Resources Engineering Facility, 9105 116th St, Edmonton, Alberta T6G 2W2, Canada

Amit Mhalas
Technology Futures Institute, Teesside University, Middlesbrough TS1 3BA, UK

Mohamad Kassem
Technology Futures Institute, Teesside University, Middlesbrough TS1 3BA, UK

Tracey Crosbie
Technology Futures Institute, Teesside University, Middlesbrough TS1 3BA, UK

Nashwan Dawood
Technology Futures Institute, Teesside University, Middlesbrough TS1 3BA, UK

Hyojoon Bae
Bradley Department of Electrical and Computer Engineering, Virginia Polytechnic Institute and State University, Blacksburg, VA, USA

Mani Golparvar-Fard
Department of Civil and Environmental Engineering, Department of Computer Science, University of Illinois at Urbana-Champaign, Urbana, IL, USA

Jules White
Department of Electrical Engineering and Computer Science, Vanderbilt University, Nashville, TN, USA

Iva Kovacic
Department for Industrial Building and Interdisciplinary Planning, Vienna University of Technology, Karlsplatz 13, 1040 Vienna, Austria

Lars Oberwinter
Department for Industrial Building and Interdisciplinary Planning, Vienna University of Technology, Karlsplatz 13, 1040 Vienna, Austria

Christoph Müller
Department for Industrial Building and Interdisciplinary Planning, Vienna University of Technology, Karlsplatz 13, 1040 Vienna, Austria

Christoph Achammer
Department for Industrial Building and Interdisciplinary Planning, Vienna University of Technology, Karlsplatz 13, 1040 Vienna, Austria

Sara Rankohi
The Department of Civil Engineering, University of New Brunswick, Fredericton, New Brunswick E3B 5A3, Canada

Lloyd Waugh
The Department of Civil Engineering, University of New Brunswick, Fredericton, New Brunswick E3B 5A3, Canada

Xiangyu Wang
Australasian Joint Research Centre for Building Information Modelling, School of Built Environment, Curtin University, Perth, Australia

Phillip S Dunston
Purdue University, School of Civil Engineering, West Lafayette, India, USA

SangUk Han
Department of Civil and Environmental Engineering, University of Illinois at Urbana-Champaign, Urbana, IL 61801, USA

Madhav Achar
Department of Computer Science, University of Michigan, Ann Arbor, MI 48109, USA

SangHyun Lee
Department of Civil & Environmental Engineering, University of Michigan, Ann Arbor, MI 48109, USA

Feniosky Peña-Mora
Department of Civil Engineering and Engineering Mechanics, Earth and Environmental Engineering, and Computer Science, Columbia University, New York, NY 10027, USA

Bhzad Sidawi
Architecture Department, College of Architecture and Planning, University of Dammam, 31451 Dammam, P O Box 2397, Kingdom of Saudi Arabia

Abdulsalam Alsudairi
Architecture Department, College of Architecture and Planning, University of Dammam, 31451 Dammam, P O Box 2397, Kingdom of Saudi Arabia

Jacob Je-Chian Lin
Computer-Aided Engineering Group, Department of Civil Engineering, National Taiwan University, 188, Sec. 3, HsinHai Rd, Taipei 10668, Taiwan

Chi-En Yang
Computer-Aided Engineering Group, Department of Civil Engineering, National Taiwan University, 188, Sec. 3, HsinHai Rd, Taipei 10668, Taiwan

Wei-Han Hung
Computer-Aided Engineering Group, Department of Civil Engineering, National Taiwan University, 188, Sec. 3, HsinHai Rd, Taipei 10668, Taiwan

Shih-Chung Kang
Computer-Aided Engineering Group, Department of Civil Engineering, National Taiwan University, 188, Sec. 3, HsinHai Rd, Taipei 10668, Taiwan

Huan-Ting Chen
Department of Civil Engineering, National Taiwan University, Taipei, Taiwan

Si-Wei Wu
Department of Civil Engineering, National Taiwan University, Taipei, Taiwan

Shang-Hsien Hsieh
Department of Civil Engineering, National Taiwan University, Taipei, Taiwan

Weilun Lee
Moh and Associates, Inc., BIM Mgmt. & Engr. Integration Center, Taipei, Taiwan

Szumin Kang
Moh and Associates, Inc., BIM Mgmt. & Engr. Integration Center, Taipei, Taiwan

Richard Moh
Moh and Associates, Inc., BIM Mgmt. & Engr. Integration Center, Taipei, Taiwan

Robert Wu
Moh and Associates, Inc., Construction Supervision Dep., Taipei, Taiwan

Hungchang Hsieh
Taiwan High Speed Rail Corporation, Taipei, Taiwan

Zhehua Shu
Taiwan High Speed Rail Corporation, Taipei, Taiwan

Donghoon Lee
Department of Architectural Engineering, Kyung Hee University, 1732 Deogyeong-dearo, Giheung-gu, Yongin-si, Gyeonggi-do, Korea

Jun Young Park
Department of Architectural Engineering, Kyung Hee University, 1732 Deogyeong-dearo, Giheung-gu, Yongin-si, Gyeonggi-do, Korea

Jongkwan Ho
Department of Q-HSE Management Division, Technical Advisory Team, Samsung C & T Corporation, Seocho2-Dong, Seocho-Gu Seoul 1321-20, Korea

Sunkuk Kim
Department of Architectural Engineering, Kyung Hee University, 1732 Deogyeong-dearo, Giheung-gu, Yongin-si, Gyeonggi-do, Korea

Farzad Pour Rahimian
Centre for Sustainable Development, The Grenfell-Baines School of Architecture, Construction and Environment, University of Central Lancashire, Preston PR1 2HE, UK

Tomasz Arciszewski
Civil, Environmental and Infrastructure Engineering Department, Volgenau School of Engineering, George Mason University, Fairfax, Virginia, USA

Jack Steven Goulding
Centre for Sustainable Development, The Grenfell-Baines School of Architecture, Construction and Environment, University of Central Lancashire, Preston PR1 2HE, UK

Gabriel B Dadi
Department of Civil Engineering, University of Kentucky, 151C Raymond Building, Lexington, KY 40506-0281, USA

Paul M Goodrum
Department of Civil,Environmental, and Architectural Engineering, University of Colorado Boulder,428UCB, 1111 Engineering Drive, Boulder, CO, USA

Timothy RB Taylor
Department of Civil Engineering, University of Kentucky, 151A Raymond Building, Lexington, KY, USA

William F Maloney
Department of Civil Engineering, University of Kentucky, 151B Raymond Building, Lexington, KY, USA

Xing Su
School of Civil Engineering, Purdue University, 550 Stadium Mall Drive, West Lafayette, IN 47907, USA

Sanat Talmaki
Department of Civil and Environmental Engineering, University of Michigan, 2340 G.G. Brown, 2350 Hayward, Ann Arbor, MI 48109, USA

Hubo Cai
School of Civil Engineering, Purdue University, Civil Engineering Building, Room 1235, 550 Stadium Mall Drive, West Lafayette, IN 47907, USA

Vineet R Kamat
Department of Civil and Environmental Engineering, University of Michigan, 2340 G.G. Brown, 2350 Hayward, Ann Arbor, MI 48109, USA

Ying-Hao Yu
Department of Electrical Engineering, National Chung Cheng University, No. 168, University Rd, Min-Hsiung Township, Chiayi County 62102, Taiwan

Tsu-Tian Lee
Department of Electrical Engineering, Tamkang University, No. 151, Yingzhuan Rd, Tamsui Dist, New Taipei City 25137, Taiwan
Department of Electrical Engineering, Chung Yuan Christian University, 200 Chung Pei Rd, Chung Li City 32023, Taiwan

Ngai-Ming Kwok
School of Mechanical and Manufacturing Engineering, The University of New South Wales, Sydney, NSW 2052, Australia

Pei-Yin Chen
Department of Computer Science and Information Engineering, National Cheng Kung University, No. 1, University Rd, Tainan City 701, Taiwan

Michael Negnevitsky
School of Engineering and ICT, University of Tasmania, Tasmania, Australia

Koon Wong
School of Engineering and ICT, University of Tasmania, Tasmania, Australia

Hossein Jadidi
Department of Construction Engineering and Management, Science and Research Branch, Islamic Azad University, Tehran, Iran

Mehdi Ravanshadnia
Department of Construction Engineering and Management, Science and Research Branch, Islamic Azad University, Tehran, Iran

Mujtaba Hosseinalipour
Department of Construction Management, Shahid Beheshti University, Tehran, Iran

Fatemeh Rahmani
Department of Information Technology Engineering, College of e-learning, Shiraz University, Shiraz, Iran

Minkyu Shin
Department of Interior Architecture Design, Hanyang University, 222Wangsimni-ro, Seongdong-gu, Seoul 133-791, Republic of Korea

Sang-ik Lee
Department of Interior Architecture Design, Hanyang University, 222Wangsimni-ro, Seongdong-gu, Seoul 133-791, Republic of Korea

Hyunsoo Lee
Department of Interior Architecture Design, Hanyang University, 222Wangsimni-ro, Seongdong-gu, Seoul 133-791, Republic of Korea

Jin-Kook Lee
Department of Interior Architecture Design, Hanyang University, 222Wangsimni-ro, Seongdong-gu, Seoul 133-791, Republic of Korea